Recent Advances in Analytical Techniques

(Volume 2)

Novel Developments in Pharmaceutical and Biomedical Analysis

Edited by

Prof. Atta-ur-Rahman, *FRS*
Honorary Life Fellow, Kings College, University of Cambridge, UK

Prof. Sibel A. Ozkan
Faculty of Pharmacy, Ankara University, Turkey

&

Dr. Rida Ahmed
Postdoctoral Fellow, TCM and Ethnomedicine Innovation & Development Laboratory, School of Pharmacy, Hunan University of Chinese Medicine, Changsha 410208, P.R. China

Recent Advances in Analytical Techniques

Volume # 2

Novel Developments in Pharmaceutical and Biomedical Analysis

Editors: Atta-ur-Rahman, *FRS*, Sibel A. Ozkan and Rida Ahmed

ISSN (Online): 2542-5625

ISSN (Print): 2542-5617

ISBN (Online): 978-1-68108-574-6

ISBN (Print): 978-1-68108-575-3

General:

1. Any dispute or claim arising out of or in connection with this License Agreement or the Work (including non-contractual disputes or claims) will be governed by and construed in accordance with the laws of the U.A.E. as applied in the Emirate of Dubai. Each party agrees that the courts of the Emirate of Dubai shall have exclusive jurisdiction to settle any dispute or claim arising out of or in connection with this License Agreement or the Work (including non-contractual disputes or claims).
2. Your rights under this License Agreement will automatically terminate without notice and without the need for a court order if at any point you breach any terms of this License Agreement. In no event will any delay or failure by Bentham Science Publishers in enforcing your compliance with this License Agreement constitute a waiver of any of its rights.
3. You acknowledge that you have read this License Agreement, and agree to be bound by its terms and conditions. To the extent that any other terms and conditions presented on any website of Bentham Science Publishers conflict with, or are inconsistent with, the terms and conditions set out in this License Agreement, you acknowledge that the terms and conditions set out in this License Agreement shall prevail.

Bentham Science Publishers Ltd.
Executive Suite Y - 2
PO Box 7917, Saif Zone
Sharjah, U.A.E.
Email: subscriptions@benthamscience.org

BENTHAM SCIENCE

CONTENTS

PREFACE

"Recent Advances in Analytical Techniques Vol. 2, "Novel Developments in Pharmaceutical and Biomedical Analysis" presents important recent developments in various analytical methods covering such fields as electrochemistry, optical sensor arrays for pharmaceutical and biomedical analysis, FTIR, high performance liquid chromatographic analysis, chiral separation and other techniques.

The first three chapters cover separation techniques including chromatographic assay of solid dosage forms of drugs, UHPLC method for the estimation of drug active compounds; and HILIC based LC/MS for metabolite analysis. The next chapter presents *in vitro* methods for the evaluation of oxidative stress; while chapter 5 describes the applications of vibrational spectroscopy in studies of structural polymorphism of drugs. Electrochemical biosensors based on conductive polymers and their applications in biomedical analysis are presented in chapter 6 while chapter 7 discusses new trends in electrochemical sensors modified with carbon nanotubes and graphene for pharmaceutical analysis. Chapter 8 describes the use of electrochemical nanobiosensors in pharmaceutical analysis while the next chapter discusses optical sensor arrays for pharmaceutical and biomedical analyses. The applications of ionic liquids in chemical science are discussed in chapter 10 whereas new trends in enantioanalysis of pharmaceutical compounds using electrochemical sensors are presented in the last chapter.

We are deeply grateful to all the authors for their excellent contributions which should be of wide interest to the readers. We are also grateful to Mr. Mahmood Alam (Director Publications) and his excellent team comprising Mr. Shehzad Naqvi (Senior Manager Publications) and Mr. Omer Shafi (Assistant Manager Publications) for their untiring efforts.

Atta-ur-Rahman, *FRS*
Honorary Life Fellow
Kings College
University of Cambridge
UK

Sibel A. Ozkan
Faculty of Pharmacy
Department of Analytical Chemistry
Ankara University
06560 Yenimahalle/Ankara
Turkey

&

Rida Ahmed
TCM and Ethnomedicine Innovation & Development Laboratory
School of Pharmacy
Hunan University of Chinese Medicine
Changsha 410208
P.R. China

List of Contributors

Abdulrahman Alwarthan	Department of Chemistry, College of Science, King Saud University, Riyadh, Kingdom of Saudi Arabia
Arben Merkoçi	Nanobioelectronics and Biosensors Group, Catalan Institute of Nanoscience and Nanotechnology (ICN2), CSIC and The Barcelona Institute of Science and Technology, Campus UAB, Barcelona, Spain ICREA, Pg. Lluís Companys, Barcelona, Spain
Ayhan Savaser	Department of Pharmaceutical Technology, University of Health Sciences, Gulhane Campus, Etlik, Turkey
Bengi Uslu	Faculty of Pharmacy, Department of Analytical Chemistry, Ankara University, Tandogan, Ankara, Turkey
Bożena Nejman-Faleńczyk	Department of Molecular Biology, University of Gdansk, Gdansk, Poland
Burcin Bozal-Palabiyik	Faculty of Pharmacy, Department of Analytical Chemistry, Ankara University, Ankara, Turkey
Burcu Dogan-Topal	Faculty of Pharmacy, Department of Analytical Chemistry, Ankara University, Ankara, Turkey
Emirhan Nemutlu	Faculty of Pharmacy, Department of Analytical Chemistry, Hacettepe University, Ankara, Turkey
Gabriela Wiergowska	Department of Pharmaceutical Chemistry, Faculty of Pharmacy, Poznan University of Medical Sciences, Poznań, Poland PozLab sp. z o.o. (Contract Research Organization) Parkowa 2, Poznań, Poland
Grzegorz Węgrzyn	Department of Molecular Biology, University of Gdansk, Gdansk, Poland
Hande Gürer Orhan	Faculty of Pharmacy, Department of Toxicology, University of Ege, Izmir, Turkey
Hassan Y. Aboul-Enein	Pharmaceutical and Medicinal Chemistry Department, Pharmaceutical and Drug Industries Research Division, National Research Centre, Giza, Egypt
Imran Ali	Department of Chemistry, Jamia Millia Islamia (Central University), New Delhi, India
Joanna Goscianska	Laboratory of Applied Chemistry, Faculty of Chemistry, Adam Mickiewicz University in Poznań, Poznań, Poland
Judyta Cielecka-Piontek	Department of Pharmaceutical Chemistry, Faculty of Pharmacy, Poznan University of Medical Sciences, Poznań, Poland
Kornelia Lewandowska	Polish Academy of Science, Mariana Smoluchowskiego, Institute of Molecular Physics, Poznań, Poland
Luciano Saso	Faculty of Pharmacy and Medicine, Department of Physiology and Pharmacology "Vittorio Erspamer", Rome, Italy
Mehmet Gokhan Caglayan	Faculty of Pharmacy, Department of Analytical Chemistry, Ankara University, Ankara, Turkey

Mehmet Gumustas	Faculty of Pharmacy, Department of Analytical Chemistry, Ankara University, Tandogan, Ankara, Turkey Faculty of Arts & Sciences, Department of Chemistry, Hitit University, Corum, Turkey
Ozgur Esim	Department of Pharmaceutical Technology, Gulhane Campus, University of Health Sciences, Etlik, 06018 Ankara, Turkey
Pavel Anzenbacher	Center for Photochemical Sciences and Department of Chemistry, Bowling Green State University, Bowling Green, OH, USA
Paweł Krzyczmonik	Department of Inorganic and Analytical Chemistry, University of Łódź, Łódź, Poland
Przemysław Zalewski	Department of Pharmaceutical Chemistry, Faculty of Pharmacy, Poznan University of Medical Sciences, Poznań, Poland
Raluca-Ioana Stefan-van Staden	Laboratory of Electrochemistry and PATLAB Bucharest, National Institute of Research for Electrochemistry and Condensed Matter, Bucharest, Romania
Sedef Kır	Faculty of Pharmacy, Department of Analytical Chemistry, Hacettepe University, Ankara, Turkey
Sevinc Kurbanoglu	Faculty of Pharmacy, Department of Analytical Chemistry, Ankara University, Tandogan, Ankara, Turkey Nanobioelectronics and Biosensors Group, Catalan Institute of Nanoscience and Nanotechnology (ICN2), CSIC and The Barcelona Institute of Science and Technology, Campus UAB, Barcelona, Spain
Sibel A. Ozkan	Faculty of Pharmacy, Department of Analytical Chemistry, Ankara University, Tandogan, Ankara, Turkey
Sibel Suzen	Faculty of Pharmacy, Department of Pharmaceutical Chemistry, University of Ankara, Tandogan, Ankara, Turkey
Sławomira Skrzypek	Department of Inorganic and Analytical Chemistry, University of Łódź, Łódź, Poland
Tamás Bálint Csont	Faculty of General Medicine, Department of Biochemistry, University of Szeged, Szeged, Hungary
Yalcin Ozkan	Department of Pharmaceutical Technology, University of Health Sciences, Gulhane Campus, Etlik, Turkey
Zeid A. Alothman	Department of Chemistry, College of Science, King Saud University, Riyadh, Kingdom of Saudi Arabia

Recent Advances in Analytical Techniques

(Volume 2)

Novel Developments in Pharmaceutical and Biomedical Analysis

CHAPTER 1

Advances in Validated Chromatographic Assay of Solid Dosage Forms and Their Drug Dissolution Studies

Sevinc Kurbanoglu[1], Ozgur Esim[2], Ayhan Savaser[2], Sibel A. Ozkan[1,*] and **Yalcin Ozkan[2]**

[1] *Ankara University, Faculty of Pharmacy, Department of Analytical Chemistry, Tandogan, 06100 Ankara, Turkey*

[2] *University of Health Sciences, Department of Pharmaceutical Technology, Gulhane Campus, Etlik, 06018 Ankara, Turkey*

Abstract: Solid dosage forms are the most common drug delivery systems because they provide reproducible and convenient delivery and they are cost effective. It is possible to use immediate, controlled or extended release systems for therapy using solid dosage form such as tablets, capsules, powders, suppositories and lozenges. Solid dosage forms depend on physical properties of the active substance and excipients. To design an effective system and to enlighten the effectiveness, it is important to determine the critical parameters both in pharmacopeia analysis and scientific studies. These critical parameters are various from active substance stability and purity to its *in vivo* profile in dosage form. The primary objective to identify these parameters is developing a fast and fully validated method. Liquid chromatographic techniques are very suitable and accurate way to determine the content of a pharmaceutical ingredient and its stability both in *in vitro* and in *in vivo* systems. Mobil phase composition, flow rate & column choice directly affect the quality of separation in pharmaceutical analysis. In validation of chromatographic methods, validation parameters should be reported in detail. In this chapter, we will discuss solid dosage forms analyses using high performance chromatographic techniques, in terms of their validation parameters and system suitability tests.

Keywords: Analysis, Dissolution, Dosage, Dosage Form, HPLC, Liquid Chromatography, Mobile Phase, Optimization, Oral Solid Drugs, Pharmaceutical Technology, Pharmacopeial, Profile, Pharmaceutical, Solid Dosage Forms, Validation.

[*] **Corresponding author Sibel A. Ozkan:** Ankara University, Faculty of Pharmacy, Department of Analytical Chemistry, Tandogan, 06100 Ankara, Turkey; Tel: +90 203 3175; E-mail: ozkan@pharmacy.ankara.edu.tr

Atta-ur-Rahman, Sibel A. Ozkan & Rida Ahmed (Eds.)

INTRODUCTION

Solid oral dosage forms comprise approximately 80% of the drugs available in the market and are the most convenient and patient-accepted drug delivery type in the world [1]. Not only are they cost-effective, but they are also extremely stable in both chemical and physical aspects [2]. They are usually intended for systemic effects resulting from drug absorption from the gastrointestinal (GI) tract; however, some oral dosage forms are used to produce local effects, thus they are not designed to be absorbed. The physical and chemical stability of solid oral dosage forms are generally better than that of other dosage forms. The main disadvantages resulting from the oral application route include irregular absorption due to the GI environment, slow onset of action, decomposition of some drugs in the stomach and insufficient permeation of high-molecular weight drugs. Due to these limitations of the oral route, different dosage forms are prepared. For instance, rapid onset of action can be provided by formulating the drug as a sublingual tablet or its decomposition in stomach can be prevented by enteric coating or using gastro-resistant dosage forms. Therefore, oral dosage forms require careful pharmaceutical formulation. The physical and chemical properties of active substances (*e.g.*, molecular weight, stability, and decomposition pH) are also important for the selection of the dosage form for oral administration [1 - 3].

Different types of dosage forms can be used for immediate or controlled drug delivery, and they can be applied to produce both local and systemic effects. With the development of different manufacturing techniques, it is now possible to use solid oral dosage forms for many different targets. For instance, by adhering a tablet to buccal mucosa, both local (dental or gingival disorders) or systemic (absorption through the gingiva) effects can be provided or drug release can be hindered in stomach by enteric coating. There are several types of solid oral dosage forms in pharmacopeias. Like other dosage forms, solid oral dosage forms also differ in each pharmacopeia. For instance, in the European Pharmacopeia (EP), chewing gums are considered to be in the solid oral dosage form whereas they are semisolid according to the United States Pharmacopeia (USP) (Table **1**) [4 - 9].

Classification of Solid Oral Dosage Forms

Extracts

Extracts are prepared by evaporating the extractives of crude drugs. There are two kinds of extracts; (I) viscous and (II) dry. In Japanese and Korean Pharmacopeias, extracts are considered as a dosage form (Table **1**) [4 - 9].

Table 1. Solid Oral Dosage Forms Listed in Pharmacopeias (EP- European Pharmacopeia, JP- Japanese Pharmacopeia, USP- United States Pharmacopeia, IP- Indian Pharmacopeia, JP- Japanese Pharmacopeia).

		European Pharmacopeia 8.0	Japanese Pharmacopeia (2006)	Chinese Pharmacopoeia 2010	USP (35)	Indian Pharmacopeia 2010	Korean Pharmacopoeia 9th
Capsules		√	√	√	√	√	√
	Hard Capsules	√	√	√	√	√	√
	Soft Capsules	√	√	√	√	√	√
	Modified-Release Capsules	√			√	√	
	Delayed-Release Capsules				√		
	Extended-Release Capsules				√		
	Sustained Release Capsules			√			
	Controlled Release Capsules			√			
	Enteric Coated Capsules			√			
	Gastro-Resistant Capsules	√				√	
	Cachets	√					
*Chewing Gums, Medicated		√					
Granules		√	√	√	√		√
	Effervescent Granules	√		√	√		
	Coated Granules	√					
	Suspended Granules			√			
	Sustained Release Granules			√			
	Controlled Release Granules			√			
	Modified-Release Granules	√					
	Enteric Coated Granules		√	√			
	Extended Release Granules		√				
	Gastro-Resistant Granules	√					
Oromucosal Preparations		√					
	Pastilles	√					
	Compressed Lozenges	√					
	Oromucosal Capsules	√					
	Orodispersible Films	√					

(Table 1) contd.....

		European Pharmacopeia 8.0	**Japanese Pharmacopeia (2006)**	**Chinese Pharmacopoeia 2010**	**USP (35)**	**Indian Pharmacopeia 2010**	**Korean Pharmacopoeia 9th**
Powders		√	√	√	√	√	√
	Effervescent Powders	√					
Tablets		√	√	√	√	√	√
	Uncoated Tablets	√				√	
	Coated Tablets	√				√	
	Film Coated Tablets					√	
	Hypodermic Tablets				√		
	Bolus Tablets				√		
	Buccal Tablets	√*			√		
	Tablet Triturates				√		
	Gastro-Resistant Tablets	√					
	Modified-Release Tablets	√			√	√	
	Delayed-Release Tablets				√		
	Extended-Release Tablets				√		
	Enteric Coated Tablets		√	√		√	
	Effervescent Tablets	√		√	√	√	
	Sustained Release Tablets			√			
	Extended Release Tablets		√				
	Prolonged Release Tablets					√	
	Soluble Tablets	√		√		√	
	Press Coated Tablets		√				
	Multilayer Tablets		√				
	Dispersible Tablets	√		√		√	
	Orodispersible Tablets	√					
	Orally Disintegrating Tablet				√		
	Sublingual Tablets	√*		√	√		
	Dental Patches			√			
	Controlled Release Tablets			√			
	Tablets Use in the Mouth					√	
	Chewable Tablets	√		√	√		
	Lozenges	√*		√	√**		
	Sugar and Film Coated Tablets		√				
	Oral Lyophilisates	√					

(Table 1) contd.....

		European Pharmacopeia 8.0	Japanese Pharmacopeia (2006)	Chinese Pharmacopoeia 2010	USP (35)	Indian Pharmacopeia 2010	Korean Pharmacopoeia 9th
Pills			√	√	√		√
	Dripping Pills			√			
	Sugar Pills			√			
	Pellets			√	√**		
Extracts			√				√
	Dry Extracts		√				√
Troches			√				√
Pellicles				√			

* Buccal Tablets, Sublingual Tablets and Lozenges are the subtopics of Oromucosal Preparations in European Pharmacopeia.
** Lozenges and Pellets are general topics in USP.

Powders

These finely divided mixtures of solids are intended for both internal and external use. Powder is finer than granules. Being in the pharmaceutical solid oral dosage form, powders may contain one or more active substances and excipients. By adding appropriate excipients, effervescent powder can be prepared.

Granules

Granules can generally be defined as small particle agglomerates. Granules can be administered either directly as powder or indirectly as suspension. Granular dosage forms allow compounding pharmacists to blend drugs in pharmacies. Types of granules as solid oral dosage form in pharmacopeias are shown in Table **1** [4 - 9].

Pellicles

Preparations in pellicular form are processed using the active substance with pellicular material. They are intended for oral or mucosal membrane usage. Pellicles are mentioned as a type of dosage form only in the Chinese pharmacopeia. (Table **1**) [4 - 9].

Pills

Pills are a small solid oral dosage form generally with a round or spherical shape. They are separated from tablets in that they are prepared by a wet massing and molding technique while tablets are typically formed by compression. In EP, there is no monograph for pills while the Chinese Pharmacopeia mentions three types of pellets; dripping pills, sugar pills, and pellets (Table **1**) [4 - 9].

Capsules

Capsules are a solid dosage form consisting of a soluble container or shell, into which drug and excipients are filled. The shell layer of a capsule can be composed of a single (soft shell) or two pieces (hard shell) referred to as a body and a cap. The difference between hard and soft shells is related to the level of plasticizer in the composition of the shell layer, which makes the latter more flexible than the former. Shells are generally made from gelatin; thus, some pharmacopoeias (*e.g.*, (Indian) refer to them as gelatin capsules. Capsule types in pharmacopoeias are similar but have certain differences (Table **1**) [4 - 9].

Lozenges, Troches, Pastilles

Lozenges are a solid oral dosage form designed to release the drug slowly in the mouth. They may contain one or more drug substances. They are often flavored and sweetened, and are generally used for local action in the oral cavity or the throat but they can also be intended to create a systemic effect. Molded lozenges are called pastilles. Lozenges, also known as troches, can be prepared by compression or by stamping or cutting from a uniform bed of paste (Table **1**) [4 - 9].

Oromucosal Preparations

This type of dosage form is only classified by EP. General forms in this class are included in other types (lozenges, buccal and sublingual tablets) in different pharmacopeias. Oromucosal preparations are solid, semi-solid or liquid preparations, containing one or more active substances intended for administration to the oral cavity and/or the throat to obtain a local or systemic effect. Solid oral oromucosal preparations are shown in Table **1** [4 - 9].

Tablets

Tablets are the most widely used dosage form in the world. Tablets are generally intended for oral administration with some being swallowed as a whole, some being chewed, and others being dissolved or dispersed in water before being administered. There are also tablet forms that are retained in the mouth where the active substance is liberated. Tablets can be produced by compressing the same volumes of particles or using another suitable manufacturing technique, such as extrusion, molding or freeze-drying (lyophilization) in a variety of sizes, shapes and surface markings. One or more active substances can be added to tablet formulation. Release of active substance can be delayed or extended, or the active substance and incompatible material can be separated using specialized tablet formulations. Types of tablets in pharmacopeias generally close but different.

General types of tablets are shown in Table **1** [4 - 9].

Pellets

Pellets are a solid dosage form also known as beads due to their uniform shape. They offer several advantages, such as allowing the physical and chemical separation of incompatible materials and modification of drug release (extended or delayed release). Oral pellets are generally contained within hard gelatin capsules for administration. Only in USP, pellets are given as a type of dosage form. In other pharmacopeia (*e.g.*, Chinese), they are considered a subtype of pills (Table **1**) [4 - 9] (Fig. **1**).

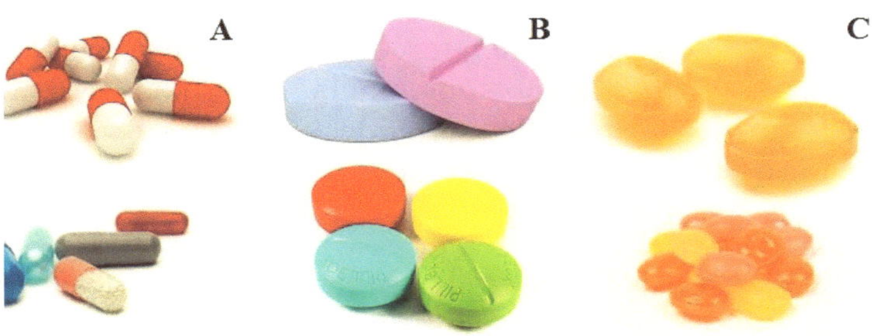

Fig. (1). Examples of solid dosage forms A) capsules B) tablets C) pastilles.

Quality Control of Solid Oral Dosage Forms

The specifications for dosage forms often provide information about the most important characteristics of drugs, which ensure their effectiveness. Finished drug products are tested for quality by assessing whether they meet the requirements for the regulatory purpose. In order to design and develop a robust solid oral dosage form, quality control studies need to be conducted routinely and appropriately. The goal of quality control studies is to evaluate and understand the critical properties, and generate a thorough understanding of dosage form stability under various processing and *in vivo* conditions, leading to an optimal drug delivery system [10, 11].

Compendial quality control tests for solid oral dosage forms are basically the same. Only specialized dosage forms that act in an intended manner require different tests or testing conditions [10, 11].

Uniformity of Dosage Units

Content Uniformity and Mass Uniformity

Especially in single solid oral dosage forms, it is very important for the patient to take the amount of drug specified on the label. It is clearly improbable to expect each single dosage form to include exactly the same amount of the active substance. Hence, pharmacopeial standards and specifications have been established to provide limits for acceptable variations in the total drug of dosage forms. Content uniformity tests are used to ensure the homogeneous distribution of the active content in a production batch. The calculated value reflects the mean drug content in the batch. However, different methodologies and specifications are still prescribed in different official pharmacopeias, such as EP and USP. Also, USP monographs have been published for each pharmaceutical dosage form individually, and EP monographs not only include the active pharmaceutical ingredient (API) individually but also provide a general description for the finished dosage forms. Statistical methods are generally used to calculate content uniformity [12, 13].

Content uniformity is established if the value of the first 10 dosage units is within the specified range. If this value is greater than specified, the next 20 samples should be considered to determine whether the final value meets this criterion. The individual contents must be within the specified range. Analysis of more tablets provides more reliable results than only using 10 dosage units. The acceptance value is calculated as the sum of the difference between the observed mean and the reference value, and the width of the tolerance interval. Therefore, the decision on accepting/rejecting a batch depends not only on the width of this interval but also on the shift of the mean from the nominal value [14, 15].

Solid oral dosage forms are generally composed of an active substance and excipient mixture. If this mixture is close to ideal, total mass is proportional to drug dose. Poor weight uniformity of the drugs results in patient dosing to vary [7].

Disintegration

Disintegration test is a useful performance test for different solid oral dosage forms. This test does not determine drug release but it is a prerequisite for drug dissolution. Conducting a disintegration test rather than a dissolution test might be appealing because the latter is more complicated and time consuming. However, if disintegration is used as a quality control test, then it must be reproducible within the specifications defined. The disintegration test is not only a quality control test but also a critical parameter for drug release [16].

Before dissolution tests became official in USP in the early 1960s, disintegration tests were the only official *in vitro* tests used to predict *in vivo* release and product performance. Even if disintegration test is not directly related to drug bioavailability, this test remains to be one of the most important tests for the pharmaceutical industry in assessing the quality and performance of any conventional oral solid dosage form. This is probably because this test is cost-effective, provides fast results, and does not require skilled personnel [17, 18].

Disintegration test is conducted using apparatuses described in pharmacopeias. In USP, two apparatuses are described; apparatus A that is the basket-rack assembly with six observation cylinders (Chapter 701), and apparatus B containing three observation cylinders with a larger diameter (Chapter 2040). The latter is used for dosage forms with larger diameters; *e.g.*, a bolus tablet. Bolus basket assembly is also mentioned in EP, but not the Japanese Pharmacopeia No. 14 [16, 19 - 21].

Disintegration test is performed to determine whether tablets or capsules disintegrate within the prescribed time when placed in a liquid medium at 37°C using the disintegration apparatus and experimental conditions proposed by a given pharmacopeia [19 - 21]. In several pharmacopeias, disintegration is defined as "a state in which any residue of the unit, except for the fragments of insoluble coating or capsule shell, remaining on the screen of the test apparatus or adhering to the lower surface of the disk, if used, is a soft mass having no palpable core" [19 - 24]. Compliance with the limits on disintegration in the individual monograph is required except where the label states that the tablets or capsules are intended for use as troches, are to be chewed, or are designed as extended release or delayed release dosage forms. The apparatus consists of the basket-rack assembly, a 1000 mL low-form beaker, a thermostatic water bath and a device for raising and lowering the basket in the immersion fluid at a constant frequency. Disintegration tests are performed with water or USP-simulated gastric fluid as the immersion fluids, except when evaluating enteric coated tablets, in which case USP-simulated gastric fluid is used for 1 h followed by USP-simulated intestinal fluid for the time period specified in each monograph [19 - 24].

Dissolution

In the pharmaceutical industry, dissolution testing is a requirement for all solid oral dosage forms in drug development and quality control. It is a key test used for providing information about the characteristics of both the active substance and the formulation. In the dissolution process, active substance is extracted from the solid dosage form and placed in a solution within a medium that mimics the desired body fluid. For efficacy, active ingredient in solid oral dosage forms must be dissolved in body fluids and absorbed into the systemic circulation. This multi-step process must be taken into account in the development of the dissolution method [25].

In vitro dissolution testing provides valuable information concerning the drug development process. Formulation scientists use dissolution not only to assess the dissolution properties of the drug itself but also to select appropriate excipients for the formulation. Dissolution testing can be used to support formulation development by indicating the suitable dosage form with the suitable and reproducible release profile [26].

Technological advancements in drug delivery research and importance of *in vivo* predictability of therapeutic effect by means of *in vitro* tests have resulted in the revival of interest in dissolution tests. Monographs for all oral solid dosage forms and reports on the formulation and development of any solid oral dosage form start with dissolution testing [18].

Studies on dissolution started at the end of the 19th century and first focused on the laws for the description of the dissolution process, not the drugs themselves. In 1897, Noyes and Whitney published an article about the dissolution of two sparingly soluble compounds; benzoic acid and lead chlorine. They attributed the mechanism of dissolution to a thin diffusion layer formed around the solid surface. Then, in 1900, Erich Brunner and Stanislaus von Tolloczko published an article about a series of experiments and showed that the rate of dissolution depends on the exposed surface, stirring rate, temperature, surface structure and apparatus. In 1904, Nerst and Brunner published a paper based on the diffusion layer concept and Fick's second law. Dissolution studies continued with Hixson and Crowell that explained the dependence of reaction velocity upon surface and agitation in 1931. Alternative models are also explained by Danckwerts (1951), Higuchi (1961), *etc* [27, 28].

Despite these advances in the *in vitro* dissolution process, the concept was not used in pharmaceutical sciences until the early 1950s. Furthermore, dissolution tests were first adopted by pharmacopeias in the 1970s [27, 28].

Dissolution tests were first developed to quantify the amount and extent of drug release from solid oral dosage forms including immediate/sustained release tablets and capsules. More recently, dissolution has become important in testing drug release of dosage forms, such as powders, chewable tablets, buccal and sublingual tablets, chewing gums, soft gelatin capsules, suppositories, transdermal patches, aerosols, and semisolids. Different apparatuses have been built for testing drug release in these formulations. Each apparatus has characteristic features and critical points, but only some have been accepted and classified by pharmacopeias [25 - 29].

Selection of suitable test parameters and apparatuses that simulate *in vivo* conditions can lead to successful formulation development. However, optimal compendial conditions for quality control purposes do not always mimic *in vivo* conditions [25 - 29].

In cases where compendial methods are not employed for *in vitro* drug release testing, guidelines published by authorities; *e.g.*, the International Federation of Pharmaceutical Sciences and US Food and Drug Administration are used. Furthermore, researchers also refer to some published papers; for instance, a position paper by the FIP Dissolution Working group on dissolution/drug release testing for special/novel dosage forms lists all apparatuses for the dissolution of these forms (Table **2**) [30].

Table 2. Suggested apparatus for drug release testing of solid oral dosage forms [30].

Solid Oral Dosage Form Type	Suggested Release Method
Conventional Solid Oral Dosage Forms	Basket apparatus, Paddle apparatus, Reciprocating cylinder, Flow through cell
Oral Disintegrating Tablets	Paddle apparatus, Disintegration method
Chewable Tablets	Basket apparatus, Paddle apparatus, Reciprocating cylinder
Powders and Granules	Flow through cell (powder/granule sample cell)
Thin Dissolvable Films	Basket apparatus, Disintegration method
Microparticulate Formulations	Modified flow through cell
Implants	Modified flow through cell

Compendial Tests

Different dissolution techniques are chosen depending on the characteristics of the dosage form and route of administration. Standard dissolution techniques mentioned in pharmacopeias for solid oral dosage forms are USP Apparatus 1 (basket), 2 (paddle), 3 (reciprocating cylinders), 4 (flow-through-cell), 5 (paddle over disk), 6 (rotating cylinder) and 7 (reciprocating holder) [25 - 29].

USP Apparatus 1 (Basket Apparatus)

This apparatus was first described in 1968 by Pernarowski *et al.* [31]. The basket method for evaluating dissolution first appeared in the 13th edition of the USP in early 1970. Hence, this method is also known as the USP basket method. It can also be referred to as a "closed-system" method due to the use of a fixed volume of a dissolution medium [32].

Apparatus basically consists of a basket which has a stirring ability in a vessel with 500-1000 mL of fluid in a temperature–controlled water bath. This method is simple, robust, and easily standardized. Generally, dissolution of immediate-release solid oral dosage forms is performed *via* a basket apparatus (Fig. **2**) [28].

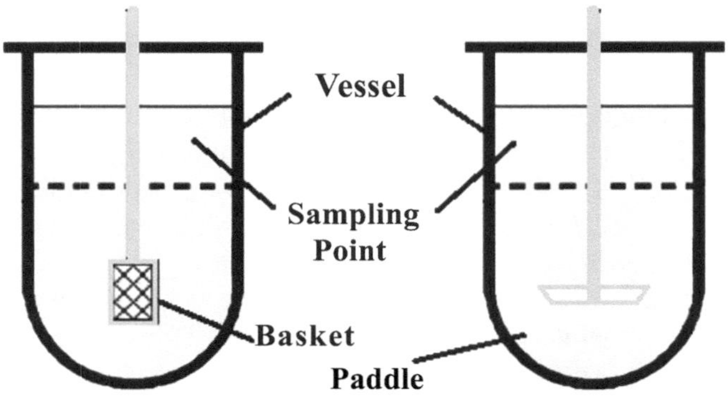

Fig. (2). Schematic Representation of Basket and Paddle Apparatus. (Sizes are approximate).

USP Apparatus 2 (Paddle Apparatus)

Levy and Hayes [33] may be considered the forerunners of the paddle method. The paddle apparatus they proposed has a different structure consisting of a 400 ml beaker and a three-blade, centrally placed polyethylene stirrer rotated at 59 rpm in 250 ml of dissolution fluid. The mechanism of the paddle apparatus is very similar to the basket method but there are slight differences, such as the replacement of the basket with a paddle as the source of agitation. As with the basket apparatus, the shaft should be positioned no more than 2 mm at any point from the vertical axis of the vessel and rotate without any significant wobble. This method was first adopted in pharmacopeias in 1978 [34].

Similarities of USP Apparatuses 1 and 2 can also be seen in their advantages and disadvantages. They are both standardized, easy to operate, robust, and widely adopted (*i.e.*, there is broad experience); thus, they are the first choice in dissolution studies of solid oral dosage forms. However, they have several disadvantages including the limited volume, difficulty of changing the properties of the dissolution media, and unpredictable hydrodynamic conditions due to shaft wobble, location, centering, and coning. Moreover, the cone formation and position of dosage form is often hard to maintain during testing [28].

Generally, the paddle and basket methods can be used for all solid oral dosage forms (Fig. **2**). Immediate release tablets can be tested using either apparatus without additional hardware; however, basket apparatus is usually preferred for capsules since during testing with the paddle method, capsules require a sinker to be kept in the medium. For enteric-coated products, using a basket apparatus is easier when the medium needs to be changed. Even if an enteric coated (EC) product consists of pellets, this does not create a problem for studies conducted with the basket method studies. The paddle and basket methods are also suitable for testing modified-release dosage forms if the formulation is robust to changes in the physiology as its proceeds through the GI tract. Due to the difficulty of changing the medium, simulating the behavior of formulation in the GI tract is harder using the paddle or basket method [25].

USP Apparatus 3 (Reciprocating Cylinder)

The design of this apparatus based on a disintegration tester was adopted by pharmacopeias for extended release products in 1991. The reciprocating cylinder consists of a cylindrical glass vessel, a glass reciprocating inner cylinder, and stainless steel fittings and screens. Dissolution is provided *via* the up- and down-agitation of the inner tube inside the outer tube. On the up position of inner tube, the dosage form contacts with the inner tube, and in the down position, dosage form floats within the inner tube [28].

Placing the dosage form inside a tube allows changing the dissolution medium at a specified time. Owing to compatibility of automated testing over longer periods, the reciprocating cylinder is mainly used for extended release and bead-type dosage forms [28].

In principle, the reciprocating cylinder can be used for a wide variety of oral dosage forms. However, since the operating volume per vessel is quite low, it may be difficult to generate sink conditions, and therefore this type of equipment is not as widely applicable for quality control of immediate-release dosage forms as the paddle or basket methods. On the other hand, for development purposes, low volumes may simulate the actual release conditions better than the volumes required for the standard paddle and basket experiments. The reciprocating cylinder has been used successfully to examine the release from lipid-filled capsules and clearly demonstrated the benefits of the reciprocating action in keeping the lipid material adequately dispersed in the dissolution medium compared to the paddle method [35]. The reciprocating cylinder may also be used for analyzing the release characteristics from enteric-coated products since the change in medium can be achieved by simply moving the cylinder into the next vessel. A particular benefit for EC products coated with polymers dissolving at higher pH is that the possibility of premature re-lease can be checked at pH values relevant to the upper small intestine as well as the stomach using three (or more) rows of vessels, each with a different pH. Examples of using the reciprocating cylinder method can be found in Klein *et al.* [36] and Li *et al.* [37], who utilized a similar setup for acquiring the release profiles of formulations with multiple pH-sensitive coating layers. Using the same approach of multiple rows of vessels to represent conditions in various parts of the GI tract, the reciprocating cylinder method can also be implemented for modified release dosage forms. Ramos Pezzini and Gomes Ferraz [38] reported the results of a study using this test design. Methods using the reciprocating cylinder have great relevance for *in vivo* experiments and are appealing for Quality-by-Design (QbD) purposes; however, as described previously, they may need some modification to be viable in a quality control testing paradigm [25].

Apparatus 4 (Flow Through Cell)

With the need for a different release testing methodology, flow-through systems were experimented with in drug release testing of oral dosage forms in the 1950s. Since then, various types of flow through systems have been proposed. Flow-through cells have been recommended as an alternative apparatus for *in vitro* drug release testing by the Section for Official Laboratories and Medicines Control Services and the Section of Industrial Pharmacists of the FIP in 1981, but they were adopted by USP only after 1995 [28].

USP Apparatus 4 has various types of application in open or closed system modes, different flow rates, and different temperatures. Furthermore, having different cell types, this apparatus can be utilized for a wide range of dosage forms, including tablets, powders, suppositories, or hard and soft gelatin capsules. Apparatus 4 is method of choice for examining the dissolution characteristics of modified release dosage forms and poorly soluble products as the single dosage form can be exposed to the different conditions across the GI tract. This apparatus consists of a pump that provides continuous flow, cells into which fresh medium is continuously pumped, and a fraction collector. Additionally, a dual sampling rack designed for diluting samples during collection and online HPLC or a UV spectrophotometer can be incorporated into this system. Nearly all solid oral dosage forms (tablets, capsules, implants, powder, granules, and hard and soft gelatin capsules) can be tested using this apparatus with optimal equipment [28].

Advantages of the apparatus incorporating the new methodology include: i) unlimited and multiple media usage, ii) providing more suitability for low soluble drugs by enhancing the sink condition in the open loop, iii) gentle hydrodynamic conditions, iv) simulation of real physiologic conditions, and v) suitability for most solid oral dosage forms. However, pump precision may influence the results, and fractioned primary data in the open loop may lead to greater experimental errors when cumulative profiles are constructed [28].

Non-compendial Tests

Formulation science offers solutions to overcome solubility and dissolution rate problems which can compromise drug absorption. However, development in the apparatus concept is not evolving as fast as dosage form administration. The simple basket or paddle (USP apparatuses I and II) systems provide a well-stirred environment but this closed system is limited due to the absence of absorptive sink conditions and inadequate hydrodynamics. While hydrodynamic conditions produced by USP apparatus III is more favorable, this approach may not accurately reproduce the physical aspects because the dissolution medium cannot be changed during analysis. The flow-through apparatus is reported not only to provide hydrodynamic conditions but also to assess the performance of controlled release dosage forms by allowing the replacement of media. However, USP Apparatus 4 is a useful tool to study extended release formulations with poorly soluble drug substances; however, dissolution prediction of basic drugs is challenging due to the precipitation risk at intestinal pH.

A variety of non-compendial dissolution models have been developed to predict *in vivo* dissolution rate which is affected by many factors. In these models, the first objective is to mimic gastric emptying and potential precipitation in the intestinal compartment in flow-through cells. Configuration of these systems allows transportation of gastric media content to intestinal media. For this type, the artificial stomach-duodenum (ASD) model is the best example. In this model, formulated drug is transferred from the stomach chamber to the duodenum part at a controlled rate. The presence of different media in the two chambers causes continuous variation of drug concentration. Although the ASD model is a convenient method for immediate-release formulations, its use in controlled release formulations is limited since this model cannot effectively mimic the lower gastrointestinal region. In addition, bioavailability of drugs with limited permeability or metabolism cannot be correlated directly. Yet, being easy to use and having bio-relevant fluid transfer properties, the ASD model can be considered powerful in providing an understanding of the dynamic dissolution of drugs [39].

Developments in *in vivo* imaging of dosage forms during gastrointestinal transit have elucidated the effects of physical forces, such as the contraction of gastrointestinal tract on the dissolution of drugs. Dissolution stress test apparatuses have been developed to mimic this pattern of movement in dosage forms during gastric emptying and intestinal transit. Using this equipment, it is possible to see multiple plasma concentration peaks *in vivo* after dosing an extended release formulation. This apparatus appears to have the ability to determine the mechanical robustness of extended release formulations and their ability to resist gastric forces and passage through the pylorus or ileocecal junction [40].

Drug dissolution occurs under sink conditions. For drugs with poor water solubility, combined dissolution-absorption models can be used to determine whether the bioavailability of a drug is limited by the dissolution rate and whether drug precipitation is a contributory factor for the poor oral bioavailability. An example of these models is the FloVitro system, which comprises gastric and small intestinal chambers, as well as an additional absorptive compartment connected to the intestinal chamber. Systems that are more complicated can be set up using cell-based membranes between the compartments to demonstrate the effect of permeation on the dissolution rate in the same assay. The major drawback of these systems is the limited hydrodynamics in simple buffer transport systems and when handling food materials. Furthermore, even when biological membranes are used, the biological system cannot be exactly mimicked due to the absence of other factors, such as the mucus layer of gut epithelium [41].

Although the above-mentioned models are useful tools in terms of solving the problems related to hydrodynamics and media composition during the dissolution process, the generated dissolution profiles generated are limited by the simulation of digestive processes. A complex *in vitro* digestion model is generally used to understand the performance of self-emulsifying drug delivery systems in the following phases: (i) the pellet, which contains a calcium soap of fatty acid and precipitated drug, (ii) the aqueous phase, and (iii) the oil phase. Lipolysis models are useful to provide an understanding of the lipolytic digestion process but they do not show the interaction between food-drug formulation and other digestive enzymes. Furthermore, they cannot fully simulate the gastric and intestinal processes (Fig. **3**) [42].

More complex systems; *e.g.*, the dynamic gastric model (DGM) and the TNO gastro intestinal model (TIM) have also been proposed to simulate gastric processes and the effect of digestive properties on dosage forms. DGM is composed of two stages, of which the first mimics gastric mixing and the dynamic secretory profiles of the stomach and the second replicates shear forces in the antral region. This two-stage system can provide an accurate simulation of gastric behaviors and generate a complete profile of digestion and dissolution using a small intestine simulator. TIM is a computer-controlled system that can also simulate physiological processes that occur in the stomach and the small intestine. In this system, absorption processes are simulated *via* dialysis membranes, but active transport, efflux and intestinal wall metabolism are not modeled. To overcome this limitation, researchers have suggested combining TIM with a CaCO-2 cell permeability assay. However, more information is required before this combined technique can be used as an effective alternative to other models (Fig. **4**) [39].

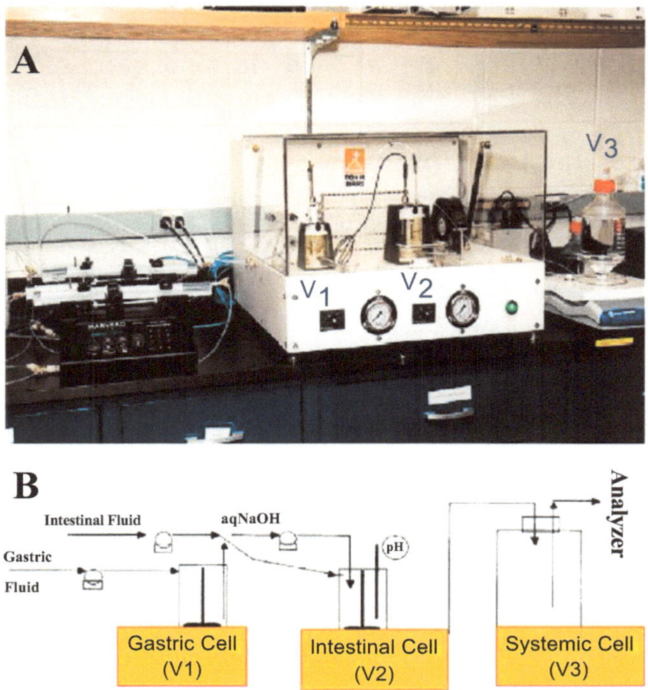

Fig. (3). A) FloVitro equipment showing gastric (V1), intestinal (V2) and absorptive compartments (V3). **B**) FloVitro equipment schematic representation. Reprinted with permission [42].

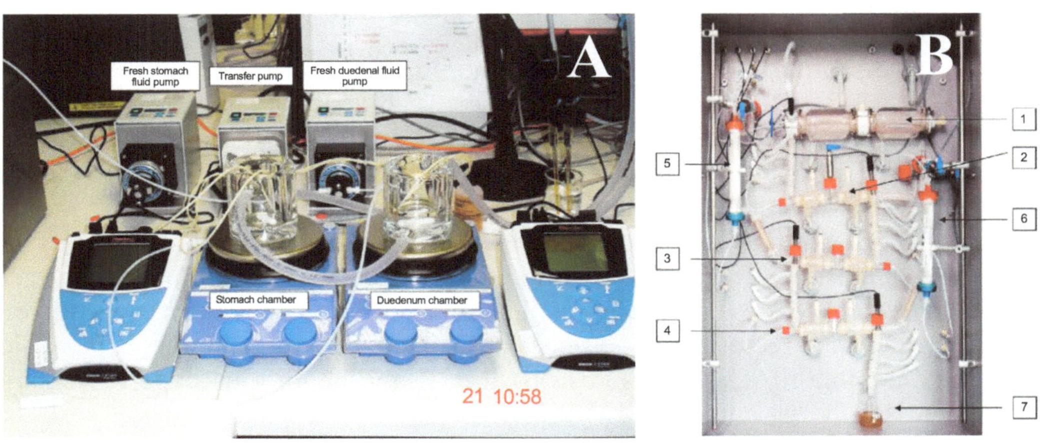

Fig. (4). A) Equipment showing gastric and duodenal compartments and transfer pumps. **B**) TNO TIM-1 apparatus: (1) gastric compartment;(2) duodenum; (3) jejunum; (4) ileum; (5) jejunal dialysis cartridge; (6) ileal dialysis cartridge; (7) ileal eluent. Reprinted with permission [42].

Despite the availability of many types of dosage forms for transmucosal drug delivery, only a few dissolution methods are defined in pharmacopeias. Several studies have been performed for the dissolution of these dosage forms using small media volumes and different apparatus. However, none of the compendial dissolution apparatuses in use correlates with the amount of saliva available for *in vivo* dissolution, and thus they cannot accurately reflect this profile [25].

Importance of Dissolution in Pharmacy

The dissolution test has an essential role in the development of drug products. The dissolution test and specifications are intended to show that the product is bioequivalent to pivotal clinical lots and critical manufacturing variables, and additional clinical studies are taken into account to implement post-approval changes and stability. However, as methods do not mimic GI conditions sufficiently, the dissolution test has been criticized for not being predictive of bioavailability. This lack of prediction may result from the erroneous selection of acceptance criteria or specific analytical conditions.

Dissolution testing can also be used to demonstrate the performance stability of a drug product throughout its shelf life. The dissolution test gives information about crystallinity, glass transition temperature, the pore structure of polymeric excipients [43], polymorphism [44], cross-linking of gelatin capsules [45], and the moisture content [46] of the formulation. Identification of these characteristics can be used to make an informed decision on the final formulation, manufacturing process, and packaging [47].

Comparison of Dissolution Profiles

Comparing dissolution profiles allows monitoring the differences between formulations and determining the stability and bioequivalence of a product; however, for this to be effective, it is important to choose the best methods to compare. For example, dissolution profiles are generally (except flow-through systems) plotted as cumulative percentage of drug released *versus* time, but *in vivo* data is not cumulative. Furthermore, the use of exploratory data analysis methods is sometimes problematic as they overlap only at certain points, not all points. To overcome these problems, some comparison methods were investigated by appropriate *in vitro* tests, which need to be designed as repeatable as possible and to realize conditions as close as possible to the conditions experienced by pharmaceuticals in the human body [48].

Mathematical comparisons based on difference (f1) or similarity (f2) factors allow quantification of profiles and can be used to utilize the differences between reference and test products. The similarity factor (f2) is a regulatory requirement and can be easily calculated. Even though the similarity factor is sensitive to the number of dissolution time points, there is no information about formulation variability. This technique is a simple statistical model, but the probability of type I (rejecting similar profiles as dissimilar) and type II (accepting dissimilar profiles as being similar) error is not defined [47].

Another technique used to compare dissolution profiles is fitting dissolution data to mathematical functions. Some of the examples of this type are the zero-order [49], first-order [50], Hixson– Crowell [51], Higuchi [51, 51], quadratic [50], [47], Weibull [52], Gompertz [50 - 52], Probit [52], exponential [52], and logistic [50 - 52] models. The advantage of these methods is not only taking into account the variance and covariance of datasets, but also handling different sampling time points for reference and test profiles. Nevertheless, finding a model that fits the data is not always possible. Poorly chosen models may produce confusing and inaccurate results. Hence, it is important to run a lack-of-fit test on the reference data before comparing model parameters. Among solid oral dosage forms, the most studied classes are dissolving forms (powders and tablets) and matrices systems including hydrogels. Powder dissolution is described by the Noyes-Whitney equation and modifications thereof. According to these models, the dissolution phenomenon of a solid particle in a liquid medium reflects a surface action. Drug release from matrix systems, especially that of hydrogels is more complicated. The release of water soluble and poorly soluble drugs incorporated in semi-solid and/or solid matrices is generally explained by the Higuchi equation. Dissolution of several types of modified-release pharmaceutical dosage forms and matrix tablets containing water soluble drugs is also generally described by the Higuchi model, in which release kinetics are proportional to the square root of time. However, this equation is not appropriate for matrix systems that show diffusivity in the presence of a solvent concentration subjected to swelling and erosion. The drawbacks of fitting these types of drug dissolution profiles to the Higuchi model have been overcome by a semi-empirical model proposed by Peppas [53], which describes the phenomena of water diffusion, swelling, drug diffusion, and polymer erosion layer by layer from the external toward the interior of the tablet. Dissolution from matrices made of polymers and drugs with different shapes has been modeled using pure hydroxypropyl methylcellulose (HPMC) [54, 55]. Diffusion problems of tablets with complex geometries (*e.g.*, convex tablets, hollow cylinders, doughnuts, and inwards hemispheres) have been resolved by finite element methods [56]. Furthermore, a model based on drug balance in the dissolution medium has been proposed in view of the resistance to release due to a layer of enteric coating [57]. Drug dissolution from solid oral dosage forms that do not disaggregate and that release the drug slowly can be modeled using zero-order kinetics. This reaction is generally used to describe the dissolution of modified-release pharmaceutical dosage forms, matrix tablets containing poorly soluble drugs in coated forms, and osmotic tablets. There are also many other methods, such as the Baker–Lonsdale model, which explains the linearization of release data from several formulations of microcapsules or microspheres, and the Hopfenberg model, which is used to describe time-dependent diffusional resistances internal or external to the eroding matrix.

Statistical methods based on analysis of variance (ANOVA) have also been used for the comparison of *in vitro* dissolution profiles. These methods do not depend on curve-fitting procedures, and the analysis is able to demonstrate the differences between native profiles in level and shape. Determining the shape of a profile is essential to learning about differences in the dissolution mechanism. An advantage of ANOVA-based methods is that they can be used to estimate type I and type II errors [58].

In Vitro and In Vivo Relationships and Bioequivalence Challenges in Dissolution Method Development

Dissolution rate, aqueous solubility, and gastrointestinal permeability are key parameters controlling the efficacy of a drug. To clarify the drug absorption mechanism, a classification system was proposed in 1995 by Amidon *et al*. Drug substances were classified into four groups: class 1 (high solubility and high permeability), class 2 (low solubility and high permeability), class 3 (high solubility and low permeability), and class 4 (low solubility and low permeability). According to this classification, drugs are considered highly soluble when the highest dose strength of the drug substance is soluble in less than 250 mL of water over a pH range of 1–6.8 whereas highly permeable drugs are those for which absorption in humans is determined to be greater than 90% of the administered dose [59].

Solid oral dosage forms are not immediately absorbed due to absorption from biological system only occurring in the solution form. *In vitro* dissolution tests offer information about the amount of drug released per unit time in a given dissolution medium, and based on this data, *in vitro* release tests can be used as a sensitive and reliable predictor of *in vivo* performances and offer a meaningful indication of physiological availability [48].

For the *in vitro* prediction of *in vivo* performance of drug products, a correlation has been proposed between *in vitro* dissolution tests and *in vivo* drug concentration based on mathematical models. This is called an *in vitro*/*in vivo* correlation (IVIVC). IVIVC is used to reduce development time, cost, and regulatory burden. Currently, there are four levels of IVIVC: Level A (the point-to-point relationship between *in vitro* dissolution and *in vivo* pharmacokinetic data), Level B (the relationship between the mean *in vitro* dissolution time and the mean *in vivo* residence time obtained by considering the full profile of *in vitro* dissolution and that of *in vivo* plasma evolution), Level C (the single- or several-point relationship between a dissolution parameter and a pharmacokinetic parameter), and Level D (a qualitative rank-order correlation).

Only Level A correlation is accepted by regulatory boards for scale-up and post-approval changes, and the requirement for additional human studies can be eliminated by establishing a Level A correlation. Level A correlation is expected from a class I drug formulated as an extended-release dosage form with site-independent characteristics of absorption and permeability. Level C correlation does not reflect the entire dissolution profile; hence, it is considered the lowest correlation level. However, Level C correlation can still provide useful information in early formulation development. For a class I drug, if the permeability is site-dependent, a level C correlation is expected [47].

Liquid Chromatography in Dissolution Testing

Chromatographic methods date back to 1855, when a German Chemist Friedrich Ferdinand Runge Ninja suggested the use of reactive impregnated filter paper in the identification of stains. In 1860, Christian Friedrich Schönbein and his student Friedrich Goppelsroeder reported that materials were drifting at different speeds with the solvent due to the capillary effect on the filter paper. These studies were followed by the research of a Russian botanist Mikhail S. Tsvet color pigments in 1906. Tsvet observed the color separations of many plant pigments, including chlorophyll and xanthophyll, a method which was given the name "chromatographie" (chromatography), combining the words "chroma" meaning color and "graphein" meaning writing. In recent years, many new chromatographic techniques have been developed, and chromatographic applications have increased dramatically as a result of the need for better techniques to separate complex mixtures [60 - 62].

Efficient separation of constituents of a sample and the acceptance of chromatographic measurements are largely dependent on the elution rates of substances. These rates are determined by the magnitude of the equilibrium constants of reactions causing the material to diffuse between the moving and stationary phases [63 - 65].

In general, the analyte in the mobile phase is separated from the other compounds in the resultant mixture so that the propagation rates in the matrix in which the stationary phase is present are different. At this point, it is necessary to classify chromatographic methods according to the composition of the surface and the mobile phase to which the stationary phase is attached [63 - 65].

Chromatographic methods are divided into two as "layer chromatography" and "column chromatography" according to the surface to which the stationary phase is attached. On the other hand, according to the composition of the mobile phase, they are classified as gas, liquid or supercritical fluid chromatography, depending on the molecular size of the working sample [63 - 65].

In the high-pressure liquid chromatography method, the increased flow rate shortens the analysis period but reduces the efficiency of separation, resulting in practical problems. It has been found that the efficiency of fillers can be increased by reducing the particle size. With the developing technology, the particle size of column fillers has been reduced to 3 - 10 µm, which in turn has increased column yield. This method has been found to have other advantages over other liquid chromatography methods in terms of applicability to complex solutions and quantitative analyses, and therefore it is now known as "high-performance liquid chromatography" (HPLC). This technique is divided into five classes according to the stationary phase used; adsorption, partition, ion exchange, size elimination and affinity chromatography [63 - 66].

Adsorption chromatography, in which the stationary phase is solid and the mobile phase is liquid, is the most widely used method in liquid chromatography. Here, solvent substances are adsorbed on the surface of a solution based on their polarity differences [63 - 66].

Partition chromatography is further divided into liquid - liquid and bound-phase chromatography. Unlike adsorption chromatography, in partition chromatography, the stationary phase is liquid and is attached to the surface of the column packing material by physical adsorption. The distinction is based on the fact that the solubility of the substance to be analyzed is different in both phases [63 - 66].

Ion exchange chromatography is based on the principle that a charged substance is held in a solid stationary phase with an overloaded charge. The stationary phase usually contains acid or base functional groups. It is an effective method for qualitative and quantitative analysis of ions.

In size elimination chromatography, molecules belonging to a substance to be analyzed are separated according to their size. Columns filled with filler material containing very different porosities are used for size elimination chromatography. Thus, pores in different diameters act like sieves, holding materials depending on their size (diameters) [63 - 66].

In affinity chromatography, the retaining mechanism is unique to the material and biological materials suitable to the key-lock model are used as the filling material. The application area of affinity chromatography is very limited due to its high cost [63 - 66].

HPLC can be equipped with various detectors, such as UV/Vis, photodiode array, fluorescence, conductivity, refractive index, electrochemical, mass and evaporative light scattering detectors. In dissolution studies carried out by HPLC, generally the UV/Vis detector is used to measure different types of pharmaceuticals absorbable at a wavelength range of 190-1100 nm. Many pharmaceuticals containing one or more double bonds in the structure and unpaired electrons can be absorbed at the specified wavelength range. Ultraviolet detectors are actually the simplest LC detectors consisting of filtered photometers that use mercury lamps as a beam source. The most common of these is the intense beam of 254 nm wavelength. The strongest absorbance detectors are photodiode series detectors [62 - 71].

Electrochemical detectors operate based on the measurement of the oxidation and reduction potentials of the electroactive substances. For electrochemical detectors, amperometry, polarography and coulometry methods are used. Although their application areas are not as wide as optical detectors, their use and sensitivity are increasing with the development of microelectrode technology [72, 73]. Fluorescence detectors for HPLC are designed similar to spectrofluorometers. The measurement of emissions emitted by certain wavelength-excited substances is performed against time [74, 75] as they return to the ground. Another widely used chromatographic technique involves the separation of species are separated from the column depending on the mass of ions. They are then passed through an electrical field in the detector to be assigned to different mass/charge ratios. From here, it is possible to make assumptions about their molecular weight and shape. Refractive index is another type of detector, in which the mobile phase used passes through a half chamber of the cell on the way of the arm, and the eluent then passes through the other chamber. If these two compartments are different from each other in terms of the refractive index of the two solutions, the incident beam is divided into two with a glass plate placed at an appropriate angle to be broken. In recent years, photodiodes have been developed as a new detector for HPLC. Deviation from the path of the incoming beam to the surface causes the output signal to change. A sample chromatogram is obtained by amplifying and recording this change. An important advantage of breakdown index determinants is that they respond to almost all materials. photodiodes is accepted as more sensitive than other refractive index-type detectors [76 - 82].

Validation in Chromatographic Analysis

"Validation" phenomena were first come by the European Federation of Pharmaceutical Industries and Associations – EFPIA which was hosted and hosted in 1990 in Belgium / Brussels, in the congress named, International Conference on Harmonization of Technical Requirements for Registration of Pharmaceuticals for Human Use – ICH). ICH is a legal entity created by European Medicines Agency –EMA; European Federation of Pharmaceutical Industries and Associations – EFPIA which contains 45 drug company in the body, Ministry of Health, Labor and Welfare, Japan – MHLW, Japan Pharmacy Manufacturers Association, Food and Drug Administration (FDA) and Pharmaceutical Research and Manufacturers of America – PhRMA [83 - 87].

A validation can be defined as a collection of processes for a device, a method and a product, the results obtained to identify defined parameters or requirements. Processes carried out at drug production and quality control stages are very important in terms of validation and interpretation of applied analytical methods, accuracy and reliability of the work done [83 - 87].

As a result of validation that is performed correctly and reliably with the selected separation method, the quality of the analysis results can be guaranteed. The variability in analyzes is most reducible, the correctness, precision and reproducibility of the process can be proved, the efficient use of the analysis period can be achieved. It is possible to establish well-controlled, reliable areas, the education of the person who will make the analysis and the information about this subject can be realized at the highest level. The aim and validation parameters of the analytical method and the required values must be specified at the beginning of the process. Moreover,

- Properties of the substance(s) to be analyzed,
- The level of concentration planned for analysis,
- Composition of the medium in which the sample is located,
- Qualitative and quantitative details,
- Required diagnostic and determination limits,
- Targeted approximate concentration range,
- Estimated accuracy and precision,
- Stability and consistency of the developed method,
- Other validation parameters required for the system being run, should also be specified at the beginning of the validation process [83 - 90].

Analytical validation is a collection of parameters that show the validity of the results obtained from analytical applications and the feasibility of the method. All validation results are expressed in terms of analytical parameters. These parameters are valid for both *in vitro* and *in vivo* studies. Only the limits of acceptability vary [83 - 90].

The following analytical method characteristics should be determined and calculated during method validation:

1. System Suitability Test Parameters
2. Linearity and Range
3. Limit of Detection
4. Limit of Quantification
5. Accuracy
6. Precision
   - In day repeatability
   - Between days repeatability
   - Reproducibility
   - Ruggedness
7. Selectivity, Specificity
8. Sensitivity
9. Robustness
10. Stability

System Suitability Tests

System suitability test parameters are the values that should be checked before calculating the other validation parameters, before the validation of the device and the developed analytical method. These are resolution (R_s; R_s >2.0), peak tailing factor (T; T≤ 2) asymmetry factor (A_s; A_s =0.95-1.2), theoretical plate number (N; N> 2000), capacity factor (κ′; κ′ = 2-10), selectivity factor (α;α>1), repeatability of the peak and peak height (RSD % <1.5) [89].

Linearity and Range

The linearity parameter analytical procedure should be displayed within the measurement range. Linearity can be studied directly on the active ingredients (diluted stock standard solutions) and / or synthetic mixtures prepared according to the recommended procedure by weighing from the components of the preparation. Visual monitoring of response signals as a function of analyte concentration provide method linearity information. If a linear relationship is found, the test results are evaluated with an appropriate statistical method (*e.g.*, plotting the regression line with the least squares method and computing the corresponding correlation coefficient and the right equation). At least five different concentration values for the control of linearity are suggested by ICH. The response of the analyte changes proportionally against concentration, in the developed method, and the response in the plotted graph lies on a straight line. Correlation (r) and / or determination (definition) (r^2) are parameters that give linearity. An analytical procedure is the determination interval; (including top and bottom concentrations) between the upper and lower concentrations that can be determined in the accuracy and precision parameters defined in the sample analyte. The analytical procedure shows that it has an appropriate level of accuracy, precision and linearity [89 - 98].

Limit of Detection

The lowest concentration that is not in the range but can be found either from calculations used or directly observed, is called as limit of detection. It can be calculated form signal to noise ratio, or by observation or using the standard deviation (σ) and slope (m) of the calibration equation with the equation of LOD: $3.3\sigma/m$ [89 - 98].

Limit of Quantification

The level at which the quantity of the material being analyzed can be determined precisely and accurately at an acceptable level, is outside the linearity range or forms the lowest limit of the range. It can be calculated form signal to noise ratio, or by observation or using the standard deviation (σ) and slope (m) of the calibration equation with the equation of LOQ: $10\sigma/m$ [89 - 98].

Accuracy

Accuracy is the defined as the resulting values, or the measure of the closeness to the true value. Accuracy gives an idea of any systematic error or deviations of the values obtained by the developed method from the correct value. The value obtained should be as close as possible to the correct value. Accuracy measures the effect of sample preparation. The calibration equation used to calculate the recovery must be the same as the equation given as linearity in the original analysis. The medium shows where the added analyte can be recovered from the analyzing medium. Accuracy can be found by comparing the results obtained with another valid and independent method of analysis, by application to the actual sample to be analyzed, or by addition of the active substance to the drug sample or matrix environment by standard addition and recovery. Results can be given as recovery %, with RSD% and Bias % values [89 - 98].

Precision

It is the measure of the reproducibility of the liquid chromatographic method developed for the analysis under actual operating conditions. It is a measure of the closeness of the results of the developed liquid chromatographic method. The results are expressed as RSD% and Bias % values. This validation parameter can be achieved by calculating RSD% in day, between days and reproducibility data.

Ruggedness also a precision value, which is a measure of the reproducibility of the method under actual use conditions. Changing the normal test parameters such as making different reagents and solvents at different temperatures, by changing the reagents and solvents in similar devices in different laboratories or by different analysts in the same laboratory or also replacing column in LC method (with a new column with the same brand and model), ruggedness can be found [89 - 98].

Selectivity

The selectivity of a liquid chromatographic method is the ability to measure the desired analyte as well as other components present in the sample to be analyzed and capable of interfering with the analyte. Other performance parameters of a chromatographic analysis method with no or insufficiently selective are also meaningless. In particular, stability is a parameter that must be done in the work. If selectivity is not sufficient, there is also a doubt about the accuracy and linearity of the method. Selectivity should be done during diagnostic testing and identification of impurities [89 - 98].

Sensitivity

Sensitivity is the slope of the linear equation in the range where linearity is valid. It is also described as LOD in many literature data and books. Smaller LOD lead to more sensitive method [89 - 98].

Robustness

The capacity of an improved chromatographic method to remain unaffected by small changes in the analysis parameters is called as robust. These parameters include the percentage of organic solvent present in the mobile phase composition, pH, ionic strength, temperature, *etc*. Robustness is a measure of the reliability of the method in routine use [89 - 98].

Stability

It is not a mandatory validation parameter to be made by the American Pharmacopoeia but is a recommended validation step to ensure reproducible and reliable results. It can be achieved by freeze-thawing stability, short term stability and long term stability [89 - 98].

In order to be able to use commonly used liquid chromatographic methods routinely today and use them effectively and reliably in analysis, validation tests must be completed. It is essential that the validation tests, which have been proven effective and reliable internationally, are carried out completely and regularly during all analysis processes, especially in the research and development and quality control departments of the health field and in the analysis and research units of the universities. The importance of ensuring the validity of chromatographic methods like other analytical methods has become increasingly important in recent years. Adequate validity tests are tried to be applied considering the approaches applied to analyze in accordance with the principles determined by various international institutions and organizations [89 - 98].

Development of Method for Drug Dissolution Testing

For the development of the HPLC method to be used, the aim and the requirements of the analysis should be prioritized. In this context, the physical and chemical properties of the analyte are examined and the detector, mobile phase, stationary phase and HPLC system to be used are determined. On the other hand; the number of samples to be investigated per unit time, possible analyte degradations (hydrolysis, photolysis, dehydration, thermolysis *etc.*), and degradation products should be investigated. Finally, the validation parameters including limit of detection, limit of quantification, linearity, accuracy and precision must be determined as a requirement of validity. The choice of the dissolution media is also important because media should meet the requirements for both solubility of the active substance and mimicking physiological conditions. Sink conditions must be provided in all study otherwise the *in vivo* conditions cannot be simulated. Also it has to be proven that used media must be compatible with chemicals used in formulation and HPLC method. Some selected researches about drug dissolutions tests are summarized in Table **3** in terms of drug dosage form, dissolution apparatus, dissolution media, condition, method conditions, linearity range, LOD, LOQ values.

Table 3. Some selected researches about drug dissolutions.

Analyte		Dissolution Conditions		HPLC Conditions			Ref
Drug	**Dosage Form**	**Dissolution Apparatus**	**Dissolution Media and Rate**	**Method and Column**	**Mobil Phase, Flow Rate and Wavelength**	**Linearity Range/ LOD/ LOQ**	
Amlodipine besylate	Tablet	USP II (Paddle)	900 mL 0.1 M HCl; 50 rpm	RP-HPLC (isocratic) Interasil Ods - 3v, 250 mm × 4.6 mm × 5 μm	Meth:ACN 50:35:15 with pH 3.0 adjusted with PA 2.0 mL.min^{-1} 217 nm	Linearity Range 1.66-16.66 ppm	[100]
Amlodipine besylate Atorvastatin calcium	Tablet	USP II (Paddle)	900mL of 0.1N HCl with 3%sls; 75 rpm	RP-HPLC (isocratic) Phenomenex Luna RP-C18 (250 mm x 4.6 mm, 5 μm)	ACN-MeOH-TEA buffer pH 4.0 (ratio 52:18:30 v/v/v) 1 mL.min^{-1} 245 nm	LOD: 0.32 μg.ml^{-1} 0.36 μg.ml^{-1} LOQ: 1.11 μg.ml^{-1} 1.20 μg.ml^{-1}	[101]

(Table 3) contd.....

Analyte		Dissolution Conditions		HPLC Conditions			Ref
Drug	**Dosage Form**	**Dissolution Apparatus**	**Dissolution Media and Rate**	**Method and Column**	**Mobil Phase, Flow Rate and Wavelength**	**Linearity Range/ LOD/ LOQ**	
Camptothecin	Controlled release Tablet	USP II (Paddle)	900 mL of PB pH 6.8; 100 rpm	RP-HPLC (isocratic) Inertsil-ODS-C18 (250mm x 4.6mm, 5μm)	15mM ammonium acetate (pH 6.5): ACN (60:40); 1mL.min^{-1} 254 nm	LOD: 0.022 μg.ml-1 LOQ: 0.077 μg.ml-1	[102]
Chlorzoxazone Aceclofenac	Film Coated Tablet	USP II (Paddle)	900 mL of 0.1 N HCl for 2 h and pH 6.8 PB; 50 rpm	RP-HPLC (isocratic); NS	ACN-PA (pH 3; 0.05 M) (65:35); 1.5 mL.min^{-1} 276 nm	NS	[103]
Chlorzoxazone	Orodispersible Tablet	USP II (Paddle)	900 mL of SSF (pH 6.8); 50 rpm	RP-HPLC (isocratic) Xterra PR18, 5mm, 4.6 mm 250 mm	Meth and 0.05M dihydrogen phosphate (50:50); 1.5 mL.min^{-1} 280 nm	NS	[104]
Daclatasvir	Tablet	USP II (Paddle)	1000 mL PB, pH 6.8 with 0.75% brij 35; 75 rpm	RP-HPLC (isocratic); Zorbax Eclipse Plus C-18 (100 x 4.6 mm, 3.5μm)	Water and Meth (20:80); 1.5 mL.min^{-1} 315 nm	Linearity Range 10-160 μg.ml^{-1}	[105]
Meloxicam	Tablet	USP IV (Flow Through Cell)	PB pH 7.5; 75 rpm	RP-HPLC (isocratic) C18-Symmetry 5 μm, 3.9 cm X 150 mm	ACN: deionized water [50:50 (% v/v)] pH 3 with GAA; 1 mL.min^{-1} 360 nm	Linearity Range 0.005 – 3 μg.ml^{-1}	[106]
Nevirapine	Sustained-Release Tablet	USP III (Reciprocating Cylinder)	250 mL 50 mM PB (pH 6.8) containing 2% SLS; 50 rpm	RP-HPLC (isocratic) Phenomenex C18 (5 μm, 250 × 4.6 mm)	ACN and water (44:56) 1 mL.min^{-1} 284 nm	Linearity Range 1–240 μg.ml^{-1}	[107]
Paracetamol and Caffeine	Tablet	USP II (Paddle)	900 mL of water 100 rpm	RP-HPLC (isocratic); ACE 5–C18, 5-μm LC (250 × 4.6 mm)	Meth/water (40:60 v/v) 1 mL.min^{-1} 270 nm	NS	[108]
Rivaroxaban	Tablet	USP II (Paddle)	900 mL 0.6% SLS solution in AB 10 mM (pH 4.5) 75 rpm	RP-HPLC (isocratic); Nova-Pak C8, 4 μm, 150 mm × 3.9 mm	KH_2PO_4 50 mM (pH 3.0) and ACN (60:40, v/v) 1 mL.min^{-1} 270 nm	LOQ: 0.58 μg.mL^{-1} LOD: 0.19 μg.mL^{-1}	[109]
Sofosbuvir Ledipasvir	Tablet	USP II (Paddle)	900 mL 1.5% polysorbate 80 in 10 mM PB with 0.0075 mg/ mL bht, pH 6.0 ± 0.5; 75 rpm	RP-HPLC (isocratic) Luna analytical 250 × 4.6 mm, 5 μm, C8	AB pH 7.0 and ACN 35:65 0.7 mL.min^{-1} 245 nm	LOD: 0.485 μg.mL^{-1} 0.175 μg.mL^{-1} LOQ: 1.619 μg.mL^{-1} 0.586 μg.mL^{-1}	[110]
Amlodipine besylate	Oral Film	USP II (Paddle)	500 mL of 0.01 M HCl 50 rpm	RP-HPLC (isocratic) Hypersil® BDS C18, Zorbax® Eclipse XDB-C18	(0.7% aqueous TEA adjusted to pH 3.0 with PA and Meth 40:60 (v/v) 1.0 mL.min^{-1} 237-241 nm	20-150 μg.mL^{-1}	[111]

(Table 3) contd.....

Analyte		Dissolution Conditions		HPLC Conditions			Ref
Drug	**Dosage Form**	**Dissolution Apparatus**	**Dissolution Media and Rate**	**Method and Column**	**Mobil Phase, Flow Rate and Wavelength**	**Linearity Range/ LOD/ LOQ**	
Etoricoxib	Tablet	USP II (Paddle)	Simulated gastric fluid pH1.2, 900 mL; 50 rpm	RP-HPLC (isocratic) Inertsil ODS-4, (250-mm x 4.6-mm), 5μm	Sodium perchlorate monohydrate (0.01M):ACN 48:52 v/v 1.5 mL.min^{-1} 235 nm	70-130 μg.ml^{-1}	[112]
Lamivudine	Tablet	USP II (Paddle)	(900 mL) (pH 1.2, 4.5, 6.8) and water; 50 rpm	RP-HPLC (isocratic) C18 (Ace 250 × 4.60 mm, 5 μm)	ACN/water (10:90) pH adjusted to pH 2.5 with PA 1 mL.min^{-1} 270 nm	LOD 0.5 μg.ml^{-1} LOQ 1 μg.ml^{-1}	[113]
Lansoprazole	Pellet	USP II (Paddle)	500 mL 0.1 HCl; 75 rpm	RP-HPLC (isocratic) LiChro-CART RP-18 (250 mm 4 mm i.d., 5 μm	ACN/ 0.01 M Na_2HPO_4 (45:55, v/v, pH 7.0 with PA 1 mL.min^{-1} 285 nm	1.0-350 μg.ml^{-1}	[114]
Memantine hydrochloride	Tablet	USP II (Paddle)	900 mL of 0.1 N HCl; 50 rpm	RP-HPLC (isocratic) C18 (Shi-seido, 25094.6 mm, 5 μm)	50 mM PB (pH 4, adjusted with orthoPA) and ACN (20:80, v/v) 2.0 mL.min^{-1} 265 nm	LOQ 1.0 μg.ml^{-1}	[115]
Metoprolol succinate and cilnidipine	Tablet	USP II (Paddle)	900mL of 0.1N HCl; 100 rpm	RP-HPLC (isocratic) ELIPSE Phenomenax C18,50X4.5mm, 5μm	PB (pH 3.0): Meth (70:30%v/v) 1 mL.min^{-1} 285 nm	Linearity Range 50-150 μg.ml^{-1}	[116]
Paracetamol Chlorpheniramide maleate Phenylephrine hydrochloride	Tablet	USP II (Paddle)	900 mL 0.1 M HCl; 50 rpm	RP-HPLC (isocratic); Eclipse XDB "C18" (4.6 mm 150 mm, 5μm)	PB (KH2PO4, 30 mM): ACN (60:40, v/v) with pH 3 1 mL.min^{-1} 265 nm	0.55-17.6, 1.4-44.8, 1.6-51.2 μg.ml^{-1}, LOD: 0.41, 1.19, 1.47 μg.ml^{-1} LOQ: 1.31, 3.51, 5.77 μg.ml^{-1}	[117]
Perindopril Indapamide	Tablet	USP II (Paddle)	500 mL of PB of pH 6.0 containing 0.5% tween 80; 100 rpm	RP-HPLC (isocratic) LiChrospher® 100 RP18 (125 mm×4.0 mm 5 μm)	ACN, Meth and PB of pH 3.0 (50:3:47, V/V/V) 0.7 mL.min^{-1} 210 nm	LOD: 3.57 μg.ml^{-1} 0.38 μg.ml^{-1} LOQ: 10.83 μg.ml^{-1} 1.15 μg.ml^{-1}	[118]
Valsartan/amlodipine	Tablet	USP II (Paddle)	900 mL pH 6.8 PB, pH 4.5 PB, and 0.1 N HCl.; 50 rpm	RP-HPLC (isocratic) Inertsil ODS-3V RP18e (250 mm × 4.6 mm, 5-μm)	(Meth/ACN/buffer [175:75:250; v/v/v]) and (water/ACN/GAA [300:700:1; v/v/v]) 1:1 1 mL.min^{-1} 220 nm	NS	[119]

(Table 3) contd.....

Analyte		**Dissolution Conditions**		**HPLC Conditions**			**Ref**
Drug	**Dosage Form**	**Dissolution Apparatus**	**Dissolution Media and Rate**	**Method and Column**	**Mobil Phase, Flow Rate and Wavelength**	**Linearity Range/ LOD/ LOQ**	
Memantine hydrochloride	Tablet	USP I (Basket)	0.1 m HCl and 2 g /L NaCl at pH 1.2; 100 rpm	RP-HPLC (isocratic) Nova-Pak C18 4 µm (250 mm×4.6 mm	ACN and sodium dihydrogenphosphate (pH 2.5; 0.05 M) (70: 30, v/v) 1.5 $mL.min^{-1}$ 360 nm	1–12 $\mu g.mL^{-1}$	[120]
Metolazone and losartan potassium	Tablet	USP II (Paddle)	900 mL of 2.0% sls; 50 rpm	RP-HPLC (isocratic) 150 4.6 mm i.d., 5µm	Meth and 0.02M PB (65:35, v/v) at pH 3.0 1 $mL.min^{-1}$ fluorescence detection at 410 nm excitation at 230nm	LOD 0.22 ng/mL 4.52 ng/mL LOQ: 0.68 ng/mL 13.70 ng/mL	[121]
Metformin and glimepiride	Tablet	USP II (Paddle)	500 mL 0.2% SLS, water and 0.1 N HCl 100 rpm	RP-HPLC (isocratic) Promocil C)8 5p 100A 4.6x150mm	ACN 55%, Ammonium acetate buffer pH 3.0 45% 1 $mL.min^{-1}$ 270 nm	NS	[122]
Nebivolol	Tablet	USP II (Paddle)	500 and 900 mL HCl 0.1 N, AB pH 4.5, PB pH 6.8; 50, 75 rpm	RP-HPLC (isocratic) ; Hypersil C18 (150 x 4.0 mm, 5 µm)	0.1% (v/v) TFAA in water: ACN (60:40, v/v) 1.25 $mL.min^{-1}$ 281 nm	1-12 $\mu g.ml^{-1}$	[123]
Pantoprazole	Enteric Coated Tablet	USP II (Paddle)	0.1N HCl, pH 1.2, PB (pH 7.4)	RP-HPLC (isocratic) l-Bondapak C18(250mm,4.6mm	Sodium hydrogen phosphate solution (0.01M)=ACN (60=40v=v) 1.5 $mL.min^{-1}$ 285nm	LOD 10 $ng.mL^{-1}$ LOQ 25$ng.mL^{-1}$	[124]
Repaglinide	Tablet	USP II (Paddle)	0.1 mol/L HCl (pH 1.0), AB (pH 4.5), PB (pH 6.8 and pH 7.4) and distilled water 50rpm	RP-HPLC (isocratic) ; C18 (150 mm 4.6 mm, 5µm)	Methe0.05 mol/l ammonium acetate (pH 4.0, 80:20, v/v) 1.0 $mL.min^{-1}$ 240 nm	NS	[125]
Simvastatin	Tablet	USP II (Paddle)	900 mL pH 7.0, 0.5% SDS in 0.01 M PB; 50 rpm	RP-HPLC (gradient) InertsilODS - 3 V (5µm, 150mm×4.6mm i.d.)	(pH=4.2, 12 mM AB and ACN 1.7 $mL.min^{-1}$ 238 nm	2.5 -16.66 $\mu g.ml^{-1}$	[126]
Valsartan Amlodipine	Immediate Release Tablet	USP I (Basket)	900 mL pH 6.8 buffer solution; 75 rpm	RP-HPLC (isocratic) C18(kinetex) (50x4.6) mm, 5µm	ACN, MeOH, d water and FA (60:30:10:0.1 v/v/v/v) 1 $mL.min^{-1}$	10-10000 $ng.ml^{-1}$	[127]

(Table 3) contd.....

Analyte		Dissolution Conditions		HPLC Conditions			Ref
Drug	**Dosage Form**	**Dissolution Apparatus**	**Dissolution Media and Rate**	**Method and Column**	**Mobil Phase, Flow Rate and Wavelength**	**Linearity Range/ LOD/ LOQ**	
Candesartan cilexetil	Tablet	USP II (Paddle)	900 mL of purified water, 0.1 N HCl, AB (pH 1.2–2.2) and PB (pH 4.5–7.2) 50 rpm	RP-HPLC (isocratic); Zorbax C8 (150·4.6 mm, 5µm)	15:85 (v/v) mixture of PB (pH 2.5) and ACN 1 mL.min^{-1} 215 nm	2-20 µg.ml^{-1}	[128]
Gabapentin	Tablet, Capsule	USP II (Paddle)	0.06N HCl; 50 rpm	RP-HPLC (isocratic) Phenomenex Luna Cyano	Meth–ACN–20 mM KH_2PO_4 (pH 2.2) (5:5:90, v/v/v) 1.25 mL.min^{-1} 210 nm	0.05–0.65 mg.mL^{-1}	[129]
Atorvastatin Amlodipine	Tablet	USP II (Paddle)	900 mL of triple distilled water; 50 rpm	RP-HPLC (isocratic) Lichrospher ® 100 C18, 5mm, 250 mm 4.0 mm	ACN and 50 mM PB (60:40, v/v), apparent pH adjusted to 3 0.1with 10% PA 1 ml.min^{-1} 254 nm	LOD: 0.4 µg.ml^{-1} 0.6 µg.ml^{-1} LOQ: 1 µg.ml^{-1} 1 µg.ml^{-1}	[130]
Abacavir, Lamivudine Zidovudine	Tablet	USP II (Paddle)	900 mL of 0.1M HCl; 75 rpm	RP-HPLC (isocratic) Zorbax C18 (150·4.6 mm ID)	Meth:water:phosphate buffer at pH 5.65 (80:10:10;v/v/v) 0.6 mL.min^{-1} 275 nm	LOD: 21.87 ng.mL^{-1} 63.03 ng.mL^{-1} 110.32 ng.mL^{-1} LOQ: 72.90 ng.mL^{-1} 210.09 ng.mL^{-1} 367.74 ng.mL^{-1}	[131]
Olmesartan Hydrochlorothiazide	Tablet	USP II (Paddle)	900 mL of PB (pH 6.8); 50 rpm	RP-HPLC (isocratic) ACE 5 CN (4.6 mm I.D·200 mm, 5µm	PA: Meth 50:50 (v/v) 1.0 mL.min^{-1} 260 nm	LOD: 0.061 gmL^{-1} 0.031 gmL^{-1} LOQ: 0.21 gmL^{-1} 0.11 gmL^{-1}	[132]
Nicotine	Chewing Gum	EP Chewing Apparatus	40 ml of artificial saliva; 60 chews/min	RP-HPLC (isocratic) Waters Xterra RP-18 4.6 x150mm	70% PB (pH 8.5): 30% ACN 1 mL.min^{-1} 260 nm	NS	[133]
Benzocaine Tyrothricin	Lozenges	Non compedial apparatus	15 ml MP; Sonication	RP-HPLC (isocratic) fused-silica 10 m × 0.53 mm i.d. capillary of 2.65-~tm	Meth-10mM PB (pH 3.31) (75:25, v/v) 1 mL.min^{-1} 270 nm	10-250 µg.ml^{-1} 100-1000µg.ml^{-1}	[134]

AB: Acetate buffer, ACN: Acetonitrile, GAA: glacial acetic acid, Meth: Methanol, MP: Mobile phase, NS. Not stated, PA: Phosphoric acid, PB: Phosphate buffer, SDS: Sodium dodecyl sulfate, TEA: triethylamine, TFAA: trifluoroacetic acid

In their study, Yan *et al.*, 2015 developed a dissolution test for Lacidipine. The authors tried to simulate the human small intestinal fluids, including fasted state simulated intestinal fluid (FaSSIF) and fed state simulated intestinal fluid (FeSSIF) (Fig. **5**). They suggested that Tween 80 solution is an alternative medium to FaSSIF and FeSSIF. Dissolution profiles of two Lacidipine formulations in FaSSIF and sodium dodecyl sulphate (SDS) solutions were studied and within 30 min. a complete dissolution was obtained (Fig. **5a** and **b**). Using tween 80 the solubility of Lacidipine was increased and faster dissolution rate was obtained (Fig. **5c**) [99].

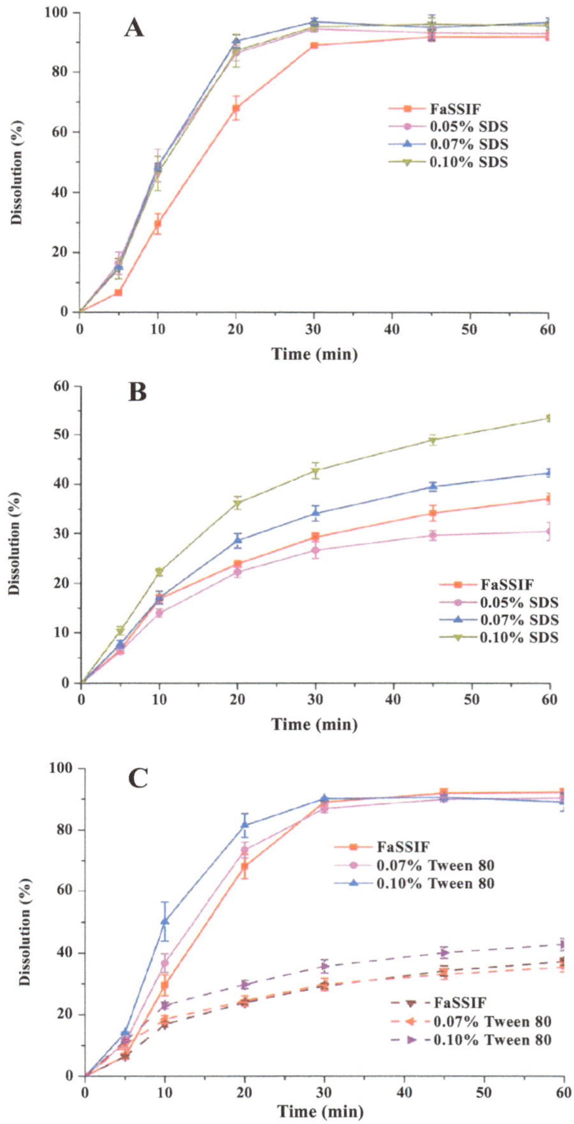

Fig. (5). A) Dissolution profiles of Lacidipine formulation A in FaSSIF and SDS solutions, **B**) Dissolution profiles of Lacidipine formulation B in FaSSIF and SDS solutions, **C**) Dissolution profiles of two Lacidipine formulations in FaSSIF and Tween 80 solutions. (Reprinted with permission [99].

CONCLUSION

Solid oral dosage forms are the most convenient and patient-accepted drug delivery type in the world. The physical and chemical stability of solid oral dosage forms are generally better than those of other dosage forms. However, they have several disadvantages resulting from the application route, including irregular absorption due to the GI environment, slow onset of action, decomposition of some drugs in the stomach, and insufficient permeation of high-molecular weight drugs. Due to these limitations, different dosage forms are prepared.

It is possible to use immediate, controlled or extended-release systems in solid dosage forms, such as tablets, capsules, powders, suppositories, and lozenges. The preparation of a solid dosage form depends on the physical properties of the active substance and excipients. To design an effective system and test the effectiveness, it is important to determine critical parameters both in pharmacopeia analysis and scientific studies. Liquid chromatography is the most common technique used in pharmaceutical analysis. System suitability tests involving parameters such as peak tailing, capacity factor, plate numbers, resolution between peaks, and selectivity relative to preceding peaks should be performed to validate chromatographic methods. Other validation parameters that should be reported in detail include linearity and range, limit of detection, limit of quantification, accuracy, precision (within-day repeatability, between-day repeatability, reproducibility, and ruggedness), selectivity, specificity, sensitivity, robustness and stability. The choice of dissolution medium is also important for meeting the requirements of the active substance concerning both solubility and physiological conditions. Sink conditions must be provided in all studies in order to simulate *in vivo* conditions. Lastly, the medium must be compatible with chemicals used in the formulation and the HPLC method.

CONSENT FOR PUBLICATION

Not applicable.

CONFLICT OF INTEREST

The authors declares no conflict of interest, financial or otherwise.

ACKNOWLEDGEMENTS

Declared none.

REFERENCES

[1] Treffer, D; Graz, I; Austria, H J Injection Moulding of Solid Oral Dosage Forms. **2016**, 1369-1372.

[2] Andrews, G.P. Advances in solid dosage form manufacturing technology. Philosophical Transactions of the Royal Society of London A: Mathematical. *Physical and Engineering Sciences,* **1861**, *2007*(365), 2935-2949.

[3] Denk, R. Forms-Review of ISPE's Baseline Guide for Oral Solid-Dosage Forms. *Pharmaceutical Technology,* **2010**, *52*, S6.

[4] Reynolds, J.E. *Martindale: the extra pharmacopoeia*; The Pharmaceutical Press: London, UK, **1982**.

[5] Lachman, L.; Lieberman, H.A.; Kanig, J.L. *The Theory and Practice of Industrial Pharmacy,* 3rd ed; Publications, Lea Febiger, Ed.; **1986**.

[6] Wermuth, C.G. *The Practice of Medicinal Chemistry,* 2nd ed; Elsevier Publications, **2003**.

[7] Carstensen, J.T. *Pharmaceutical Principles of Solid Dosage Forms,* 1st ed; Technomic Publishing Company, **1993**.

[8] Tenjarla, S. Microemulsions: an overview and pharmaceutical applications. *Crit. Rev. Ther. Drug Carrier Syst.,* **1999**, *16*(5), 461-521. [http://dx.doi.org/10.1615/CritRevTherDrugCarrierSyst.v16.i5.20] [PMID: 10635455]

[9] Cole, E.T. Liquid filled hard gelatin capsules. *Pharm Tech,* **1989**, *13*(9), 124-140.

[10] Yu, L.X. Pharmaceutical quality by design: product and process development, understanding, and control. *Pharm. Res.,* **2008**, *25*(4), 781-791. [http://dx.doi.org/10.1007/s11095-007-9511-1] [PMID: 18185986]

[11] Görög, S. Drug safety, drug quality, drug analysis. *J. Pharm. Biomed. Anal.,* **2008**, *48*(2), 247-253. [http://dx.doi.org/10.1016/j.jpba.2007.10.038] [PMID: 18082992]

[12] Sampson, C.B.; Breunig, H.L. Some Statistical Aspects of Pharmaceutical Content Uniformity. *J. Qual. Technol.,* **1971**, *3*(4), 170-178.

[13] Williams, R.L.; Adams, W.P.; Poochikian, G.; Hauck, W.W. Content uniformity and dose uniformity: current approaches, statistical analyses, and presentation of an alternative approach, with special reference to oral inhalation and nasal drug products. *Pharm. Res.,* **2002**, *19*(4), 359-366. [http://dx.doi.org/10.1023/A:1015114821387] [PMID: 12033365]

[14] Bánfai, B.; Ganzler, K.; Kemény, S. Content uniformity and assay requirements in current regulations. *J. Chromatogr. A,* **2007**, *1156*(1-2), 206-212. [http://dx.doi.org/10.1016/j.chroma.2006.10.067] [PMID: 17109871]

[15] Martins, J.M.; Farinha, A. Uniformity of dosage units--comparative study of methods and specifications between Eur. Pharm. 3rd and USP 23. *J. Pharm. Biomed. Anal.,* **1998**, *18*(4-5), 487-495. [http://dx.doi.org/10.1016/S0731-7085(98)00269-6] [PMID: 9919948]

[16] Donauer, N.; Löbenberg, R. A mini review of scientific and pharmacopeial requirements for the disintegration test. *Int. J. Pharm.,* **2007**, *345*(1-2), 2-8. [http://dx.doi.org/10.1016/j.ijpharm.2007.08.045] [PMID: 17935916]

[17] Cohen, J.L.; Hubert, B.B.; Leeson, L.J.; Rhodes, C.T.; Robinson, J.R.; Roseman, T.J.; Shefter, E. The development of USP dissolution and drug release standards. *Pharm. Res.,* **1990**, *7*(10), 983-987. [http://dx.doi.org/10.1023/A:1015922629207] [PMID: 2281043]

[18] Zahirul, M.; Khan, I. Dissolution testing for sustained or controlled release oral dosage forms and correlation with *in vivo* data: challenges and opportunities. *Int. J. Pharm.,* **1996**, *140*(2), 131-143. [http://dx.doi.org/10.1016/0378-5173(96)04561-9]

[19] Us Pharmacopoeia. *XXI, Pharmacopoeial Convention,* **1985**, Rockville, Md.

[20] Japanese Pharmacopoeia Editorial Committee. *The Japanese Pharmacopoeia,* 15th ed; The Society of

Japanese Pharmacopoeia: Tokyo, Japan, **2006**.

[21] European Directorate for the Quality Of Medicines. *European Pharmacopoeia*; **2008**.

[22] China Pharmacopoeia Committee. *Chinese Pharmacopoeia*; China Medica Science Press: Beijing, China, **2010**.

[23] Pharmacopoeia, Indian. *The Indian Pharmacopoeia Commision*; **2007**.

[24] Food, Korea, And Drug Adminitration. *The Korean Pharmacopoeia*; **2007**.

[25] Kostewicz, E.S.; Abrahamsson, B.; Brewster, M.; Brouwers, J.; Butler, J.; Carlert, S.; Dickinson, P.A.; Dressman, J.; Holm, R.; Klein, S.; Mann, J.; McAllister, M.; Minekus, M.; Muenster, U.; Müllertz, A.; Verwei, M.; Vertzoni, M.; Weitschies, W.; Augustijns, P. *In vitro* models for the prediction of *in vivo* performance of oral dosage forms. *Eur. J. Pharm. Sci.,* **2014**, *57*, 342-366. [http://dx.doi.org/10.1016/j.ejps.2013.08.024] [PMID: 23988843]

[26] Garbacz, G.; Klein, S. Dissolution testing of oral modified-release dosage forms. *J. Pharm. Pharmacol.,* **2012**, *64*(7), 944-968. [http://dx.doi.org/10.1111/j.2042-7158.2012.01477.x] [PMID: 22686342]

[27] Dressman, J.B.; Amidon, G.L.; Reppas, C.; Shah, V.P. Dissolution testing as a prognostic tool for oral drug absorption: immediate release dosage forms. *Pharm. Res.,* **1998**, *15*(1), 11-22. [http://dx.doi.org/10.1023/A:1011984216775] [PMID: 9487541]

[28] Uddin, R.; Saffoon, N.; Sutradhar, K.B. Dissolution and dissolution apparatus: a review. *Int J Cur Biomed Phar Res,* **2011**, *1*(4), 201-207.

[29] Dokoumetzidis, A.; Macheras, P. A century of dissolution research: from Noyes and Whitney to the biopharmaceutics classification system. *Int. J. Pharm.,* **2006**, *321*(1-2), 1-11. [http://dx.doi.org/10.1016/j.ijpharm.2006.07.011] [PMID: 16920290]

[30] Siewert, M.; Dressman, J.; Brown, C.K.; Shah, V.P.; Aiache, J.M.; Aoyagi, N.; Crison, J. FIP; AAPS. FIP/AAPS guidelines to dissolution/*in vitro* release testing of novel/special dosage forms. *AAPS PharmSciTech,* **2003**, *4*(1), E7. [http://dx.doi.org/10.1208/pt040107] [PMID: 12916916]

[31] Pernarowski, M.; Woo, W.; Searl, R.O. Continuous flow apparatus for the determination of the dissolution characteristics of tablets and capsules. *J. Pharm. Sci.,* **1968**, *57*(8), 1419-1421. [http://dx.doi.org/10.1002/jps.2600570832] [PMID: 5677352]

[32] Uddin, R.; Saffoon, N.; Sutradhar, K.B. Dissolution and dissolution apparatus: a review. *Int J Cur Biomed Phar Res,* **2011**, *1*(4), 201-207.

[33] Levy, G.; Hayes, B.A. Physicochemical basis of the buffered acetylsalicylic acid controversy. *N. Engl. J. Med.,* **1960**, *262*, 1053-1058. [http://dx.doi.org/10.1056/NEJM196005262622102] [PMID: 14416389]

[34] Dyas, A.M.; Shah, U.U. Dissolution and Dissolution Testing.*Encyclopedia of Pharmaceutical Technology*; Swarbrick, J., Ed.; Informa Healthcare USA Inc.: USA, **2007**, pp. 908-928.

[35] Jantratid, E.; Janssen, N.; Chokshi, H.; Tang, K.; Dressman, J.B. Designing biorelevant dissolution tests for lipid formulations: case example--lipid suspension of RZ-50. *Eur. J. Pharm. Biopharm.,* **2008**, *69*(2), 776-785. [http://dx.doi.org/10.1016/j.ejpb.2007.12.010] [PMID: 18226882]

[36] Klein, S.; Stein, J.; Dressman, J. Site-specific delivery of anti-inflammatory drugs in the gastrointestinal tract: an in-vitro release model. *J. Pharm. Pharmacol.,* **2005**, *57*(6), 709-719. [http://dx.doi.org/10.1211/0022357056172] [PMID: 15969925]

[37] Li, J.; Yang, L.; Ferguson, S.M.; Hudson, T.J.; Watanabe, S.; Katsuma, M.; Fix, J.A. In vitro evaluation of dissolution behavior for a colon-specific drug delivery system (CODES) in multi-pH media using United States Pharmacopeia apparatus II and III. *AAPS PharmSciTech,* **2002**, *3*(4), E33. [http://dx.doi.org/10.1208/pt030433] [PMID: 12916927]

[38] Ramos Pezzini, B.; Gomes Ferraz, H. Bio-dis and the paddle dissolution apparatuses applied to the release characterization of ketoprofen from hypromellose matrices. *AAPS PharmSciTech,* **2009**, *10*(3), 763-771.
[http://dx.doi.org/10.1208/s12249-009-9262-x] [PMID: 19495988]

[39] McAllister, M. Dynamic dissolution: a step closer to predictive dissolution testing? *Mol. Pharm.,* **2010**, *7*(5), 1374-1387.
[http://dx.doi.org/10.1021/mp1001203] [PMID: 20698545]

[40] Cassilly, D.; Kantor, S.; Knight, L.C.; Maurer, A.H.; Fisher, R.S.; Semler, J.; Parkman, H.P. Gastric emptying of a non-digestible solid: assessment with simultaneous SmartPill pH and pressure capsule, antroduodenal manometry, gastric emptying scintigraphy. *Neurogastroenterol. Motil.,* **2008**, *20*(4), 311-319.
[http://dx.doi.org/10.1111/j.1365-2982.2007.01061.x] [PMID: 18194154]

[41] Buch, P.; Langguth, P.; Kataoka, M.; Yamashita, S. IVIVC in oral absorption for fenofibrate immediate release tablets using a dissolution/permeation system. *J. Pharm. Sci.,* **2009**, *98*(6), 2001-2009.
[http://dx.doi.org/10.1002/jps.21576] [PMID: 18855916]

[42] Dahan, A.; Hoffman, A. The effect of different lipid based formulations on the oral absorption of lipophilic drugs: the ability of *in vitro* lipolysis and consecutive ex vivo intestinal permeability data to predict *in vivo* bioavailability in rats. *Eur. J. Pharm. Biopharm.,* **2007**, *67*(1), 96-105.
[http://dx.doi.org/10.1016/j.ejpb.2007.01.017] [PMID: 17329087]

[43] Fitzpatrick, S.; McCabe, J.F.; Petts, C.R.; Booth, S.W. Effect of moisture on polyvinylpyrrolidone in accelerated stability testing. *Int. J. Pharm.,* **2002**, *246*(1-2), 143-151.
[http://dx.doi.org/10.1016/S0378-5173(02)00375-7] [PMID: 12270617]

[44] Phadnis, N.V.; Suryanarayanan, R. Polymorphism in anhydrous theophylline--implications on the dissolution rate of theophylline tablets. *J. Pharm. Sci.,* **1997**, *86*(11), 1256-1263.
[http://dx.doi.org/10.1021/js9701418] [PMID: 9383736]

[45] Marchais, H.; Cayzeele, G.; Legendre, J.Y.; Skiba, M.; Arnaud, P. Cross-linking of hard gelatin carbamazepine capsules: effect of dissolution conditions on *in vitro* drug release. *Eur. J. Pharm. Sci.,* **2003**, *19*(2-3), 129-132.
[http://dx.doi.org/10.1016/S0928-0987(03)00070-8] [PMID: 12791415]

[46] Rohrs, B.R.; Thamann, T.J.; Gao, P.; Stelzer, D.J.; Bergren, M.S.; Chao, R.S. Tablet dissolution affected by a moisture mediated solid-state interaction between drug and disintegrant. *Pharm. Res.,* **1999**, *16*(12), 1850-1856.
[http://dx.doi.org/10.1023/A:1018951309506] [PMID: 10644073]

[47] Gray, V.; Kelly, G.; Xia, M.; Butler, C.; Thomas, S.; Mayock, S. The science of USP 1 and 2 dissolution: present challenges and future relevance. *Pharm. Res.,* **2009**, *26*(6), 1289-1302.
[http://dx.doi.org/10.1007/s11095-008-9822-x] [PMID: 19165579]

[48] Grassi, M.; Lamberti, G.; Cascone, S.; Grassi, G. Mathematical modeling of simultaneous drug release and *in vivo* absorption. *Int. J. Pharm.,* **2011**, *418*(1), 130-141.
[http://dx.doi.org/10.1016/j.ijpharm.2010.12.044] [PMID: 21237258]

[49] Sathe, P.M.; Tsong, Y.; Shah, V.P. In-vitro dissolution profile comparison: statistics and analysis, model dependent approach. *Pharm. Res.,* **1996**, *13*(12), 1799-1803.
[http://dx.doi.org/10.1023/A:1016020822093] [PMID: 8987074]

[50] Polli, J.E.; Rekhi, G.S.; Shah, V.P. Methods to compare dissolution profiles. *Drug Inf. J.,* **1996**, *30*(4), 1113-1120.
[http://dx.doi.org/10.1177/009286159603000428]

[51] Costa, P. An alternative method to the evaluation of similarity factor in dissolution testing. *Int. J. Pharm.,* **2001**, *220*(1-2), 77-83.

[http://dx.doi.org/10.1016/S0378-5173(01)00651-2] [PMID: 11376969]

[52] Tsong, Y.; Hammerstrom, T.; Chen, J.J. Multipoint dissolution specification and acceptance sampling rule based on profile modeling and principal component analysis. *J. Biopharm. Stat.,* **1997**, *7*(3), 423-439.
[http://dx.doi.org/10.1080/10543409708835198] [PMID: 9252835]

[53] Siepmann, J.; Peppas, N.A. Hydrophilic matrices for controlled drug delivery: an improved mathematical model to predict the resulting drug release kinetics (the “sequential layer” model). *Pharm. Res.,* **2000**, *17*(10), 1290-1298.
[http://dx.doi.org/10.1023/A:1026455822595] [PMID: 11145237]

[54] Chirico, S.; Dalmoro, A.; Lamberti, G.; Russo, G.; Titomanlio, G. Analysis and modeling of swelling and erosion behavior for pure HPMC tablet. *J. Control. Release,* **2007**, *122*(2), 181-188.
[http://dx.doi.org/10.1016/j.jconrel.2007.07.001] [PMID: 17706830]

[55] Barba, A.A.; d’Amore, M.; Chirico, S.; Lamberti, G.; Titomanlio, G. A general code to predict the drug release kinetics from different shaped matrices. *Eur. J. Pharm. Sci.,* **2009**, *36*(2-3), 359-368.
[http://dx.doi.org/10.1016/j.ejps.2008.10.006] [PMID: 19022380]

[56] Wu, X.Y.; Zhou, Y. Finite element analysis of diffusional drug release from complex matrix systems: II. Factors influencing release kinetics1Presented in part at the Conference on Advances in Controlled Delivery, Controlled Release Society, Inc. Proceedings, pp. 169–170 (August, 1996, Baltimore, USA), and at the Annual Meeting of the American Association of Pharmaceutical Scientists (October, 1996, Seattle, USA). 1. *J Cont Rel,* **1998**, *51*(1), 57-71.
[http://dx.doi.org/10.1016/S0168-3659(97)00159-4]

[57] Grassi, M.; Zema, L.; Sangalli, M.E.; Maroni, A.; Giordano, F.; Gazzaniga, A. Modeling of drug release from partially coated matrices made of a high viscosity HPMC. *Int. J. Pharm.,* **2004**, *276*(1-2), 107-114.
[http://dx.doi.org/10.1016/j.ijpharm.2004.02.016] [PMID: 15113619]

[58] Yuksel, N.; Kanik, A.E.; Baykara, T. Comparison of *in vitro* dissolution profiles by ANOVA-based, model-dependent and -independent methods. *Int. J. Pharm.,* **2000**, *209*(1-2), 57-67.
[http://dx.doi.org/10.1016/S0378-5173(00)00554-8] [PMID: 11084246]

[59] Amidon, G.L.; Lennernäs, H.; Shah, V.P.; Crison, J.R. A theoretical basis for a biopharmaceutic drug classification: the correlation of *in vitro* drug product dissolution and *in vivo* bioavailability. *Pharm. Res.,* **1995**, *12*(3), 413-420.
[http://dx.doi.org/10.1023/A:1016212804288] [PMID: 7617530]

[60] Tsvet, M. Physikalisch-Chemische Studien Über Das Chlorophyll. *Die Adsorptionen. Ber Dtsch Botan Ges,* **1906**, *24*, 316-323.

[61] Tsvet, M. Adsorptionsanalyse Und Chromatographische Methode. *Anwendung Auf Die Chemie Des Chlorophylls. Ber Dtsch Botan Ges,* **1906**, *24*, 384-393.

[62] Robards, K.; Robards, K.; Haddad, P.R.; Jackson, P.E. *Principles and practice of modern chromatographic methods*; Academic Press, **1994**.

[63] Colin, H.; Guiochon, G. Introduction to reversed-phase high-performance liquid chromatography. *J. Chromatogr. A,* **1977**, *141*(3), 289-312.
[http://dx.doi.org/10.1016/S0021-9673(00)93537-4]

[64] Engelhardt, H. *High performance liquid chromatography*; Springer Science Business Media, **2012**.

[65] Swartz, M.E. UPLC: An Introduction and Review. *Journal of LJ Liq Chromatogr R T,* **2005**, *28*, 1253-1263.
[http://dx.doi.org/10.1081/JLC-200053046]

[66] Gumustas, M.; Kurbanoglu, S.; Uslu, B.; Ozkan, S.A. UPLC *versus* HPLC on drug analysis: advantageous, applications and their validation parameters. *Chromatographia,* **2013**, *76*(21-22), 1365-1427.

[http://dx.doi.org/10.1007/s10337-013-2477-8]

[67] Zotou, A. An overview of recent advances in HPLC instrumentation. *Cent. Eur. J. Chem.*, **2012**, *10*(3), 554-569.

[68] Shen, Y.; Zhang, R.; Moore, R.J.; Kim, J.; Metz, T.O.; Hixson, K.K.; Zhao, R.; Livesay, E.A.; Udseth, H.R.; Smith, R.D. Automated 20 kpsi RPLC-MS and MS/MS with chromatographic peak capacities of 1000-1500 and capabilities in proteomics and metabolomics. *Anal. Chem.*, **2005**, *77*(10), 3090-3100. [http://dx.doi.org/10.1021/ac0483062] [PMID: 15889897]

[69] Huber, J.F. Evaluation of detectors for liquid chromatography in columns. *J. Chromatogr. Sci.*, **1969**, *7*(3), 172-176. [http://dx.doi.org/10.1093/chromsci/7.3.172]

[70] Novotny, M. Recent advances in microcolumn liquid chromatography. *Anal. Chem.*, **1988**, *60*(8), 500A-510A. [http://dx.doi.org/10.1021/ac00159a718] [PMID: 3369701]

[71] Barba, A.O.; Hurtado, M.C.; Mata, M.S.; Ruiz, V.F.; De Tejada, M.L. Application of a UV–vis detection-HPLC method for a rapid determination of lycopene and β-carotene in vegetables. *Food Chem.*, **2006**, *95*(2), 328-336. [http://dx.doi.org/10.1016/j.foodchem.2005.02.028]

[72] Sittampalam, G.; Wilson, G.S. Surface-modified electrochemical detector for liquid chromatography. *Anal. Chem.*, **1983**, *55*(9), 1608-1610. [http://dx.doi.org/10.1021/ac00260a039]

[73] Riggin, R.M.; Kissinger, P.T. Determination of catecholamines in urine by reverse-phase liquid chromatography with electrochemical detection. *Anal. Chem.*, **1977**, *49*(13), 2109-2111. [http://dx.doi.org/10.1021/ac50021a052] [PMID: 907164]

[74] Her, N.; Amy, G.; McKnight, D.; Sohn, J.; Yoon, Y. Characterization of DOM as a function of MW by fluorescence EEM and HPLC-SEC using UVA, DOC, and fluorescence detection. *Water Res.*, **2003**, *37*(17), 4295-4303. [http://dx.doi.org/10.1016/S0043-1354(03)00317-8] [PMID: 12946913]

[75] Smalley, M.B.; McGown, L.B. Fluorescence detectors in HPLC. *Adv. Chromatogr.*, **1997**, *37*, 29-72.

[76] Christie, W.W. Rapid separation and quantification of lipid classes by high performance liquid chromatography and mass (light-scattering) detection. *J. Lipid Res.*, **1985**, *26*(4), 507-512. [PMID: 4009068]

[77] Covey, T.R.; Lee, E.D.; Bruins, A.P.; Henion, J.D. Liquid chromatography/mass spectrometry. *Anal. Chem.*, **1986**, *58*(14), 1451A-1461A. [http://dx.doi.org/10.1021/ac00127a001] [PMID: 3789400]

[78] Binder, H. Separation of monosaccharides by high-performance liquid chromatography: comparison of ultraviolet and refractive index detection. *J. Chromatogr. A*, **1980**, *189*(3), 414-420. [http://dx.doi.org/10.1016/S0021-9673(00)80322-2]

[79] Haddad, P.R.; Heckenberg, A.L. High-performance liquid chromatography of inorganic and organic ions using low-capacity ion-exchange columns with indirect refractive index detection. *J. Chromatogr. A*, **1982**, *252*, 177-184. [http://dx.doi.org/10.1016/S0021-9673(01)88409-0]

[80] Zuo, Y.; Chen, H.; Deng, Y. Simultaneous determination of catechins, caffeine and gallic acids in green, Oolong, black and pu-erh teas using HPLC with a photodiode array detector. *Talanta*, **2002**, *57*(2), 307-316. [http://dx.doi.org/10.1016/S0039-9140(02)00030-9] [PMID: 18968631]

[81] Viñas, P.; López-Erroz, C.; Marín-Hernández, J.J.; Hernández-Córdoba, M. Determination of phenols in wines by liquid chromatography with photodiode array and fluorescence detection. *J. Chromatogr. A*, **2000**, *871*(1-2), 85-93.

[http://dx.doi.org/10.1016/S0021-9673(99)01087-0] [PMID: 10735289]

[82] Nogata, Y.; Ohta, H.; Yoza, K.I.; Berhow, M.; Hasegawa, S. High-performance liquid chromatographic determination of naturally occurring flavonoids in Citrus with a photodiode-array detector. *J. Chromatogr. A,* **1994**, *667*(1-2), 59-66. [http://dx.doi.org/10.1016/0021-9673(94)89051-X]

[83] Aboul-Enein, H.Y.; Stefan, R-I.; Baiulescu, G-E. *Quality and Reliability in Analytical Chemistry*; Crc Press: Boca Raton, Fl, **2000**. [http://dx.doi.org/10.1201/9781420038576]

[84] Peters, F.T.; Maurer, H.H. Bioanalytical Method Validation and Its Implications for Forensic and Clinical Toxicology – A Review.*Validation in Chemical Measurement*; De Bievre, P.; Günzler, H., Eds.; Springer: Heidelberg, **2002**, pp. 1-9. [http://dx.doi.org/10.1007/3-540-27034-5_1]

[85] Shah, V.P.; Midha, K.K.; Dighe, S.; Mcgilveray, I.J.; Skelly, J.P.; Yacobi, A.; Layloff, T.; Viswanathan, C.T.; Cook, C.E.; Mcdowall, R.D.; Pittman, K.A.; Spector, S. Analytical Methods Validation: Bioavailability, Bioequivalence And Pharmacokinetic Studies. *Conference Report. Pharm Res,* **1992**, *9*, 588-592.

[86] Pickering, W.F. General Strategies for Speciation.*Chemical Speciation in the Environment*; Ure, A., Ed.; Blackwell Science Ltd.: Oxford, **2002**, pp. 9-29. [http://dx.doi.org/10.1002/9780470988312.ch2]

[87] Hartmann, C.; Smeyers-Verbeke, J.; Massart, D.L.; McDowall, R.D. Validation of bioanalytical chromatographic methods. *J. Pharm. Biomed. Anal.,* **1998**, *17*(2), 193-218. [http://dx.doi.org/10.1016/S0731-7085(97)00198-2] [PMID: 9638572]

[89] ICH Guidelines. *Validation of Analytical Procedures: Text and Methodology,* **1994**, *Q2*(R1).

[90] ICH Guidelines. *Stability Testing of New Drug Substances and Products,* **2003**, *Q1A*(R2).

[91] USP 31. Validation of Compendial Methods. In: *The United States Pharmacopeia XXXI*; The United States Pharmacopeial Convention, Inc.: Rockville, MD, **2008**.

[92] Riley, C.M. Statistical Parameters and Analytical Figures of Merit.*Pharmaceutical and Biomedical Analysis: Development and Validation of Analytical Methods*; Riley, C.M.; Rosanske, T.W., Eds.; Elsevier: Oxford, **1996**.

[93] Me, S. *Krull I. Handbook of Analytical Validation*; Crc Press: Boca Raton, Fl, **2012**.

[94] Özkan, S.A. *Electroanalytical Methods in Pharmaceutical Analysis and Their Validation*; Hnb Pub.: USA, **2012**.

[95] Özkan, S.A.; Kauffmann, J.M.; Zuman, P. *Electroanalysis in Biomedical and Pharmaceutical Sciences (Main Title), (Voltammetry, Amperometry, Biosensors, Applications)*; Springer-Verlag Berlin Heidelberg, **2015**. [http://dx.doi.org/10.1007/978-3-662-47138-8]

[96] Srinubabu, G.; Jaganbabu, K.; Sudharani, B.; Venugopal, K.; Irizasankar, G. Rao Jvlns. Development and Validation for the Determination of an Experimental Design. *Chromatographia,* **2006**, *64*, 95-100. [http://dx.doi.org/10.1365/s10337-006-0829-3]

[97] Heyden, Y.V.; Massart, Dl. Review of the Use of Robustness and Ruggedness in Analytical Chemistry.*Robustness of Analytical Chemical Methods and Pharmaceutical Technologies Products*; Hendricks, M.M.; De Boer, J.H.; Smilde, A.K., Eds.; Elsevier: Amsterdam, **1996**, pp. 79-148. [http://dx.doi.org/10.1016/S0922-3487(96)80016-5]

[98] FDA. *Analytical Procedures and Methods Validation for Drugs and Biologics Guidance for Industry,* **2015**.

[99] Yang, B.; Wu, C.; Ji, B.; Ai, X.; Kuang, X.; Wu, M.; Sun, J. The biorelevant concentration of Tween 80 solution is a simple alternative medium to simulated fasted state intestinal fluid. *RSC Advances,*

2015, *5*(127), 104846-104853.
[http://dx.doi.org/10.1039/C5RA17674C]

[100] Kumar, G.; Mittal, A. Development and Validation of an Analytical Method for the Estimation of Amlodipine Besylate Tablets by Reverse Phase HPLC. *Int J Multidis Res Inn,* **2012**, *2*, 1-20.

[101] Toan, P.D.; Minh, P.P. A, Ly DTT. Novel HPLC-UV Method Using Volatile Buffer for Simultaneous Determination of Amlodipine Besylate and Atorvastatin Calcium. *Int J. Pharm. Pharm. Sci.,* **2016**, *8*(7)

[102] Nalluri, B.N.; Valluru, S.; Bonthu, C. Development Of Validated Stability Indicating RP-HPLC-PDA Method For Camptothecin Analysis. *J Appl Pharmaceut Sci Vol,* **2016**, *6*(09), 140-146.
[http://dx.doi.org/10.7324/JAPS.2016.60921]

[103] Jain, S.; Jain, A.; Jain, A.; Shrivastava, S.; Jain, A.K. Development and Evaluation of Film Coated Aceclofenac and Chlorzoxazone Tablet with Enhanced Dissolution Rate. *J Pharma Investig,* **2016**, *46*(5), 467-474.

[104] Moqbel, H.A.; ElMeshad, A.N.; El-Nabarawi, M.A. A pharmaceutical study on chlorzoxazone orodispersible tablets: formulation, in-vitro and in-vivo evaluation. *Drug Deliv.,* **2016**, *23*(8), 2998-3007.
[http://dx.doi.org/10.3109/10717544.2016.1138340] [PMID: 26828616]

[105] Chakravarthy, V.A.; Sailaja, B.B. Method Development And Validation Of Assay And Dissolution Methods For The Estimation Of Daclatasvir In Tablet Dosage Forms By Reverse Phase HPLC. *European J Pharm And Med Res,* **2016**, *3*(7), 356-364.

[106] Emara, L. H.; Emam, M. F.; Taha, N. F.; Raslan, H. M.; El-Ashmawy, A. A. A Simple and Sensitive HPLC/Uv Method For Determination Of Meloxicam In Human Plasma For Bioavailability And Bioequivalence Studies. *J Appl Pharmaceut Sci,* **2016**, *6*(7), 12-19.

[107] Mwila, C.; Khamanga, S.M.; Walker, R.B. Development And Assessment Of A Usp Apparatus 3 Dissolution Test Method For Sustained-Release Nevirapine Matrix Tablets. *Dissolut. Technol.,* **2016**, *23*(3), 22-30.
[http://dx.doi.org/10.14227/DT230316P22]

[108] Akgeyik, E.; Kaynak, M.S.; Çelebier, M.; Altınöz, S.; Şahin, S. Evaluation of Pharmaceutical Quality Of Conventional Dosage Forms Containing Paracetamol And Caffeine Available In The Turkish Drug Market. *Diss Tech,* in press

[109] Souri, E.; Mottaghi, S.; Zargarpoor, M.; Ahmadkhaniha, R.; Jalalizadeh, H. Development Of A Stability-Indicating HPLC Method And A Dissolution Test For Rivaroxaban Dosage Forms. *Acta Chromatogr.,* **2016**, 1-15.

[110] Zaman, B.; Siddique, F.; Hassan, W. RP-HPLC Method For Simultaneous Determination Of Sofosbuvir And Ledipasvir In Tablet Dosage Form And Its Application To *In Vitro* Dissolution Studies. *Chromatographia,* **2016**, *79*(23-24), 1605-1613.
[http://dx.doi.org/10.1007/s10337-016-3179-9]

[111] Pawar, H.A.; Yadav, A. Development And Application of RP-HPLC Method For Dissolution Study Of Oral Formulations Containing Amlodipine Besylate. *Pharm. Anal. Acta,* **2015**, *6-10*, 1-6.

[112] Singh, B.; Santhakumar, R.; Bala, I.; Prasad, S.B.; Verma, S. Development and Validation of RP-HPLC Method For The Dissolution And Assay Of Etoricoxib In Pharmaceutical Dosage Forms. *Int J Pharmaceut Qual Ass,* **2015**, *6*, 1-7.

[113] Ozturk, N.; Kaynak, M.S.; Sahin, S. Comparison of Dissolution Profiles Of Commercially Available Lamivudine Tablets. *Dissolut. Technol.,* **2015**, *22*(4), 38-43.
[http://dx.doi.org/10.14227/DT220415P38]

[114] Tsai, T.R.; Cham, T.M.; Wu, Y.T. HPLC Determination of Lansoprazole and Method Application for the Formulation Development of Enteric-Coated Lansoprazole Pellets. *J Chinese Chem Soci,* **2015**, *62*(6), 513-520.

[http://dx.doi.org/10.1002/jccs.201500003]

[115] Maeng, H.J.; Choi, S.U.; Jang, D.J.; Lee, D.W.; Ahn, B.N.; Choi, M.K.; Cho, K.H. Validation And Application Of A Simple Reverse Phase HPLC Method For *In Vitro* Dissolution Studies Of Memantine Hydrochloride Tablet. *J. Pharm. Investig.,* **2015**, *45*(5), 415-421. [http://dx.doi.org/10.1007/s40005-015-0184-1]

[116] Rao, D.S.; Venkateswarlu, T.; Krishna, G.R. Development and Validation of HPLC Method of Dissolution Test For Metoprolol Succinate And Cilnidipine. *Int. J. Pharm. Chem. Biol. Sci.,* **2015**, *5*(4), 971-981.

[117] Upadhyay, Y.; Sharma, N.; Sarma, G.S.; Rawal, R.K. Application of RP–HPLC Method In Dissolution Testing And Statistical Evaluation By Nassam For Simultaneous Estimation Of Tertiary Combined Dosages Forms. *J. Pharm. Anal.,* **2015**, *5*(5), 307-315. [http://dx.doi.org/10.1016/j.jpha.2014.11.001]

[118] Gumieniczek, A.; Mączka, P.; Komsta, Ł.; Pietraś, R. Dissolution profiles of perindopril and indapamide in their fixed-dose formulations by a new HPLC method and different mathematical approaches. *Acta Pharm.,* **2015**, *65*(3), 235-252. [http://dx.doi.org/10.1515/acph-2015-0024] [PMID: 26431103]

[119] Zaid, A.N.; Qaddomi, A.; Ghanem, M.; Shehadeh, L.; Abualhasan, M.; Natur, S.; Khammash, S. Development of A Dissolution Method To Compare Tablet Formulations Containing Valsartan/Amlodipine. *Dissolut. Technol.,* **2015**, *22*(3), 32-38. [http://dx.doi.org/10.14227/DT220315P32]

[120] Jalalizadeh, H.; Raei, M.; Tafti, R.F.; Farsam, H.; Kebriaeezadeh, A.; Souri, E. A Stability-Indicating HPLC Method for the Determination of Memantine Hydrochloride in Dosage Forms through Derivatization with 1-Fluoro-2,4-dinitrobenzene. *Sci. Pharm.,* **2013**, *82*(2), 265-279. [http://dx.doi.org/10.3797/scipharm.1310-09] [PMID: 24959398]

[121] Walash, M.I.; El-Enany, N.M.; Eid, M.I.; Fathy, M.E. Simultaneous determination of metolazone and losartan potassium in their binary mixtures using high-performance liquid chromatography with fluorimetric detection: application to combined tablets and spiked human plasma. *Luminescence,* **2014**, *29*(5), 462-470. [http://dx.doi.org/10.1002/bio.2570] [PMID: 23996979]

[122] Ahmed, R. A simple and convenient method for the simultaneous *in vitro* study of metformin and glimepiride tablets. *Pak. J. Pharm. Sci.,* **2014**, *27*(6), 1939-1943. [PMID: 25362617]

[123] Szabo, Z.I.; Szabo, T.; Emőke, R.; Sipos, E. Validated HPLC Method For Determination Of Nebivolol In Pharmaceutical Dosage Form And *In Vitro* Dissolution Studies. *Stud. Univ. Babes-Bolyai Chem.,* **2014**, *59*(4), 195-203.

[124] Emami, J.; Rezazadeh, M.; Kalani, M. Quantification of Pantoprazole by High Performance Liquid Chromatography (HPLC) Method: *In Vitro* And *In Vivo* Applications. *J. Liq. Chromatogr. Relat. Technol.,* **2014**, *37*(5), 681-695. [http://dx.doi.org/10.1080/10826076.2012.758137]

[125] Zhu, Z.; Yang, T.; Zhao, Y.; Gao, N.; Leng, D.; Ding, P. A Simple Method To Improve The Dissolution Of Repaglinide And Exploration Of Its Mechanism. *Asian J Pharmaceut Sci,* **2014**, *9*(4), 218-225. [http://dx.doi.org/10.1016/j.ajps.2014.06.004]

[126] Karbane, M.E.; Ramlib, Y.; Al-Kamarany, M.A.; Bouchfra, H.; Radi, M.; Karrouchi, K.; Azougagh, M. Development and Validation of HPLC Dissolution Assay of Simvastatin Tablets under Normal and Accelerated Conditions. *J. Chem. Pharm. Res.,* **2014**, *6*(6), 886-893.

[127] Zaid, A.N.; Natur, S.; Qaddomi, A.; Abualhasan, M.; Al-Ramahi, R.; Shraim, N.; Khammash, S.; Jaradat, N. Formulation and bioequivalence of two Valsartan/Amlodipine immediate release tablets after a single oral administration. *Pak. J. Pharm. Sci.,* **2014**, *27*(4), 755-762.

[PMID: 25015436]

[128] Revathi, R.; Ethiraj, T.; Marreddy, J.L.; Ganeshan, V. Development And Validation Of A Dissolution Test For Candesartan Cilexetil İn Tablet Forms Using Reverse Phase-High Performance Liquid Chromatography. *J Pharmaceut Educ Res,* **2011**, *2*(2), 71.

[129] Gupta, A.; Ciavarella, A.B.; Sayeed, V.A.; Khan, M.A.; Faustino, P.J. Development and application of a validated HPLC method for the analysis of dissolution samples of gabapentin drug products. *J. Pharm. Biomed. Anal.,* **2008**, *46*(1), 181-186. [http://dx.doi.org/10.1016/j.jpba.2007.08.023] [PMID: 17935923]

[130] Chaudhari, B.G.; Patel, N.M.; Shah, P.B. Stability indicating RP-HPLC method for simultaneous determination of atorvastatin and amlodipine from their combination drug products. *Chem. Pharm. Bull. (Tokyo),* **2007**, *55*(2), 241-246. [http://dx.doi.org/10.1248/cpb.55.241] [PMID: 17268096]

[131] Savaşer, A.; Goraler, S.; Taşöz, A.; Uslu, B.; Lingeman, H.; Özkan, S.A. Determination of Abacavir, Lamivudine And Zidovudine In Pharmaceutical Tablets, Human Serum And In Drug Dissolution Studies By HPLC. *Chromatographia,* **2007**, *65*(5-6), 259-265. [http://dx.doi.org/10.1365/s10337-006-0166-6]

[132] Sagirli, O.; Önal, A.; Toker, S.E.; Şensoy, D. Simultaneous HPLC Analysis of Olmesartan and Hydrochlorothiazide in Combined Tablets and *In Vitro* Dissolution Studies. *Chromatographia,* **2007**, *66*(3-4), 213-218. [http://dx.doi.org/10.1365/s10337-007-0304-9]

[133] Morjaria, Y.; Irwin, W.J.; Barnett, P.X.; Chan, R.S.; Conway, B. *In Vitro* Release of Nicotine from Chewing Gum Formulations. *Dissolut. Technol.,* **2004**, *11*(2), 12-15. [http://dx.doi.org/10.14227/DT110204P12]

[134] Ortiz-Boyer, F.; Tena, M.T.; Luque de Castro, M.D.; Valcárcel, M. Development and validation of chromatographic methods (HPLC and GC) for the determination of the active components (benzocaine, tyrothricin and menthol) of a pharmaceutical preparation. *J. Pharm. Biomed. Anal.,* **1995**, *13*(11), 1297-1303. [http://dx.doi.org/10.1016/0731-7085(95)01557-2] [PMID: 8634346]

CHAPTER 2

Novel Validated UHPLC Method for the Estimation of Drug Active Compounds

Mehmet Gumustas, Bengi Uslu and **Sibel A. Ozkan***

Ankara University, Faculty of Pharmacy, Department of Analytical Chemistry, 06100, Tandogan, Ankara, Turkey

Abstract: Recently, pharmaceutical companies have been driven to create cheaper, simpler, more novel, and more efficient tools for the discovery, development, delivery, and monitoring of pharmaceutical active ingredients. In this contest, the development of sensitive, rapid, and more effective chromatographic methods is vital for analytical quality control and R&D laboratories. Liquid chromatography (LC) with different kinds of detection modes are widely used in the pharmaceutical analysis area for the quantification and identification of active pharmaceutical ingredients in varied matrices. A completely new version of LC design with improved technology has been developed, namely ultra-high performance liquid chromatography (UHPLC), which has evolved from conventional HPLC. UHPLC can be regarded as a new direction for LC. It is a rather new technique providing new possibilities in liquid chromatography, especially reducing solvent consumption and decreasing the run times. UHPLC improves three areas of LC — namely the speed, efficiency, and sensitivity of analysis by using sub 2 μm particles as a stationary phase. The system is designed as stable in order to overcome high system pressures while the analysis is performed with no negative influence on either the analytical column or the other components of the system. During these days, the quality control of pharmaceutical preparations in industry is moving one step forward with transferring their methods from HPLC to UHPLC. The application of UHPLC methods in the pharmaceutical industry introduces a powerful tool not only for routine analysis but also for clinical approaches. Due to the high throughput investigation like therapeutic drug monitoring studies, special attention is paid to the various detection possibilities. In this chapter, the principles, differences, advantages, and disadvantages of UHPLC are discussed in detail. As a consequence, pharmaceutical applications *via* UHPLC are also summarized, based on a survey of the literature written since 2012.

Keywords: Active Pharmaceutical Ingredient, Analysis, Dosage Forms, Liquid Chromatography, Pharmaceutical, Separation, UPLC, UHPLC, Validation.

* **Corresponding author Sibel A. Ozkan:** Ankara University, Faculty of Pharmacy, Department of Analytical Chemistry, 06100, Tandogan, Ankara, Turkey; Tel: +90 203 3175; E-mail: ozkan@pharmacy.ankara.edu.tr

Atta-ur-Rahman, Sibel A. Ozkan & Rida Ahmed (Eds.)

INTRODUCTION

Pharmaceutical assay is an indispensable step for quality control of the drugs in the pharmaceutical industry. For this reason, analytical methods are used to illustrate stability, pharmacokinetic properties, and release of the pharmaceutically active compounds for years [1 - 3]. Official monographs in pharmacopoeias present the most commonly used methodologies for the analysis of pharmaceuticals [4 - 6] where titrimetric, optic, as well as separation techniques are predominant. Generally, pharmacopeial procedures require time-consuming sample preparation steps and long analysis times.

In recent years, strict regulations for the quality control of pharmaceutical dosage forms led to increasing demands for the automation of analytical assays carried out in appropriate control laboratories. The applicability of separation techniques in the pharmaceutical industry for the analysis of drugs and biological samples has become the mainstay of pharmaceutical research development and independent research laboratories since the 1970s. The recent past has witnessed an extensive development of automation instead of manual operation in the analysis of active pharmaceutical ingredients (API). This automation was motivated by the need for short assay of a high number of samples in the areas of food, environmental, clinical, and pharmaceutical chemistry [7 - 11]. Automation is also a significant demand of modern pharmaceutical investigations for each step, such as routine quality control processes, validation of the manufacturing process, and development of a new product. Typical quality control tests that involve the analysis of many samples include:

- Assay of drug dosage form;
- Dissolution studies during production and validation of manufacturing processes;
- Dosage uniformity tests and content uniformity tests for all types of (solid, semi-solid, *etc.*) dosage forms.

Flow processing is characterized by its versatility and ease of automation, as the apparatus often comprises only two mechanical components, a pump and a valve. Modern approaches in the automated analysis of pharmaceuticals are based on liquid flow analytical techniques that operate under hydrodynamically controlled conditions. In flow analytical techniques, an analyte solution is moved by the flowing stream from the place of sampling to the waste through the hypothetically separated stages of the analysis process. Flow analytical techniques involve dispersion, *i.e.*, the broadening of a discrete analyte zone during its travel through the flow-through system. These techniques comprise those involving separation, which is called as liquid chromatography (LC) [7 - 12].

Nowadays, both analytical laboratories and the pharmaceutical industry are looking for contemporary ways to decrease run times and the cost for analysis of pharmaceuticals. They are also searching for new ways to improve the quality of their products. Over time, considerable technological advances occurred like enhanced particle chemistry performance, improved detector design, data processors, and a variety of chromatographic techniques. As a result of harmonization, the outstanding performance of ultra-high performance liquid chromatography (UHPLC) occurred, based on the same principle as the high performance liquid chromatography (HPLC) technique. One of the major disadvantages of the current conventional HPLC technology is the lack of sufficient pressure as compared to the UHPLC technique, due to this, Waters designed a new instrumentation called ultra-pressure liquid chromatography (UPLC™). The term "UPLC", in fact, is a trademark of the Waters® Company [13]. However, this technique is mostly referred to as UPLC and sometimes UHPLC. Table **1** represents some commercially available UHPLC instruments and specifications [14]. It is a relatively new technique which offers better separation capabilities in comparison with conventional LC. The additional benefits of reduced run time, as well as waste organic solvent consumption, increasing the efficiency and selectivity of analysis.

Table 1. Features of some commercially available UHPLC instruments.

Brand name	Vendor	Highest Pressure (psi)	Flow range (mL min^{-1})
Acquity UPLC I-Class	Waters	18 000	0.01–2
Acquity UPLC H-Class	Waters	15 000	0.01–2
Acquity UPLC	Waters	15 000	0.01–2
Ultimate 3000 RSLC	Dionex	15 000	0.1–8
1290 Infinity LC	Agilent	17 400	0.001–5
Accela High Speed LC	ThermoFisher	15 000	0.001–5
Alliance 2695	Waters	5 000	Up to 10
Nexera	Shimadzu	18 800	0.0001–5
Flexar FX-15	PerkinElmer	15 000	Unknown
X-LC	Jasco	15 000	Up to 5
Platin Blue	Knauer	15 000	Up to 5

HPLC has been the predominant technology used in laboratories worldwide during the past 35 years for various pharmaceutical applications. One of the major promoters for the growth of this technique has been the evolution of particle size inside the column, which is one of the variables that has been investigated more often in order to enhance the chromatographic performance during pharmaceutical analysis. Fig. (**1**) represents the increasing number of UHPLC related studies on pharmaceuticals from 2007-2014 according to the Scopus data.

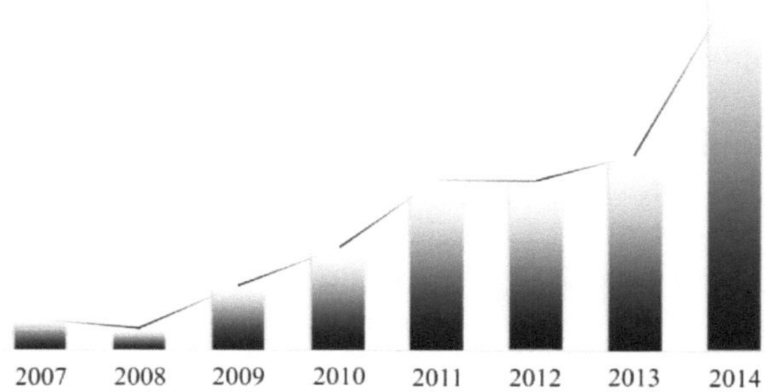

Fig. (1). UHPLC related studies on pharmaceuticals from 2007-2014 according to the Scopus data.

Analytical chemists have tried to develop quicker ways of delivering quality data across a range of needs. Particularly, many approaches are being employed to increase the throughput of separation. The "need for speed"is due to the large number of specimens in the discovery and drug analysis labs. Furthermore, the aim of the pharmaceutical industry is to increase productivity and put new pharmaceutical formulations onto the showcase in a shorter time period [15 - 18].

As underlined above, UHPLC improves three areas — namely the sensitivity, speed and resolution of an analysis *via* sub 2 µm particles. As the system operates at high pressures, the mobile phase is able to run at higher linear velocities in comparison with conventional HPLC. The separation on the UHPLC instrument is performed normally up to 15000 psi pressure without having any negative effects on either the guard and analytical columns or other components of the instrument, thereby decreasing the run time and waste solvents compared with traditional HPLC.

The aim of this chapter is to underline the new trends for the analysis of pharmaceutical active compounds by using UHPLC instruments when compared to conventional liquid chromatographic techniques. In this chapter, prior insight is given on the changes that have been made concerning both techniques (UHPLC and HPLC), focusing also on the implementation of the UHPLC technique in different areas of interest. Some selected applications published in the last 5 years are also reviewed in this chapter, along with other relevant characteristics, such as column types, chemistry, and validation parameters.

Liquid Chromatographic Techniques

The term "chromatography" originated from the Greek words meaning "color" and "to write". Chromatographic techniques have been developed over the past century and have major input in many areas of science [1 - 9]. Chromatography represents one of the greatest techniques with an extremely fruitful output for future prospects. Liquid chromatography was defined in 1903 by the work of the Russian botanist, Mikhail S. Tswett. His pioneering studies focused on separating pigments of leaves in the early 20th century [19]. He used a rudimentary form of chromatographic separation to purify a mixture of plant pigments as pure constituents. The "inventors" of modern chromatography, Martin and Synge, mentioned that the stationary phase requires small particles, hence, a high pressure needs to force the mobile phase through the column. Similar to this theory, the acronym of high performance liquid chromatography (HPLC), given by Prof. Csaba Horváth in his 1970 Pittcon paper, originally indicated that high pressure was used to generate the flow required for liquid chromatography in packed columns in order to separate the sample (that contains a single analyte and/or mixture), exploiting its breakdown between two phases. The stationary phase is either a solid, porous, surface-active material in small-particle form or a thin film of liquid coated on a solid support or column wall. The mobile phase is a liquid in all types of LC. In LC methods, the sample is injected into a mobile liquid phase, which is fed to a column packed with a stationary phase. The sample components are separated based on differences in their distribution between the stationary and mobile phases. There are two well-known types of separation modes available on liquid chromatography: reversed phase and normal phase. Apart from these techniques, a rather new separation mode named hydrophilic interaction liquid chromatography (HILIC) has been used since the 1970s, later improved by Alpert in the 1990s [20]. Among these separation modes, reverse phase LC is the most commonly used one, especially in the pharmaceutical area. An HPLC and/or UHPLC instrument includes 5 main modules called pump and solvent reservoir, sample manager, column, and detector. The chromatographic process has been significantly improved over the years: it gained speed, higher separation efficiency, and versatility. The separation efficiency of a column can be expressed in terms of the number of the theoretical plates in the column. Theoretical plate number (N) defines the ability of the column to produce sharp, narrow peaks for achieving good resolution of band pairs with the acceptable selectivity (α) values. A large number of plates is necessary for high efficiency. For an existing LC method, the number of the theoretical plates of a new column should be determined for a particular sample analyte, using designated conditions for the separation, because the plate number depends on experimental factors [19, 21 - 29]. One of the well-known basic concerns for the growth of chromatographic techniques is the packing material, which directly affects the separations. The van Deemter equation showed, for a packed chromatographic column, that the peak broadening is the sum of several experimental independent factors. According to this equation, the decrease in particle size causes an increase in the separation efficiency. On the other hand, the efficiency diminishes according to higher linear velocities or flow rates. Smaller particles like sub 2.0 micron is caused to increase speed and the number of resolved peaks per unit time. High pressure and columns with a very small particle size are the key features of UHPLC methodology, but despite the high speed, low dispersion detectors that allow high sensitivity detection in a small cell volume should not neglected. The detectors will be highlighted in the following parts of this chapter. The UHPLC is based on the use of the stationary phase consisting of sub-2-µm particles, while classical LC columns are typically filled with 2.5-10 µm particles. In general, stationary phases are classified based on four distinguishing features: particle type, geometry (pore diameter, particle shape, surface area, *etc.*), surface chemistry, and substrate type (polymer, silica, *etc.*). The UHPLC method is based on the van Deemter relationship which explains the correlation between flow rate and plate height [1 - 7]. According to the van Deemter equation, the higher flow rate of the mobile phase pumped, the more the molecule on the packing material is inclined to lag behind molecules in the mobile phase. This term is inversely proportional to linear velocity [30, 31].

Fig. (**2**) illustrates the effect of particle size on plate height (HETP) using the van Deemter plot. The figure highlights a significant increase in the efficiency using particle size <2 µm.

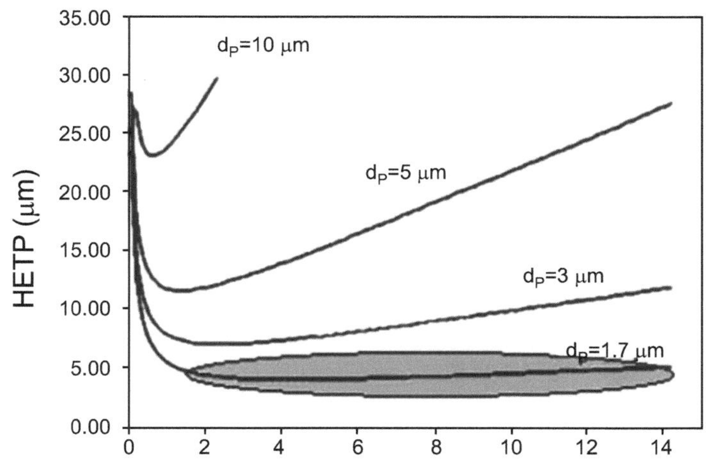

Fig. (2). The effect of particle size on plate height (HETP) using the van Deemter plot over the last decades [32].

Stationary Phases (Columns)

In modern LC, almost all reversed phase separations are performed on chemically modified adsorbents. The column can be stated as the "heart" of the chromatographic system, and it is the only device where actual separation of the analyte mixture takes place (Fig. **3**) [33 - 57].

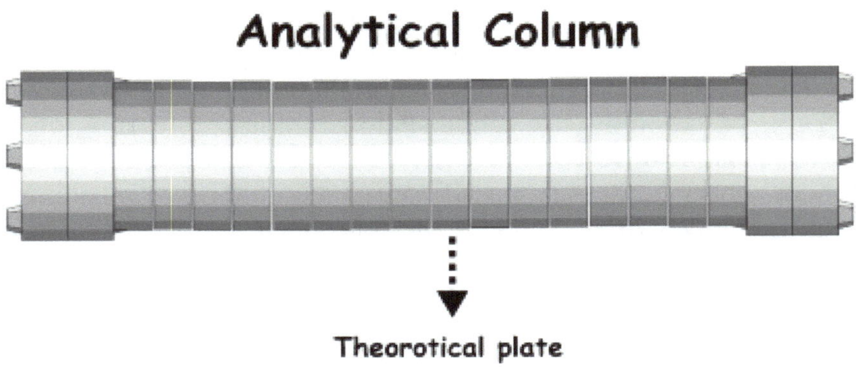

Fig. (3). Schematic representation of an analytical column with illustration of theoretical plates.

Historical milestones of column development are represented in Scheme **1** [58]. However, current LC columns in most cases are a stainless-steel tube packed with very small (1.3-10 µm) particles of rigid porous material. Several thousands of different columns are commercially available. Stainless steel or titanium frits have a pore size ranging from 0.2-0.5 µm. Commercial columns C8, C18, cyano, phenyl, amine, *etc.*, could vary widely in their separation power from supplier to supplier. Most commercial adsorbents reflect their surface chemistry in their names while the base material used usually is not specified, although their properties have high importance.

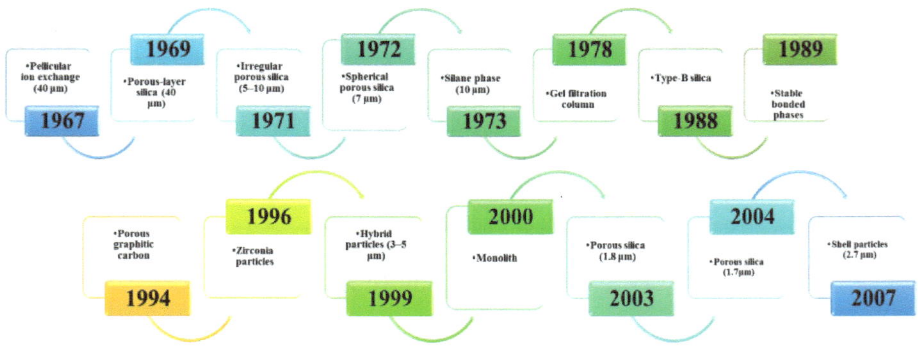

Scheme 1. Historical milestones of column development [58].

Analysis speed and separation efficiency may also be increased using superficially porous particles, usually termed as fused core and/or core-shell particles as illustrated in Fig. (**4**). The advantage of using core-shell silica material was their greater efficiency in high speed analysis without the generation of high back-pressures, which is typical in UHPLC. As the UHPLC column filled with small particles (<2 μm), the role and the effect of small particles and their sizes in LC methods have been mentioned following parts. When the particle size decreases, efficiency and subsequently, resolution (Rs) between the analyte signals slightly increases. Sharper peaks are desirable for many applications, such as environmental and biological matrices. Having a UHPLC system that can both reliably deliver the requisite pressures and maintain the separation efficiency of the small particles with tightly managed volumes when analysis speed is the primary objective.

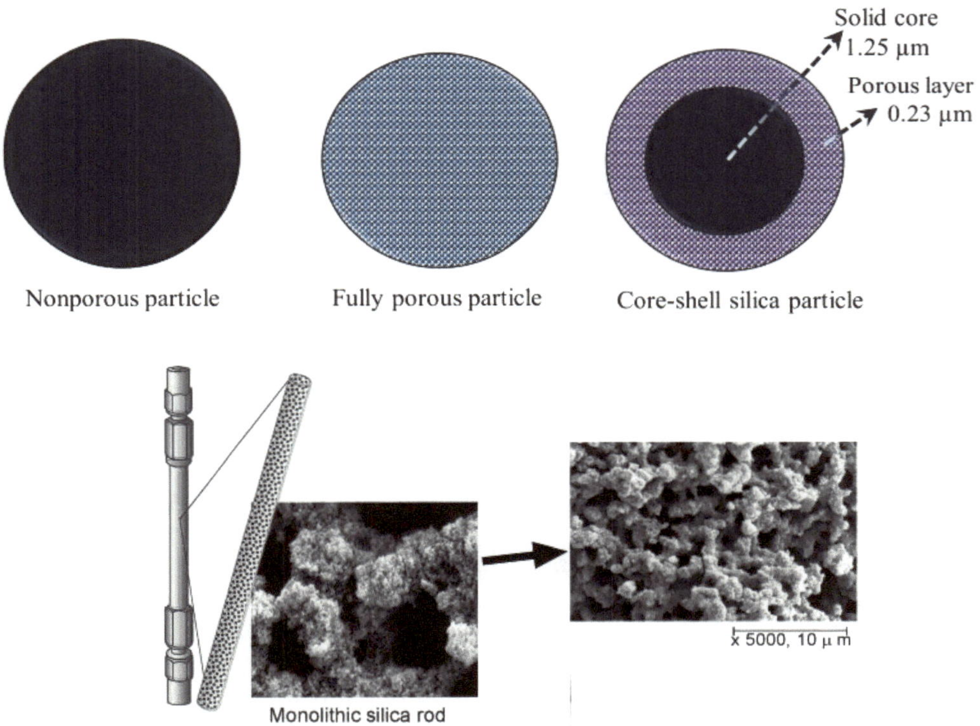

Fig. (4). The illustration of the type of particles used as column packing material in UHPLC. "Reprinted (adapted) with permission from [59]; [Nagase, K., Kobayashi, J., Kikuchi, A., Akiyama, Y., Kanazawa, H.,Okano, T. 2014. Monolithic Silica Rods Grafted with Thermoresponsive Anionic Polymer Brushes for High-Speed Separation of Basic Biomolecules and Peptides. Biomacromolecules, 15, 1204-1215]. Copyright (2014) American Chemical Society."

Monolith sorbent usage (Fig. **4**) has become popular, especially in bioanalytical approaches, during the recent years. This stationary phase does not need to have a long equilibration time while a gradient mobile phase is used. They also allow easy flow-rate gradients and the coupling of several columns together for increasing the efficiency of the separation. On the contrary, a limited number of monolith technology stationary phases are commercially available (Table **2**) [43 - 53].

Table 2. Some of the current commercial monolith stationary phases commercially available.

Vendor	Chemistry	Product	Separation Mode	Monolith Shape
BIA Separations	Modified polymethacrylate or polystyrene copolymers	CIM disc	Ion exchange, hydrophobic interaction, reverse phase, bioaffinity	Disc
BIA Separations	Modified polymethacrylate	CIM tube	Ion exchange	Tube
Merck	Modified silica	Chromolith	Reverse phase	Cylinder
Conchrom	Modified silica	CB silica plate	Reverse phase, normal phase	Disc
LC Packings	Polystyrene copolymer	Monoliths	Reverse phase	Cylinder
ISCO	Modified polymethacrylate or polystyrene copolymers	Swift	Ion exchange, reverse phase	Cylinder
Sepragen	Modified cellulose	SepraSorb	Ion exchange	Disc
BioRad	Polyacrylamide based copolymers	UNO	Ion exchange	Cylinder

Monolith technology allows a significant decrease in sample run time and has been proven to be suitable for bioanalytical applications in particular. Furthermore, Chromolith type monolith columns have been used for routine analyses of biological fluids (urine, plasma, whole blood, and serum). Typically, a 100-mm column length is used in order to enhance separation efficiency. Monoliths are mostly used in connection to ultraviolet (UV) or fluorescence detection (FD) [47 - 55].

All LC techniques operate under hydrodynamically controlled conditions [19, 21 - 27]. In UHPLC, the dispersion of analyte arises at first from the processes occurring in the column, such as:

1. Slow analyte distribution between the stationary phase and the mobile phase,
2. Convection of the analyte due to local turbulences of the tortuous liquid flow paths through the stationary phase,
3. Axial diffusion.

UHPLC is designed for the separation of each of the components of the sample. In order to obtain the highest possible resolution of the component zones, the dispersion in every part of the UHPLC instrument should be suppressed as much as possible. A pump pushes the mobile phase mixture through the stationary phase. A detector shows the retention times of the different molecules while the mobile phase is passing through the analytical column. In this particular case, retention time, abbreviated as t_r, is differentiated depending on the interactions between the mobile phase (solvent composition), the stationary phase, and the analyzed compound. When the speed of analysis is the primary goal, higher efficiency with better resolution can be leveraged even further. Therefore, the length of the column can be reduced by the same factor as the particle size without loss of selectivity. For instance, increasing the flow rate two times higher due to the sub 2 micron particles and shortening the column by half of the current length, the separation is completed four times faster while maintaining resolution. So, if sample capacity, speed, and throughput is a concern, in theory, can be further leveraged to get much higher throughput. So, it is possible to take nearly complete advantage of chromatographic principles to run analysis by using higher flow rates and shorter columns, increasing the speed with better resolution by using sub 2 μm particles. In this context, the smaller particles provide not only increased efficiency of an analysis, but also the ability to study with higher velocity [15 - 18].

For the columns, new hybrid material was developed. The new material advantages are based on the theories behind LC. In general, the resolution is increased with the increase of the efficiency. Efficiency and optimal flow rate are inversely proportional to particle size. Hence, a decrease in the particle size will increase efficiency and expedite the flow rate. The task to decrease particle size was not easy. A commercially available, non-porous, small particle has poor loading capacity and retention due to low surface area. The novel porous particles that can withstand high pressures are used to maintain retention and capacity. In the beginning of 2000, the hybrid of silica and polymeric column was introduced. Hybrid columns, which are mechanically strong, consist of classical sol-gel synthesis that incorporates carbon in the form of methyl groups. These columns can be operated at a wide range of pH and are highly efficient. In order to provide the kind of enhanced mechanical stability required for UHPLC, a second-generation bridged Ethylene hybrid technology was developed [53] These ≤ 2.5 μm particles have become mechanically stable by bridging the methyl groups in the silica matrix. In spite of the good mechanical strength, silica-based particles have tailing for basic analytes and limited pH range. Columns that have polymeric materials may overcome limitations in pH. On the other hand, they have poor mechanical strength and limited loading capacities, as well as low efficiencies. Waters Corp. has introduced a hybrid silica material, the XTerra® brand. It is announced as a first generation of hybrid chemistry. It has high efficiency, has mechanical stability, and can perform at a wide range of pH [53].

Classical sol-gel synthesis is used to incorporate carbon in the form of methyl groups. However, in order to improve the mechanical stability for these particles, bridged ethylene hybrid technology was developed as a second-generation material. Furthermore, packed bed uniformity is also critical for maintaining the resolution *via* shorter columns [47 - 53].

Some other types of stationary phases produced through various technologies are: charged surface hybrid, high strength silica, particle technologies, and peptide separation technology that was also marketed by the various companies. Among these technologies, charged surface hybrid particles is the newest one for the development of hybrid materials. In this technology, low-level surface charged particles for the enhancement of the efficiency, peak shapes (especially for basic compounds), and the selectivity under extreme pH conditions, were used.

The advent of UHPLC has demanded the development of a new instrumental system for LC, which can take advantage of the separation performance by reducing dead volumes and keep its stability against high pressure (about 8000 to 15.000 Psi) [15 - 18].

Many pharmaceutical companies try to reduce the research and development costs associated with the time by using faster separation methods. Hence, this type of separation can save the time for method development both in quality control and R&D laboratories. UHPLC is a very attractive tool for the pharmaceutical development laboratory because of high resolution obtained in extremely short period of times, as it provides high resolution, high productivity, and high throughput. These categories have specific applications on the pharmaceutical area and distinct separation goals. Now these goals have been achieved by utilizing UHPLC with typical column dimensions and particle sizes. Owing to its sensitivity and speed, UHPLC has gained considerable attention in recent years for pharmaceutical and biomedical analysis [15 - 18, 60]. One of the best solutions in the quest to increase performance in liquid chromatography is decreasing the particle sizes. Due to the low mass transfer resistance, optimal separations are also achieved at higher and wider linear velocities. By reducing the particle size, better resolution and shorter analysis time could be attained. These factors have been taken advantage of through the increasingly used technique of UHPLC. It is obvious that increasing the peak capacity caused an improvement in the data quality. On the other hand, solvent consumption decreased dramatically with the decrease in both run time and column equilibration time. From the viewpoint of injection volume, UHPLC can obtain nearly ten times smaller volumes than in HPLC, resulting in low carry over effects and better peak shapes. It has to be adapted to work in an ultra-fast mode with the reduction in column diameters, limiting frictional heating and substantially reducing solvent consumption. In this technology, system dwell time changes in accordance with the mobile phase flow rate. In this particular case, the restrictions of the limitations are essential, especially the mobile phase changing rates can be limited by dwell volume and the accuracy of the pump. This system needs a small gradient delay volume because of shorter run times.

The requirements for the UHPLC system can be summarized as: fast injection cycles; a robust pumping and injection modules; small gradient delay volume; high detector sampling rate; fast detection constant; low volumes for columns, detection cell, and tubing; and appropriate columns and stationary phases [15 - 18, 60, 61]. UHPLC increases productivity by giving more data per unit of work, thereby improving the quality of the data.

UHPLC Over HPLC

HPLC and UHPLC are both liquid chromatographic techniques that are used for separating the mixtures in samples. UHPLC sometimes defined as a special version of HPLC because the theory and application process are so similar and easily understood. HPLC is the most preferable technique for the identification, quantification, and separation of components in a sample. High pressure is used for pushing mobile phase solvents through the column. It is an ideal way for the identification, separation, and quantification of active pharmaceutical ingredients: proteins, amino acids, hydrocarbons, nucleic acids, pesticides, carbohydrates, antibiotics, anticancer, antivirals, and steroids, as well as the countless other inorganic and organic substances. HPLC has many advantages over many types of analytical methods. It allows for higher resolution and columns can be used several times without regeneration and repacking if the system suitability parameter results are acceptable [19 - 29, 62 - 69].

In UHPLC, sub 2 μm particles are used in order to obtain better resolution with better peak shapes in comparison with 5 μm particle sizes that are used in conventional HPLC. In UHPLC, mobile phase viscosity must be minimized or the chromatographic system should be redesigned to overcome high backpressure when compared with HPLC. Whereas the pump pressure in HPLC is up to 400 bar, this pressure can go up to 1000-1200 bar in UHPLC, which makes this technique efficient and attractive [15 - 18, 32, 60].

UHPLC allows faster method development with maintaining resolution, dynamic range, selectivity, and sensitivity of HPLC analysis. As it reduces analysis of cycle times, more samples can be analyzed with existing resources. Related to these examples, the advantage of UHPLC over traditional HPLC is the capability to increase the speed without sacrificing efficiency. In comparison to classical HPLC, the research shows that UHPLC can decrease run times up to 7-10 times that of HPLC. It assures quality results with a decrease in run time and cost, and it increases the sensitivity of sample analysis along with less solvent consumption [15 - 19, 21 - 29, 60, 62 - 69].

UHPLC methods are more effective than HPLC methods because of the following reasons:

- Peak capacity is the number of peaks that can be resolved in a specific amount of time.
- Peak capacity is proportional to the inverse of the square root of the number of theoretical plates (N): N = L/H.
- Lower plate heights generate a smaller number of plates.
- Plate heights are correlated through the van Deemter equation.

A comparison between UHPLC and HPLC can be summarized in Table **3**.

Table 3. UHPLC *vs* HPLC.

Attribute	HPLC	UHPLC
Pressure	6000 Psi	100000 Psi
Particle Size	3-5 μm	1.3-2.0 μm
Flow Rate	mL per minute	μL per minute
Injection Volume	5-50 μL	0.2-10 μL
Max Resolution	Relatively low	Relatively high
Detection Speed	40 Hz	80-120 Hz
Solvent Consumption	Relatively high	Relatively low

As a result, the advantages of UHPLC methods over the HPLC methods can be ordered as:

- Better resolution (separation efficiency),
- Faster chromatography,
- Better sensitivity (sharper and higher peaks),
- Fewer solvent requirements,
- Environmental friendly method,
- Withstand high back pressure system.

In spite of these advantages, UHPLC has some disadvantages. The performance of UHPLC may be similar or even higher that demonstrated by using sub 2 μm stationary phases without the adverse effects of high pressure. The phases less than 2 µm are generally non-regenerable and thus, have limited use as pressure increases; it requires more maintenance and reduces the life of the columns [32, 40 - 53]. Other disadvantages can be ordered as:

- High cost of system and spare parts,
- Detector and data collection system may not cope with sharper peaks (data acquisition rate effect),
- Only binary pump systems can be used. This problem may make the method transfer not flat out,

Increased resolution in shorter run times can generate more information faster without sacrifices in UHPLC. Nowadays, higher sample throughput with more information per sample may decrease the time to market, an important driving force in the pharmaceutical industry. The corresponding HPLC separation takes an excess of 12 min; UHPLC accomplishes the same separation in under 30 s.

These achievements enable method development to be more efficient, allowing products to be brought to market faster. Hence use of such UHPLC systems will become the option of choice for the development of fast LC methods in drug development and analysis. As a result of this knowledge, it can be said that UHPLC out classes HPLC in all aspects and is expected to replace HPLC in the near future.

Detector Types for UHPLC

A detector is a device for continuous registration of specific physical or chemical properties of the column effluent that monitors the quantity of the analytes in the sample leaving from the column. The electrical signal, called peak that is provided has to be proportional to quantity. In this particular case, detectors should be stable, sensitive, and precise. Furthermore, they are capable of constituting both the concentration of effluents in the mobile phase stream and identity.

The attractive specialties of a detector include low cell volumes for reducing the memory effects, sensitivity towards solute over mobile phase, low noise, large linear dynamic range, and lower limit of detection, while selective, precise, and applicable detectors are used. As mentioned above, the cell volume in which the measurement is realized have to be as low as possible.

UHPLC ensures that the separation sensitivity is two to three times more than that of the conventional HPLC, which is also based on the method that is employed for the detection. A wide range of detectors that meet different requirements (Table **4**) are commercially available. Specific detectors, such as electrochemical, respond to a particular compound and/or the compounds that have special properties (*e.g.*, electroactive groups. In this case, the response is independent from the mobile phase composition or the pH. Furthermore, the response of bulk property detectors is related to changes in the composition of the mobile phase. Detectors in UHPLC can be divided into three groups:

Table 4. Detector types and some of their basic properties.

Type of the Detector	Application range	Sensitivity
UV-Vis, PDA	Selective and universal	Nanograms
Refractive Index	Universal	Micrograms
Conductivity	Selective	High level of Nanograms
Light Scattering	Universal	High level of Nanograms
Charged Aerosol	Universal	Low level Nanograms
Fluorescence	Highly selective	Picograms
Electrochemical	Highly selective	Femtograms to Nanograms

Specific property detectors: These types of detectors monitor the quantity of the analytes by measuring a property that is special to the analytes in question. This can be fluorescence or UV-Visible absorbance of a compound.

Bulk property detectors: A typical example of this type is the Refractive index detector. It measures the general changes in a property that is typical of the solvent and the mobile phase.

The other types and measurement types of detectors such as Evaporative Light Scattering (ELSD) and Mass Spectrometry (MS) can also be used.

As a result of these classifications, the choice of detectors for use with UHPLC varies in a wide range, including Tunable UV-VIS (TUV), Photodiode Array (PDA), Fluorescence (FLR), Evaporative Light Scattering (ELSD), MS, and most recently, refractive index (RI). The system can be configured with ELSD, FLR, PDA detectors, or a combination of them [32, 70 - 72].

In general, the ideal detector should be:

- React in a quick way for getting sharper peak signals
- Easily operated and manipulated
- Precise and robust (not affected by small changes in the conditions like temperature, mobile phase composition)
- Have small cell volumes (to avoid band broadening)

- Tend to trace samples
- Sensitive enough for the analytes

Based on above necessities some selected detector types can be briefly explained as follows:

UV based detectors like UV-Vis or PDA (Fig. **5** and **6**) are the most commonly used type of detectors in UHPLC. It has some significant attributes like, universal, easy to use, reliable, cheap, gradient compatible, available for spectral library searches and identification, for nondestructive to sample. On the other hand, some limitations may have occurred, such as the compound should have a chromophore group and used solvents must be transparent.

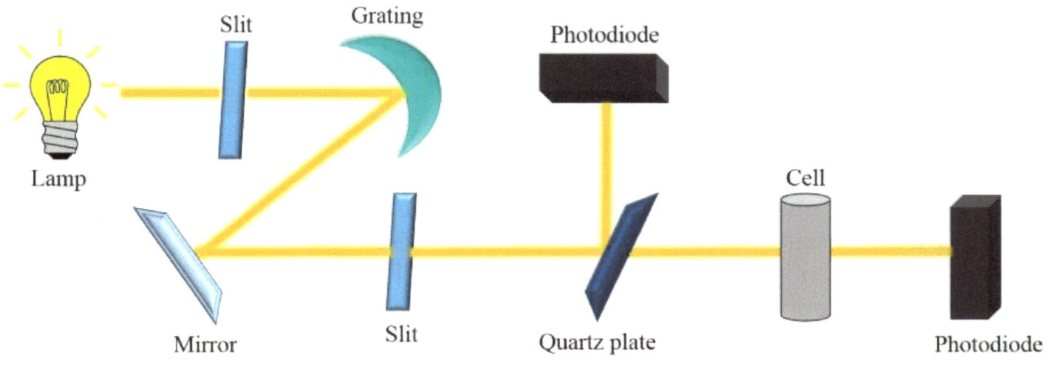

Fig. (5). Basic principle of UV detector.

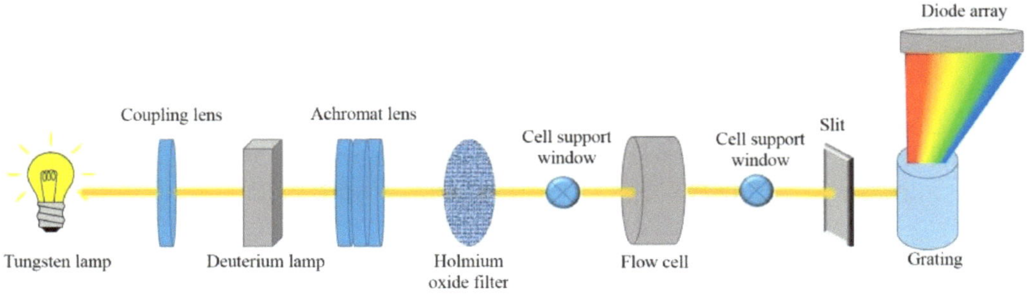

Fig. (6). Optical system of the PDA detector.

UV-VIS Detectors

They records compounds that absorb UV or visible light. Absorption takes place at a wavelength above 200 nm, provided that the molecule has at least a chromophore group, such as carbonyl group, two conjugated double bonds, iodine, sulfur, bromine, or an aromatic ring. If an analyte has no absorbance, it can give a suitable response after derivatization procedure before the injection (post column derivatization) to introduce light absorbing entities [58, 72, 73].

These detectors separated two different types:

Tunable UV (TUV) Detector

It is a tunable, dual wavelength UV detector. It is especially designed for the entire line of the UHPLC system. This device allows for decreased sensitivity to variable changes like humidity and temperature. In general, it has 80 points/s sampling rate.

Photo Diode Array (PDA) Detector

The introduction of diode array detector technology to liquid chromatography in the early 1980s was announced as the beginning of a major new era in analytical HPLC. Multiwavelength detection, with a methods' operating range from 190 nm to 950s nm for helping establish peak homogeneity, method development, compound verification, and other techniques have been performed using modern multiwavelength detection. It allows detection of multiwavelength simultaneously because of the incorporation of a large number of diodes. Indeed, things were certainly looking up for the chromatographers. This specification provides us reduction of cost and time due to the usage of expensive solvents [58, 72, 73].

Fluorescence (FLR) Detector

It is stated as multi wavelength detector, which has an emission wavelength range from 210-900 and excitation wavelength range from 200 to 890 nm. It actually helps to develop methods easily due to its 3D scanning capability. Fluorescence detection works well with gradient system, and also, it offers higher sensitivity than UV-Vis-based detectors. It is not uncommon to achieve femtomole detection limits with linearity over three to four orders of magnitude. Fluorescence utilizes two distinct wavelengths as opposed to one in absorbance (Fig. **7**). However, using two wavelengths decreases the chance that interfering (co-eluting) peaks will be detected. However, there are not many natural fluorescent compounds available in comparison to UV-Vis light absorbing ones. For this reason, it can be a disadvantage in the field of chromatography. This situation overcame by post column derivatization techniques. Furthermore, quenching and cost for performance can be the other disadvantages for this detection system.

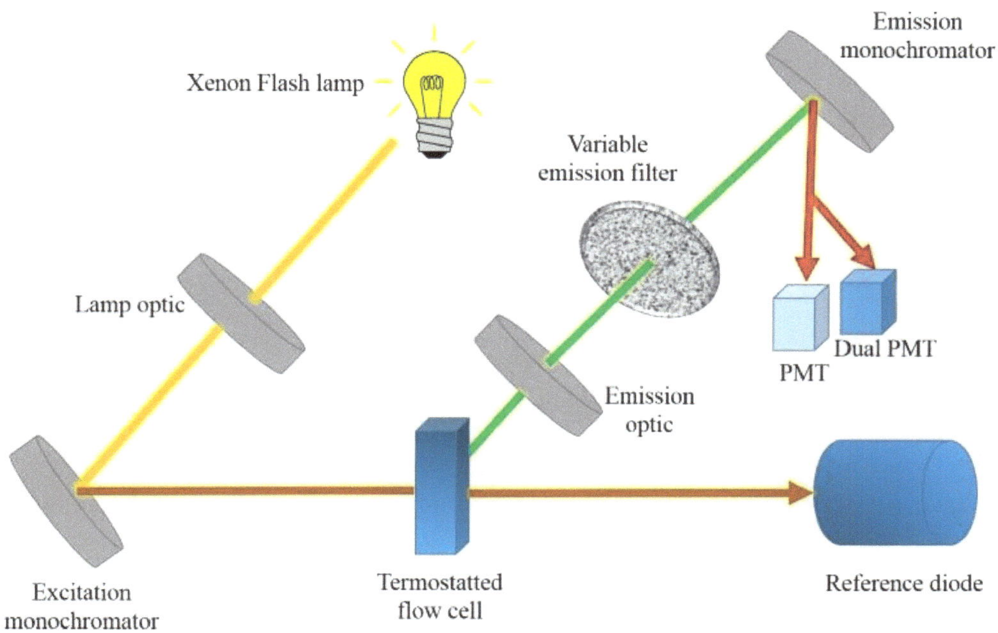

Fig. (7). Schematic representation of fluorescence detector.

Refractive Index (RI) Detector

RI detectors were among the first commercial detectors for HPLC. Two types, the Fresnel type (Watson 1968) and the beam deflection type (Waters 1970), were commercially introduced in the late 1960s. The major advantage of this RI detector is the universal nature of its response; in contrary to this major advantage, disadvantage is its lack of sensitivity when compared with other type of detectors. On the other hand, it is possible to see both positive and negative peaks in the chromatogram, and this detector is not capable with the gradient mode. In most cases, RI used for the analytes that do not have strong chromophore, electrochemical, and fluorophore moieties like carbohydrates, lipids, *etc.* In this particular case, the response is based on changes in RI of effluents in the mobile phase. RI detectors record all eluting zones which have an RI different to that of the mobile phase. The difference between the refractive indices of the eluent and the analyte becomes greater while the signal becomes more intense. RI detectors are near a thousand-fold less sensitive than UV based ones; LOD approximately 5 10^{-7} g of sample per mL, in contrast to UV based detectors senses 5 10^{-10} g mL^{-1}. One of the primers for this detection system is the temperature control is essential. Due to this reason, in this type of detectors, cells should be very well controlled thermostatically (Fig. **8**) [58, 72, 73].

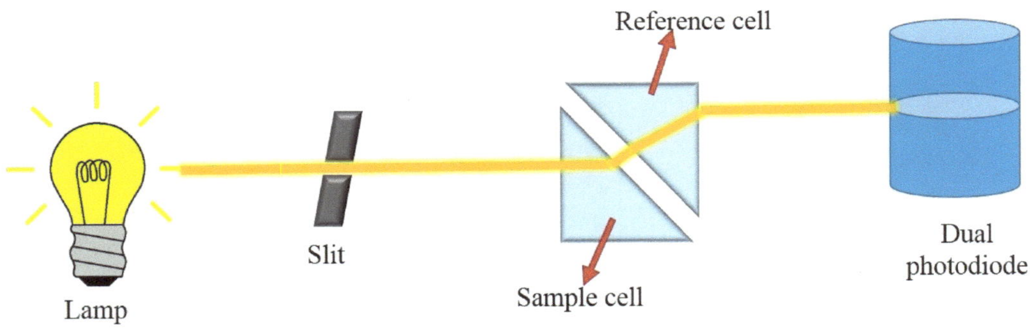

Fig. (8). Schematic representation of RI detector.

Evaporative Light Scattering (ELS) Detector

ELSD (Fig. **9**) is used for nonselective detection of nonvolatile compounds. It is also useful for detection of the analytes that have especially high molecular weights. This type of detector is available for working in gradient conditions. Depending on the analyte, it has also better sensitivity than RI detection (10 to 100 times). On the other hand, as a limitation, it requires the use of volatile buffers (including formic acid and trifluoroacetic acid), and it has poor reproducibility, as well as limited dynamic range. The principal of this technique is based on nebulization of column eluate in a stream of inert gas (nitrogen, compressed air, or helium). After that, with the evaporation of liquid droplets, solid particles pass through a laser or polychromatic light beam. Finally, scattered light is registered by a photomultiplier [74].

Mass Spectrometer (MS) Detector

The biggest challenge for the MS detection is its high cost. However, mass spectrometer detection instantly improves results for UHPLC technology. For this reason, it provides an ideal profile for an ideal detector. Another limitation is the need of higher level experienced or well-trained analyst than that needed for an absorbance based detector. It combines optimum sensitivity with maximum identification capability. The point in LC-MS is the connection of the MS to the chromatographic system. Generally, this spectrometer for UHPLC includes 3 parts: interface (where the ions are generated), mass analyzer, and detector. A number of different approaches have been used in interfacing LC and MS (Fig. **10**). The earliest attempts focused on methods of overcoming the incompatibility of the liquid flow rate and maintenance of the mass spectrometer high vacuum. Recently, more attention has been focused on the practical use of ionization techniques that do not require sample volatilization. Nowadays, there are two main ionization techniques available in MS such as atmospheric pressure chemical ionization and electrospray ionization. The final ions can be sent to more than one mass analyzer in order to acquire more detailed information about the structure of a sample. This type of technique is called MS/MS (if it has two steps) [72, 75 - 77].

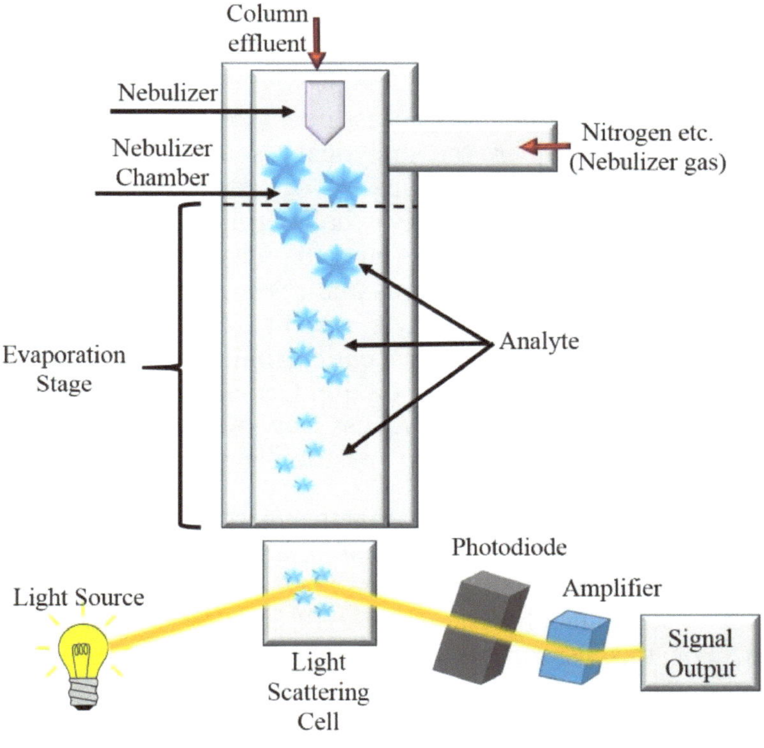

Fig. (9). Schematic representation of ELS detector.

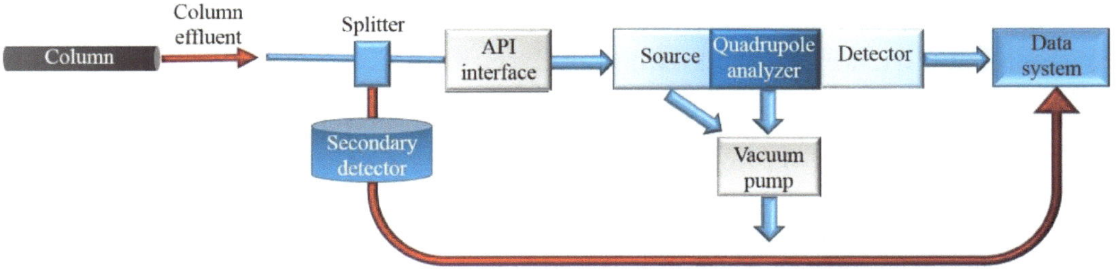

Fig. (10). Schematic representation of LC/MS system.

Application on Drug Assay

In recent years, the pharmaceutical industry started to apply quality by design strategies on different tasks in drug manufacturing, quality control, and the R&D laboratories [4 - 6, 78], which states that, to have a high-quality product, quality must be designed into the product, not tested into it. Furthermore, drug discovery and the development process have undergone significant changes, particularly in the last 80 years. The process continues to evolve in response to new technologies, new discoveries, and the increasing demand to bring more drugs into the market in an efficient way.

The availability of an instrument is necessary for testing pharmaceutically active ingredients. Laboratories where the analytical research is carried out should seek optimum efficiency since a high quality of results are anticipated and the processes involved are significantly challenging. As a consequence, high throughput systems, like UHPLC tandem mass spectrometry, have been developed to ensure that the operations and research in the pharmaceutical area are properly implemented. Nowadays, stability of compounds is one of the essential topics, especially in pharmaceutical industry, since it determines the effect that such compounds may have on life. Generally, this process is monitored by using chromatography-like techniques. Nevertheless, these methods include the disadvantages of narrow resolution spectrum and long time periods.

The combined utilization of UHPLC together with the detectors, such as photo diode array, fluorescence, electrochemical, and mass spectrometry, facilitates an extended range of results and provides more accurate detections. UHPLC is known as one step ahead version of HPLC for the assay of drug active compounds. For this reason, fully validated UHPLC methods use different detection systems in order to identify and quantify the pharmaceutical preparations, their impurities and metabolites from the biological media.

The application of UHPLC methods dramatically reduced the run time and decreased organic solvent consumption with higher efficiency in comparison to the traditional HPLC methods. Hence, it can be described as greener and more environmental friendly method than HPLC.

With recent developments and improvement in column packing materials and the utilization of smaller particles like sub 2 μm, the process has overcome the problems with significant results. The key changes allowed this system to gain ultra-performance and facilitate a higher flow rate of mobile phase without affecting the efficiency that is reached by using larger particle analytical columns.

UHPLC combined with MS detection has been used for routine clinical analysis. UHPLC-MS provides quick analysis of biological samples since the volume required is small. Based on the developments in software and instrument control, it facilitates clear and baseline chromatographic separation associated with the powerful detection and measurement capabilities of MS for clinical testing laboratory.

The new developments of the column technology make UHPLC a reliable and practical technique for the specific analysis of some vitamins, analgesics, antibiotics, anticholinesterase, antihypertensive, hormones, and other types of drugs. Some selected application examples are given in Table **5** and **6** with their details [79 - 201].

Table 5. Optimized conditions for pharmaceutical applications.

Compound	Column Supplier and Brand	Column Chemistry	Dimensions (mm x mm)	P.S. (µm)	Mobile phase (v/v)	pH	Tem (°C)	Detector	Ref
Bosentan monohydrate and related impurities	Phenomenex Kinetex XB-C18	C18	50x3	1.7	Gradient MeOH:0.1%TEA; 40:60, v/v and ACN	2.5	35	UV	79
Salidroside Rhodiosin Rhodionin Herbacetin Kaempferol Quercetin	Zorbax Eclipse Plus	C18	100x2.1	3.5	Gradient Water (0.2% formic acid):ACN	-	25	MS/MS	80
Gallic Acid Chlorogenic Acid Vanillic Acid Caffeic Acid Paeoniflorin Ferulic Acid Liquiritin Verbascoside	Waters Acquity BEH	C18	100x2.1	1.7	Gradient ACN (0.1% formic acid):water (0.1% formic acid)	-	35	MS/MS	81
4-Methoxy-5-hydroxycanthin-6-one Andrographolide Dehydroandrographolide Neoandrographolide Rosmarinic acid	Waters Acquity BEH	C18	100x2.1	1.7	Gradient ACN: water (0.1% formic acid)	-	30	DAD, MS	82
Ardoside, Shanzhiside Deacetyl asperulosidic acid methyl ester Gardenoside Genipin-1-β-D-gentiobioside Chlorogenic acid Gardenoside.	Waters Acquity HSS T3	C18	-	-	Gradient ACN: water (0.05% phosphoric acid)	-	30	DAD	83
Rivaroxaban and degradation products	Agilent Zorbax Eclipse Plus	C18	50x2.1	1.8	Ammonium hydroxide (0.1% formic acid): ACN (0.1% formic acid); 63:37, v/v	4	30	MS/MS	84
Fenofibric acid	Phenomenex Kinetex	C18	50x2.1	2.6	Gradient 2 mM ammonium acetate:methanol	-	RT	MS/MS	85
Mesaconitine calycosin-7-O-β-d-glucopyranoside Hypaconitine Aconitine Benzoylmesaconitine Benzoylaconitine Benzoylhypaconitine Liquiritin Ammonium glycyrrhizinate Tetrahydropalmatine Paeoniflorin Sinomenine	Waters Acquity BEH	C18	100x2.1	1.7	10 mM ammonium acetate (0.1% acetic acid):ACN	-	30	MS/MS	86
Nebivolol	Waters Acquity BEH	C18	50x2.1	1.7	Potassium di hydrogen phosphate:ACN; 25:75, v/v	-	30	DAD	87
Clozapine and degradation products	Zorbax Eclipse Plus	C18	50x2.1	1.8	Gradient MeOH:Water (0.1% Formic acid)	-	35	DAD, MS	88

(Table 5) contd.....

Compound	Column Supplier and Brand	Column Chemistry	Dimensions (mm x mm)	P.S. (µm)	Mobile phase (v/v)	pH	Tem (°C)	Detector	Ref
Acetaminophen Phenylephrine HCl Guaifenesin Dextromethorphan HBr	-	C18	100x2.1	1.7	(18.3mM phosphoric acid and 4.94 mM 1-heptane sulfonic acid sodium salt) and water in ratio 95:5, v/v, and, a buffer (mixture of 18.3mM phosphoric acid and 4.94mM 1-heptane sulfonic acid sodium salt) and water in ratio 20:80	-	45	DAD	89
Carfilzomib and degradation products	Microsolv Silica type C Cogent Bidentate	C18	100x2.1	2.2	Gradient 2 mM ammonium format:ACN	-	30	DAD	90
Conjugated parabens	Thermo Scientific Syncronis aQ column	C18	100x2.1	1.7	Gradient Water: ACN	-	40	MS/MS	91
Acetazolamide	-	-	-	-	Gradient 0.5 M anhydrous sodium acetate in water (95% v/v) and ACN: MeOH; 60:40 (v/v)	4	21	DAD	92
Telmisartan Clinidipine	Waters Acquity BEH	C18	100x2.1	1.7	Gradient ACN: 0.01M sodium phosphates monobasic dihydrate (68:32, v/v)	3	-	DAD	93
Imatinib Mesylate and related impurities	Waters Acquity CSH	C18	50x2.1	1.7	Gradient 0.1% g/mL ammonium formate buffer:ACN	8.57	41.88	MS/MS	94
Nevirapine and related impurities	Agilent Poroshell 120	C18	100x2.1	2.7	Gradient Water:ACN	-	30	DAD	95
Hederacoside C	Waters Acquity BEH	C18	100x2.1	1.7	Gradient Water (0.02% acetic acid):ACN (0.02% acetic acid)	-	35	MS/MS	96
Methyl Paraben Propyl Paraben Loratadine	Waters Acquity BEH	C18	100x2.1	1.7	Gradient 0.01 M $KHPO_4$ and a mixture of MeOH:ACN (50:50)	-	25	DAD	97
Agomelatine	Waters Acquity BEH Shield RP 18	C18	50x2.1	1.7	Gradient ACN:MeOH, 50:50, v/v and Ammonium acetate	9.5	30	DAD	98
Bambuterol and degradation products	Waters Acquity BEH	C18	100x2.1	1.7	Gradient Water (TFA, 0.025%): ACN (TFA, 0.025%); 30:70, v/v	-	40	DAD	99
Oleracone	Shimadzu Shimpack ODS	C18	50x2.0	1.6	MeOH:Water, 44:56, v/v	-	40	DAD	100
Amlodipine Rosuvastatin	Waters Acquity HSS	C18	50x2.1	1.8	ACN:Water (0.1% Phosphoric acid)	3	25	DAD	101
Bosentan and related impurities	Waters Acquity BEH	C18	100x2.1	1.7	Gradient Water (0.1% Acetic acid):MeOH	-	30	UV	102
Efavirenz Lamivudine	Waters Acquity BEH Shield RP 18	C18	50x3.0	1.7	Gradient ACN:MeOH:10mM Phosphate Buffer	4	35	DAD	103
Dimenhydrinate and related impurities	Waters X Select HSS T3	C18	100x2.1	1.8	Gradient Phosphate buffer;ACN and Phosphate buffer:MeOH	2.5	30	DAD	104
Cefetamet pivoxil	Waters Acquity BEH	C18	100x2.1	1.7	Water (0.1% Formic acid):ACN, 40:60, v/v	-	30	DAD	105
Bisoprolol, Hydrochlorothiazide and related impurities	Waters Acquity BEH	C18	100x2.1	1.7	Gradient Water (5mM monobasic potassium phosphate):ACN:MeOH	3.6	29	DAD	106
Amoxicillin	Waters Acquity BEH	C18	100x2.1	1.7	Phosphate buffer:MeOH 95:5, v/v	5	35	DAD	107

(Table 5) contd.....

Compound	Column Supplier and Brand	Column Chemistry	Dimensions (mm x mm)	P.S. (µm)	Mobile phase (v/v)	pH	Tem (°C)	Detector	Ref
Diclofenac and related impurities	Waters Acquity HSS T3	C18	100x2.1	1.8	Gradient Water (Phosphoric acid):MeOH	2.3	45	DAD	108
Diloxanide Furoate Metronidazole Benzoate Methyl Paraben Propyl Paraben	Waters Acquity BEH Shield RP 18	C18	100x2.1	1.7	Gradient Water (10 mM monobasic potassium phosphate):ACN:MeOH	-	40	DAD	109
Trans-resveratrol	Waters Acquity HSS	C18	100x2.1	1.8	(Water, containing 0.1% formic acid and 10% ACN):(ACN, containing 2% water)	-	40	DAD	110
Propranolol	Waters Acquity BEH	C18	30x2.1	1.7	Gradient Trifluoroacetic acid (0.1%):ACN, 80:20, v/v	-	35	DAD	111
Naproxen Fluvastatin Ibuprofen	Waters Acquity BEH	C18	100x2.1	1.7	MeOH:Water (0.1% formic acid), 50:50, v/v	-	25	MS	112
Valsartan Hydrochlorothiazide Amlodipine and related impurities	Agilent Zorbax Eclipse Plus RRHD	C8	100x3.0	1.8	Gradient Water (1.15 g/L ammonium dihydrogen phosphate):ACN	2.5	30	DAD	113
Meropenem	Waters Acquity HSS	C18	150x2.1	1.8	Gradient Water (20 mM KH_2PO_4):ACN	3.23	40	DAD	114
Rifampicin	Waters Acquity BEH	C18	100x2.1	1.7	Water (Formic acid):ACN, 40:60, v/v	4.5	-	DAD	115
Cinacalcet and related impurities	Waters Acquity BEH Shield RP 18	C18	100x2.1	1.7	Gradient Phosphate buffer:ACN	6.6	35	DAD	116
Losartan Potassium Chlorthalidone	Waters Acquity BEH HSS	C18	100x2.1	1.8	Water (20 mM KH_2PO_4):MeOH. 56:44, v/v and ACN:MeOH, 90:10, v/v	3	RT	DAD	117
Chloramphenicol Dexamethasone Tetryzoline	Phenomenex Kinetex	C18	50x2.1	1.7	ACN:Phosphate buffer, 15:85, v/v	3.5	35	DAD	118
Tacrolimus Tacrolimus diol Tacrolimus C-10 epimer	Waters Acquity BEH	C18	100x2.1	1.7	Water (0.1% phosphoric acid):ACN: methyl tert-butyl ether	-	65	DAD	119
R-Telaprevir	Waters Acquity BEH	C18	100x2.1	1.7	Gradient KOH and MeOH:ACN, MeOH, 2-propanol and Ethanol	11.5	40	DAD	120
Cabazitaxel and related degradation products	Waters Acquity BEH Shield RP 18	C18	100x2.1	1.7	Gradient Water (20 mM KH_2PO_4):MeOH:ACN and ACN:Water	-	30	DAD	121
Tolterodine tartarate and related impurities	Waters Acquity BEH Shield RP 18	C18	100x2.1	1.7	Gradient	3.5	50	DAD	122
Tolperisone HCl Paracetamol	Waters Acquity BEH	C18	50x2.1	1.7	Gradient Water (0.1% Phosphoric acid):ACN, 70:30, v/v	-	35	DAD	123
Isoconazole Diflucortolone	Waters Acquity HSS	C18	50x2.1	1.8	MeOH:Water, 61:39, v/v	-	40	DAD	124
Emtricitabine and related impurities	Waters Acquity BEH	C18	100x2.1	1.7	Gradient 0.01 M ammonium acetate buffer and MeOH:Buffer, 80:20, v/v	4.8	40	DAD	125
Sitagliptin Metformin Vildagliptin Diphenhydramine Sitagliptin Imp-1 Sitagliptin Imp-2	Thermo Scientific Hypersil Gold	C18	50x2.1	1.9	Gradient ACN:Water (0.2% Formic acid)	-	25	MS/MS	126
Chloroquine Primaquine	Thermo Scientific Hypersil	C18	50x2.1	1.9	Gradient ACN:Water(0.1% TEA)	3	25	DAD	127

(Table 5) contd.....

Compound	Column Supplier and Brand	Column Chemistry	Dimensions (mm x mm)	P.S. (µm)	Mobile phase (v/v)	pH	Tem (°C)	Detector	Ref
Benzalkonium chloride	Advanced Chromatography Technologies Ltd Excel C18 AR	C18	50x2.1	2.0	Gradient 10 mM ammonium phosphate buffer and ACN:MeOH,15:85,v/v	3.3	45	DAD	128
Sodium cromoglicate Tetryzoline and related degradants	Phenomenex Kinetex	C18	50x2.1	1.7	MeOH:Water (1% Phosphoric acid), 65:35, v/v	-	-	UV	129
Hydrochlorothiazide Valsartan Amlodipin	Waters Acquity BEH	C18	100x2.1	1.7	MeOH:Water (0.05M phosphate buffer), 70:30, v/v	3	-	DAD	130
Ritonavir and related impurities	Waters Acquity BEH Shield RP18	C18	100x2.1	1.7	Gradient Water (10 mM KH_2PO_4):ACN	3.5	50	DAD	131
Vitamin C	Waters Acquity BEH	C18	100x2.1	1.7	Gradient MeOH:Water (5 mM KH_2PO_4)	2.65	-	DAD	132
Duloxetine Mecobalamin	Waters Acquity BEH	C18	100x2.1	1.7	MeOH:Water, 45:55, v/v	-	40	UV	133
Sitagliptin Metformin	Waters Symmetry	C18	100x2.1	2.2	MeOH:Buffer, 20:80, v/v	3.5	50	DAD	134
Repaglinide	Phenomenex Kinetex	C18	50x2.1	1.3	ACN:Water (0.05% TFA) 50:50, v/v	3	25	DAD	135
Metformin Linagliptin Empagliflozin	Waters Symmetry Acclaim RSLC	C18	100x2.1	2.2	MeOH:Water (KH_2PO_4), 50:50, v/v	4	RT	DAD	136
5 Fluorouracil	Thermo Scientific Hypersil Gold	C18	50x2.1	1.9	MeOH:Water, 80:20, v/v	-	-	DAD	137
Domperidone and related impurities	Agilent Hypersil Zorbax eXtra Densely Bonded	C18	30x4.6	1.8	Gradient 0.06 M ammonium acetate:MeOH	-	40	DAD	138
Rabeprazole Levosulpride	Waters Acquity BEH	C18	100x2.1	1.7	Gradient Phosphate buffer:ACN	6.5	-	DAD	139
Azithromycin	Waters Acquity BEH	C18	100x2.1	1.7	Gradient ACN:Ammonia acetate	-	-	MS/MS	140
Sitagliptin Simvastatin	Waters Symmetry BEH	C18	100x2.1	1.7	Phosphate Buffer:ACN, 30:70, v/v	4	-	DAD	141
Cefuroxime	Waters Acquity BEH	C18	100x2.1	1.7	1-octane sulfonic acid sodium salt (5.78 × 10^{-4} M):ACN, 50:50, v/v	-	25	MS/MS	142
Ampicillin and degradation products	Waters Acquity BEH	C18	100x2.1	1.7	Water (0.001% acetic acid):MeOH, 30:70, v/v	-	-	MS	143
Carbamazepine and related impurities	Waters Acquity CSH	C18	100x2.1	1.7	Gradient Water (0.2% phosphoric acid):ACN	2.4	60	DAD	144
Alfuzosin	GL Sciences Inertsil ODS-3	C18	50x3.0	2.0	Tetrahydrofuran:ACN:Water:Perchloric acid, 10:220:770:1, v/v/v/v	-	25	DAD	145
Mebeverine	Waters Symmetry	C18	75x4.6	3.5	Buffer:ACN, 60:40, v/v	5.2	50	DAD	146
Cetirizine	Silica	-	33x4.6	3.0	Sulfuric acid:Water:ACN, 4:66:930, v/v/v	-	25	DAD	147
Fingolimod HCl and degradation products	Waters Acquity BEH	C18	100x2.1	1.7	Gradient ACN:Water (20 mM KH_2PO_4, Triethylamine)	6.5	40	DAD	148
Milnacipran and degradation products	Waters Acquity BEH	C18	100x2.1	1.7	Gradient ACN:Water (10 mM KH_2PO_4)	3.5	27	DAD	149
Dronedarone and degradation products	Waters Acquity BEH	C18	100x2.1	1.7	Gradient ACN:Water (10 mM KH_2PO_4)	-	40	DAD	150
Perindopril L-Arginine	Agilent Poroshell 120	HILIC	150x4.6	2.7	ACN:Water (0.1% Formic acid), 20:80, v/v	-	25	DAD	151

(Table 5) contd.....

Compound	Column Supplier and Brand	Column Chemistry	Dimensions (mm x mm)	P.S. (μm)	Mobile phase (v/v)	pH	Tem (°C)	Detector	Ref
Moxonidine and degradation products	Thermo Scientific Hypersil Gold aq.	C18	100x2.1	1.9	MeOH:Ammonium acetate buffer (10 mM), 09:99.1, v/v	3.4	50.13	DAD	152
Talinolol	Waters Acquity HSS	C18	50x2.1	1.8	ACN:Phosphate Buffer, 30:70, v/v	4.5	-	DAD	153
Telmisartan, Chlortalidone and related impurities	Waters Acquity BEH Shield RP18	C18	100x2.1	1.7	Gradient Buffer:ACN	4.5	25	DAD	154
Bisoprolol Fumarate Hydrochlorothiazide	Waters Acquity BEH	C18	50x2.1	1.7	ACN:Phosphate buffer (20 mM)	3	40	DAD	155
Nilotinib and related impurities	Shimadzu Shim-pack XR-ODS II	C18	75x30	1.8	Gradient Water (0.1% Phosphoric acid):ACN	-	40	DAD	156
Lopinavir, ritonavir and related impurities	Waters Acquity BEH	C18	50x2.1	1.7	Gradient ACN:MeOH (0.1% Phosphoric acid)	2.2	30	DAD	157
Entacapone	Waters Acquity HSS	C18	50x2.1	1.8	ACN:Water (0.1% Phosphoric acid), 43:57, v/v	3	40	DAD	158
Telmisartan and related impurities	Waters Acquity BEH	C18	100x2.1	1.7	Gradient Phosphate buffer (10 mM):MeOH:ACN:Water	2.5	30	DAD	159
Maraviroc and related impurities	Waters Acquity BEH Shield RP18	C18	100x2.1	1.7	0.01 M Ammonium acetate: ACN, 63:37, v/v	-	40	DAD	160
Omeprazole Esomeprazole Lansoprazole Pantoprazole Rabeprazole	Waters Acquity BEH	C18	50x3.0	1.7	Gradient Water:ACN	-	35	DAD	161
Famotidine	Waters Symmetry	C18	50x2.1	1.7	Water (KH_2PO_4):ACN, 40:60, v/v	7	-	DAD	162
Enalapril Hydrochlororthiazide	Waters Acquity BEH	C18	100x2.1	1.7	Gradient Water (Phosphoric acid):ACN	2.5	60	DAD	163
Metoclopramide and related impurities	Waters X-terra RP18	C18	150x4.6	3.5	Gradient Phosphate buffer:Pentane-1-sulphonic acid:ACN	3	35	DAD	164
Rabeprazole and related impurities	Waters Acquity RP18	C18	100x2.1	1.7	Gradient Phosphate buffer:ACN	6.4	25	DAD	165
Epalrestat	Agilent	C18	50x2.1	1.8	0.5% Ammonium acetate:MeOH:ACN, 40:40:20, v/v/v	3	-	DAD	166
Ebastine and related impurities	Waters Acquity BEH	C18	50x2.1	1.7	ACN:2-propanol, 50:50, v/v	6.2	60	DAD	167
Tadalafil and related impurities	Waters Acquity HSS T3	C18	150x2.1	1.8	Gradient MeOH:Ammonium acetate	4	-	DAD	168
Lansoprazole and related impurities	Waters Acquity BEH	C18	50x2.1	1.7	Gradient Phosphate buffer:MeOH:ACN	7	40	DAD	169
Aciclovir and related impurities	Thermo Scientific Syncronis	C18	100x3.5	1.7	Gradient Ammonium acetate:ACN:Water, 90:6:4, v/v	5.5	60	DAD	170
Darifenacin HBr and related impurities	Waters Acquity BEH	C18	100x2.1	1.7	Gradient Phosphate buffer (20 mM) and triethylamine: Buffer and ACN	6.5	30	DAD	171
Lacosamide and related impurities	Waters Acquity HSS	C18	100x2.1	1.8	Phosphate buffer:ACN, 85:15, v/v	2	-	DAD	172
Atropine sulphate	Merck Hiber HR Purospher Star	C18	100x2.1	2.0	Phosphate buffer:ACN, 87:13, v/v	3.5	30	DAD	173
Acyclovir	Waters Acquity BEH	C18	50x2.1	1.7	Water (0.25% Formic acid)	-	50	DAD	174

(Table 5) contd.....

Compound	Column Supplier and Brand	Column Chemistry	Dimensions (mm x mm)	P.S. (μm)	Mobile phase (v/v)	pH	Tem (°C)	Detector	Ref
(Bisphenols) Cl-BPA Cl_2-BPA Cl_3-BPA Cl_4-BPA	Phenomenex Gemini	C 18	100x2.0	3.0	Gradient Ammonia: MeOH	-	30	MS	175
Imatinib and impurities	Waters Acquity BEH	C18	50x2.1	1.7	Gradient Ammonium acetate:MeOH: ACN	-	-	DAD	176
Aliskiren ADPA	Waters XBridge	HILIC	150x4.6	3.5	K_2HPO_4: ACN (26:74; v/v)	7.2	-	FLD	177
Meta-cresol Parathyroid hormone	-	C18	100x2.1	3.0	Gradient	-	-	UV	178
Erythropoietin HSA	-	C18	50x2.1	1.7	Gradient Water (0.1% TFA):ACN	-	-	UV	179
Phenylpropanolamine Cathine Ephedrine Pseudoephedrine Methylephedrine	Waters Acquity BEH	C18	50x2.1	1.7	Ammonium bicarbonate: ACN (90:10; v/v)	10	-	DAD	180
Yohimbine	Waters Acquity BEH	C18	100x2.1	1.7	Gradient Water (0.1% NH_4OH): MeOH(0.1% NH_4OH)	-	-	DAD	181
Sibutramin Modafinil Ephedrine Norefedrine Metformine	Grace Vision HT C18-P	C18	100x2.0	1.5	Gradient 0.025 M ammonium acetate:ACN	5	-	DAD	182
Zolmitriptan and impurities	Sigma Aldrich Ascentis Express	Phenyl Hexyl	100x3.0	2.7	Gradient 20 mM KH_2PO_4,5 mM sodium 1-hexansulfonate:ACN	2	-	DAD	183
Esomeprazole and impurities	Waters Acquity BEH	C18	50x2.1	1.7	Gradient 0.04M glycine in water:ACN	9	-	DAD	184
Genistein Daidzein Glicitein Genistin Daidzin Glycitin	Waters Acquity BEH	C18	50x2.1	1.7	Gradient Water (%0.1 formic acid):ACN (%0.1 formic acid)	-	-	UV	185
Cyanindin-3-glucoside Cyanindin-3-rutinoside	Waters Acquity HSS	C18	100x2.1	1.8	Gradient Formic acid in water:ACN	-	-	DAD	186
Mesembrenol Mesembranol Mesembrenone Mesembrine	Waters Acquity BEH	C18	150x2.1	1.7	Gradient 0.1% Ammonia: MeOH	-	-	DAD	187
Ganciclovir	Waters Acquity HSS T3	C18	50x2.1	1.8	Gradient 0.02 M KH_2PO_4:1% ACN	2	-	DAD	188
Domperidon Omeprazole	Waters Acquity BEH	C18	50x2.1	1.7	Phosphate buffer:ACN (68:32; v/v)	5	-	DAD	189
Isoleucine Leucine Phenylalanine Valine	Waters Acquity BEH	C18	100x2.1	1.7	Gradient Phosphate buffer:ACN	3	-	UV	190
Olmesartan and 17 Impurities	Waters Acquity BEH	Phenyl	100x2.1	1.7	Gradient Phosphate buffer:ACN	2.5	-	DAD	191
Rivastigmine and related impurities	Waters Acquity BEH	Phenyl	100x2.1	1.7	ACN:Na_2HPO_4:TEA (10:90:0.1; v/v/v)	7.5	-	DAD	192
Norisoboldine	Waters Acquity BEH	C18	50x2.1	1.7	Gradient ACN:Ammonium acetate	11.1	-	DAD	193
Ramelteon and related impurities	Waters Acquity UPLC BEH Shield RP18	C18	100x2.1	1.7	Gradient 0.1% TFA mixture of water:ACN	-	-	DAD	194

(Table 5) contd.....

Compound	Column Supplier and Brand	Column Chemistry	Dimensions (mm x mm)	P.S. (µm)	Mobile phase (v/v)	pH	Tem (°C)	Detector	Ref
Paliperidon and related impurities	Waters Acquity BEH	C18	50x2.1	1.7	KH_2PO_4: 1.5 mL of TEA	6.5	-	DAD	195
Raloxifene	Waters Acquity BEH	C18	50x2.1	1.7	Gradient 0.1% formic acid:ACN	-	-	DAD	196
Gemicitabine and related impurities	Waters Acquity HSS T3	C18	100x2.1	1.8	KH_2PO_4:MeOH (90:10; v/v)	2.5	-	DAD	197
Astaxanthin	Waters Acquity BEH	C18	50x2.1	1.7	Gradient ACN:MeOH (ammonium acetate):dichloromethane (75:20:5; v/v/v)	-	-	DAD	198
Homocysteine L-Cysteine	Sihma Aldrich Supelcosil	C18	150x4.6	3.0	KH_2PO_4:ACN	2.1	-	FLD	199
Glucosamines	Waters Acquity HSS T3	C18	50x2.1	1.7	Gradient water,MeOH: ACN	6.5	-	DAD	200
Rabeprazole and related impurities	Waters Acquity BEH	C18	100x2.1	1.7	Gradient KH_2PO_4:ACN	6	-	DAD	201

ACN: Acetonitrile; P.S: Particle Size; Tem: Temperature; DAD: Diode Array Detector; IS: Internal standard; TFA: Trifluoroacetic acid; PFP: Pentafluorophenyl; TEA: Triethylamine; FLD: Fluorescence detector; Imp: Impurity; HILIC: Hydrophilic Interaction Liquid Chromatography; MeOH: Methanol; MS: Mass spectrometry; UV: Ultraviolet; RT: Room temperature.

Table 6. Validation parameters of the pharmaceutical applications.

Compound	Analytical Column	Run Time (min)	System Suitability				Validation Parameters					Ref
			Rs	α	N	T or A_s	Linearity (µg/mL)	LOD (µg/mL)	LOQ (µg/mL)	Precision (RSD %)	Accuracy %	
Imp-E Imp-D Imp-B Bosentan Imp-A Imp-C	Kinetex XB-C18 100A (50x3.0 mm; 1.7 µm)	14	- 3.91 17.55 1.75 17.18 31.87	-	-	1.11 1.15 1.00 1.04 1.19 1.08	0.52-3.12	0.045	0.151	<2.7	97.0-103.0	79
Salidroside Rhodiosin Rhodionin Herbacetin Kaempferol Quercetin	Zorbax Eclipse plus C18 (100x2.1 mm; 3.5 µm)	-	-	-	-	-	0.01-20 0.01-4 0.01-4 0.005-2 0.01-2 0.005-2	-	-	4.05-5.07	99.27-108.91	80
Gallic Acid Chlorogenic Acid Vanillic Acid Caffeic Acid Paeoniflorin Ferulic Acid Liquiritin Verbascoside	Acquity BEH C18 (100 x 2.1 mm, 1.7 µm)	13	-	-	-	-	4.4-139.5 3.3-104.5 0.12-3.86 1.3-41.6 3.3-106.5 0.19-6.1 5.31-170 0.41-13.1	0.227 0.551 0.113 0.143 0.158 0.157 1.330 0.065	0.767 1.280 0.324 0.441 0.472 0.359 3.540 0.231	1.26 1.76 0.55 0.82 1.77 2.32 1.85 1.25	98.7 99.1 102.3 102.6 97.0 98.4 99.1 98.0	81
4-Methoxy-5-hydroxycanthin-6-one Andrographolide Dehydroandrographolide Neoandrographolide Rosmarinic acid	Acquity BEH C18 (100 x 2.1 mm, 1.7 µm)	30	-	-	-	-	0.50-32.16 10-160.02 40-200.7 17.3-122 1.91-61.3	-	0.50 10.0 40.13 17.03 1.91	2.94 0.11 0.20 2.14 0.30	102.4 100.1 103.3 97.2 100.9	82
Fenofibric acid	Kinetex (50x2.1 mm; 2.6 µm)	3	-	-	-	-	0.005-10	-	0.005	11.3<	94.9-105.2	85

(Table 6) contd.....

Compound	Analytical Column	Run Time (min)	System Suitability				Validation Parameters					Ref
			Rs	α	N	T or A_s	Linearity (μg/mL)	LOD (μg/mL)	LOQ (μg/mL)	Precision (RSD %)	Accuracy %	
Mesaconitine calycosin-7-O-β-d-glucopyranoside Hypaconitine Aconitine Benzoylmesaconitine Bbenzoylaconitine Benzoylhypaconitine Liquiritin Ammonium glycyrrhizinate Tetrahydropalmatine Paeoniflorin Sinomenine	Acquity BEH C18 (100x2.1 mm, 1.7 μm)	36	-	-	-	-	0.115–11.79 0.582–23.30 0.002–0.127 0.013–1.72 0.002–0.050 0.094–4.71 0.0331–1.66 0.0337–2.70 2.292–114.6 2.498–199.9 0.313–15.6 4.604–736.6 30.78–153.9	65.51 0.2330 0.0325 0.0344 0.0255 0.0379 0.0132 0.0135 0.4585 0.1999 0.0313 0.03683 3.078	92.12 1.165 0.0814 0.0860 0.0636 0.190 0.0662 0.0675 0.9170 0.9993 0.313 0.3683 6.156	2.57 2.68 2.86 2.37 2.61 1.31 2.98 2.50 2.65 1.95 1.61 2.73 1.14	101.68 100.45 100.99 100.76 100.26 99.35 102.77 104.44 102.64 103.79 99.04 100.24 100.68	86
Nebivolol	Acquity BEH C18 (50x2.1 mm, 1.7 μm)	0.6	-	-	3218	1.3	10-50	2.95	9.97	0.58	100.6	87
Clozapine and degradation products	Zorbax Eclipse plus C18 (50x2.1 mm; 1.8 μm)	7					0.4-14	0.07	0.23	0.63	98.64	88
Acetaminophen Phenylephrine HCl Guaifenesin Dextromethorphan HBr	C18 (100x2.1 mm; 1.7 μm)	13	-	-	-	1.2 1.0 1.0 1.1	25-112.5 16-74 16-74 32-144	2.3 3.3 3.5 4.9	6.2 11.1 11.9 14.7	0.4 0.8 1.0 0.7	100.1 99.7 99.4 99.9	89
Carfilzomib and degradation products	Microsolv Silica type C Cogent Bidentate C18 (100x2.1 mm; 2.2 μm)	6	-	-	-	-	10-250	-	10	2.0<	95.4-100.4	90
Methyl Paraben Ethyl Paraben isopropyl Paraben nPr Paraben sec Butyl Paraben iso Butyl Paraben nButyl Paraben Benzyl Paraben	Syncronis aQ column (100x2.1 mm; 1.7 μm)	-	-	-	-	-	0.100–100 0.100–100 0.100–250 0.100–100 0.100–250 0.100–250 0.100–250 0.100–100	-	0.50 0.21 0.13 0.10 ND 0.10 0.10 0.10	1.9 2.5 4.4 0.9 1.8 8.1 6.2 4.9	118 103 95 107 132 99 100 105	91
Acetazolamide	-	2.5	+	+	+	+	2-10	1.36	4.13	0.71	102.4	92
Telmisartan Clinidipine	Acquity BEH C18 (100x2.1 mm, 1.7 μm)	13	+	+	3628 3290	1.24 1.46	40-160 10-40	+	+	1.67 0.36	99.2 98.11	93
Imp-A Imp-B Imp-C1 Imp-D Imp-C2 Imatinib Mesylate Imp-E	Acquity CSH C18 (50x2.1 mm, 1.7 μm)	9	- 9.6 5.00 3.63 2.98 7.16 4.51	+	+	1.21 1.04 1.20 1.19 1.08 0.99 1.13	LOQ-200%	0.64 1.78 1.63 0.91 0.16 0.088 0.96	1.93 5.41 4.93 2.74 0.50 0.27 2.91	1.46 1.06 1.09 1.96 0.98 1.08 2.08	-	94
Nevirapine and related impurities	Agilent Poroshell 120 C18 (100x2.1 mm, 2.7 μm)	16	+	+	+	+	8-88	0.092	0.174	0.57-1.72	98-102	95
Hederacoside C	Acquity BEH C18 (100x2.1 mm, 1.7 μm)	4	-	-	-	-	0.001-1	-	0.001	4.7	95.5-101.3	96
Methyl Paraben Propyl Paraben Loratadine	Acquity BEH C18 (100x2.1 mm, 1.7 μm)	2.5	- 7.67 19.32	1.33 1.52 1.75	15075 17067 17695	1.07 0.99 0.94	4-12 8-16 0.8-1.2	0.07 0.07 0.21	0.23 0.23 0.69	0.76 0.72 0.76	99.0 98.5 100.5	97
Nitrile Dimer Amide Agomelatine Acid Amine AgoSalt	Acquity BEH Shield RP 18 (50x2.1 mm, 1.7 μm)	5	6.12 16.53 27.35 12.33 - 8.68 10.81	-	-	1.24 1.06 1.38 1.33 1.27 1.18 1.34	0.1-25	0.03 0.03 0.03 0.03 0.03 0.03 0.03	0.1 0.1 0.1 0.1 0.1 0.1 0.1	1.9 2.3 0.4 2.0 1.2 0.7 1.0	102.7 99.5 100.9 97.1 97.3 98.4 98.3	98

(Table 6) contd.....

Compound	Analytical Column	Run Time (min)	System Suitability				Validation Parameters					Ref
			Rs	α	N	T or A_s	Linearity (µg/mL)	LOD (µg/mL)	LOQ (µg/mL)	Precision (RSD %)	Accuracy %	
Bambuterol and degradation products	Acquity BEH C18 (100x2.1 mm, 1.7 µm)	1.3	-	1.9	12590	-	7-13	0.11	0.36	0.19	99.2	99
Oleracone	Shimadzu Shimpack ODS (50x2.0 mm, 1.6 µm)	4.0	-	-	-	-	0.15-30	0.042	0.141	3.2-8.7	86.2-89.1	100
Internal Standard Amlodipine Rosuvastatin	Waters Acquity HSS (50x2.1 mm, 1.8 µm)	1.5	- 4.96 5.04	- 2.9 1.8	1830 4140 10200	1.99 1.86 1.31	0.5-100	0.011 0.034	0.016 0.049	0.35 0.52	99.8 101.2	101
BO-2 BO-1 BO-3 Imp-D Imp-E Bosentan	Waters Acquity (100x2.1 mm, 1.7 µm)	14	18.6 9.2 5.4 2.6 13.2 -	+	+	1.3 1.2 1.2 1.2 1.1 1.1	0.03-20 0.03-20 0.5-40 0.03-1.5 0.03-1.5 0.03-1.5	0.1 0.1 0.05 0.1 0.01 0.1	0.3 0.3 0.3 0.3 0.3 0.3	0.86 0.36 0.51 0.63 0.49 0.69	99.91-99.82	102
Efavirenz Lamivudine	Waters Acquity BEH Shield RP 18 (50x3 mm, 1.7 µm)	3	- 3.35	+	12448 10275	1.32 1.47	10-60 5-30	0.12 0.04	0.37 0.13	0.81 1.27	100.79 100.33	103
Dimenhydrinate Chloroteophylline Theophylline Imp-F Benzhydrol Benzophenone	Waters X Select HSS T3 (100x2.1 mm, 1.8 µm)	5	+	+	+	+	0.2-11 0.3-6	0.01 0.01 0.02 0.01 0.02 0.02	0.02 0.02 0.04 0.03 0.04 0.04	0.5-6.4	90.6-100	104
Bisoprolol Hydrochlorothiazide and related impurities	Waters Acquity (100x2.1 mm, 1.7 µm)	22	+	+	+	+	LOQ-150%	3-4	10-12	0.4-1	98.4-104	105
Cefetamet pivoxil	Waters Acquity (100x2.1 mm, 1.7 µm)	2	4.92	+	+	1.34	10-240	2.08	6.29	2<	97.8-102	106
Amoxicillin	Waters Acquity BEH (100x2.1 mm, 1.7 µm	2.6	3.19		12076	1.27	0.024-0.096	+	+	1.8	99.7	107
Diclofenac and related impurities	Waters Acquity HSS T3 (100x2.1 mm, 1.8 µm)	7	+	+	+	+	LOQ-200%	+	+	3<	99-101	108
Methyl Paraben Propyl Paraben Metronidazole Benzoate Diloxanide Furoate	Waters Acquity BEH Shield RP-18 (100x2.1 mm, 1.7 µm)	3.6	- 11.9 15.9 10.4	- 1.68 1.85 1.50	15075 17067 17695 16765	1.07 0.99 0.94 0.94 1.1	8-16 0.8-1.6 8-40 8-40	0.07 0.08 0.22 0.28	0.23 0.27 0.73 0.93	0.88 0.76 1.12 1.23	99.0 98.5 100.5 101.5	109
Trans-resveratrol	Waters Acquity HSS (100x2.1 mm, 1.8 µm)	13	7.5	+	+	+	0.5-15	0.14	0.43	0.58	101.2	110
Propranolol	Waters Acquity BEH (30x2.1 mm, 1.7 µm	2.5	-	-	-	-	10-50	-	-	0.72	96.7-98.7	111
Naproxen Fluvastatin Ibuprofen	Waters Acquity BEH (100x2.1 mm, 1.7 µm)	6	+	+	+	+	0.018-0.27 0.015-0.225 0.02-0.15	0.0013 0.0014 0.0019	0.0038 0.0041 0.0057	3.24 0.70 3.94	100.32 99.99 100.86	112

(Table 6) contd.....

Compound	Analytical Column	Run Time (min)	System Suitability				Validation Parameters					Ref
			Rs	α	N	T or A_s	Linearity (μg/mL)	LOD (μg/mL)	LOQ (μg/mL)	Precision (RSD %)	Accuracy %	
Hydrochlorothiazide Amlodipine at 225 nm Amlodipine at 360 nm Valsartan Imp-A(HCT) Imp-B (HCT) Imp-C (HCT) Imp-D (Amlodipine) Imp-C (Valsartan)	Agilent Zorbax Eclipse Plus RRHD C8 (100x3.0 mm, 1.8 μm)	12	>2.2	+	+	+	0.05–0.30 0.10–0.30 0.10–0.30 0.05–0.30 0.05–1.20 0.05–1.20 0.10–1.24 0.10–0.66 0.05–0.41	0.028 - 0.093 0.047 0.056 0.040 0.028 0.103 0.044	0.086 - 0.283 0.141 0.171 0.120 0.084 0.313 0.135	0.59 0.96 2.3 1.5 1.4 0.8 3.7 1.3 0.54	91.8-103.7	113
Meropenem	Waters Acquity HSS (150x2.1 mm, 1.8 μm	3.4	-	-	-	-	47-9120	-	-	-	-	114
Rifampicin	Waters Acquity BEH (100x2.1 mm, 1.7 μm)	5	+	+	2231	1.27	5-80	0.03	0.09	1.02-1.52	97.9-101	115
Cinacalcet and related impurities	Waters Acquity BEH Shield RP 18 (100x2.1 mm, 1.7 μm)	30	3.5	+	113576	1.2	0.22-1.65	-	0.022-0.078	0.3-2.2	95-108	116
Losartan Potassium Chlorthalidone	Waters Acquity BEH HSS (100x2.1 mm, 1.8 μm)	2	+	+	11040 8325	1.36 1.25	12.5-125 3.125-31.25	0.5	1.5	0.51 0.37	99.8-100.3	117
Tetryzoline Chloramphenicol Dexamethasone	Kinetex C18 (50x2.1 mm, 1.8 μm)	3.3	- 4.58 8.79	- 1.88 2.23	2009 2800 4019	1.01 0.91 0.95	4-30 1-15 1-12	-	-	0.77 0.91 0.82	100.3 100.3 100.5	118
R-Telaprevir	Waters Acquity BEH (100x2.1 mm, 1.7 μm)	10	2.1	1.09	-	-	LOQ-5.25%	0.15	0.5	0.3	101.6	120
Imp-C Imp-D Imp-B Imp-A Imp-E Cabazitaxel Imp-F	Waters Acquity BEH Shield RP 18 (100x2.1 mm, 1.7 μm)	12	>2.0	-	-	1.5<	5-50 5-50 5-50 5-50 5-50 50-200 5-50	0.002 0.002 0.001 0.001 0.002 0.002 0.001	0.005 0.006 0.003 0.003 0.003 0.006 0.003	4.5 4.2 4.9 4.6 4.3 3.5 2.6	104.5 102.8 101.8 103.4 103.9 - 99.7	121
Tolterodine tartarate Des-N-N-diisopropyl TL Imp-A Imp-B Imp-C Imp-D Imp-E	Waters Acquity BEH Shield RP 18 (100x2.1 mm, 1.7 μm)	7	>1.5	+	>5000	2<	LOQ-200%	0.011 0.003 0.009 0.012 0.009 0.015 0.015	0.034 0.01 0.032 0.035 0.033 0.045 0.045	0.9 1.1 1.5 1.2 1.4 1.0 1.5	101.5 102.5 101.5 100.2 99.2 99.9 100.2	122
Tolperisone HCl Paracetamol	Waters Acquity BEH (50x2.1 mm, 1.7 μm)	5	>2	+	6676	1.13	6-54 20-180	0.2 0.06	0.5 0.15	1.43 1.09	100.5 100.0	123
Isoconazole Diflucortolone	Waters Acquity HSS C18 (50x2.1 mm, 1.8 μm)	2	1.66	1.41	8220 38500	+	1-200 1-200	0.11 0.09	0.35 0.27	0.93 0.68	98.37 96.49	124
Emtricitabine Carboxylic acid S-Oxide impurity Lamivudine Des amino impurity	Waters Acquity BEH (100x2.1 mm, 1.7 μm)	8	-	-	18189	1.1	+	3.5 3.1 3.2 3.3	11.3 9.6 13 11.3	0.4-0.8	97.7-107	125

(Table 6) contd.....

Compound	Analytical Column	Run Time (min)	System Suitability				Validation Parameters					Ref
			Rs	α	N	T or A_s	Linearity (μg/mL)	LOD (μg/mL)	LOQ (μg/mL)	Precision (RSD %)	Accuracy %	
Sitagliptin Vildagliptin Metformin Diphenhydramine (IS) Sitagliptin Imp-1 Sitagliptin Imp-2	Hypersil Gold C18 (50x2.1 mm, 1.9 μm)	3	-	-	-	-	0.005-0.1 0.005-0.5 0.001-0.4	0.0015 0.0015 0.003	0.005 0.005 0.001	0.71 0.81 0.83	99.24 100.27 100.59	126
Chloroquine Primaquine	Hypersil Gold C18 (50x2.1 mm, 1.9 μm)	1.5	-	-	-	-	90-210 60-140	10 7	30 22	1.1 1.04	99.22 99.67	127
Benzalkonium chloride	ACE Excel C1-AR (50x2.1 mm, 2.0 μm)	2	-	-	-	-	25-75	-	25	0.16	98	128
Sodium cromoglicate Tetryzoline Deg-1 Deg-2 Deg-2	Kinetex C18(50x2.1 mm, 1.7 μm)	6	3.83 7.45 - 2.91 4.72	1.33 1.52 - 1.44 1.76	9606 7084 2932 3138 4505	1.01 1.21 1.09 1.14 1.10	25-100 0.5-10	0.83 0.15	2.51 0.46	1.14 0.91	99.56 100.55	129
Hydrochlorothiazide Amlodipine Besylate Valsartan	Waters Acquity BEH (100x2.1 mm, 1.7 μm)	1	- 3.1 3.8	-	2039 2180 2830	-	6.5-15 5-12 76.5-178.5	0.011 0.023 0.06	0.037 0.077 0.22	0.11-0.95	98-103	130
Imp-1 Imp-2 Imp-3 Imp-4 Imp-5 Imp-6 Imp-7 Imp-8 Imp-9 Ritonavir Imp-10 Imp-11 Imp-12 Imp-13	Waters Acquity BEH Shield RP18 (100x2.1 mm, 1.7 μm)	20	- 10.7 12.2 6.7 3.5 14.5 9.5 16 6.6 2.0 3.0 5.0 21.2 14.5	+	2,504 3834 11684 15813 15144 24615 21762 41446 62164 58922 71709 112080 143742 171696	1.1 1.1 1.1 1.1 1.1 1.1 1.0 1.0 1.0 1.1 1.1 1.1 1.0 0.9		0.014 0.078 0.102 0.125 0.037 0.084 0.108 0.137 0.165 0.112 0.113 0.138 0.131 0.122	0.050 0.265 0.349 0.419 0.125 0.286 0.370 0.471 0.570 0.381 0.381 0.470 0.444 0.416	4.16 2.48 3.03 2.45 1.77 2.32 3.27 4.84 6.11 6.70 5.43 6.71 4.60 5.30	95.9 101.5 100.4 101.2 103.8 105.8 100.5 101.7 103.8 - 96.4 102.9 97.3 99.2	131
Vitamin C	Waters Acquity BEH (100x2.1 mm, 1.7 μm)	4	-	-	-	-	5-50	0.024	0.073	0.3	99.3-100	132
Duloxetine Mecobalamin	Waters Acquity BEH (100x2.1 mm, 1.7 μm)	10	- 3.75	+	3594 3334	1.04 0.92	1.5-5.25 37.5-131.25	0.048 0.045	0.145 1.363	0.79 0.89	99.74 99.82	133
Sitagliptin Metformin	Waters Symmetry (100x2.1 mm, 2.2 μm)	5	- 8.1		1241 1847	1.0 1.2	2-12 5-35	0.12 1.64	0.36 4.92	0.24 0.27	100.2 99.8	134
Repaglinide	Kinetex (50x2.1 mm, 1.3 μm)	2.2	6.09	3.65	16400	1.05	0.2-300	0.012	0.035	1.18	100.26	135
Metformin Linagliptin Empagliflozin	Waters Symmetry Acclaim RSLC 120 (100x2.1 mm, 2.2 μm)	5	- 2.08 4.81	-	1584 2614 2874	1.02 1.01 1.0	1-100 0.5-16 1-32	0.21 0.12 0.26	0.63 0.36 0.78	0.16 0.22 0.19	100.65 98.88 99.81	136
5 Fluorouracil	Hypersil Gold C18 (50x2.1 mm, 1.9 μm)	0.8	-	-	-	-	1-50	0.50	1.5	1.25-1.77	101.2	137

(Table 6) contd.....

Compound	Analytical Column	Run Time (min)	System Suitability				Validation Parameters					Ref
			Rs	α	N	T or A_s	Linearity (μg/mL)	LOD (μg/mL)	LOQ (μg/mL)	Precision (RSD %)	Accuracy %	
Imp-A Imp-B Imp-C Imp-D Imp-E Imp-F Domperidone Droperidol	Agilent Hypersil Zorbax eXtra Densely Bonded (30x4.6 mm, 1.8 μm)	7.5	-	-	-	-	15-55	0.11 0.04 0.04 0.05 0.08 0.08 0.04 0.05	0.30 0.11 0.12 0.14 0.22 0.24 0.12 0.15	0.25 0.30 0.31 0.29 0.29 0.35 0.63 0.65	98-100.9	138
Levosulpride Rabeprazole	Waters Acquity BEH (100x2.1 mm, 1.7 μm)	2	29	-	4193 30972	1.7 1.2	4-40 15-150	-	-	0.1 0.4	99.8-100	139
Azithromycin	Waters Acquity BEH (100x2.1 mm, 1.7 μm)	1.5	-	-	-	-	0.025-0.5	-	0.025	7.98	99.5-103	140
Sitagliptin Simvastatin	Waters Symmetry (100x2.1 mm, 1.7 μm)	1.8	-	-	2556 2318	1.3 1.1	500-900 200-360	0.18 0.17	0.61 0.56	0.39 0.95	99.8 99.4	141
Cefuroxime	Waters Acquity BEH (100x2.1 mm, 1.7 μm)	1	+	+	+	+	0.5-5	0.1	0.303	1.19	100.16	142
Ampicillin and degradation products	Waters Acquity BEH (100x2.1 mm, 1.7 μm)	1.5	+	+	+	+	0.25-3	0.016	0.049	1.98-2.67	99.5-100.9	143
Alfuzosin	GL Sciences Inertsil ODS-3 (50x3.0 mm, 2.0 μm)	2	-	-	-	1.5	10-300	-	-	0.5	97-103	145
Mebeverine	Waters Symmetry C18 (75x4.6 mm, 3.5 μm)	1.5	-	-	5263	1.19	50-400	-	-	0.23	100.5	146
Cetirizine	Silica (33x4.6 mm, 3.5 μm)	1.5	-	-	2693	1.5	10-300	-	-	0.69	97-103	147
Fingolimod HCl Imp-A Imp-B Imp-C Imp-D Imp-E Imp-F	Waters Acquity BEH (100x2.1 mm, 1.7 μm)	16	>2	+	+	1.5<	250-750	0.0004 0.0003 0.0002 0.0003 0.0004 0.0003 0.0004	0.0012 0.0010 0.0006 0.0009 0.0012 0.0009 0.0012	2.5 1.8 4.5 4.7 3.1 3.6 2.4	99.2 100.2 99.1 103.4 99.1 98.6 99.1	148
Imp-12 Imp-1 Imp-2 Imp-11 Milnacipran Imp-3 Imp-4 Imp-5 Imp-6 Imp-7 Imp-8 Imp-9 Imp-10	Waters Acquity BEH (100x2.1 mm, 1.7 μm)	12	- 20.59 16.79 6.35 2.25 4.11 4.35 6.20 7.09 9.17 6.07 41.20 2.92	+	+	1.00 0.99 1.03 1.05 1.19 1.03 1.17 1.10 1.05 1.00 1.02 1.05 1.07	LOQ-200%	0.015 0.045 0.045 0.035 0.043 0.045 0.035 0.035 0.035 0.035 0.020 0.030 0.020	0.045 0.140 0.135 0.110 0.130 0.140 0.105 0.105 0.105 0.110 0.065 0.085 0.060	0.2-1.4	94.7-103.4	149

(Table 6) contd.....

Compound	Analytical Column	Run Time (min)	System Suitability				Validation Parameters					Ref
			Rs	α	N	T or A_s	Linearity (µg/mL)	LOD (µg/mL)	LOQ (µg/mL)	Precision (RSD %)	Accuracy %	
Imp-1	Waters Acquity BEH (100x2.1 mm, 1.7 µm)	10	-	+	+	0.99	LOQ-200%	0.035	0.10	0.2-2.5	98.7-102.4	150
Imp-2			3.96			1.10		0.045	0.14			
Imp-3			17.49			1.02		0.045	0.13			
Imp-4			3.33			1.16		0.050	0.16			
Imp-5			4.84			1.01		0.045	0.13			
Imp-6			3.22			1.03		0.050	0.16			
Dronedarone			4.84			1.02		0.035	0.111			
Imp-7			7.96			1.02		0.050	0.16			
Imp-8			2.03			0.99		0.050	0.15			
Imp-9			12.23			0.98		0.030	0.095			
Imp-10			7.56			1.01		0.025	0.075			
Imp-11			10.92			1.03		0.025	0.075			
Perindopril L-Arginine	Poroshell 120 (150x4.6 mm, 2.7 µm)	2		-								151
Isomer I			1.68		34693	1.83	0.40-1.40	0.15	0.46	0.9-1.91	95.6-96.3	
Isomer II			1.47		28716	1.51	0.40-2.40	0.036	0.11	1.1-1.98	97.1-97.9	
Moxonidine	Hypersil Gold aq. (100x2.1 mm, 1.9 µm)	5.2	+	+	+	+	50-15	0.04	0.14	0.03	101.14	152
Imp-A							0.05-0.75	0.02	0.05	0.90	93.31	
Imp-B							0.05-0.75	0.02	0.05	0.49	91.05	
Imp-C							0.1-1.5	0.03	0.08	0.28	97.30	
Imp-D							0.1-1.5	0.04	0.10	0.90	97.63	
Talinolol	Waters Acquity BEH (50x2.1 mm, 1.8 µm)	2	-	-	2829	1.05	0.5-100	0.14	0.50	2.58	99.7-101.1	153
Telmisartan (TS) Chlortalidone (CT) Impurities	Waters Acquity Shield RP18 (100x2.1 mm, 1.7 µm)	16	-	-	-	-	0.05-6.81	~0.010	~0.030	0.5-4.0	95.4-98.6	154
Hydrochlorothiazide	Waters Acquity BEH (50x2.1 mm, 1.7 µm)	1.5	6.8	4.2	-	1.3	0.5-150	0.01	0.03	0.01	99.72	155
Bisoprolol Fumarate						1.5	0.5-250	0.07	0.21	0.01	100.12	
Imp-4	Shimadzu Shim-pack XR-ODS II (75x3.0 mm, 1.8 µm)	7	-	-	-	1.10	LOQ-150%	0.004	0.01	1.52	89-116.5	156
Imp-3			8.41			1.12		0.010	0.02	2.42		
Nilotinib			6.91			1.39		0.004	0.01	1.11		
Imp-2			2.11			1.01		0.004	0.01	1.33		
Imp-1			8.67			1.00		0.002	0.005	1.25		
2,6-dimethyl Imp	Waters Acquity BEH (50x2.1 mm, 1.7 µm)	11	-	-	-	1.06	LOQ-150%	0.014	0.576	0.29-0.64	90.1-108	157
Imp-F			20.11			1.04		0.371	0.927			
Imp-I			9.70			1.05		0.314	0.785			
Imp-A			6.62			1.10		0.382	0.954			
Ritonavir			4.46			1.11		0.488	1.220			
Imp-L			2.32			1.10		0.371	1.240			
Lopinavir			5.03			1.06		0.332	0.831			
Imp-B			8.27			1.09		0.541	1.353			
N-Acyl Imp			34.15			1.11		0.170	0.425			
O-Acyl Imp			8.32			1.10		0.214	0.536			
Entacapone	Waters Acquity HSS (50x2.1 mm, 1.8 µm)	0.7	2.43	2.36	9060	1.54	0.5-100	0.03	0.08	0.11	99.2-101.5	158
Imp-A	Waters Acquity BEH (50x2.1 mm, 1.7 µm)	7	—	-	-	1.13	LOQ-200%	0.0012	0.0033	0.55	101.2	159
Imp-B			10.71			1.20		0.0016	0.0031	0.62	102.5	
Imp-C			8.70			1.02		0.0012	0.0031	0.52	99.5	
Imp-D			7.19			1.19		0.0013	0.0031	0.81	99.2	
Telmisartan			3.06			1.45		0.0010	0.0033	0.71	95.2	
Imp-E			3.56			1.25		0.0016	0.0036	0.51	94.6	
Imp-F			13.43			1.08		0.0014	0.0031	0.75	103.1	
Imp-G			11.01			1.16		0.0013	0.0032	0.69	102.6	
Imp-H			6.02			1.23		0.0015	0.0031	0.60	102.2	
Imp-1	Waters Acquity Shield RP18 (100x2.1 mm, 1.7 µm)	10	—	-	-	1.103	LOQ-200%	0.20	0.50	3.45	100.1	160
Imp-4			5.758			1.117		0.20	0.50	2.95	97.6	
Maraviroc			9.256			1.388		0.10	0.50	4.11	-	
Imp-2			10.052			1.007		0.20	0.50	1.98	88.7	
Imp-6			3.878			1.001		0.35	1.0	2.89	90.8	
Imp-3			8.637			1.013		0.20	0.50	3.45	95.8	
Imp-5			14.378			1.011		0.35	1.0	5.02	96.2	

(Table 6) contd.....

Compound	Analytical Column	Run Time (min)	System Suitability				Validation Parameters					Ref
			Rs	α	N	T or A_s	Linearity (μg/mL)	LOD (μg/mL)	LOQ (μg/mL)	Precision (RSD %)	Accuracy %	
Omeprazole Esomeprazole Lansoprazole Pantoprazole Rabeprazole	Waters Acquity BEH (50x3.0 mm, 1.7 μm)	2.5	-	-	-	-	1.78-200 1.42-200 0.75-200 1.08-200 1.00-200	0.59 0.47 0.23 0.36 0.33	1.78 1.42 0.71 1.08 1.00	0.17 0.14 0.13 0.21 0.11	99.08 101.70 101.58 98.68 101.87	161
Famotidine	Symmetry C18 (50x2.1 mm, 1.7 μm)	1	-	-	8896	1.48	5-20	0.006	0.02	0.9	81.4-93.4	162
Hydrochlorothiazide Enalapril IS	Waters Acquity BEH (100x2.1 mm, 1.7 μm)	2	- 22.1 5.2	- 2.06 1.20	14411 18820 56097	-	0.07-249 0.27-399	0.0029 0.0028	0.095 0.092	0.06 0.05	100.06 100.15	163
Metoclopramide and related impurities	Waters X-terra RP18 (150x4.6 mm, 3.5 μm)	18	+	-	-	-	0.06-3.04	0.022-0.037	0.067-0.112	1.4-4.9	100-104	164
Imp-A Imp-I Imp-II Imp-III Imp-B Imp-C Imp-D Rabeprazole Imp-E Imp-F	Waters Acquity RP18 (100x2.1 mm, 1.7 μm)	18	- 2.65 8.59 11.21 33.57 10.29 22.40 1.72 10.83 29.43	-	-	1.43 1.87 1.35 1.37 1.07 1.09 0.86 0.94 1.01 1.08	LOQ-200%	0.028 0.023 0.038 0.042 0.029 0.040 0.067 0.061 0.038 0.053	0.085 0.071 0.117 0.127 0.087 0.121 0.201 0.185 0.116 0.162	1.3 1.3 1.1 0.7 1.1 1.4 1.0 1.2 1.3 1.3	95-105	165
Epalrestat	Agilent C18 (50 x 2.1 mm, 1.8 μm)	-	-	-	-	-	250-1500	-	-	0.07	109.78	166
Ebastine	Waters Acquity BEH (50x2.1 mm, 1.7 μm)	2.5	-	-	-	-	LOQ-125%	0.061	0.090	3.2<	98.2-100.3	167
Imp-A Imp-B Imp-C Imp-D Imp-E Tadalafil	Waters Acquity HSS T3 (150x2.1 mm, 1.8 μm)	10	+	+	+	+	0.118-1000	0.039 0.040 0.038 0.076 0.039 0.040	0.11 0.12 0.12 0.11 0.012 0.12	0.202 0.200 0.206 0.206 0.207 0.600	99.3-100.3	168
Lansoprazole Imp-1 Imp-2 Imp-3 Imp-4 Imp-5	Waters Acquity BEH (50x2.1 mm, 1.7 μm)	10	>2.0	-	-	1.4<	LOQ-150%	0.008 0.006 0.007 0.010 0.010 0.009	0.03 0.02 0.02 0.025 0.03 0.02	0.2-2.3	93.5-106.5	169
Aciclovir Guanine Diguanine Diacetlyaciclovir MP Benzoate PP Benzoate	Syncronis C18 (100x3.0mm, 1.7 μm)	5	>2.0	-	-	0.8-1.5	400-600 0.04-5.25 0.03-0.75 0.04-1.50 8-12 4-6	- 0.01 0.01 0.01 0.3 0.3	- 0.04 0.03 0.04 1.0 1.0	0.38 0.13 1.13 0.89 0.47 0.77	100.17 98.31 96.64 100.43 101.22 100.94	170
Darifenacin and related impurities	Waters Acquity BEH (100x2.1 mm, 1.7 μm)	13	>2.0	+	+	+	25-100 LOQ-1	0.003-0.016	0.01-0.49	0.3-2.5	93.8-99.8	171
Lacosamide and related impurities	Waters Acquity BEH (100x2.1 mm, 1.8 μm)	5	>2.0	-	-	1.17	20-600	0.03-1.3	0.08-3.9	0.49	88.8-110	172
Atropine sulphate	Hiber HR Purospher Star (100x2.1 mm, 2.0 μm)	3.5	-	-	11802	1.49	2.5-17.5	0.0033	0.0102	0.08-0.61	100.61	173
Acyclovir	Waters Acquity BEH (50x2.1 mm, 1.7 μm)	1.2	-	-	5069	-	1-400	0.3	1.0	0.47-6.17	96.1-100.9	174

(Table 6) contd.....

Compound	Analytical Column	Run Time (min)	System Suitability				Validation Parameters					Ref
			Rs	α	N	T or A_s	Linearity (μg/mL)	LOD (μg/mL)	LOQ (μg/mL)	Precision (RSD %)	Accuracy %	
(Bisphenols) Cl-BPA Cl_2-BPA Cl_3-BPA Cl_4-BPA	Gemini C18 (100x 2mm, 3μm)	7	-	-	-	-	ng/g 14-1000 14-1000 8-1000 11-1000	ng/g 4 4 2 3	ng/g 14 14 8 11	0.90 0.70 0.40 1.30	100.20 98.40 100.60 99.00	175
Imp-A Imp-B Imp-C Imp-D Imp-E Imp-F Imp-G Imatinib mesylate Imp-H	Acquity BEH C18 (50 x 2.1 mm, 1.7 μm)	8	– 4.02 1.77 1.63 1.59 3.16 4.69 1.57 5.59	-	-	1.01 1.08 1.14 1.20 1.50 1.36 1.03 1.35 0.94	LOQ-200%	0.06 0.05 0.02 0.05 0.09 0.05 0.03 0.06 0.05	0.20 0.15 0.07 0.15 0.30 0.17 0.10 0.20 0.16	0.0 0.0 1.5 0.0 3.2 2.2 0.0 0.0 2.6	90 -110	176
Aliskiren ADPA	XBridge HILIC (150 x 4.6 mm, 3.5 μm)	12	≥2.0	+	> 5000	≤ 1.5	+	0.006	0.020	1.10 0.90	+	177
Meta-cresol Parathyroid hormone	C18 (100 x 2.1 mm, 3 μm)	6	+	+	+	1.24	50-300	-	-	<1.0	95 -105	178
Erythropoietin HSA	C18 (50 x 2.1 mm, 1.7 μm)	4	5.35	+	+	1.57 5.63	2.5-150	-	-	<1.5	97-103	179
Phenylpropanolamine Cathine Ephedrine Pseudoephedrine Methylephedrine	Acquity BEH C18 (50 x 2.1 mm, 1.7 μm)	6	+	+	+	+	-	-	-	+	-	180
Yohimbine	Acquity BEH C18 (100 x 2.1 mm, 1.7 μm)	36	-	-	-	-	0.11-214.10	$26.8x10^{-3}$	$80.4x10^{-3}$	1.22	99.59	181
Sibutramin Modafinil Ephedrine Norefedrine Metformine	Vision HT C18-P (100 x 2.0 mm, 1.5 μm)	-	-	-	-	100-1000; 125-1000; 250-2000; 250-2000; 30-250	-	-	0.842 0.617 0.614 1.080 0.617	102.10 98.90 99.40 100.00 101.10	-	182
Zolmitriptan and impurities	Ascentis Express Phenyl-Hexyl (100 x 3.0 mm, 2.7 μm)	5	4.56	+	+	1.36	LOQ-200%	0.057	0.173	<2.0	103.90	183
Imp-6 Imp-2 Imp-1 Imp-3 Imp-4 Esomeprazole Imp-7 Imp-5	Acquity BEH C18 (50 x 2.1 mm, 1.7 μm)	18	- 11.1 19.8 10.8 10.6 - 4.2 16.9	0.43 0.58 0.81 0.88 0.96 1.00 1.19 1.39	-	1.0 1.1 1.1 1.1 1.1 1.1 1.1 1.1	LOQ-200%	0.044 0.016 0.024 0.024 0.024 0.020 0.028 0.032	0.136 0.044 0.064 0.080 0.064 0.060 0.072 0.092	1.2-1.9	98.90 98.60 98.90 101.90 99.90 99.90 98.10 99.30	184
Genistein Daidzein Glicitein Genistin Daidzin Glycitin	Acquity BEH C18 (50 x 2.1 mm, 1.7 μm)	8	-	-	-	-	0.1-5.0 0.1-5.0 0.1-5.0 0.1-10.0 0.1-10.0 0.1-5.0	mg/Kg 0.04 0.02 0.03 0.05 0.04 0.02	mg/Kg 0.14 0.07 0.09 0.17 0.12 0.08	6.40 5.47 7.24 3.71 3.59 6.13	101.50 97.50 98.00 97.00 97.40 95.80	185
Cyanindin-3-glucoside Cyanindin-3-rutinoside	Acquity HSS C18 (100 x 2.1 mm, 1.8 μm)	10	-	-	-	-	1.0-49.00 1.0-48.60	0.32 0.12	1.07 0.97	4.63 3.10	+	186
Mesembrenol Mesembranol Mesembrenone Mesembrine	Acquity BEH C18 (150 x 2.1 mm, 1.7 μm)	8	-	-	-	-	5-100	0.91 1.04 1.15 1.21	2.72 3.25 3.60 3.71	2.11 2.62 2.85 1.78	80.67 81.33 80.00 80.33	187
Ganciclovir	Acquity HSS T3 (50 x 2.1 mm, 1.8 μm)	2	-	-	-	-	0.5-30	0.50	0.50	1.02	102.90	188

(Table 6) contd.....

Compound	Analytical Column	Run Time (min)	System Suitability				Validation Parameters					Ref
			Rs	α	N	T or A_s	Linearity (μg/mL)	LOD (μg/mL)	LOQ (μg/mL)	Precision (RSD %)	Accuracy %	
Domperidon Omeprazole	Acquity BEH C18 (50 x 2.1 mm, 1.7 μm)	2	+	+	+	+	5-40	-	-	0.50 0.30	98.20 99.00	189
Isoleucine Leucine Phenylalanine Valine	Acquity BEH C18 (100 x 2.1 mm, 1.7 μm)	4	-	-	-	-	50-400	0.1 0.1 0.1 0.1	1.0 1.0 1.0 1.0	0.41 0.35 0.34 0.36	98.90 98.00 98.15 98.00	190
Olmesartan and 17 Impurities	Acquity UPLC BEH phenyl (100x 2.1 mm, 1.7 μm)	8	>1.5	+	+	<1.4	LOQ–150%	0.0005	0.018	1.3-3.6	98.5-101.8	191
Imp-1 Imp-2 Imp-3 Imp-4 Imp-5 Imp-6 Imp-7 Imp-8 Rivastigmine Imp-9 Imp-10 Imp-11	Acquity UPLC BEH phenyl (100 x 2.1 mm, 1.7 μm)	10	- 7.49 3.99 20.2 2.93 2.99 13.5 2.54 2.34 16.5 12.1 7.18	+	+	1.25 1.20 1.15 1.05 1.20 1.08 1.09 1.01 1.05 1.00 1.02 1.01	LOQ–200%	0.022 0.015 0.017 0.021 0.020 0.018 0.016 0.021 0.022 0.018 0.015 0.013	0.072 0.050 0.055 0.073 0.070 0.060 0.050 0.073 0.073 0.060 0.050 0.043	2.7 2.5 2.2 2.1 2.5 2.4 2.6 2.3 2.3 2.5 2.6 2.3	95.5-103.7	192
Norisoboldine	Acquity BEH C18 (50 x 2.1 mm, 1.7 μm)	4	+	+	11000	1.05	0.302-301.8	0.151	0.302	1.1	97.3	193
Ramelteon Imp-A Imp-B Imp-C Imp-D	Acquity UPLC BEH Shield RP18 (100 x 2.1 mm, 1.7 μm)	5	>2.0	+	+	<1.2	0.038-0.225	0.018 0.022 0.014 0.025 0.021	0.05 0.07 0.04 0.08 0.06	- 3.90 0.87 1.41 0.81	87.0-105	194
Paliperidon Imp-A Imp-B Imp-C Imp-D Imp-E Imp-F Imp-G Imp-H	Acquity BEH C18 (50 x 2.1 mm, 1.7 μm)	10	>2.0	+	+	<2.0	LOQ-150%	0.0001 0.0001 0.0001 0.0002 0.0002 0.0002 0.0002 0.0002 0.0002	0.0005 0.0003 0.0003 0.0005 0.0005 0.0005 0.0005 0.0005 0.0005	5.8 2.9 5.6 4.9 3.8 5.9 5.0 3.8 2.9	99.9 102.1 105.4 94.0 99.7 98.0 100.6 92.4 104.3	195
Raloxifene	Acquity BEH C18 (50 x 2.1 mm, 1.7 μm)	6	+	+	51191	1.1	0.1-50	0.033	0.100	1.08	102.5	196
Imp-A Imp-B Gemicitabine Imp-C	Acquity UPLC HSS T3 (100 x 2.1 mm, 1.8 μm)	5	4.0	1.2	8650	1.1	LOQ-200%	+	+	2.0	100.1	197
Astaxanthin	Acquity BEH C18 (50 x 2.1 mm, 1.7 μm)	3	+	-	-	-	0.16-2.10	0.054	0.160	3.59	90.5	198
Homocysteine L-Cysteine	Supelcosil LC-18-DB (150 x 4.6 mm, 3 μm)	10	9.5 9.5	7.0 2.0	3243 2459	1.3 0.9	μM 1-60 150-300	μM 0.05 0.15	μM 0.24 0.80	2.22 2.66	93.6 92.4	199
Glucosamines	Acquity UPLC HSS T3 (50 x 2.1 mm, 1.7 μm)	8	+	-	-	-	10-500	-	10	3.50	82.9-101.1	200

(Table 6) contd.....

Compound	Analytical Column	Run Time (min)	System Suitability				Validation Parameters					Ref
			Rs	α	N	T or A_s	Linearity (μg/mL)	LOD (μg/mL)	LOQ (μg/mL)	Precision (RSD %)	Accuracy %	
Imp-A	Acquity BEH C18 (100 x 2.1 mm, 1.7 μm)	18	-	-	-	1.43	LOQ-200%	0.028	0.085	0.5	95-105.0	201
Imp-I			2.65			1.87		0.023	0.071	0.2		
Imp-II			8.59			1.35		0.038	0.120	0.3		
Imp-III			11.2			1.37		0.130	0.130	1.0		
Imp-B			33.5			1.07		0.029	0.087	1.0		
Imp-C			10.2			1.09		0.040	0.121	0.5		
Imp-D			22.4			0.86		0.066	0.201	2.5		
Rabeprazole			1.72			0.94		0.061	0.185	1.1		
Imp-E			10.8			1.01		0.038	0.116	1.7		
Imp-F			29.4			1.08		0.053	0.162	0.6		

IS: Internal standard; PFP: Pentafluorophenyl; Imp: Impurity; HILIC: Hydrophilic Interaction Liquid Chromatography; Rs: Resolution; α: Selectivity; N: Theoretical plate number; T: Tailing; A: Asymmetry; LOD: Limit of Detection; LOQ: Limit of Quantification.

CONCLUDING REMARKS

Ultra high-performance liquid chromatography provides simple, fast, accurate, precise, sensitive, and reproducible analytical methods for the routine analysis of active pharmaceutical ingredients. According to the wide range of usage for this technique, scientists still continue to see performance gains and benefits such as efficient plate numbers, higher sample throughput, better sensitivity, and increased resolution by novel advancements in this technology. The general consensus in the literature is that positive characteristics are beneficial with future advancements even higher operational pressures will be reached, especially in UHPLC instruments. Thus, it has revolutionized the science of LC. In a special manner, pharmaceutical and the chemical industries need to work efficiently and rapidly for achieving the highest levels of quality.

From this point of view, development of this technique will open new doors for those working with liquid chromatography. It has also provided reduction in the workload, difficulties, and errors faced by analysts. The literature survey shows that it has been safely utilized in the pharmaceutical industry, as well as in the research and development laboratories in academia, with significant results. Furthermore, it demonstrated the capacity to be easily coupled with high throughput systems in order to achieve much more sensitive results.

UHPLC supplied very fast method development. It needs less time for validation, and high-throughput quality control analysis. This system is designed in a special way to withstand high system back-pressures and allows the significant reduction of analysis time in comparison to the conventional HPLC system. The main advantages include short turnaround time, method reliability, sensitivity, and specificity for various groups of the pharmaceutically active compounds.

In practical terms, changes in conditions are allowed within the design space to meet system suitability requirements. The design space encompasses the range of allowed values of various conditions, such as percentage of organic phase, pH, temperature, and flow rate. Changes in any combination of these variables are allowed. This approach provides the flexibility to adjust a method to restore performance. Various types of detection modes are applied to UHPLC techniques and PDA detection is one of the most commonly used in the analysis of drugs.

It would appear that the detector requirements commonly cited for HPLC are the same for UHPLC. The application of highly selective detectors enables the determination of medicinal substances without derivatization.

Drug active compounds can therefore be detected selectively and determined sensitively by UHPLC methods. This selectivity and sensitivity depend on the accessible potential range, the number of compounds which are active in this range, and on the half-widths of the individual signals. UHPLC is commonly used in many pharmaceutical related studies and biochemical laboratories with various detectors.

FUTURE DEMANDS

Drug discovery and development strategies are changing with increasing speed. This is also reflected in an increased chemical diversity of the targeted active pharmaceutical molecules. With the increasing speed, these instruments need the high data acquisition rates of detectors because of narrow peaks produced by new generation columns. In this context, sometimes the peaks eluting below 1 second and conventional detectors could not show sufficient quantitative performance for the analysis of pharmaceuticals. Nowadays, the spectroscopic detectors designed especially (up to 200 Hz acquisition rates) for UHPLC systems will improve the quantitative performance significantly. On the other hand, the MS based detectors still need to improve in order to meet the requirements especially in biomedical analysis. However, instrument companies and scientists will overcome this problem by high resolution mass detection in the near future.

New developments in both column chemistries and decreasing the size of the column packing materials are another promising point for the new generation UHPLC systems, but it can create the back-pressure problems. To overcome this problem, current technology will be modified to be able to work up to 1400 bar. It is also not possible with the narrow bore columns that we used in UHPLC. Miniaturization of the columns to the micro or nano scale may solve the pressure problems. It will also safeguard the environment.

The coupling of this technology with the high acquisition rate of MS based detectors will carry UHPLC technology to the next step, especially in the fields of biopharmaceuticals, proteomics, metabolomics, fluxomics, drug residue screening, and drug metabolism studies.

CONSENT FOR PUBLICATION

Not applicable.

CONFLICT OF INTEREST

The authors declares no conflict of interest, financial or otherwise.

ACKNOWLEDGEMENTS

Declared none.

REFERENCES

[1] Watson, D.G. *Pharmaceutical Analysis,* 2nd ed; Elsevier: New York, **2005**.

[2] Ohannesian, L.; Streeter, A.J., Eds. *Handbook of Pharmaceutical Analysis*; Marcel Dekker: New York, **2002**.

[3] Ahuja, S.; Scypinski, S., Eds. *Handbook of Modern Pharmaceutical Analysis*; Academic Press: London, **2001**, Vol. 3, .

[4] British Pharmacopoeia Commission. *British Pharmacopoeia*; TSO: London, **2016**.

[5] Eur, Ph. *Council of Europe,* 9th ed; European Pharmacopoeia: Strasbourg, **2016**.

[6] USP 39 [NF 34]. *U.S. Pharmacopoeia-National Formulary.,* **2016**,

[7] Christian, G.D., Ed. *Analytical Chemistry,* 6th ed; Wiley: New Jersey, **2004**.

[8] Kissinger, P.T.; Heineman, W.R., Eds. *Laboratory Techniques in Electroanalytical Chemistry,* 2nd ed; Marcel Dekker: New York, **1996**.

[9] Adamovics, J.A., Ed. *Chromatographic Analysis of Pharmaceuticals,* 2nd ed; Marcel Dekker: New York, **1997**.

[10] Ruzicka, J.; Hansen, E.H. *Flow Injection Analysis,* 2nd ed; John Wiley: New York, **1988**.

[11] Snyder, L.R.; Kirkland, J.J.; Glajch, J.L., Eds. *Practical HPLC method development,* 2nd ed; Wiley: New York, **1997**.
[http://dx.doi.org/10.1002/9781118592014]

[12] Vakh, C.; Falkova, M.; Timofeeva, I.; Moskvin, A.; Moskvin, L.; Bulatov, A. Flow Analysis: A Novel Approach For Classification. *Crit. Rev. Anal. Chem.,* **2016**, *46*(5), 374-388.
[http://dx.doi.org/10.1080/10408347.2015.1087301] [PMID: 26364745]

[13] Guillarme, D.; Veuthey, J.L.; Smith, R.M. *UHPLC in Life Sciences*; Royal Society of Chemistry, **2012**.
[http://dx.doi.org/10.1039/9781849735490]

[14] Fountain, K.J.; Iraneta, P.C. *Instrumentation and Columns for UHPLC Separations. UHPLC in Life Sciences*; The Royal Society of Chemistry, **2012**.

[15] Russo, R.; Guillarme, D.; T-T Nguyen, D.; Bicchi, C.; Rudaz, S.; Veuthey, J.L. Pharmaceutical applications on columns packed with sub-2 microm particles. *J. Chromatogr. Sci.,* **2008**, *46*(3), 199-208.
[http://dx.doi.org/10.1093/chromsci/46.3.199] [PMID: 18334085]

[16] Mazzeo, J.R.; Neue, U.V.; Marianna, K.; Plumb, R.S. Advancing LC performance with smaller particles and higher pressure. *Anal. Chem.,* **2005**, *77*, 460A-467A. [http://dx.doi.org/10.1021/ac053516f]

[17] de Villiers, A.; Lestremau, F.; Szucs, R.; Gélébart, S.; David, F.; Sandra, P. Evaluation of ultra performance liquid chromatography. Part I. Possibilities and limitations. *J. Chromatogr. A,* **2006**, *1127*(1-2), 60-69. [http://dx.doi.org/10.1016/j.chroma.2006.05.071] [PMID: 16797562]

[18] Wren, S.A.; Tchelitcheff, P. Use of ultra-performance liquid chromatography in pharmaceutical development. *J. Chromatogr. A,* **2006**, *1119*(1-2), 140-146. [http://dx.doi.org/10.1016/j.chroma.2006.02.052] [PMID: 16564533]

[19] Tswet, M.S. On the new form of adsorption phenomena and its application in biochemical analysis. *Proc. Warsaw Natural Biol. Soc.,* **1903**, *14*, 20-39.

[20] Alpert, A.J. Hydrophilic-interaction chromatography for the separation of peptides, nucleic acids and other polar compounds. *J. Chromatogr. A,* **1990**, *499*, 177-196. [http://dx.doi.org/10.1016/S0021-9673(00)96972-3] [PMID: 2324207]

[21] Tswet, M.S. Chromophils in Animal and Plant World. *Doctor of Science dissertation, Warsaw,* **1910**.

[22] Dong, M.W. *Modern HPLC for Practicing Scientists*; John Wiley & Sons, Ltd: Chichester, **2006**. [http://dx.doi.org/10.1002/0471973106]

[22] Moldoveanu, S.C.; David, V. *Sample Preparation in Chromatography*; Elsevier: Amsterdam, **2002**.

[23] Dyson, N. *Chromatographic Integration Methods,* 2nd ed; Royal Society of Chemistry: London, **1998**.

[24] Kromidas, S. *HPLC Made to Measure*; Wiley-VCH: Weinheim, **2006**. [http://dx.doi.org/10.1002/9783527611973]

[26] Snyder, L.R.; Dolan, J.W. *High-Performance Gradient Elution*; Wiley-Interscience: Hoboken, **2007**.

[27] Snyder, L.R.; Kirkland, J.J.; Glajch, J.L. *Practical HPLC Method Development,* 2nd ed; Wiley-Interscience: New York, **1997**. [http://dx.doi.org/10.1002/9781118592014]

[28] Heftmann, E., Ed. *Chromatography, Part A: Fundamentals and Techniques, Part B: Applications,* 6th ed; Elsevier: Amsterdam, **2004**.

[29] Miller, J.M. *Chromatography – Concepts and Contrasts,* 2nd ed; John Wiley & Sons, Ltd: Chichester, **2009**. [http://dx.doi.org/10.1002/9780471980582]

[30] Knox, J.H.; Scott, H.P. B and C terms in the Van Deemter equation for liquid chromatography. *J. Chromatogr. A,* **1983**, *282*, 297-313. [http://dx.doi.org/10.1016/S0021-9673(00)91609-1]

[31] van Deemter, J.J.; Zuiderweg, F.J.; Kinkenberg, A. Longitudinal diffusion and resistance to mass transfer as causes of nonideality in chromatography. *Chem. Eng. Sci.,* **1956**, *5*, 271-289. [http://dx.doi.org/10.1016/0009-2509(56)80003-1]

[32] Cielecka-Piontek, J.; Zalewski, P.; Jelińska, A.; Garbacki, P. UHPLC: The Greening Face of Liquid Chromatography. *Chromatographia,* **2013**, *76*, 1429-1437. [http://dx.doi.org/10.1007/s10337-013-2434-6] [PMID: 24273332]

[33] Nováková, L.; Matysová, L.; Solichová, D.; Koupparis, M.A.; Solich, P. Comparison of performance of C18 monolithic rod columns and conventional C18 particle-packed columns in liquid chromatographic determination of Estrogel and Ketoprofen gel. *J. Chromatogr. B Analyt. Technol. Biomed. Life Sci.,* **2004**, *813*(1-2), 191-197. [http://dx.doi.org/10.1016/j.jchromb.2004.09.045] [PMID: 15556533]

[34] Vancea, S.; Imre, S.; Donáth-Nagy, G.; Béla, T.; Nyulas, M.; Muntean, T.; Borka-Balás, R.

Determination of free captopril in human plasma by liquid chromatography with mass spectrometry detection. *Talanta,* **2009**, *79*(2), 436-441.
[http://dx.doi.org/10.1016/j.talanta.2009.04.003] [PMID: 19559901]

[35] Lu, S.; Tran, B.N.; Nelsen, J.L.; Aldous, K.M. Quantitative analysis of mitragynine in human urine by high performance liquid chromatography-tandem mass spectrometry. *J. Chromatogr. B Analyt. Technol. Biomed. Life Sci.,* **2009**, *877*(24), 2499-2505.
[http://dx.doi.org/10.1016/j.jchromb.2009.06.024] [PMID: 19577523]

[36] Lee, H.W.; Ji, H.Y.; Kim, H.Y.; Park, E.S.; Lee, K.C.; Lee, H.S. Determination of metoclopramide in human plasma using hydrophilic interaction chromatography with tandem mass spectrometry. *J. Chromatogr. B Analyt. Technol. Biomed. Life Sci.,* **2009**, *877*(18-19), 1716-1720.
[http://dx.doi.org/10.1016/j.jchromb.2009.04.027] [PMID: 19442594]

[37] Guitton, J.; Coste, S.; Guffon-Fouilhoux, N.; Cohen, S.; Manchon, M.; Guillaumont, M. Rapid quantification of miglustat in human plasma and cerebrospinal fluid by liquid chromatography coupled with tandem mass spectrometry. *J. Chromatogr. B Analyt. Technol. Biomed. Life Sci.,* **2009**, *877*(3), 149-154.
[http://dx.doi.org/10.1016/j.jchromb.2008.11.040] [PMID: 19095507]

[38] Neue, U.D. *HPLC Columns: Theory, Technology, and Practice*; JohnWiley&Sons, Inc.: New York, **1997**.

[39] Posch, H.; Trathnigg, B. *HPLC of Polymers*; Springer: Berlin, Heidelberg, **1998**.

[40] Iler, R.K. *Chemistry of Silica*; Wiley-Interscience: New York, **1979**.

[41] Unger, K.K. *Porous Silica, Its Properties and Use as Support in Column Liquid Chromatography, Journal of Chromatography Library*; Elsevier: Amsterdam, **1979**, Vol. 16, .

[42] Song, W.; Pabbisetty, D.; Groeber, E.A.; Steenwyk, R.C.; Fast, D.M. Comparison of fused-core and conventional particle size columns by LC-MS/MS and UV: application to pharmacokinetic study. *J. Pharm. Biomed. Anal.,* **2009**, *50*(3), 491-500.
[http://dx.doi.org/10.1016/j.jpba.2009.05.011] [PMID: 19540084]

[43] Mallett, D.N.; Ramírez-Molina, C. The use of partially porous particle columns for the routine, generic analysis of biological samples for pharmacokinetic studies in drug discovery by reversed-phase ultra-high performance liquid chromatography-tandem mass spectrometry. *J. Pharm. Biomed. Anal.,* **2009**, *49*(1), 100-107.
[http://dx.doi.org/10.1016/j.jpba.2008.09.041] [PMID: 19019615]

[44] Wilson, N.S.; Dolan, J.W.; Snyder, L.R.; Carr, P.W.; Sander, L.C. Column selectivity in reversed-phase liquid chromatography III. The physico-chemical basis of selectivity. *J. Chromatogr. A,* **2002**, *961*(2), 217-236.
[http://dx.doi.org/10.1016/S0021-9673(02)00658-1] [PMID: 12184620]

[45] Leinweber, F.C.; Tallarek, U. Chromatographic performance of monolithic and particulate stationary phases. Hydrodynamics and adsorption capacity. *J. Chromatogr. A,* **2003**, *1006*(1-2), 207-228.
[http://dx.doi.org/10.1016/S0021-9673(03)00391-1] [PMID: 12938887]

[46] Halasz, I.; Naefe, M. Influence of column parameters on peak broadening in high-pressure liquid chromatography. *Anal. Chem.,* **1972**, *44*, 76-84.
[http://dx.doi.org/10.1021/ac60309a021]

[47] Majors, R.E. High-performance liquid chromatography on small particle silica gel. *Anal. Chem.,* **1972**, *44*(11), 1722-1726.
[http://dx.doi.org/10.1021/ac60319a007] [PMID: 22324582]

[48] Dewaele, C.; Versele, M. Influence of the particle size distribution of the packing material in reversed-phase high-performance liquid chromatography. *J. Chromatogr. A,* **1983**, *260*, 13-21.
[http://dx.doi.org/10.1016/0021-9673(83)80002-8]

[49] Lisichkin, G.V., Ed. *Chemistry of Bonded Surface Compounds*; Fizmatlit: Moscow, **2003**.

[50] Wyndham, K.D.; O'Gara, J.E.; Walter, T.H.; Glose, K.H.; Lawrence, N.L.; Alden, B.A.; Izzo, G.S.; Hudalla, C.J.; Iraneta, P.C. Characterization and evaluation of C18 HPLC stationary phases based on ethyl-bridged hybrid organic/inorganic particles. *Anal. Chem.,* **2003**, *75*(24), 6781-6788. [http://dx.doi.org/10.1021/ac034767w] [PMID: 14670036]

[51] Cheng, Y-F.; Walter, T.H.; Lu, Z.; Iraneta, P.; Alden, B.A.; Gendreau, C.; Neue, U.D.; Grassi, J.M.; Carmody, J.L.; O'Gara, J.E.; Fisk, R.P. Hybrid organic-inorganic particle technology: Breaking through traditional barriers of HPLC separations. *LC GC,* **2000**, *18*, 1162-1173.

[52] Chan, F.; Yeung, L.S.; LoBrutto, R.; Kazakevich, Y.V. Characterization of phenyl-type HPLC adsorbents. *J. Chromatogr. A,* **2005**, *1069*(2), 217-224. [http://dx.doi.org/10.1016/j.chroma.2005.02.022] [PMID: 15830948]

[53] Waters. *XBridge HPLC columns, USA,* **2015**. http://www.waters.com/webassets/cms/library/docs/720001255en.pdf

[54] Tanaka, N.; Araki, M. Polymer-Based Packing Materials for Reversed-Phase Liquid Chromatography In: *Advances in Chromatography*; Giddings, J.C.; Grushka, E.; Brown, P.R., Eds.; Marcel Dekker: New York, **1989**; 30, pp. 181-122.

[55] Nevejans, F.; Verzele, M. On the structure and chromatographic behaviour of polystyrene phases. *J. Chromatogr. A,* **1987**, *406*, 325-342. [http://dx.doi.org/10.1016/S0021-9673(00)94040-8]

[56] Tanaka, N.; Ebata, T.; Hashizume, K.; Hosoya, K.; Araki, M. Polymer-based parking materials with backbones for reversed-phase liquid chromatography. Performance and retention selectivity. *J. Chromatogr. A,* **1989**, *475*, 195-208. [http://dx.doi.org/10.1016/S0021-9673(01)89675-8]

[57] Dawkins, J.V.; Lloyd, L.; Warner, F.P. Chromatographic characteristics of polymer-based high-performance liquid chromatography packings. *J. Chromatogr. A,* **1986**, *352*, 157-167. [http://dx.doi.org/10.1016/S0021-9673(01)83376-8]

[58] Snyder, L.R.; Dolan, J.W. Milestones in the Development of Liquid Chromatography A2 - Fanali, S.*Liquid Chromatography*; Haddad, P.R.; Poole, C.F.; Schoenmakers, P.; Lloyd, D., Eds.; Chapter 1. Elsevier: Amsterdam, **2013**.

[59] Nagase, K.; Kobayashi, J.; Kikuchi, A.; Akiyama, Y.; Kanazawa, H.; Okano, T. Monolithic silica rods grafted with thermoresponsive anionic polymer brushes for high-speed separation of basic biomolecules and peptides. *Biomacromolecules,* **2014**, *15*(4), 1204-1215. [http://dx.doi.org/10.1021/bm401779r] [PMID: 24552290]

[60] Leandro, C.C.; Hancock, P.; Fussell, R.J.; Keely, B.J. Comparison of ultra-performance liquid chromatography and high-performance liquid chromatography for the determination of priority pesticides in baby foods by tandem quadrupole mass spectrometry. *J. Chromatogr. A,* **2006**, *1103*(1), 94-101. [http://dx.doi.org/10.1016/j.chroma.2005.10.077] [PMID: 16321395]

[61] Bezhitashvili, L.; Bardavelidze, A.; Ordjonikidze, T.; Chankvetadze, L.; Chity, M.; Farkas, T.; Chankvetadze, B. Effect of pore-size optimization on the performance of polysaccharide-based superficially porous chiral stationary phases for the separation of enantiomers in high-performance liquid chromatography. *J. Chromatogr. A,* **2017**, *1482*, 32-38. [http://dx.doi.org/10.1016/j.chroma.2016.12.055] [PMID: 28049582]

[62] Poole, C.F. *The Essence of Chromatography*; Elsevier: Amsterdam, **2002**.

[63] Robards, K.; Jackson, P.E.; Haddad, P.R. *Principles and Practice of Modern Chromatographic Methods*; Academic: London, San Diego, **1995**.

[64] Dolan, J.W.; Snyder, L.R. *Troubleshooting LC Systems*; Aster: Chester, **1989**. [http://dx.doi.org/10.1007/978-1-59259-640-9]

[65] Snyder, L.R.; Kirkland, J.J.; Dolan, J.W. *Introduction to Modern Liquid Chromatography,* 3rd ed; John Wiley & Sons Ltd.: Chichester, **2010**.

[66] Horvath, C. *My Focus on Chromatography over 40 Years, Journal of Chromatography Library*; Elsevier: Amsterdam, **2001**, Vol. 64, pp. 238-247.

[67] Ettre, L.S.; Zlatkis, A., Eds. *75 Years of Chromatography: A Historical Dialogue, Journal of Chromatography Library*; Elsevier: Amsterdam, **1979**, Vol. 17, p. 155.

[68] Sadek, P.C. *Troubleshooting LC Systems*; JohnWiley & Sons, Inc.: New York, **1999**.

[69] Martin, A.J.; Synge, R.L. A new form of chromatogram employing two liquid phases: A theory of chromatography. 2. Application to the micro-determination of the higher monoamino-acids in proteins. *Biochem. J.,* **1941**, *35*(12), 1358-1368.
[http://dx.doi.org/10.1042/bj0351358] [PMID: 16747422]

[70] Gumustas, M.; Kurbanoglu, S.; Uslu, B.; Ozkan, S.A. UPLC *versus* HPLC on Drug Analysis: Advantageous, Applications and Their Validation Parameters. *Chromatographia,* **2013**, *76*, 1365-1427.
[http://dx.doi.org/10.1007/s10337-013-2477-8]

[71] Parriott, D. *A Practical Guide to HPLC Detection*; Academic Press: San Diego, **1993**.

[72] Meyer, V.R. *Detectors. Practical High-Performance Liquid Chromatography*; John Wiley & Sons, Ltd., **2010**.
[http://dx.doi.org/10.1002/9780470688427]

[73] Wickham, D. Photodiode Array Absorbance Detection. Chapter 4. In: *A Practical Guide to HPLC Detection*; Parriott, D., Ed.; Academic Press: San Diego, **1993**.

[74] Mourey, T.H.; Oppenheimer, L.E. Principles of operation of an evaporative light-scattering detector for liquid chromatography. *Anal. Chem.,* **1984**, *56*, 2427-2434.
[http://dx.doi.org/10.1021/ac00277a039]

[75] Downard, K. *Mass Spectrometry: A Foundation Course*; Royal Society of Chemistry: Cambridge, **2004**.

[76] Rossi, D.T.; Sinz, M. *Mass Spectrometry in Drug Discovery*; CRC Press: New York, **2001**.
[http://dx.doi.org/10.1201/9781420002478]

[77] Sparkman, O.D. *Mass Spectrometry Desk Reference*; Global View Pub.: Pittsburgh, **2000**.

[78] ICH (International Council on Harmonisation). *Q2(R1) Validation of Analytical Procedures: Text and Methodology*; , **2005**.

[79] Zielińska, A.; Wicherkiewicz, S.; Łuniewski, W.; Sidoryk, K.; Pesta, E.; Kutner, A. The comparison of the stability indicating two HPLC methods and their application for the determination of bosentan in coated tablets. Acta Poloniae Pharmaceutica -. *Drug Res.,* **2016**, *73*(4), 865-874.

[80] Zhao, L.; Qi, T.; Ge, B.K.; Xu, P.X.; Xue, M. Quantification of six active ingredients in Rhodiola extract by UPLC-MS/MS. *Journal of International Pharmaceutical Research.,* **2016**, *43*(5), 975-979.

[81] Yao, W.; Li, H.; Liu, Q.; Gao, Y.; Dai, J.; Bao, B.; Zhang, L.; Ding, A.; Tang, Y.; Duan, J. Simultaneous quantitation and comparison of eight components in Jiao-ai decoction and Si-wu decoction by ultra high performance liquid chromatography with triple quadrupole tandem mass spectrometry. *J. Sep. Sci.,* **2016**, *39*(17), 3311-3317.
[http://dx.doi.org/10.1002/jssc.201600456] [PMID: 27384525]

[82] Yang, N.; Xiong, A.; Wang, R.; Yang, L.; Wang, Z. Quality evaluation of traditional Chinese medicine compounds in xiaoyan lidan tablets: Fingerprint and quantitative analysis using UPLC-MS. *Molecules,* **2016**, *21*(2), 83.
[http://dx.doi.org/10.3390/molecules21020083] [PMID: 26805803]

[83] Wu, Y.C.; Yang, W.J.; Zhang, L.; Zhao, W.S.; Zhang, W.S.; Ye, Z.L. Determination and quality

evaluation of seven chemical components of Gardenia jasminoides ellis. *Chung Kuo Yao Hsueh Tsa Chih,* **2016**, *51*(10), 841-847.

[84] Wingert, N.R.; dos Santos, N.O.; Nunes, M.A.; Gomes, P.; Müller, E.I.; Flores, É.M.; Steppe, M. Characterization of three main degradation products from novel oral anticoagulant rivaroxaban under stress conditions by UPLC-Q-TOF-MS/MS. *J. Pharm. Biomed. Anal.,* **2016**, *123*, 10-15. [http://dx.doi.org/10.1016/j.jpba.2016.01.053] [PMID: 26855380]

[85] Wei, X.; Li, P.; Liu, M.; Du, Y.; Wang, M.; Zhang, J. Absolute oral bioavailability of fenofibric acid and choline fenofibrate in rats determined by ultra-performance liquid chromatography tandem mass spectrometry. *Biomed. Chromatogr.,* **2016**. [PMID: 27594083]

[86] Wang, C.; Xie, Y.; Xiang, Z.; Zhou, H.; Liu, L. Simultaneous determination of thirteen major active compounds in Guanjiekang preparation by UHPLC-QQQ-MS/MS. *J. Pharm. Biomed. Anal.,* **2016**, *118*, 315-321. [http://dx.doi.org/10.1016/j.jpba.2015.10.043] [PMID: 26588049]

[87] Sujana, K.; Jagadish, P.; Venu, S.; Priyanka, T.P. Stability indicating validated UPLC method for determination of Nebivolol HCl in bulk and pharmaceutical dosage form. *Research Journal of Pharmacy and Technology.,* **2016**, *9*(5), 533-537.

[88] Skibiński, R.; Trawiński, J.; Komsta, Ł.; Bajda, K. Characterization of forced degradation products of clozapine by LC-DAD/ESI-Q-TOF. *J. Pharm. Biomed. Anal.,* **2016**, *131*, 272-280. [http://dx.doi.org/10.1016/j.jpba.2016.09.007] [PMID: 27611099]

[89] Siddareddy, K.; Reddy, M.R.; Sreeramulu, J. Simultaneous estimation of acetaminophen, phenylephrine HCl, guaifenesin and dextromethorphan HBr in reverse phase ultra performance liquid chromatography. *Res. J. Pharm. Biol. Chem. Sci.,* **2016**, *7*(6), 2274-2281.

[90] Sestak, V.; Roh, J.; Klepalova, L.; Kovarikova, P. A UHPLC-UV-QTOF study on the stability of carfilzomib, a novel proteasome inhibitor. *J. Pharm. Biomed. Anal.,* **2016**, *124*, 365-373. [http://dx.doi.org/10.1016/j.jpba.2016.03.015] [PMID: 26994320]

[91] Schlittenbauer, L.; Seiwert, B.; Reemtsma, T. Ultrasound-assisted hydrolysis of conjugated parabens in human urine and their determination by UPLC-MS/MS and UPLC-HRMS. *Anal. Bioanal. Chem.,* **2016**, *408*(6), 1573-1583. [http://dx.doi.org/10.1007/s00216-015-9265-2] [PMID: 26753983]

[92] Santoveña, A.; Suárez-González, J.; Martín-Rodríguez, C.; Fariña, J.B. Formulation design of oral pediatric Acetazolamide suspension: dose uniformity and physico-chemical stability study. *Pharm. Dev. Technol.,* **2016**, 1-7. [PMID: 27109383]

[93] Rupareliya, R.H.; Joshi, H.S.; Ram, V.R.; Dave, P.N.; Khosla, E. Stability indicating simultaneous validation of telmisartan and cilnidipine with forced degradation behavior study by RP-UPLC in tablet dosage form. *International Journal of Pharmaceutical Quality Assurance.,* **2016**, *7*(3), 39-45.

[94] Ren, Z.; Zhang, X.; Wang, H.; Jin, X. Using an innovative quality-by-design approach for the development of a stability-indicating UPLC/Q-TOF-ESI-MS/MS method for stressed degradation products of imatinib mesylate. *RSC Advances,* **2016**, *6*(16), 13050-13062. [http://dx.doi.org/10.1039/C5RA26456A]

[95] Reis, N.F.; de Assis, J.C.; Fialho, S.L.; Pianetti, G.A.; Fernandes, C. Stability-indicating UHPLC method for determination of nevirapine in its bulk form and tablets: identification of impurities and degradation kinetic study. *J. Pharm. Biomed. Anal.,* **2016**, *126*, 103-108. [http://dx.doi.org/10.1016/j.jpba.2016.05.005] [PMID: 27179642]

[96] Rehman, S.U.; Choi, M.S.; Kim, I.S.; Kim, S.H.; Yoo, H.H. An ultra-high-performance liquid chromatography-tandem mass spectrometric method for the determination of hederacoside C, a drug candidate for respiratory disorder, in rat plasma. *J. Pharm. Biomed. Anal.,* **2016**, *129*, 90-95. [http://dx.doi.org/10.1016/j.jpba.2016.06.039] [PMID: 27411171]

[97] Qadir, M.A.; Ahmed, M.; Tahir, M.S.; Shakoor, A.; Shafiq, M.I.; Shahzad, S. Sensitive and selective UPLC method for simultaneous determination of methyl paraben, propyl paraben and loratadine in antihistamine suspension and biological samples: Stability indicating assay. *Lat. Am. J. Pharm.*, **2016**, *35*(5), 869-876.

[98] Plachká, K.; Chrenková, L.; Douša, M.; Nováková, L. Development, validation and comparison of UHPSFC and UHPLC methods for the determination of agomelatine and its impurities. *J. Pharm. Biomed. Anal.*, **2016**, *125*, 376-384. [http://dx.doi.org/10.1016/j.jpba.2016.04.020] [PMID: 27131147]

[99] Noreen, S.; Ahmed, M.; Qadir, M.A.; Shahzad, S.; Shafiq, M.I.; Mumtaz, A.M. Stability-indicating UPLC-PDA method for the determination of bambuterol hydrochloride: Application to pharmaceutical dosage form and biological samples. *Lat. Am. J. Pharm.*, **2016**, *35*(5), 937-944.

[100] Meng, Y.; Ying, Z.; Xiang, Z.; Hao, D.; Zhang, W.; Zheng, Y.; Gao, Y.; Ying, X. The anti-inflammation and pharmacokinetics of a novel alkaloid from Portulaca oleracea L. *J. Pharm. Pharmacol.*, **2016**, *68*(3), 397-405. [http://dx.doi.org/10.1111/jphp.12526] [PMID: 26888212]

[101] Karadas-Bakirhan, N.; Gumustas, M.; Uslu, B.; Ozkan, S.A. Simultaneous determination of amlodipine besylate and rosuvastatin calcium in binary mixtures by voltammetric and chromatographic techniques. *Ionics,* **2016**, *22*(2), 277-288. [http://dx.doi.org/10.1007/s11581-015-1534-8]

[102] Jatczak, M.; Sidoryk, K.; Kossykowska, M.; Łuniewski, W.; Zagrodzka, J.; Lipiec-Abramska, E. Development and Validation of a UHPLC UV Method for the In-Process Control of Bosentan Monohydrate Synthesis. *Chromatographia,* **2016**, *79*(17), 1131-1141. [http://dx.doi.org/10.1007/s10337-016-3124-y] [PMID: 27616782]

[103] Induri, M.; Mantripragada, B.R.; Yejella, R.P. Development and validation of UPLC method for simultaneous estimation of Efavirenz and Lamivudine in pharmaceutical formulations. *J. Appl. Pharm. Sci.*, **2016**, *6*(3), 29-33. [http://dx.doi.org/10.7324/JAPS.2016.60305]

[104] Girish, B.V.; Praveen, S.; Kathyayini, N. Analytical quality by design for development of a sensitive, rapid and stability indicating method for estimation of impurities of dimenhydrinate in its pharmaceutical formulation. *Res. J. Pharm. Biol. Chem. Sci.*, **2016**, *7*(4), 1309-1322.

[105] Garbacki, P.; Cielecka-Piontek, J.; Zalewski, P.; Oszczapowicz, I.; Jelińska, A. A simple and sensitive stability-indicating UHPLC-DAD method for the determination of cefetamet pivoxil hydrochloride. Acta Poloniae Pharmaceutica -. *Drug Res.*, **2016**, *73*(3), 621-626. [PMID: 27476279]

[106] Ganipisetty, V.N.; Jalandhar, D.; Gnanadev, G.; Manoj, P.; Venkata Nadh, R. A novel reversed-phase liquid chromatographic method for the simultaneous determination of potential impurities of bisoprolol fumarate and hydrochlorothiazide in a fixed dosage form. *Separation Science and Technology (Philadelphia),* **2016**, *51*(8), 1362-1369. [http://dx.doi.org/10.1080/01496395.2016.1154873]

[107] Eid, M.A.; Yusof, N.A.; Faruq, M.; Abdullah, J.; Sulaiman, Y. Quantitative measurement of amoxicillin in Ibuprofen tablets using UPLC. Measurement. *Journal of the International Measurement Confederation.*, **2016**, *93*, 465-472. [http://dx.doi.org/10.1016/j.measurement.2016.07.039]

[108] Azougagh, M.; Elkarbane, M.; Bakhous, K.; Issmaili, S.; Skalli, A.; Iben Moussad, S.; Benaji, B. A novel stability-indicating UPLC method development and validation for the determination of seven impurities in various diclofenac pharmaceutical dosage forms. *Ann. Pharm. Fr.*, **2016**, *74*(5), 358-369. [http://dx.doi.org/10.1016/j.pharma.2016.07.001] [PMID: 27475309]

[109] Ahmad, S.A.; Ahmed, M.; Qadir, M.A.; Shafiq, M.I.; Safa, Y.; Awan, N. Simultaneous determination of diloxanide furoate,metronidazole benzoate, methyl paraben and propyl paraben by UPLC-DAD in

pharmaceutical suspension. *Lat. Am. J. Pharm.*, **2016**, *35*(7), 1626-1633.

[110] Zupančič, Š.; Lavrič, Z.; Kristl, J. Stability and solubility of trans-resveratrol are strongly influenced by pH and temperature. *Eur. J. Pharm. Biopharm.*, **2015**, *93*, 196-204. [http://dx.doi.org/10.1016/j.ejpb.2015.04.002] [PMID: 25864442]

[111] Yadav, P.; Majee, C. Method development and validation for quantification of propranolol HCL in pharmaceutical dosage form by RP-UPLC. *Int. J. Pharm. Tech. Res.*, **2015**, *7*(1), 197-203.

[112] Wabaidur, S.M.; AlOthman, Z.A.; Siddiqui, M.R.; Mohsin, K.; Bousiakou, L.G.; Karikas, G.A. UPLC-MS method for the simultaneous determination of naproxen, fluvastatin and ibuprofen in waste water samples. *J. Ind. Eng. Chem.*, **2015**, *24*, 302-307. [http://dx.doi.org/10.1016/j.jiec.2014.09.046]

[113] Vojta, J.; Jedlička, A.; Coufal, P.; Janečková, L. A new, rapid, stability-indicating UPLC method for separation and determination of impurities in amlodipine besylate, valsartan and hydrochlorothiazide in their combined tablet dosage form. *J. Pharm. Biomed. Anal.*, **2015**, *109*, 36-44. [http://dx.doi.org/10.1016/j.jpba.2015.01.059] [PMID: 25756663]

[114] Tomasello, C.; Leggieri, A.; Cavalli, R.; Di Perri, G.; D'Avolio, A. *In vitro* stability evaluation of different pharmaceutical products containing meropenem. *Hosp. Pharm.*, **2015**, *50*(4), 296-303. [http://dx.doi.org/10.1310/hpj5004-296] [PMID: 26448659]

[115] Swamy, N.; Basavaiah, K.; Vamsikrishna, P.; Krishnamurthy, G. Development and validation of a stability-indicating ultra-performance liquid chromatographic method for the determination of rifampicin in bulk drug and capsules. *Thaiphesatchasan,* **2015**, *39*(2), 41-48.

[116] Sunil Reddy, P.; Raju, T.V.; Raju, P.S.; Varma, N.S.; Babu, K.S. Development and validation of a stability-indicating RP-UPLC method for the estimation of impurities in cinacalcet hydrochloride API and its formulation. *Sci. Pharm.*, **2015**, *83*(4), 583-598. [http://dx.doi.org/10.3797/scipharm.1502-06] [PMID: 26839840]

[117] Sirisha, T.; Gurupadayya, B.; Siddiraju, S. Optimized and validated RP-UPLC method for the determination of losartan potassium and chlorthalidone in pharmaceutical formulations. *Adv. Pharm. Bull.*, **2015**, *5*(1), 133-136. [PMID: 25789231]

[118] Salem, H.; Hassan, N.Y.; Lotfy, H.M.; Saleh, S.S. Column performance study of different variants of liquid chromatographic technique: an application on pharmaceutical ternary mixtures containing tetryzoline. *J. Chromatogr. Sci.*, **2015**, *53*(5), 708-715. [http://dx.doi.org/10.1093/chromsci/bmu109] [PMID: 25217705]

[119] Rozman Peterka, T.; Grahek, R.; Hren, J.; Bastarda, A.; Bergles, J.; Urleb, U. Solid state compatibility study and characterization of a novel degradation product of tacrolimus in formulation. *J. Pharm. Biomed. Anal.*, **2015**, *110*, 67-75. [http://dx.doi.org/10.1016/j.jpba.2015.02.047] [PMID: 25804434]

[120] Rajendra Reddy, G.; Siva Jyothi, P.; Ravindra Reddy, P.; Saidi Reddy, G.; Indira Priya, J.K.; Lasker, R. Development and validation of a stability indicating UPLC method for the estimation of R-telaprevir in drug substance and pharmaceutical dosage form of telaprevir by enhanced approach. *Int J. Pharm. Pharm. Sci.*, **2015**, *7*(5), 200-207.

[121] Rajan, N.; Anver Basha, K. A validated stability indicating UPLC method for simultaneous determination of related substances, and degradation products of cabazitaxel drug substance and its pharmaceutical injection forms. *Journal of Pharmaceutical Sciences and Research.*, **2015**, *6*(12), 411-419.

[122] Prakash, L.; Himaja, M.; Vasudev, R. Isolation, identification, and characterisation of degradation products and the development and validation of a stability-indicating method for the estimation of impurities in the tolterodine tartrate formulation. *Sci. Pharm.*, **2014**, *83*(1), 65-83. [http://dx.doi.org/10.3797/scipharm.1407-18] [PMID: 26839802]

[123] Patel, M.; Singh, B.; Shah, A. Stability indicating UPLC method for quantification of tolperisone HCL and paracetamol from muscle relaxant combination tablet. *Int J. Pharm. Pharm. Sci.*, **2015**, *7*(8), 299-304.

[124] Ozen, G.; Gumustas, M.; Satana Kara, H.E.; Goger, N.G.; Uslu, B.; Ozkan, S.A. Quantitative determination of isoconazole nitrate and diflucortolone valerate in pharmaceutical creams by uplc, hplc and improved derivative uv spectrophotometry. *Rev. Roum. Chim.*, **2015**, *60*(5-6), 527-535.

[125] Nadig, S.; Jacob, J.T. UPLC method for the quantitative estimation of emtricitabine impurities in formulated drug product. *Der Pharmacia Lettre.*, **2015**, *7*(2), 160-164.

[126] Mowaka, S.; Mohamed, D. Novel contribution to the simultaneous analysis of certain hypoglycemic drugs in the presence of their impurities and degradation products utilizing UPLC-MS/MS. *RSC Advances,* **2015**, *5*(74), 60467-60481. [http://dx.doi.org/10.1039/C5RA11448A]

[127] Miranda, T.A.; Silva, P.H.; Pianetti, G.A.; César, I.C. Simultaneous quantitation of chloroquine and primaquine by UPLC-DAD and comparison with a HPLC-DAD method. *Malar. J.,* **2015**, *14*(1), 29. [http://dx.doi.org/10.1186/s12936-015-0570-1] [PMID: 25626728]

[128] Mallik, R.; Raman, S.; Liang, X.; Grobin, A.W.; Choudhury, D. Development and validation of a rapid ultra-high performance liquid chromatography method for the assay of benzalkonium chloride using a quality-by-design approach. *J. Chromatogr. A,* **2015**, *1413*, 22-32. [http://dx.doi.org/10.1016/j.chroma.2015.08.010] [PMID: 26316034]

[129] Lotfy, H.M.; Saleh, S.S.; Hassan, N.Y.; Salem, H. Development and validation of impurity-profiling UPLC method for the determination of sodium cromoglicate and tetryzoline hydrochloride: Application on rabbit aqueous humor. *J. Chromatogr. B Analyt. Technol. Biomed. Life Sci.,* **2015**, *1006*, 121-129. [http://dx.doi.org/10.1016/j.jchromb.2015.10.020] [PMID: 26547297]

[130] Lahsini, R.; Monser, L. Development of a new UPLC method for simultaneous determination of valsartan, amlodipine besylate, and hydrochlorothiazide pharmaceutical products. *Acta Chromatogr.,* **2015**, *27*(3), 449-460. [http://dx.doi.org/10.1556/AChrom.27.2015.3.4]

[131] Koppala, S.; Panigrahi, B.; Raju, S.V.; Padmaja Reddy, K.; Ranga Reddy, V.; Anireddy, J.S. Development and validation of a simple, sensitive, selective and stability-indicating RP-UPLC method for the quantitative determination of ritonavir and its related compounds. *J. Chromatogr. Sci.,* **2015**, *53*(5), 662-675. [http://dx.doi.org/10.1093/chromsci/bmu097] [PMID: 25186277]

[132] Klimczak, I.; Gliszczyńska-Świgło, A. Comparison of UPLC and HPLC methods for determination of vitamin C. *Food Chem.,* **2015**, *175*, 100-105. [http://dx.doi.org/10.1016/j.foodchem.2014.11.104] [PMID: 25577057]

[133] Ismail, Y.; Chandrasekhar, K.B.; Gunasekaran, V. A new stability indicating UPLC method development and validation for the simultaneous estimation of duloxetine and mecobalamin in bulk and in its dosage forms. *Int J. Pharm. Pharm. Sci.,* **2015**, *7*(1), 273-279.

[134] Hendy, M.S. Improved chromatographic separation of sitagliptin phosphate and metformin hydrochloride. *Int. J. Biomed. Sci.,* **2015**, *11*(4), 190-194. [PMID: 26759536]

[135] Gumustas, M.; Coskun, G.; Ozkan, S.A. Selective and sensitive determination of repaglinide in pharmaceuticals by voltammetric and LC methods. *Rev. Roum. Chim.,* **2015**, *60*(5-6), 477-490.

[136] Ayoub, B.M. UPLC simultaneous determination of empagliflozin, linagliptin and metformin. *RSC Advances,* **2015**, *5*(116), 95703-95709. [http://dx.doi.org/10.1039/C5RA17231D]

[137] Alanazi, F.K.; Haq, N.; Radwan, A.A.; Alsarra, I.A.; Shakeel, F. Development and validation of

UHPLC-DAD method for the determination of cholesteryl-hexahydrophthaloyl-5-fluorouracil in lipid nanoemulsion. *J. Anal. Chem.*, **2015**, *70*(5), 593-599. [http://dx.doi.org/10.1134/S1061934815050056]

[138] Curtin Whelan, L.; Geary, M.; Wharton, M.; Sweetman, P. The ultra-performance liquid chromatography determination of domperidone and its process-related impurities. *J. Chromatogr. Sci.*, **2015**, *53*(2), 226-232. [http://dx.doi.org/10.1093/chromsci/bmu044] [PMID: 24904093]

[139] Thummala, V.R.; Seshadri, R.K.; Tharlapu, S.S.; Ivaturi, M.R.; Nittala, S.R. Development and validation of a UPLC method by the QbD-approach for the estimation of rabeprazole and levosulpiride from capsules. *Sci. Pharm.*, **2014**, *82*(2), 307-326. [http://dx.doi.org/10.3797/scipharm.1310-17] [PMID: 24959404]

[140] Sun, L.; Zhang, W.; Liu, X.; Sun, J. Preparation and evaluation of sustained-release azithromycin tablets *in vitro* and *in vivo*. *Asian Journal of Pharmaceutical Sciences.*, **2014**, *9*(3), 155-161. [http://dx.doi.org/10.1016/j.ajps.2014.03.003]

[141] Sravan Kumar Reddy, G.; Ashutosh Kumar, S.; Raj Kumar, V. A new, Simple, Sensitive, Accurate & rapid Analytical method development & validation for simultaneous Estimation of Sitagliptin & Simvastatin in tablet dosage form by using UPLC. *Int. J. Pharm. Tech. Res.*, **2014**, *6*(3), 880-893.

[142] Siddiqui, M.R.; Wabaidur, S.M.; Alothman, Z.A.; Khan, M.A.; Ola, M.S. Quantitative assessment of cefuroxime: A second generation cephalosporin in pharmaceutical formulations and spiked plasma using ultra performance liquid chromatography-tandem mass spectrometry. *Asian J. Chem.*, **2014**, *26*(24), 8479-8482. [http://dx.doi.org/10.14233/ajchem.2014.18060]

[143] Raza Siddiqui, M.; Alothman, Z.A.; Mohammad Wabaidur, S. Ultraperformance liquid chromatography-mass spectrometric method for determination of ampicillin and characterization of its forced degradation products. *J. Chromatogr. Sci.*, **2014**, *52*(10), 1273-1280. [http://dx.doi.org/10.1093/chromsci/bmt211] [PMID: 24474427]

[144] Schmidt, A.H.; Wess, C. A QBD with design-of-experiments approach for development of a state-o--the-art UPLC purity method for Carbamazepine. *J. Liq. Chromatogr. Relat. Technol.*, **2014**, *37*(18), 2653-2666. [http://dx.doi.org/10.1080/10826076.2013.853312]

[145] Reddy, P.R.; Reddy, V.K.; Goud, E.S.; Reddy, Y.R. Development and validation of an ultra performance liquid chromatographic method for assay of alfuzosin hydrochloride in blend sample. *Int J. Pharm. Pharm. Sci.*, **2014**, *6*(5), 72-76.

[146] Ravi Kumar Reddy, P.; Krishna Reddy, V.; Sasikiran Goud, E.; Ramachandra Reddy, Y. Development and validation of a ultra performance liquid chromatographic method for assay of mebeverine hydrochloride. *Int J. Pharm. Pharm. Sci.*, **2014**, *6* Suppl. 2, 442-445.

[147] Ravi Kumar Reddy, P.; Krishna Reddy, V.; Krishna Reddy, E.; Ramachandra Reddy, Y. Development and validation of a ultra performance liquid chromatographic method for assay of cetirizine dihydrochloride. *Int J. Pharm. Pharm. Sci.*, **2014**, *6* Suppl. 2, 430-432.

[148] Rajan, N.; Anver Basha, K. Rapid Determination of Fingolimod Hydrochloride-Related Substances and Degradation Products in API and Pharmaceutical Dosage Forms by Use of a Stability-Indicating UPLC Method. *Chromatographia,* **2014**, *77*(21-22), 1545-1552. [http://dx.doi.org/10.1007/s10337-014-2751-4]

[149] Pydimarry, S.P.; Cholleti, V.K.; Venagala, R.R. Rapid quantitative determination of related substances and degradants in milnacipran. *J. Chromatogr. Sci.*, **2014**, *52*(1), 42-51. [http://dx.doi.org/10.1093/chromsci/bms202] [PMID: 23258392]

[150] Pydimarry, S.P.; Cholleti, V.K.; Vangala, R.R. Stability-indicating UPLC method for determining related substances and degradants in dronedarone. *J. Chromatogr. Sci.*, **2014**, *52*(7), 666-675.

[http://dx.doi.org/10.1093/chromsci/bmt097] [PMID: 23863770]

[151] Paczkowska, M.; Zalewski, P.; Garbacki, P.; Talaczyńska, A.; Krause, A.; Cielecka-Piontek, J. The Development and Validation of a Stability-Indicating UHPLC-DAD Method for Determination of Perindopril l-Arginine in Bulk Substance and Pharmaceutical Dosage Form. *Chromatographia,* **2014**, *77*(21-22), 1497-1501. [http://dx.doi.org/10.1007/s10337-014-2724-7] [PMID: 25400289]

[152] Otašević, B.; Milovanović, S.; Zečević, M.; Golubović, J.; Protić, A. UPLC method for determination of moxonidine and its degradation products in active pharmaceutical ingredient and pharmaceutical dosage form. *Chromatographia,* **2014**, *77*(1-2), 109-118. [http://dx.doi.org/10.1007/s10337-013-2580-x]

[153] Mohsin, K.; Al-Qahtani, S.; Alanazi, F.K. Rapid and sensitive bioanalytical stability-indicating method for quantification of talinolol, a selective β1 adrenoceptor antagonist in lipid based formulations using ultrafast UHPLC systems. *Acta Pol. Pharm.,* **2014**, *71*(5), 737-745. [PMID: 25362802]

[154] Marineni, B.; Sreenivasulu Reddy, T. Simultaneous determination of Telmisartan impurities and Chlorthalidone impurities by UPLC. *Int. J. Pharm. Sci. Rev. Res.,* **2014**, *26*(2), 226-230.

[155] Kurbanoglu, S.; Rodriguez San Miguel, P.; Uslu, B.; Ozkan, S.A. Stability-indicating UPLC method for the determination of bisoprolol fumarate and hydrochlorothiazide: Application to dosage forms and biological sample. *Chromatographia,* **2014**, *77*(3-4), 365-371. [http://dx.doi.org/10.1007/s10337-013-2606-4]

[156] Kondra, S.B.; Madireddy, V.; Chilukuri, M.; Papadasu, N.; Jonnalagadda, L. A validated stability-indicative UPLC method for nilotinib hydrochloride for the determination of process-related and degradation impurities. *J. Chromatogr. Sci.,* **2014**, *52*(8), 880-885. [http://dx.doi.org/10.1093/chromsci/bmt134] [PMID: 24029617]

[157] Killi, G.D.; Maddinapudi, R.K.; Dinakaran, S.K.; Harani, A. A novel validated UPLC method for quantitation of lopinavir and ritonavir in bulk drug and pharmaceutical formulation with its impurities. *Braz. J. Pharm. Sci.,* **2014**, *50*(2), 301-307. [http://dx.doi.org/10.1590/S1984-82502014000200009]

[158] Gumustas, M.; Uslu, B.; Ozkan, S.A.; Aboul-Enein, H.Y. Validated Stability-Indicating HPLC and UPLC Assay Methods for the Determination of Entacapone in Pharmaceutical Dosage Forms. *Chromatographia,* **2014**. [http://dx.doi.org/10.1007/s10337-014-2758-x]

[159] Dhekale, N.H.; Bindu, K.H.; Kirankumar, K.Y.; Gore, A.H.; Anbhule, P.V.; Kolekar, G.B. Development and optimization of a multivariate RP-UPLC method for determination of telmisartan and its related substances by applying a two-level factorial design approach: Application to quality control study. *Anal. Methods,* **2014**, *6*(14), 5168-5182. [http://dx.doi.org/10.1039/c3ay42260g]

[160] Chilukuri, M.; Hussainreddy, K.; Narayanareddy, P.; Venkataramana, M. A validated stability-indicating UPLC method for the determination of impurities in Maraviroc. *J. Chromatogr. Sci.,* **2014**, *52*(7), 609-616. [http://dx.doi.org/10.1093/chromsci/bmt085] [PMID: 23825352]

[161] Addo, R.T.; Davis, K.; Ubale, R.; Owen, J.S.; Watkins, E.B. Development and validation of a UPLC method for rapid and simultaneous analysis of proton pump inhibitors. *AAPS PharmSciTech,* **2015**, *16*(1), 30-34. [http://dx.doi.org/10.1208/s12249-014-0207-7] [PMID: 25160675]

[162] Vijaya Bhaskara Reddy, T.; Sowjanya Reddy, N.; Ravi Prakash Reddy, S.; Ramu, G.; Rambabu, C. Optimized and validated stability indicating RP-UPLC method for the determination of famotidine in pharmaceutical formulations. *J. Pharm. Res.,* **2013**, *6*(8), 865-869. [http://dx.doi.org/10.1016/j.jopr.2013.08.003]

[163] Uslu, B.; Özden, T. HPLC and UPLC Methods for the Simultaneous Determination of Enalapril and Hydrochlorothiazide in Pharmaceutical Dosage Forms. *Chromatographia,* **2013**, *76*(21-22), 1487-1494.
[http://dx.doi.org/10.1007/s10337-013-2442-6]

[164] Sowjanya, P.; Shanmugasundaram, P.; Naidu, P.; Singamsetty, S.K. Novel validated stability-indicating UPLC method for the determination of Metoclopramide and its degradation impurities in API and pharmaceutical dosage form. *J. Pharm. Res.,* **2013**, *6*(7), 765-773.
[http://dx.doi.org/10.1016/j.jopr.2013.07.004]

[165] Seshadri, R.K.; Raghavaraju, T.V.; Kumar, Y.R.; Babu, B.S.; Venkatasubbaiah, B.; Chakravarthy, I.E. Development and validation of a stability indicating UPLC method, identification, and characterization of three degradant impurities in pharmaceutical dosage form of rabeprazole sodium. *J. Liq. Chromatogr. Relat. Technol.,* **2013**, *36*(16), 2261-2278.

[166] Seelam, S.; Dhanalakshmi, K.; Nagarjuna Reddy, G. Validation of Derivative spectrophotometric and UPLC methods for quantitative determination of Epalrestat in bulk and pharmaceutical dosage form. *Int J. Pharm. Pharm. Sci.,* **2013**, *5* Suppl. 3, 341-346.

[167] Schmidt, A.H.; Molnár, I. Using an innovative Quality-by-Design approach for development of a stability indicating UHPLC method for ebastine in the API and pharmaceutical formulations. *J. Pharm. Biomed. Anal.,* **2013**, *78-79*, 65-74.
[http://dx.doi.org/10.1016/j.jpba.2013.01.032] [PMID: 23454599]

[168] Satheesh, B.; Sree Ganesh, K.K.; Saravanan, D. Simultaneous determination of tadalafil and its related compounds in pharmaceutical dosage forms by UPLC. *J. Liq. Chromatogr. Relat. Technol.,* **2013**, *36*(10), 1451-1465.

[169] Rao, P.V.; Kumar, M.N.; Kumar, M.R. A novel, validated stability-indicating UPLC method for the estimation of Lansoprazole and its impurities in bulk drug and pharmaceutical dosage forms. *Sci. Pharm.,* **2013**, *81*(1), 183-193.
[http://dx.doi.org/10.3797/scipharm.1210-09] [PMID: 23641337]

[170] Pavla, P.; Petr, K.; Tomáš, N.; Jiří, K. Development and validation of rapid UHPLC method for determination of aciclovir, its impurities and preservatives in topical cream. *Anal. Methods,* **2013**, *5*(10), 2610-2614.
[http://dx.doi.org/10.1039/c3ay40282g]

[171] Murthy, M.V.; Krishnaiah, Ch.; Srinivas, K.; Rao, K.S.; Kumar, N.R.; Mukkanti, K. Development and validation of RP-UPLC method for the determination of darifenacin hydrobromide, its related compounds and its degradation products using design of experiments. *J. Pharm. Biomed. Anal.,* **2013**, *72*, 40-50.
[http://dx.doi.org/10.1016/j.jpba.2012.09.013] [PMID: 23146225]

[172] Molleti, S.; Rao, V.; Jayaveera, K.N. Stability indicating RP-UPLC method for the determination of lacosamide and its impurities in bulk drugs and its pharmaceutical dosage forms. *Pharma Chem.,* **2013**, *5*(1), 81-89.

[173] Arvadiya, A.C.; Dahivelker, P.P. Development and validation of novel RP-UPLC method for estimation of atropine sulphate in pharmaceutical dosage form. *Chem. Ind. Chem. Eng. Q.,* **2013**, *19*, 333-337.
[http://dx.doi.org/10.2298/CICEQ120319068A]

[174] Al-Amri, K.A.; Mohsin, K.; Alanazi, F.K. Development and validation of a UPLC method for quantification of antiviral agent, Acyclovir in lipid-based formulations. *Arab. J. Chem.,* **2013**.

[175] Dorival-García, N.; Zafra-Gómez, A.; Navalón, A.; Vílchez, J.L. Analysis of bisphenol A and its chlorinated derivatives in sewage sludge samples. Comparison of the efficiency of three extraction techniques. *J. Chromatogr. A,* **2012**, *1253*(0), 1-10.
[http://dx.doi.org/10.1016/j.chroma.2012.06.079] [PMID: 22794796]

[176] Nageswari, A.; Reddy, K.V.; Mukkanti, K. Stability-indicating UPLC method for determination of Imatinib Mesylate and their degradation products in active pharmaceutical ingredient and pharmaceutical dosage forms. *J. Pharm. Biomed. Anal.,* **2012**, *66*(0), 109-115. [http://dx.doi.org/10.1016/j.jpba.2012.03.022] [PMID: 22502910]

[177] Douša, M.; Břicháč, J.; Svoboda, J.; Klvaňa, R. Rapid HILIC method with fluorescence detection using derivatization reaction utilizing o-phthaldialdehyde for determination of degradation product of aliskiren. *J. Pharm. Biomed. Anal.,* **2012**, *66*(0), 359-364. [http://dx.doi.org/10.1016/j.jpba.2012.03.015] [PMID: 22475518]

[178] Rane, S.S.; Ajameri, A.; Mody, R.; Padmaja, P. Development and validation of RP-HPLC and RP-UPLC methods for quantification of parathyroid hormones (1-34) in medicinal product formulated with meta-cresol. *J. Pharm. Anal.,* **2012**, *2*(2), 136-142. [http://dx.doi.org/10.1016/j.jpha.2011.12.001]

[179] Rane, S.S.; Ajameri, A.; Mody, R.; Padmaja, P. Development and validation of RP-HPLC and RP-UPLC methods for quantification of erythropoietin formulated with human serum albumin. *J. Pharm. Anal.,* **2012**, *2*(2), 160-165. [http://dx.doi.org/10.1016/j.jpha.2011.11.006]

[180] Heaton, J.; Gray, N.; Cowan, D.A.; Plumb, R.S.; Legido-Quigley, C.; Smith, N.W. Comparison of reversed-phase and hydrophilic interaction liquid chromatography for the separation of ephedrines. *J. Chromatogr. A,* **2012**, *1228*(0), 329-337. [http://dx.doi.org/10.1016/j.chroma.2011.09.026] [PMID: 21999982]

[181] Sun, J.; Chen, P. Chromatographic fingerprint analysis of yohimbe bark and related dietary supplements using UHPLC/UV/MS. *J. Pharm. Biomed. Anal.,* **2012**, *61*(0), 142-149. [http://dx.doi.org/10.1016/j.jpba.2011.11.013] [PMID: 22221902]

[182] Deconinck, E.; Verlinde, K.; Courselle, P.; Beer, J.O. A validated Ultra High Pressure Liquid Chromatographic method for the characterisation of confiscated illegal slimming products containing anorexics. *J. Pharm. Biomed. Anal.,* **2012**, *59*(0), 38-43. [http://dx.doi.org/10.1016/j.jpba.2011.09.036] [PMID: 22075376]

[183] Douša, M.; Gibala, P.; Rádl, S.; Klecán, O.; Mandelová, Z.; Břicháč, J.; Pekárek, T. Identification, preparation and UHPLC determination of process-related impurity in zolmitriptan. *J. Pharm. Biomed. Anal.,* **2012**, *58*, 1-6. [http://dx.doi.org/10.1016/j.jpba.2011.08.043] [PMID: 21982522]

[184] Nalwade, S.U.; Reddy, V.R.; Rao, D.D.; Morisetti, N.K. A validated stability indicating ultra performance liquid chromatographic method for determination of impurities in Esomeprazole magnesium gastro resistant tablets. *J. Pharm. Biomed. Anal.,* **2012**, *57*(0), 109-114. [http://dx.doi.org/10.1016/j.jpba.2011.08.025] [PMID: 21920688]

[185] Toro-Funes, N.; Odriozola-Serrano, I.; Bosch-Fusté, J.; Latorre-Moratalla, M.L.; Veciana-Nogués, M.T.; Izquierdo-Pulido, M.; Vidal-Carou, M.C. Fast simultaneous determination of free and conjugated isoflavones in soy milk by UHPLC-UV. *Food Chem.,* **2012**, *135*(4), 2832-2838. [http://dx.doi.org/10.1016/j.foodchem.2012.06.011] [PMID: 22980879]

[186] Dias, A.L.; Rozet, E.; Chataigné, G.; Oliveira, A.C.; Rabelo, C.A.; Hubert, P.; Rogez, H.; Quetin-Leclercq, J. A rapid validated UHPLC-PDA method for anthocyanins quantification from Euterpe oleracea fruits. *J. Chromatogr. B Analyt. Technol. Biomed. Life Sci.,* **2012**, *907*(0), 108-116. [http://dx.doi.org/10.1016/j.jchromb.2012.09.015] [PMID: 23026226]

[187] Shikanga, E.A.; Kamatou, G.P.; Chen, W.; Combrinck, S.; Viljoen, A.M. Validated RP-UHPLC PDA and GC–MS methods for the analysis of psychoactive alkaloids in Sceletium tortuosum. *S. Afr. J. Bot.,* **2012**, *82*(0), 99-107. [http://dx.doi.org/10.1016/j.sajb.2012.05.004]

[188] Padullés, A.; Colom, H.; Armendariz, Y.; Cerezo, G.; Caldes, A.; Pou, L.; Torras, J.; Grinyó, J.M.; Lloberas, N. Determination of ganciclovir in human plasma by ultra performance liquid

chromatography-UV detection. *Clin. Biochem.,* **2012**, *45*(4-5), 309-314. [http://dx.doi.org/10.1016/j.clinbiochem.2011.12.014] [PMID: 22245106]

[189] Rao, P.V.; Sanjeeva Reddy, C.K.; Kumar, M.R.; Rao, D.D. Simultaneous determination of omeprazole and domperidone in capsules and *in vitro* dissolution studies by using stability indicating UPLC. *J. Liq. Chromatogr. Relat. Technol.,* **2012**, *35*(16), 2322-2332.

[190] Thakkar, R.; Saravaia, H.; Ambasana, M.; Shah, A. Ultra performance liquid chromatographic method for quantitative analysis of some keto-analogues of essential amino acid calcium salt used in severe renal failure. *J. Liq. Chromatogr. Relat. Technol.,* **2012**, *35*(15), 2125-2133.

[191] Raj, S.; Kumari, K.S.; Rao, A.N.; Reddy, I.U.; Raju, M.N. Development of a stability-indicating UPLC method for determination of olmesartan medoxomil and its degradation products in active pharmaceutical ingredient and dosage forms. *J. Liq. Chromatogr. Relat. Technol.,* **2012**, *35*(8), 1011-1026. [http://dx.doi.org/10.1080/10826076.2011.615091]

[192] Raju, T.S.; Kalyanaraman, L.; Reddy, V.V.; Swamy, P.Y. Development and validation of an UPLC method for the rapid separation of positional isomers and potential impurities of rivastigmine hydrogen tartrate in drug substance and drug product. *J. Liq. Chromatogr. Relat. Technol.,* **2012**, *35*(7), 896-911. [http://dx.doi.org/10.1080/10826076.2011.613143]

[193] Chen, J-Z.; Chou, G-X.; Yang, L.; Bligh, S.A.; Wang, Z-T. Quantification of norisoboldine in Linderae Radix by ultra-performance liquid chromatography and high-performance liquid chromatography with UV detection. *J. Liq. Chromatogr. Relat. Technol.,* **2012**, *35*(6), 788-797. [http://dx.doi.org/10.1080/10826076.2011.608236]

[194] Reddy, I.U.; Rao, P.N.; Reddy, V.R.; Satyanarayana, K. Stability-indicating UPLC method for determination of ramelteon and their degradation products in active pharmaceutical ingredients. *J. Liq. Chromatogr. Relat. Technol.,* **2012**, *35*(5), 688-699. [http://dx.doi.org/10.1080/10826076.2011.606587]

[195] Bindu, K.H.; Dhekale, N.H.; Suryanarayana, M.; Anjaneyulu, Y. A validated stability indicating UPLC method for simultaneous determination of assay, related substances, and degradation products of paliperidone palmitate active pharmaceutical ingredient and its pharmaceutical injection forms. *J. Liq. Chromatogr. Relat. Technol.,* **2012**, *35*(4), 533-546. [http://dx.doi.org/10.1080/10826076.2011.601503]

[196] Saini, D.; Baboota, S.; Ali, M.; Patel, H.; Jain, P.; Neerumulla, S. Development and validation of a stability-indicating reversed phase ultra performance liquid chromatographic method for the quantitative analysis of raloxifene hydrochloride in pharmaceutical dosage form. *J. Liq. Chromatogr. Relat. Technol.,* **2012**, *35*(1), 162-173. [http://dx.doi.org/10.1080/10826076.2011.597067]

[197] Naga Malleswararao, C.S.; Suryanaryana, M.; Krishna, K.; Mukkanti, K. Stability indicating UPLC method for the rapid separation of related components of gemcitibine hydrochloride. *J. Liq. Chromatogr. Relat. Technol.,* **2012**, *35*(18), 2511-2523.

[198] Sanches-Silva, A.; Ribeiro, T.; Albuquerque, T.G.; Paseiro, P.; Sendón, R.; de Quirós, A.B.; López-Cervantes, J.; Sánchez-Machado, D.I.; Soto Valdez, H.; Angulo, I.; Aurrekoetxea, G.P.; Costa, H.S. Ultra-high pressure LC for astaxanthin determination in shrimp by-products and active food packaging. *Biomed. Chromatogr.,* **2013**, *27*(6), 757-764. [http://dx.doi.org/10.1002/bmc.2856] [PMID: 23225623]

[199] Valente, A.; Bronze, M.R.; Bicho, M.; Duarte, R.; Costa, H.S. Validation and clinical application of an UHPLC method for simultaneous analysis of total homocysteine and cysteine in human plasma. *J. Sep. Sci.,* **2012**, *35*(24), 3427-3433. [http://dx.doi.org/10.1002/jssc.201200672] [PMID: 23225709]

[200] Sanches-Silva, A.; Ribeiro, T.; Albuquerque, T.G.; Paseiro, P.; Sendón, R.; de Quirós, A.B.; López-

Cervantes, J.; Sánchez-Machado, D.I.; Valdez, H.S.; Angulo, I.; Aurrekoetxea, G.P.; Costa, H.S. Ultra-high pressure LC determination of glucosamine in shrimp by-products and migration tests of chitosan films. *J. Sep. Sci.,* **2012**, *35*(5-6), 633-640. [http://dx.doi.org/10.1002/jssc.201100855] [PMID: 22517638]

[201] Seshadri, R.K.; Raghavaraju, T.V.; Kumar, Y.R.; Babu, B.S.; Venkatasubbaiah, B.; Chakravarthy, I.E. Development and validation of a stability indicating uplc method, identification, and characterization of three degradant impurities in pharmaceutical dosage form of rabeprazole sodium. *J. Liq. Chromatogr. Rel. Technol.,* **2013**, *36*(16), 2261-2278.

CHAPTER 3

HILIC Based LC/MS for Metabolite Analysis

Emirhan Nemutlu* and **Sedef Kır**

Faculty of Pharmacy, Department of Analytical Chemistry, Hacettepe University, Sıhhiye 06100, Ankara, Turkey

Abstract: Recent developments in column packing materials have improved analysis of a broad range of chemical groups with better chromatographic results in terms of peak shape, capacity factor *etc*. Hydrophilic interaction liquid chromatography (HILIC) has achieved significant progress especially on the analysis of polar molecules. The technique is analogous to normal-phase chromatography and an alternative approach to effectively separate small polar compounds. Researches focusing on applications of HILIC are increasing due to their unique application in targeted and untargeted metabolomic profiling, which enable the analysis of a wide range of metabolites from multiple metabolic pathways. Accordingly, improved profiling leads to individualized and predictive disease information for personalization of treatment. Combined application of advanced spectroscopic and separation technique is essential for metabolomic studies. This chapter reviews the reference information on the application of HILIC and MS related to separation of metabolites and its effect on various parameters.

Keywords: Biological matrix, Hydrophilic interaction liquid chromatography, HILIC, HILIC-MS, Metabolite analyze, Metabolites, Metabolomics, Method development steps, Organic solvent, pH, Polar compounds, Sample preparation, Separation mechanism, Stationary phases, Targeted metabolomics.

INTRODUCTION

Living cells represent an integrated and interacting network of genes, transcripts, proteins, small signaling molecules, and metabolites that define cellular phenotype and function. Traditionally the focus of biomedical researches was on individual genes, single protein targets, single metabolites, and metabolic or signaling pathways. This "molecular reductionist" paradigm was based on the assumption that identifying genetic variations and molecular components would lead to discovery of cured for diseases. However, most of the diseases are and

* **Corresponding author Emirhan Nemutlu:** Faculty of Pharmacy, Department of Analytical Chemistry, Hacettepe University, Sıhhiye 06100, Ankara, Turkey; Tel: +90 312 305 1499; E-mail: enemutlu@hacettepe.edu.tr

Atta-ur-Rahman, Sibel A. Ozkan & Rida Ahmed (Eds.)

multi-factorial and the disease phenotype is determined by the alterations of multiple genes, pathways, proteins and metabolites (at cellular, tissue, and organ levels). Therefore, an integrated "omics" approach is a more viable direction for uncovering alterations in metabolic networks, disease mechanisms, and drug effects [1 - 5].

Technological innovations and implementation of basic discoveries to clinical practice drive advances in medicine. Today's innovative technologies enable comprehensive screening of the genome, transcriptome, proteome, and metabolome. The detailed knowledge, converged in the integrated "omics" (genomics, transcriptomics, proteomics, and metabolomics), holds an immense potential for understanding mechanism of diseases, facilitating their early diagnostics, selecting personalized therapeutic strategies, and assessing their effectiveness. Metabolomics is the newest "omics" approach aimed to analyze large metabolite pools, which may be indicator of energy and metabolic imbalances as a result of genetic deficiencies, myocardial ischemia, heart failure, inflammatory and neurodegenerative disorders, *etc*. The analysis of metabolic fingerprints left by disease processes and metabolic monitoring of disease progression or treatment efficacy play a crucial role in personalized and predictive medicine. Moreover, the "reverse omic" approach or "metabolomics-informed pharmacogenomics," based on the discovery of specific metabolite changes, may help to identify discovery of genetic alterations [3, 4, 6].

Metabolomics is the study of identification and quantification of small molecule metabolites (molecular weight <1000 Da) in tissues, cells and physiological fluids in a short period of time. Differences in metabolite profiling are due to genetic or environmental factors as well as cellular and systemic differences from diseases, medicines, toxins and lifestyle [6 - 9]. The analytical methods, developed for monitoring these alterations, have to perform a comprehensive analysis of hundreds of metabolites from complex biological materials [3]. Metabolomic analyses are the study of performed as targeted and untargeted analyzes which have superiority over each other. Targeted metabolomics studies are aimed to quantification of selected biomarkers with high accuracy and reproducibility, while untargeted analyses are based on finding novel markers for disease diagnosis [10]. Since untargeted metabolomic studies are carried out by semi-quantitative analysis, the variations among laboratories are very high. Targeted metabolomic analyzes are less affected from these variations, because they are based on concentration measurements. Therefore, targeted metabolomic analyses are preferred for routine analysis [1 - 5].

For a comprehensive characterization of metabolic networks and functions, the concentrations of metabolites must be determined. Furthermore, the discovery of several metabolites is one of the foremost challenges in metabolomic studies. Simultaneous and comprehensive fingerprint analysis of a large number of metabolites is required to identify the metabolic phenotype. Analyzes of a broad range of metabolites with different physical and chemical features (polar/nonpolar, acidic/basic, *etc.*) cannot be performed by a single analytical method. Therefore, different analytical platforms must be used for comprehensive metabolomic analysis [7, 11, 12]. The most widely used methods for metabolomic analysis are nuclear magnetic resonance spectroscopy (NMR) and mass spectroscopy (MS) combined with liquid and/or gas chromatography. All three methods have different advantages over each other in terms of sensitivity, coverage and efficiency. The main advantage of the NMR based analysis is nondestructive and it requires lower cost for sample analyses, but it is slower and less sensitive as compared to the MS system [13]. In contrast, MS is very sensitive and selective for specific molecules. The combination of MS with liquid chromatography or gas chromatography becomes a very powerful technique for metabolomic analysis. In plasma-based metabolomic studies, about 30 metabolites can be determined by NMR spectroscopy, while the number of metabolites that can be identified by GC-MS is about 100 reaching about 500 with LC-MS. To achieve these performances, LC based metabolomics analyses must be carried out with samples both in positive and negative ionization mode and using reversed as well as normal phase chromatographic conditions. Reversed phase columns are widely used for the separation of non-polar and weakly polar compounds, while normal phase for highly hydrophilic, ionic, and polar molecules which having lack of adequate retention time on a reversed phase column. In this latter case, there are significant limitations occurred, due to solubility problems of polar metabolites and low ionization efficiency with organic and non-polar solvents used. These limitations are surmounted using hydrophilic interaction liquid chromatography (HILIC). The success of the analysis of polar compounds such as amino acids, lipids, nucleotides, organic acids, carbohydrates and some drug metabolites, especially in MS systems, has increased the frequency of use of the HILIC phase in metabolomic analysis [14, 15].

The retention of analytes using HILIC occurs mainly by adsorption or partition mechanism depending on the structure of the stationary phase, organic solvent type and percentages of solvents in the mobile phase [16].

The targeted analyses published in last five years with HILIC-MS are summarized in Table **1**, in terms of type of sample analyzed, number of metabolites analyzed, type of stationary phase, mobile phase composition, linearity, sensitivity and sample preparation.

Table 1. List of publications related to targeted metabolomic studies by HILIC-MS analysis.

Sample	N*	Stationary Phase	Mobile Phase/Flow Rate	Linearity range	LOQ	Sample Preparation	Ref.
Human urine	2	ZIC-HILIC (2.1 × 150 mm, 5 µm)	**A:** 50 mM Ammonium acetate: acetonitrile (95:5, v/v) **B:** 50 mM Ammonium acetate: water: acetonitrile (50:45:5, v/v/v) **Flow rate:** 0.2 mL/min	0.5 nM-5000 µM	-	Dilution with water and 3 kD molecular weight cut-off filter	[17]
Extracellular medium	5	Waters XBridge HILIC (4.6 × 150 mm,5 µm)	**A:** Acetonitrile: 10 mM ammonium acetate (pH 4.5) (70:30, v/v) **Flow rate:** 0.5 mL/min	0.5-150 nM	0.5 nM	LLE (Chloroform:methanol, 2:1, v/v)	[18]
Cell culture	80	Acquity BEH Amide (2.1 x 150 mm, 1.7 µm)	**A:** Acetonitrile + 0.1% formic acid **B:** Water + 0.1% formic acid **Flow rate:** 0.5 mL/min	0.008-4000 µg/mL	0.3-537.3 ng/mL	PP (Acetonitrile)	[19]
Grape extracts	135	Acquity BEH Amide (2.1 x 150 mm,1.7 µm)	**A:** Acetonitrile: 0.1% formic acid + 0.075% NH_4OH (95:5, v/v) **B:** Acetonitrile: 0.2%formic acid + 0.1% NH_4OH (2:98, v/v) **Flow rate:** 0.5 mL/min	0.02 - 200 µg/mL	0.02-1 µg/mL	PP (Water:methanol, 20:80, v/v)	[20]
Rhesus monkey brain and blood	5	Capillary column (0.75 x 150 mm, 5 µm)	**A:** 50 mM Ammonium formate + 1.0% formic acid (pH 2.87) **B:** Acetonitrile **Flow rate:** 3.5 µL/min	0.5 nM – 5000 µM	450 pM- 6 µM	Microdialysis	[21]
Mouse plasma	1	Kinetex HILIC (4.6 x 100 mm, 2.6 µm)	**A:** 5 mM Ammonium formate (pH 2.9) **B:** Acetonitrile **Flow rate:** 0.9 mL/min	0.01- 1.6 µM	10 nM	PP (Acetonitrile)	[22]
Human skin stratum corneum	3	Acquity BEH Amide (2.1 x 150 mm, 1.7 µm)	**A:** Water **B:** Acetonitrile **Flow rate:** 0.3 mL/min	1-250 ng/mL	0.05 ng/mL	Extraction (Phosphate buffer saline)	[23]
Rat serum	15	TSK-GEL Amide-80 (2.0 x 50 mm, 3 µm)	**A:** Acetonitrile **B:** 5 mM Ammonium acetate + 0.1% formic acid **Flow rate:** 0.15 mL/min	0.05-10 µg/mL	0.05 µg/mL	PP (Acetonitrile)	[24]

(Table 1) contd.....

Sample	N*	Stationary Phase	Mobile Phase/Flow Rate	Linearity range	LOQ	Sample Preparation	Ref.
Human, rat, mouse plasma	7	Nucleosil 100-5OH Diol (3 x 250 mm, 5 µm)	**A:** Acetonitril + 53 mM formic acid **B:** 60 mM ammonium formate + 53 mM formic acid **Flow rate:** 0.5 mL/min	0.035-34.47 µg/mL	-	PP (Isopropanol)	[25]
Human urine		Cosmosil HILIC (2.0 x 150 mm, 5 µm)	**A:** 10 mM Ammonium acetate **B:** Acetonitrile **Flow rate:**0.4 mL/min	0.2-100 nM	0.2 nM	SPE (Oasis HLB)	[26]
Human serum	1	Acquity BEH Amide (2.1 x 50 mm, 1.7 µm)	**A:** Acetonitrile **B:** 100 mM Ammonium acetate (pH 4.5) **Flow rate:** 0.4 mL/min	0.03-20 µM	0.1 µM	SPE (SAX)	[27]
Human urine	7	ZIC-HILIC (4.6 x 250 mm, 5 µm)	**A:** Acetonitrile **B:** 10 mM Ammonium acetate **Flow rate:** 0.5 mL/min	0.5-300 µg/mL	0.5-2 µg/mL	Centrifugation	[28]
Dried blood spot	17	Acquity HILIC Amide (2.1 x 50 mm, 1.7 µm)	**A:** 0.1% formic acid **B:** Acetonitrile + 10 mM ammonium (90:10, v/v) **Flow rate:** 0.3 mL/min	0.048- 1291 µM		Extraction (Methanol with 0.1% formic acid)	[29]
Pichia pastoris cell	29	ZIC-HILIC (4.6 x 150 mm, 3.5 µm)	**A:** Water:acetonitrile: formic acid (98:1:1, v/v/v) **B:** Acetonitrile: water: formic acid (98:1:1, v/v/v) **Flow rate:** 0.5 mL/min	-	-	Extraction (Ethanol)	[30]
Mouse brain and tissue	10	Atlantis HILIC silica (2.1 x 150 mm, 5 µm)	**A:** 0.1% Formic acid **B:** Acetonitrile **Flow rate:** 0.3 mL/min	2.5-20 ng/mL	2.5-20 ng/mL	Homogenization and PP (Methanol)	[31]
Cow urine	9	ZIC-HILIC (2 x 125 mm, 3 µm)	**A:** 20 mM Ammonium formate (pH 4) **B:** Acetonitrile **Flow rate:** 0.35 mL/min	2-10000 µM	1-10 µM	PP (Acetonitrile)	[32]
Human plasma-synovial fluid and urine	1	TSK Gel-Amide (2.0 x 150 mm, 5 µm)	**A:** 10 mM Ammonium acetate (pH 6.8) **B:** Acetonitrile **Flow rate:** 0.2 mL/min	1-50 µM	10 fmol	PP (Acetonitrile)	[33]

(Table 1) contd.....

Sample	N*	Stationary Phase	Mobile Phase/Flow Rate	Linearity range	LOQ	Sample Preparation	Ref.
Human urine	2	Atlantis HILIC silica (2.1 x 100 mm, 5 µm)	**A:** 0.2% formic acid: methanol + 0.2% formic acid: acetonitrile+ 0.2% formic acid (35:35:30, v/v/v) **Flow rate:** 0.2 mL/min	0.1 -19.95 µM	0.07- 0.1 µM	PP (Methanol with 0.1% formic acid)	[34]
Fungus (*Cordyceps*)	15	Waters Xbridge Amide (4.6 x 250 mm, 3.5 µm)	**A:** 0.2% Acetic acid + 10 mM ammonium acetate **B:** Acetonitrile **Flow rate:** 0.4 mL/min	0.001-10 µg/mL	0.50-33.33 ng/mL	PP (Methanol:water, 60:40 v/v)	[35]
Ziziphus plants	20	Acquity BEH Amide (2.1 x 100 mm, 1.7 µm)	**A:** 0.8% Acetic acid + 10 mM ammonium acetate **B:** Acetonitrile + 0.1% acetic acid	0.002- 9.120 µg/mL	0.29-12.48 ng/mL	Ultrasonication Extraction (Water)	[36]
Human plasma	1	Atlantis HILIC silica (2.1 x 150 mm, 5 µm)	**A:** Acetonitrile: 0.025% trifluoroacetic acid + 1% propionic acid (pH 1) (80:20, v/v) **Flow rate:** 0.25 mL/min	0.1- 10 µM	0.1- µM	PP (Formic acid + ammonium formate + acetonitrile)	[37]
Human plasma-urine	8	ZIC-HILIC (2.1 x 250 mm, 5 µm)	**A:** Acetonitrile: 100 mM ammonium acetate (80:20, v/v) (pH 4.7) **Flow rate:** 0.5 mL/min	0.1-2500 µg/mL	0.1-30 µg/mL	SPE (Anion exchange)	[38]
Human serum-urine	2	Acquity BEH HILIC (2.1 x 50 mm, 1.7 µm)	**A:** 7.5 mM ammonium formate **B:** Acetonitrile + 7.5 mM ammonium formate **Flow rate:** 0.25 mL/min	0-350 µg/mL	3.15-4.34 nM	PP (Acetonitrile + methanol)	[39]
Human plasma	12	Kinetex HILIC silica (2 x 50 mm, 2.6 µm)	**A:** Acetonitrile:water:100 mM ammonium formate (pH 3.2) (90:5:5, v/v/v) **B:** Acetonitrile:water:100 mM ammonium formate (pH 3.2) (50:45:5, v/v/v) **Flow rate:** 0.3 mL/min	0.08-152 µM		PP (Acetonitrile)	[40]
Human blood	2	SeQuant ZIC-HILIC (2.1 x 100 mm, 5 µm)	**A:** 1 mM Ammonium acetate: acetonitrile (5:95, v/v) **Flow rate:** 0.2 mL/min	0.2-2000 µM	0.4-3 µM	SPE	[41]

(Table 1) contd.....

Sample	N*	Stationary Phase	Mobile Phase/Flow Rate	Linearity range	LOQ	Sample Preparation	Ref.
Human plasma	8	Aquity BEH HILIC (2.1 × 100 mm, 1.7 μm)	**A:** 13 mM Ammonium formate + 1% formic acid (pH 2.4) **B:** Acetonitrile + 0.1% formic acid **Flow rate:** 0.55 mL/min	0.5-2000 μM	0.5 μM	PP and LLE (Chloroform-methanol)	[42]
Human plasma	3	LUNA-NH_2 (2.0 x 250 mm, 3 μm)	**A:** 100 mM HFIP + 0.5% DEA (pH 8.9) **B:** Acetonitrile + 100 mM HFIP + 0.5% DEA (pH 8.9) **Flow rate:** 0.35 mL/min	2 - 1000 ng/mL	2 ng/mL	PP (Methanol)	[43]
Human plasma	1	ZIC-HILIC (4.6 x 50 mm, 3.5 μm)	**A:** Acetonitrile + 0.1% formic acid B: Water + 0.1% formic acid **Flow rate:** 0.5 mL/min	0.5-32.5 nM	0.5 nM	PP (Acetonitrile:1% zinc sulphate, 70:30, v/v)	[44]
Calf brain tissue	2	ZIC-HILIC (2.1 x 150 mm, 3.5 μm)	A: Acetonitrile + 0.1% formic acid B: Water + 0.1% formic acid **Flow rate:** 0.2 mL/min	20 - 1500 ng/g	20 ng/g	PP (trichloroacetic acid) LLE SPE	[45]
Human urine	1	Nucleodur HILIC (2 x 100 mm, 1.8 μm)	**A:** Water **B:** Acetonitrile **C:** 200 mM Ammonium acetate (pH 6.7) **Flow rate:** 0.25 mL/min	5-500 ng/mL	5 ng/mL	PP (100 mM Ammonium acetate + acetonitrile)	[46]
Embryonal carcinoma stem cell	7	ZIC-HILIC (2.1 x 100 mm, 3 μm)	**A:**Methanol:water, 55:45 (v/v) + 20 mM ammonium formate, pH 3.0 (formic acid) **Flow rate:** 0.2 mL/min	5-2000 ng/mL	5 ng/mL	On-line microdialysis	[47]
Human, beagle dog, rat, mouse blood	1	Acquity BEH HILIC (2.1 x 50 mm, 1.7 μm)	**A:** 10 mM Ammonium acetate **B:** Acetonitrile + 10 mM ammonium acetate: water + 0.1% formic acid (98:2, v/v) **Flow rate:** 0.4 mL/min	0.2-500 ng/mL	0.2 ng/mL	PP (Acetonitrile)	[48]
Pichia pastoris cell	10	Nucleodur HILIC (3 x 100 mm, 1.8 μm)	**A:** 10 mM Ammonium formate (pH 3.25) **B:** Acetonitrile **Flow rate:** 0.3 mL/min	2-800 μM	LOD: 0.02-4 μM	Quenching (Methanol) Extraction (Ethanol)	[49]

(Table 1) contd.....

Sample	N*	Stationary Phase	Mobile Phase/Flow Rate	Linearity range	LOQ	Sample Preparation	Ref.
Arabidopsis leaves	2	Zorbax SB-C_{18} (3 x 150 mm, 3.5 µm)	**A:** 0.1% Formic acid **B:** Acetonitrile: methanol:water (1:2:4, v/v/v) **Flow rate:**0.35 mL/min	20-2000 ng/mL	20 ng/mL	PP (Acetonitrile) SPE	[50]
Invitro fertilization cell culture amniotic fluid, urine, blood serum	100	Acquity BEH Amide (2.1 x 150 mm,1.7 µm)	**A:** Acetonitrile:water (95:5, v/v) **B:** Water:acetonitrile (70:30, v/v) **Flow rate:** 0.5 mL/min	0.01-95 µg/mL	2.7 nM- 0.08 mM	PP (Acetonitrile)	[51]
Freshwater snail (*Lymnaea Stagnalis*)	20	ZIC-cHILIC (2.1X150 mm, 3 µm)	**A:** Acetonitrile:10 mM ammonium formate (pH 2.8) (90:10,v/v) **Flow rate:** 0.2 mL/min	0.05-3000 ng/mL	0.0001-0.076 ng/mL	Dilution (Water) PP (Methanol)	[52]
Peritonea mast cell- Rat plasma	1	Kinetex HILIC (2.1 x 50 mm, 2.6 µm)	**A:** Acetonitrile + 0.1% formic acid **B:** 20 mM Ammonium formate + 0.2% formic acid **Flow rate:** 0.7 mL/min	1-1000 ng/mL	15.6 ng/mL	Cell degranulation PP (Acetonitrile +0.1% formic acid)	[53]
Human plasma	1	Tosoh TSKgel Amide-80 (2.0 x 150 mm, 3 µm)	**A:** 1 mM ammonium acetate:acetonitrile (90:10, v/v) **Flow rate:** 0.2 mL/min	10-500 nM	-	PP (Acetonitrile) Ultrafiltration	[54]
Human plasma	13	Acquity HILIC (2.1 x 50 mm, 1.7 µm)	**A:** 5mM Ammonium acetate:methanol (9:1, v/v) **B:** Acetonitrile **Flow rate:** 0.5 mL/min	-	LOD: 0.015 - 0.11 µg/mL	PP and LLE (Hexane + methanol)	[55]
Bovine aortic endothelial cell- Human muscle	4	Zorbax SB-CN (4.6 x 150 mm, 5 µm)	**A:** Water + 0.1% formic acid **B:** Acetonitrile **Flow rate:** 0.6 mL/min	0.001-25 µM	LOD: 1-8 nM	PP (Acetonitrile + water, 80:20 v/v)	[56]
Human urine	1	Atlantis HILIC Silica (2.1 x 50 mm, 5 µm)	**A:** Acetonitrile + 0.1% formic acid **B:** 5 mM ammonium formate + 0.1% formic acid **Flow rate:** 0.3 mL/min	2-200 ng/mL	2 ng/mL	LLE (Methanol:chloroform:water	[57]
Animal horn	14	Acquity BEH Amide (2.1 x 100 mm, 1.7 µm)	**A:** 0.8% Acetic acid + 10 mM ammonium acetate **B:** Acetonitrile + 0.1% acetic acid **Flow rate:** 0.4 mL/min	0.001-31.050 µg/mL	0.23-4.43 ng/mL	Ultrasonication PP (methanol)	[58]

(Table 1) contd.....

Sample	N*	Stationary Phase	Mobile Phase/Flow Rate	Linearity range	LOQ	Sample Preparation	Ref.
Human urine	8	ZIC HILIC (2.1 x 100 mm, 1.7 µm)	**A:** Acetonitrile **B:** Water + 25 mM formic acid **Flow rate:** 0.5 mL/min	2-10 ng/mL	-	On line SPE	[59]
Human blood	6	ZIC-cHILIC (0.3 x 150 mm, 3.5 µm)	**A:** 70% Acetonitrile + 30% 100 mM ammonium formate (pH 3) + 300 µM ascorbic acid **Flow rate:**4 µL/min	0.05-250 nM	0.2-1 nM	On line SPE	[60]
C-reactive protein Amino acid	8	ZIC-HILIC (2.1 x 250 mm, 5 µm)	**A:** 10 mM acetic acid **B:** Acetonitrile **Flow rate:** 0.1 mL/min	50-1250 nM	0.01- 0.1 pM	Hydrolysis (Hydrochloric acid)	[61]
Cell (intracelular extracts of *E.Coli*)	50	ZIC-pHILIC (2.1 x 150 mm, 5 µm)	**A:** 10 mM Ammonium acetate + acetonitrile (10:90, v/v) **B:** 10 mM Ammonium acetate + acetonitrile (90:10, v/v) A-B: pH 5.6 (acetic acid) or pH 9.2 (25% ammonium hydroxide) **Flow rate:** 0.2 mL/min	10-5000 nM (alkaline) 20-5000 nM (acid)	LOD: 2.2-2949.3 nM	Centrifugation-Quenching-Extraction (Water)	[62]
Human plasma	2	XBridge Amide (2.1 x 100 mm, 3.5 µm) Atlatis HILIC Silica (3.0 x 50 mm, 5 µm)	**A:** Water + 0.2% acetic acid + 0.05% trifluoracetic acid **B:** Acetonitrile + 0.2% acetic acid + 0.05% trifluoraceticacid **Flow rate:** 0.8 mL/min	10-10000 ng/mL	10-25 ng/mL	PP (Acetonitrile)	[63]
Human urine-oral fluid-hair	2	Acquity BEH HILIC (2.1 x 50 mm, 1.7 µm)	**A:** 10 mM Ammonium formate (pH 3) **B:** Acetonitrile **Flow rate:** 0.6 mL/min	0.1-2000 ng/mL	0.02 -0.26 ng/mL 0.04 - 0.48 ng/mL	Extraction (Dichlorometane)	[64]
Primatone RL animal tissue hydrolysate	18	LUNA-NH_2 (2.0 x 250 mm, 5 µm)	**A:** 20 mM Ammonium acetate (pH 9.4) **B:** Acetonitrile **Flow rate:** 0.2 mL/min	0.03-200 µM	LOD 1-100 nM	PP (Acetonitrile) SPE	[65]
Cycad plant tissue - mussel tissue	5	TSK gel Amide (2.0 x 250 mm, 5 µm)	**A:** Water + 50 mM formic acid **B:** Acetonitrile + 50 mM formic acid **Flow rate:** 340 µL	0.01-6 µM	LOD: 20 ng/mL	Hydrolysis (Hydrochlorid acid)	[66]

(Table 1) contd.....

Sample	N*	Stationary Phase	Mobile Phase/Flow Rate	Linearity range	LOQ	Sample Preparation	Ref.
Medicago truncatula plant	3	ZIC-HILIC PEEK (2.1 x 150 mm, 3.5 µm)	**A:** 0.5 mM Methylphosphonic acid +0.001% piperidine **B:** Acetonitrile **Flow rate:** 0.3 mL/min	0.1-50 µM	0.1-1 µM	Extraction	[67]
Blood- urine	5	Luna HILIC (4.6 × 100 mm, 3 µm)	**A1:** Acetonitrile: 10 mM ammonium acetate (90:10, v/v) **B1:** 10 mM Ammonium acetate (pH 4) **A2:** Acetonitrile + 10 mM ammonium formate (90:10, v/v) **B2:** 10 mM Ammonium formate (pH 3) **Flow rate:** 0.75 mL/min	0.5-200 µM	0.5-2 µM	PP (Methanol)	[68]
Serum	16	Kinetex HILIC silica (2.1 x150 mm, 2.6 µm)	**A:** 75% Acetonitrile + 7.5 mM ammonium formate **B:** 0.5% formic acid (pH 3.0) **Flow rate:** 0.25mL/min	-	0.003-9.36 µM	Washed (Isopropanol)	[69]
Schizochytrium sp.S31	5	BEH HILIC (1 x100 mm, 1.7 µm)	**A:** Acetonitrile **B:** 50 mM Ammonium acetate +0.1% formic acid (pH 3.65) **Flow rate:** 0.3 mL/min	1.25-25 µg/mL	0.1-5 µg/mL	PP and LLE (Methanol+ chloroform)	[70]
Human urine	13	XBridge Amide BEH (3 x 100 mm, 3.5 µm)	**A:** 10 mM ammonium formate (pH 3.0) **B:** Acetonitrile + 10 mM ammonium formate **Flow rate:** 1mL/min	5-2000 ng/mL	10-20 ng/mL	LLE (Dichloromethane + ethanol)	[71]
Serum	74	Kinetex HILIC core shell (2.1 x 150 mm, 2.6 µm)	**A:** Water **B:** Acetonitrile + 20 µM ammonium formate (+ ionization) or ammonium acetate (- ionization) (pH 3.5) **Flow rate:** 0.1 – 0.3 mL/min	0.1-3238.3 µM	0.1- 218.6 µM	PP (Acetonitrile + methanol)	[72]
Human plasma-urine	12	XBridge Amide BEH (3 x 100 mm, 3.5 µm)	**A:** Water +10 mM ammonium formate (pH 3.0) **B:** Acetonitrile + 10 mM ammonium formate (pH 3.0) **Flow rate:** 1 mL/min	10-2000 ng/mL	10-20 ng/mL	Microextraction	[73]

(Table 1) contd.....

Sample	N*	Stationary Phase	Mobile Phase/Flow Rate	Linearity range	LOQ	Sample Preparation	Ref.
Cerebrum-cerebellum-brain stem tissue	2	BEH HILIC (2.1 x 50 mm, 1.7 µm)	**A:** 7 mM Ammonium formate +0.1% ammonia **B:** Acetonitrile + 0.1% ammonia **Flow rate:** 0.3 mL/min	10-4000 ng/mL	3.2-16 fmol	DNA extraction Hydrolysis (Formic acid)	[74]
Human plasma	36	Acquity BEH Amide (2.1 x100 mm, 1.7 µm)	**A:** Acetonitrile:10 mM ammonium formate + 0.15% formic acid (85:15, v/v) **B:** 10 mM ammonium formate + 0.15% formic acid (pH 3.0) **Flow rate:** 0.4 mL/min	0-1000 µM	0.001-0.36 µM	PP (Mobile phase A) SPE	[75]
Human urine	3	SeQuant ZIC-HILIC (2.1 x 150 mm, 3.5 µm)	**A:** 5 mM Ammonium acetate (pH 5.72) + 0.1% formic acid **B:** Acetonitrile + 0.1% formic acid **Flow rate:**0.3 mL/min	0.05-2500 µg/mL	0.05-20 µg/mL	PP (Acetonitrile)	[76]
Wheat	4	ZIC-pHILIC (2.1 x 100 mm, 5 µm)	**A:** 20 mM Ammonium acetate **B:** Acetonitrile: 20 mM ammonium acetate (pH 3.0) (96:4, v/v) **Flow rate:** 0.5 mL/min	0.6-1000 ng/mL	2-33 ng/mL	Homogenization and PP (Methanol+ water)	[77]

*N: Number of metabolites analyzed in publication, PP: Protein precipitation, LLE: Liquid-liquid extraction, SPE: Solid Phase Extraction.

Separation Mechanism

The retention mechanism in HILIC is based mainly on partitioning, hydrogen bonding and ionic interactions. Metabolites interact with the stationary phase according to their structure and the separation usually happens with the more hydrophilic compounds that more strongly retained by the HILIC stationary phase. The retention of the polar metabolites in the HILIC is enhanced by increasing the polarity of the stationary phase and decreasing polarity of the organic elution solvent [78].

The presence of polar and hydrophobic groups in the stationary phase leads to water absorption on stationary phase and it plays an important role in separation (Fig. **1**). The absorption amount varies with the type of the stationary bonded phases and affects the retention of the metabolites by a partitioning mechanism [16, 79]. The partition mechanism of HILIC is based on the distribution of metabolites between mobile phase, containing high amount of organic non-polar phase and a water-rich layer on the surface of the polar stationary phase [78, 80]. In fact, as possible to observe in Fig. (**1**), uncharged metabolites are retained for portioning mechanism, a high correlation between retentivity and number of hydroxyl groups in the molecules existing due to hydrogen bonding mechanism and finally, increase in retention related to the number of amine groups in the molecule (ionic interaction) [20].

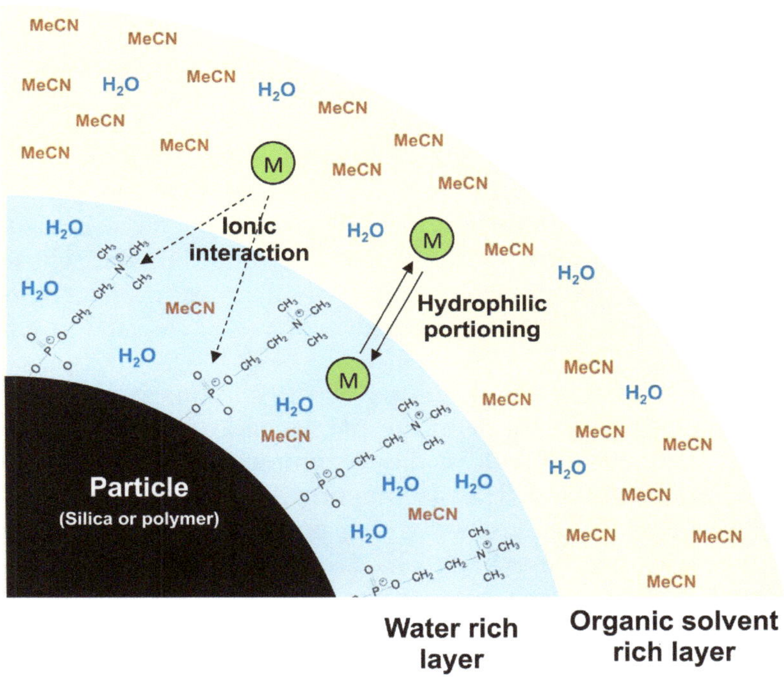

Fig. (1). Separation mechanism in hydrophilic interaction liquid chromatography (M: Metabolite, MeCN: Acetonitrile).

The organic phase (acetonitrile or some other organic solvents) ratio in the mobile phase is a predominating parameter determining the water rich layer composition. When the organic solvent ratio increases, the residual silanols on the stationary phases interact more strongly with water molecules and can create excess of adsorbed water in association with the ratio of water in the mobile phase such as about up to 13% of the pore volume of the silica phase is engaged by a water-rich layer when the acetonitrile ration in the mobile phase is between 75 and 90% [81, 82]. Likewise the mobile phase composition, the number of layers adsorbed on the bare silica depends on organic phase ratio. The water ratio in the mobile phase is lower than 20%, with water adsorption being multilayer in nature [78].

PARAMETERS EFFECTS ON SEPARATIONS

Stationary Phase Effects

The selection of the stationary phase is the key point in order to achieve a success separation and selectivity in HILIC based metabolomic analyses, because the type of interactions (hydrophilic, dipole–dipole, π–π interaction, hydrogen bonding and electrostatic) with metabolites largely diverge between stationary phases [16, 83].

HILIC columns can be classified into three categories based on the charge features of the functional groups of the stationary phases [81, 83]:

- Uncharged (neutral) phases: Neutral phases such as amide, cyano, diol, aspartamide and cyclodextrin cannot be charged in the working pH range for HILIC. Neutral groups usually are bonded directly to the silica surface *via* a propyl linker to form a monomeric phase. The amide phase, whose yield and long-term stability are better than the others, is the most preferred neutral phase in metabolomic studies. Neutral cyclodextrin phases were used to separate polar chiral compounds [83, 84].
- Charged phases: The functional groups of the stationary phases are charged positively or negatively in the working pH range and confer to the stationary phase typical ionization features. For example, phases containing amino, aminopropyl, imidazole and triazole functional groups are positively charged (cationic phases), while phases containing poly (2-sulfoethylaspartamide) and poly (aspartic acid) are negatively charged (anionic phases). The separation mechanism of charged stationary phases is mainly hydrophilic interaction; moreover, ion exchange also plays an important role in separation. For example, nucleosides and nucleobases metabolites from different matrixes are generally analyzed by amino columns [85].
- Zwitterionic phases: The sulphobetaine stationary phase is composed of 3-sulphopropyldimethylalkylammonium functional groups, which are both positively and negatively charged [81, 83, 86]. A water-rich layer occurs on sulfobetaine stationary phases by hydrogen binding and it becomes very effective in the retention mechanism [78]. Additionally, the negatively charged sulphonic acid group at the distal end of the sulphobetaine phase can electrostatically interact with charged solutes [81, 83]. Phosphocholine groups are also used to form zwitterionic stationary phases. With respect to the sulphobetaine phase, there is a positively charged quaternary ammonium group at the distal end of the phosphocholine phase [83].

The chemical structures of the bared silica, amino, diol, cyano, amide and zwitterion columns, which are frequently used in targeted metabolomics analyses are given in Fig. (**2**). Among these stationary phases, amide and zwitterion are the most preferred stationary phases while diol and cyano are the least ones (Fig. **3**). The retention mechanism on HILIC is based on the type of the stationary phases of functional groups. For example, in bare silica, stationary phases separation based on mainly adsorption, whereas partition is mainly in zwitterion columns. In this perspective, amides, sulfonates and zwitterionic groups demonstrated higher hydrophilic retention and less ion exchange interactions than phases containing hydroxy, amino groups and bare silica materials [15, 79, 87].

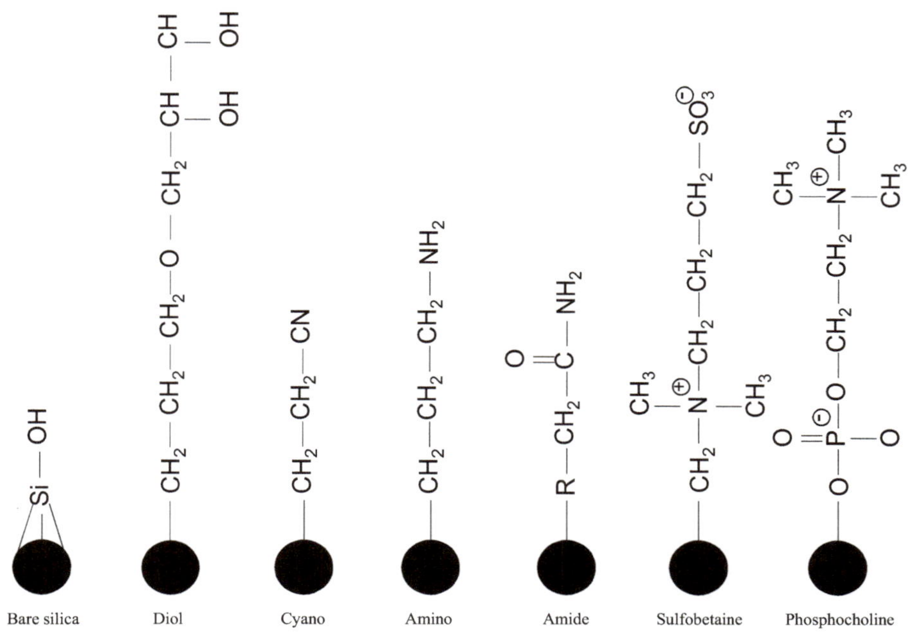

Fig. (2). Schematic illustration of HILIC stationary phases (●: Support material; silica or polymers).

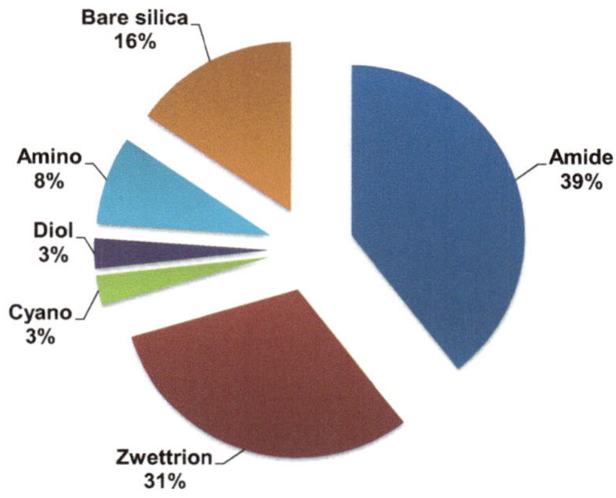

Fig. (3). Distribution of stationary phase in targeted metabolomic studies.

Stationary phases must be selected based on the metabolite physical and chemical properties of metabolites such as ionizable metabolites can be selectively separated on ion exchange phases while non-ionizable metabolites can be selectively separated in hydrophilic interaction-based phases [15, 87].

The experimental data indicate that most of the HILIC stationary phases have similar retentivity for neutral metabolites which may change for charged metabolites especially on charged stationary phases. Consequently, the stationary phase selection is based on the metabolites to be analyzed [83]. Therefore, selection of the suitable stationary phase is the key to achieve a desired separation [83]. For instance, amide or amino bonded phases for the analysis of carbohydrates, bare silica columns for basic analytes and sulfobetaine-bonded phases may be selected for specific group of metabolite analyses [85, 88].

Additionally, the type, particle size and pore size of the stationary phase may affect the retentivity and selectivity of analytes [79]. Based on the Dinh *et al.* studies (2011) on commercially available 22 hydrophilic stationary phases including zwitterionic, neutral, cationic, and anionic functional groups, they found that the column selectivity is not affecting from pore diameter of stationary phases, or the particle diameter, while retention of analytes is significantly effecting from particle pore diameters [79]. The decrease in particle size results in a longer retention of analytes [79]. However, the selectivity decreases with the increase of water content of the mobile phase. Furthermore, the ionic strength of the mobile phase should be optimized according to the metabolites analyzed, since higher salt concentration leads to a decreased selectivity [79].

Besides chromatographic parameters (peak shapes, numbers of metabolites analyzed), the column lifetime is also a critical step for column and mobile phase selection. For example, a polymer based amide stationary phase can be used for over 1,000 analyses with high reproducibility while, a silica based amine column can be used for about 500 highly reproducible analyses at pH = 9. Although amine columns have shorter lifetime, they have slightly higher metabolite peak separation and shapes especially for nucleotides than the other columns [89] (Fig. **4**).

Kloos *et al.* (2013) tested three porous HILIC columns (amino, cyano and diol) following three C_{18} columns in order to find the fastest and efficient chromatographic conditions. 54 urinary metabolites were monitored to test the performance of the columns and mobile phases (retention time, dead time, and peak height and shape). Overall, HILIC columns generate outperforms than the RPLC columns. Authors suggested the use of diol-HILIC columns for urinary metabolic profiling because of a higher number of metabolites detected and shorter analyses time [90].

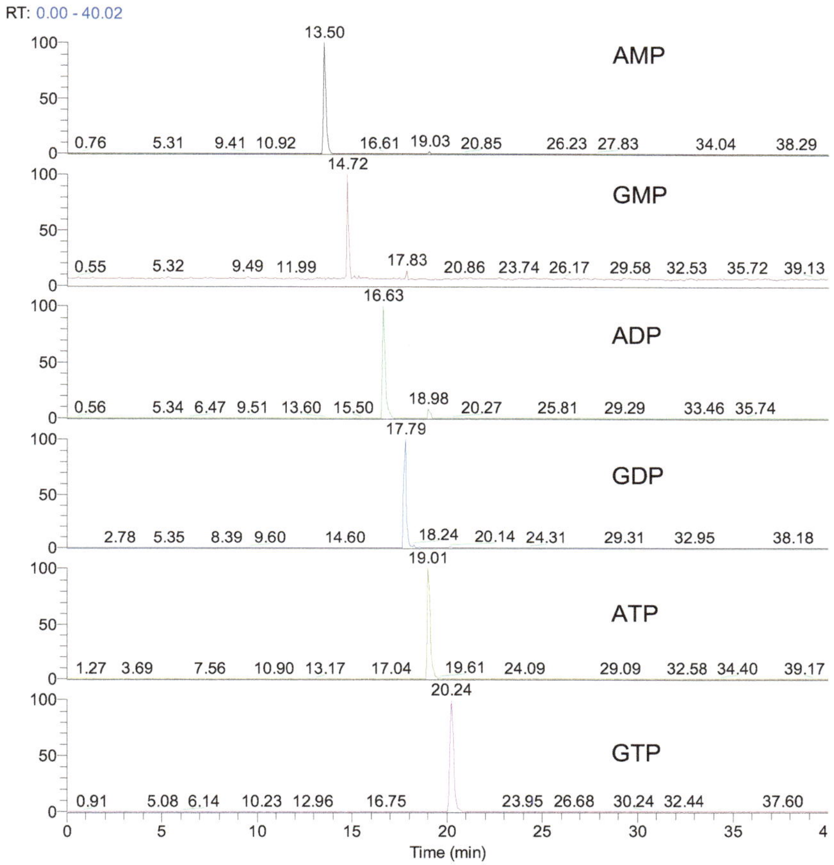

Fig. (4). Separation of adenosine and guanidine nucleotides in amino-based stationary phase. Chromatographic parameters: stationary phase: Luna NH_2 Column (100 x 2.1 mm, 3 µm); mobile phase: 5 mM ammonium carbonate (pH 9):acetonitrile, (80:20 v/v); flow rate: 0.3 mL/min; MS system: Thermo TSQ Vantage Triple Quad.

Mobile Phase Effects on Separation

Likewise the stationary phase, the mobile phase features (pH, ionic strength, organic solvent type *etc.*) are also effective in separation in terms of peak capacity, resolution, retentivity, peak shapes and sensitivity [88]. Initially, the organic solvent in mobile phase may require optimization, because it may have a significant effect on retention and selectivity. Then, the mobile phase pH and ionic strength must be optimized especially working with charged metabolites and/or stationary phases [83].

Organic Phase

The type of the organic solvent is one of the primary parameters which affects the retention in HILIC. Acetonitrile alone, acetonitrile/water mixture, methanol or isopropyl alcohol is the most preferred organic solvents in HILIC due to its moderate elution strength that gives acceptable retention times by avoiding formation of hydrogen bonds. The retention in HILIC is influenced strongly by the polarity of the mobile phase. The main component of a strong mobile phase is water, while acetonitrile is the main component of a weak mobile phase. Gradient elution in HILIC begins with a high ratio of organic solvent and ends up with a high ratio of aqueous solution. The elution strength of the organic solvents depends on the polarity (water > methanol > ethanol > acetonitrile > acetone) [78]. HILIC phases are, in general, vulnerable to fast gradients and require longer equilibration because of the aqueous layer on the stationary phase, which depends on the composition of the mobile phase. Therefore, HILIC gradient must be carried out in 5% to 95% of aqueous phase to ensure a sufficient hydration of the stationary phase.

Periat and coworkers (2013) observed minor differences in selectivity replacing the pure acetonitrile with acetonitrile/methanol or acetonitrile/isopropanol mixtures. They found that the retention order was not changed but the retention was lower with isopropanol and higher with methanol. Furthermore, the peaks were significantly broadening in the presence of isopropanol and methanol compared to pure acetonitrile [91].

Mata *et al.* (2016) used methylphosphonic acid in the mobile phase in order to prevent the interaction between metabolites and metallic surface of the chromatographic system. This improved the chromatographic parameters like reducing peak tailing and increasing resolution between sugar phosphates [67].

pH

Although some alterations have been made in selectivity by changing the nature of organic solvent, the outcome is less noticeable than changing the functional groups of the stationary phase or the mobile phase pH. Especially working with ionizable compounds, since the retention of them is strongly influenced by pH of the mobile phase [19, 91]. Furthermore, another critical parameter for determining the pH of mobile phase is the column pH working range because; this will affect the column lifetime. The pH range of the mobile phase is between 2 and 8 for silica based stationary phases while between 2 and 10 for with polymeric stationary phases [78].

One of the major challenges in HILIC results from imprecise pH measurement when a large proportion of organic modifier is used in mobile phase. This not only affects the precise measurement of mobile phase pH but also a analyte pK_a and silanols pK_a of stationary phase [20, 91]. pH value of the mobile phases generally considers in aqueous buffers. The concentrated buffer is diluted with organic solvent resulting in a desired buffer and organic solvent concentration ratio with a less acidic or basic apparent pH [20].

The pH may also subsidize to the polarity of the metabolite and affect the differential behavior in retention times. For example, the retention of saccharides on HILIC column was not significantly affected by the pH of mobile phase while amides showed increased retention in neutral pH conditions. Moreover, the pH must be adjusted to provide that the analyte will be in a single ionic form [20].

Cifkova *et al.* (2016) have optimized an HILIC method for the analysis of lipids in porcine brain and kidney samples. Nine HILIC stationary phases (unmodified silica, modified silica using diol, 2- picolylamine, diethylamine and 1-aminoanthracene and hydride silica) were compared for the separation of 18 lipids including (lyso)phosphatidylserines, (lyso)phosphatidic acids and other lipids. Besides column types, the effect of pH of the mobile phase and, the type and concentration of mobile phase additives on chromatographic behavior of lipids, were also investigated. The column type was found to be one of the most important parameters on the separation, in addition to column type, pH of the mobile phases is another important parameter affecting separation, since it is determining not only the ionic ratio of the species from acidic lipids but also free silanol groups on silica based stationary phase. For that reason, ionization degree of lipids, especially those for acidic lipids, is the most affected parameter that mostly influence the peak tailing as the ionic species from acidic lipids compete for protons derived from silanol groups on the surface of silica stationary phase. To avoid competition between analyte and stationary phase, in the study, the hydride stationary phase in which free silanol (Si-OH) groups were replaced with silicon-hydride (Si-H) bonds, were tested and a better separation was obtained for acidic lipids than the other columns used in the study [92].

Tufi *et al.* (2015) tested six different HILIC stationary phases (silica, amino, amide and two zwitterionic stationary phases) and two different buffer salts (ammonium formate and ammonium acetate) for the analysis of 20 endogenous neurotransmitters. They evaluated different chromatographic parameters and found zwitterionic phase (ZIC-cHILIC) column as the best for the separation of these metabolites. Moreover, they observed that there was no significant difference on the separation of neurotransmitters due to buffer type [52]. On the contrary, Kumar *et al.* (2013) have shown that an increase in the buffer concentration led to a decreased retention of strongly basic and quaternary compounds on different HILIC columns [93].

Based on the Periat *et al.* studies (2013), the peak asymmetry and broadening can be altered by changing pH. The effect of pH was tested at four different pH and pH 6 was found to be the most suitable condition in the terms of peak shape and selectivity, while pH 3 was the worst. However the influence of pH on peak shape and retention primarily depends on the ionization degree of the metabolites [91]. Similarly, Paglia and coworkers (2012) showed that acidic mobile phases in HILIC could be used to separate polar basic compounds such as arginine, lysine, ornithine, and cysteine. The retention times of these basic metabolites are pH dependent and the increase of pH in mobile phase leads to longer retention on stationary phase resulting in peak tailing although the effect on purines and nucleosides was less significant [19].

Sampsonidis *et al.* (2015) tested three different stationary phases (an amide, a diol and a zwitterionic phase) and two different pH values (pH 4 and 8.5) for more than 75 metabolites [88]. Data were evaluated using the number of detected compounds, retentivity, resolved critical pairs and chromatographic parameters (peak width, resolution and tailing factor) with a Derringer functions used to calculate overall desirability. Although the columns show resemblances, diol column was able to detect more compounds than the other ones. Favorable chromatographic peak shapes were obtained in zwitterionic column at basic conditions whereas in amide column at acidic conditions [88]. The pH of mobile phase has no effect on amino acids analyses, while carbohydrates elution is highly dependent on pH and column used for separation. At basic pH, carbohydrates and organic acids were retained more in zwitterion column than diol column. When the columns were compared according to peak shapes, zwitterionic phase appears to provide better peak shapes in general than diol and amide columns, but this depends on the targeted class of metabolites. Moreover, the zwitterionic and amide columns show more uniform distribution across the different metabolite groups, which is preferred for MS analyses to prevent ion suppression. Shortly, the performance of zwitterionic and amide columns may vary with mobile phase pH and ionic strength. Therefore, these parameters must be adjusted in accordance to physicochemical characteristic of targeted metabolites. The same authors have suggested to run amide column under basic conditions, since it has a capability of simultaneously analyzing higher number of metabolites [88]. Similarly in the other studies from the same group, they found amide column as the best separation tool for over 100 metabolites in *in vitro* fertilization culture [51].

Ionic Strength

The buffer ionic strength potentially affects the retention and selectivity, especially analyzing ionizable metabolites, since it is affecting adsorption, ion exchange and hydrophilic partitioning [91]. The bare silica and diol phases are most sensitive stationary phases to ionic strength, due to the strongest cation exchange interaction mechanism between free silanol groups of stationary phase and ionic metabolites. The ionic strength lowers the distance between column materials and metabolites due to shielding of electrostatic interaction. For that reasons, ionic strength can be considered a secondary parameter depending on the stationary phase type [91]. Buffer salt concentrations between 2 and 20 mM are recommended for organic content not up to 90%. Higher salts concentrations are not recommended due to insolubility in organic solutions and for impair problems with MS system [91].

Buffer salts compatible with MS, such as formic acid, ammonium acetate and ammonium formate, are typically used to control the mobile phase pH and ionic strength. Ionic strength is effective when analyzing ionic metabolites and ion exchanges control the retention. The solvation of the metabolites must be considered otherwise an asymmetric peak shape, and/or poor recovery from the stationary phase will be observed. On the other hand, there is no buffer needed to separate neutral polar metabolites [78].

Periat and coworkers (2013) experimentally showed which parameters affect the HILIC separation with 82 compounds with different polarity including acidic, basic and neutral properties. Five different columns (bare silica, hybrid silica, amide, diol and zwitterionic), four different pH (3, 4, 5 and 6), two ionic strengths (10 and 50 mM) and several organic phase modifiers (acetonitrile, methanol and isopropanol) were tested. The stationary phase and pH were primary parameters that affect the selectivity and retention time in HILIC, while ionic strength and organic phase modifiers could be considered as secondary parameters for HILIC method development [91]. Therefore, ionic additives such as formic acid, acetic acid, ammonium formate, ammonium acetate and ammonium hydroxide are used to control pH and ionic strength of the mobile phases [78].

Detectors

HILIC techniques compatible with many HPLC detectors such as ultraviolet, fluorescence, refractive index, evaporative light scattering, charged aerosol and MS. Among these detectors, MS is extensively used in the analysis of polar compounds, because it; i) is able to detect and identify a wide range of metabolites; ii) is compatible with the aqueous organic mobile phase; iii) can be operated in positive and negative ionization mode; iv) is selective, v) is speed and

vi) easy to use. Therefore HILIC-MS combined systems have wide application areas for comprehensive metabolomic analyses in different matrices (urine, serum, cerebrospinal fluid, amniotic fluid tissues, cells *etc.*) [14, 15, 78, 94, 95].

In targeted metabolomics studies, triple quadrupole system is predominantly used because of its higher sensitivity. Recently new analytical MS systems also become available for metabolomics analyses. Among them, the ion mobility spectrometry (IMS) was an extremely sensitive stand-alone device but are often coupled with mass spectrometry, gas chromatography or high-performance liquid chromatography. It is able to separate and identify ionized molecules in the gas phase based on their mobility in a carrier buffer gas. This unique system is orthogonal to LC and MS separations and allows better isomer separation. Beach *et al.* (2015) used the differential ion mobility spectrometry to separate β- N-methylamino-L-alanine from its isomers [66].

Sample Preparation

The sample preparation aims to maximize the number of metabolites from different matrixes without a prior knowledge of the metabolites present in a sample as well as to reduce metabolites degradation and ex-vivo alteration of sample composition. Sample preparation has a critical effect on experimental reproducibility. To minimize variation due to the sample preparation, a standardized methods must be used as well as an appropriate method to block the enzymatic activity.

Targeted metabolomic studies are focused on selective extraction in order to remove interfering matrix components and on enrichment of samples in order to increase the LOQ values. Therefore, targeted metabolomic studies include protein precipitation often followed by liquid-liquid extraction and/or solid-phase extraction in order to reach cleaner samples [96]. Especially lipid removal is a crucial step for MS based metabolomic studies, since lipids may interfere with MS and lead to suppression during analysis. Therefore, the combination of protein and lipid removal is often used to improve both metabolite coverage and reproducibility [96].

HILIC based metabolomic analyses were performed in different biological human or animal matrixes such as blood, plasma, serum, urine, tissue other than in plant and cell lines. The protocols are mostly based on protein precipitation with organic solvent or acid, evaporation in order to concentrate the metabolites and reconstitution with mobile phase in order to use an analysis-compatible solvent. For cell lines, the sample preparation is generally performed in three steps: quenching, harvesting and extracting of metabolites [96].

Special extraction methods for metabolomic analysis have been developed for different biological samples to avoid the matrix effect. The matrix effect may occur due to solubility of interfering substances in extraction solvent, total metabolites content of sample and co-precipitation with proteins. In order to minimize the matrix effect, special attention must be given during the optimization of sample preparation, which is very critical for reproducibility.

In targeted metabolomic analyses, single step organic solvent-based protein precipitation method is a preferred technique for sample preparation (Fig. **5**). Acetonitrile is widely used for sample preparation because of its high efficacy on protein removal [96]. Only 16% of studies have used two steps extraction, generally a protein precipitation step followed by liquid-liquid extraction or solid phase extraction (SPE). About 10% of the studies have used SPE for sample preparation. However, it requires smaller volumes and provides cleaner extracts than liquid-liquid extraction and protein precipitation methods. The other advantage of SPE is minimal ion suppression in HILIC-MS system [38].

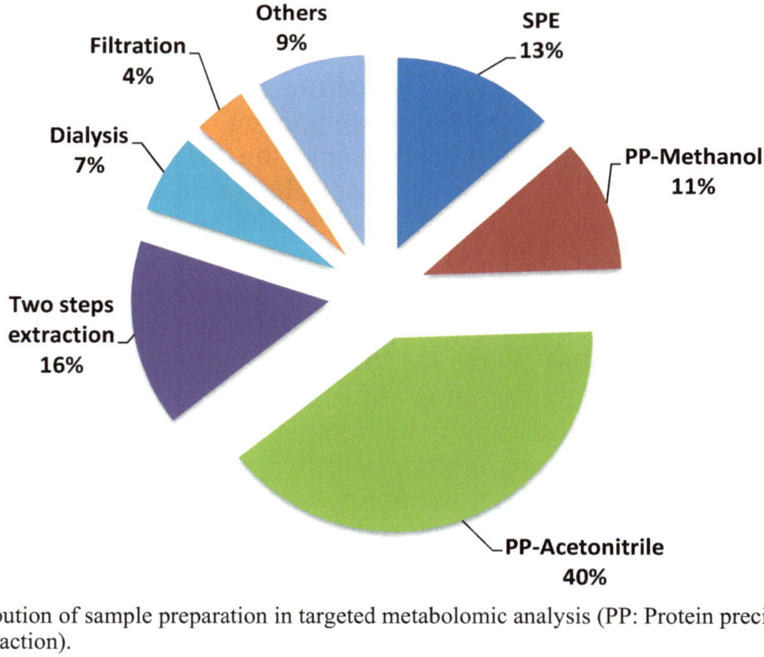

Fig. (5). Distribution of sample preparation in targeted metabolomic analysis (PP: Protein precipitation, SPE: Solid phase extraction).

Different organic phases or mixtures can be used for protein precipitation. Acetonitrile, methanol, ethanol, acetone or their mixture are frequently used for targeted metabolomic studies. Selection of organic solvent for protein precipitation must be evaluated based on the recovery, matrix effect and cost [96].

Recovery of different metabolite classes was altered by the nature of the solvent or solvent mixtures as well as solvent/sample ratio used. Paglia *et al.* (2012) have investigated the effect of extraction solvents on cells culture samples. For this purpose, the authors used seven different protocols including methanol:water (7:3) neutral, methanol:water acidic (pH 2), methanol:water basic (pH 10), hot methanol (80°C), methanol:acetonitrile:water (4:4:2), methanol:chloroform (2:1) and methanol+water two-step extraction. The best recovery was achieved with methanol:water (7:3) neutral extraction; however, this protocol did not provide a complete recovery of all metabolites. Therefore, for complete recovery, different extraction methods and orthogonal analytical platforms were suggested [97].

Sriboonvorakul *et al.* (2013) have developed a 96-well SPE system in order to analyze eight organic acids in plasma and urine samples. The high recovery (94.6–99.8%) and lack of matrix effect (−3.7 to 3.4%) confirmed that the SPE is an excellent method for targeted sample preparation method in both urine and plasma samples [38].

Development Steps of HILIC Method

Understanding the physico-chemical properties of a metabolite and column is a key parameter for successful method development. A combination of partition coefficient (log P), distribution coefficient (log D) and acid disassociation constant (pKa) data will guide the stationary phase selection process. The Log P represents a very valuable information to understand if a metabolite is suitable for HILC separation. If log P or log D is near to zero or even below, the metabolite may be retained by HILIC stationary phase. The pKa gives information about the charge of the metabolite at certain pH of the mobile phase. These parameters are important factors while selecting the right stationary phase, which is a very critical step for successful separation. After stationary phase selection, the most appropriate mobile phase starting conditions, must be selected. The pKa of the analytes and the stationary phase type are critical parameters to decide the initial conditions. Generally, a mobile phase consists of 80% acetonitrile in water or buffer, may be a good starting point. Special caution must be given during the buffer choose for mobile phase. The buffer must be compatible with MS system and also with the column working pH range. Before running initial experiments, the MS parameters such ion spray voltage, temperature, ionization mode *etc.* have to optimized with standard metabolites. Based on the initial experimental results, the mobile phase (type of buffer, pH and ionic strength) and/or stationary phase may require modifications in order to reach a good separation. Therefore, a systematic approach is suggested for HILIC method development (Fig. **6**).

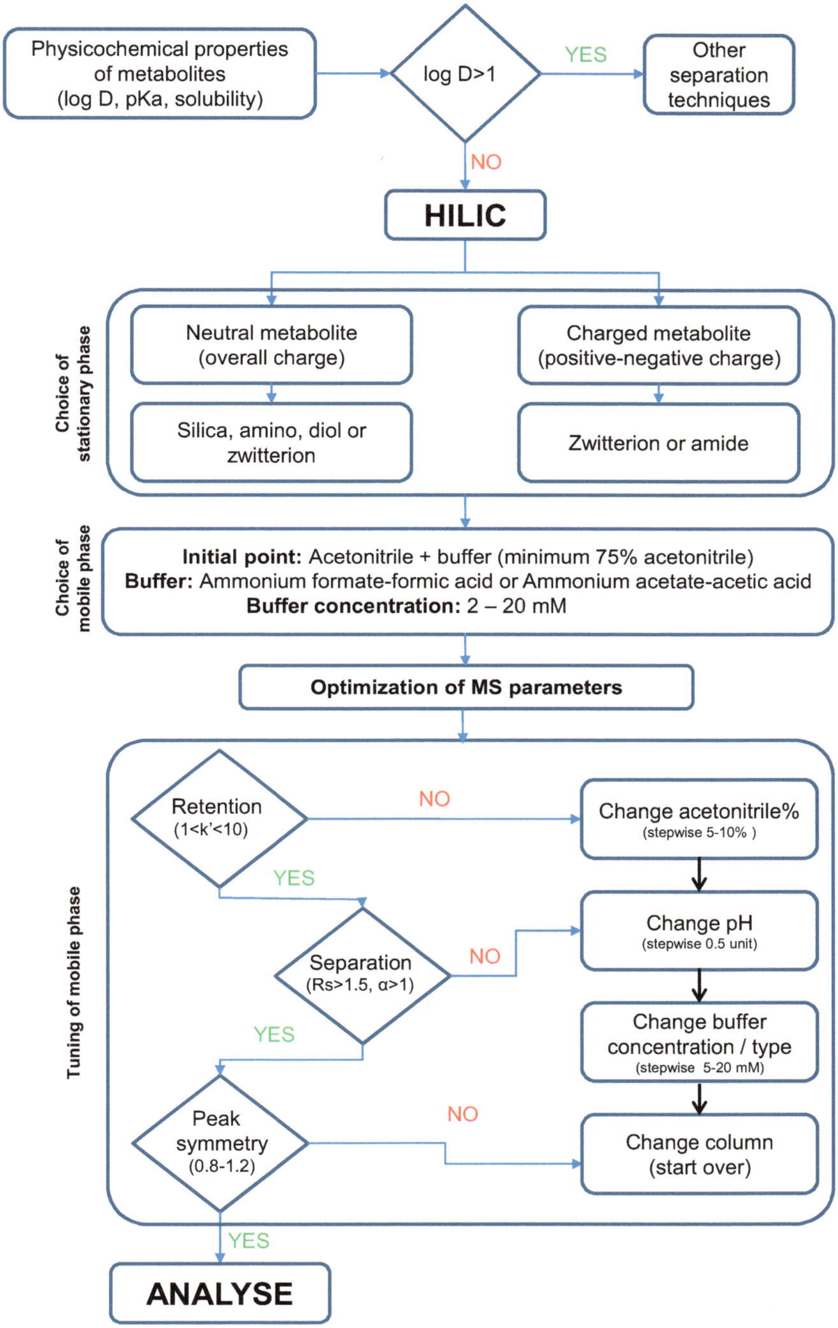

Fig. (6). Flow chart for HILIC-MS method development steps.

CONCLUSION

Results of metabolomic studies provide physicians with the benefit of obtaining detailed information at earlier stages of the diseases. Consequently, physicians are able to diagnose diseases and make arrangements for customized treatments earlier. Research process starts by determining the biomarker candidates *via* untargeted metabolite analysis, completing the targeted analysis with determination of the candidate biomarker quantities during disease, validating the results and identifying potential biomarkers. Application of HILIC tandemMS is the preferred method to separate polar phase metabolites from bio-fluid, cells, tissues, *etc*. HILIC-MS is the most preferred method of analysis considering its sensitivity and retentivity for polar metabolites. Therefore, optimization of samples preparation, chromatographic conditions and MS parameters during HILIC-MS based metabolomic studies, might lead to quantitative analysis of increased number of metabolites.

CONSENT FOR PUBLICATION

Not applicable.

CONFLICT OF INTEREST

The authors declares no conflict of interest, financial or otherwise.

ACKNOWLEDGEMENTS

Declared none.

REFERENCES

[1] Nemutlu, E.; Zhang, S.; Xu, Y.Z.; Terzic, A.; Zhong, L.; Dzeja, P.D.; Cha, Y.M. Cardiac resynchronization therapy induces adaptive metabolic transitions in the metabolomic profile of heart failure. *J. Card. Fail.,* **2015**, *21*(6), 460-469. [http://dx.doi.org/10.1016/j.cardfail.2015.04.005] [PMID: 25911126]

[2] Nemutlu, E.; Zhang, S.; Juranic, N.O.; Terzic, A.; Macura, S.; Dzeja, P. 18O-assisted dynamic metabolomics for individualized diagnostics and treatment of human diseases. *Croat. Med. J.,* **2012**, *53*(6), 529-534. [http://dx.doi.org/10.3325/cmj.2012.53.529] [PMID: 23275318]

[3] Nemutlu, E.; Zhang, S.; Gupta, A.; Juranic, N.O.; Macura, S.I.; Terzic, A.; Jahangir, A.; Dzeja, P. Dynamic phosphometabolomic profiling of human tissues and transgenic models by 18O-assisted ^{31}P NMR and mass spectrometry. *Physiol. Genomics,* **2012**, *44*(7), 386-402. [http://dx.doi.org/10.1152/physiolgenomics.00152.2011] [PMID: 22234996]

[4] Dzeja, P. P. Rearrangement of energetic and substrate utilization networks compensate for chronic myocardial creatine kinase deficiency. *J Physiol,* **2011**, *589*(Pt 21), 5193-5211. [http://dx.doi.org/10.1113/jphysiol.2011.212829]

[5] Zhou, B.; Xiao, J.F.; Tuli, L.; Ressom, H.W. LC-MS-based metabolomics. *Mol. Biosyst.,* **2012**, *8*(2), 470-481.

[http://dx.doi.org/10.1039/C1MB05350G] [PMID: 22041788]

[6] Mangalam, A.; Poisson, L.; Nemutlu, E.; Datta, I.; Denic, A.; Dzeja, P.; Rodriguez, M.; Rattan, R.; Giri, S. Profile of Circulatory Metabolites in a Relapsing-remitting Animal Model of Multiple Sclerosis using Global Metabolomics. *J. Clin. Cell. Immunol.,* **2013**, *4* [PMID: 24273690]

[7] Beckonert, O.; Keun, H.C.; Ebbels, T.M.; Bundy, J.; Holmes, E.; Lindon, J.C.; Nicholson, J.K. Metabolic profiling, metabolomic and metabonomic procedures for NMR spectroscopy of urine, plasma, serum and tissue extracts. *Nat. Protoc.,* **2007**, *2*(11), 2692-2703. [in English]. [http://dx.doi.org/10.1038/nprot.2007.376] [PMID: 18007604]

[8] Brown, M.; Dunn, W.B.; Dobson, P.; Patel, Y.; Winder, C.L.; Francis-McIntyre, S.; Begley, P.; Carroll, K.; Broadhurst, D.; Tseng, A.; Swainston, N.; Spasic, I.; Goodacre, R.; Kell, D.B. Mass spectrometry tools and metabolite-specific databases for molecular identification in metabolomics. *Analyst (Lond.),* **2009**, *134*(7), 1322-1332. [in English]. [http://dx.doi.org/10.1039/b901179j] [PMID: 19562197]

[9] Kaddurah-Daouk, R.; Kristal, B.S.; Weinshilboum, R.M. Metabolomics: a global biochemical approach to drug response and disease. *Annu. Rev. Pharmacol. Toxicol.,* **2008**, *48*, 653-683. [in English]. [http://dx.doi.org/10.1146/annurev.pharmtox.48.113006.094715] [PMID: 18184107]

[10] Aura, A-M.; Mattila, I.; Seppänen-Laakso, T.; Miettinen, J.; Oksman-Caldentey, K-M.; Orešič, M. Microbial metabolism of catechin stereoisomers by human faecal microbiota: comparison of targeted analysis and a non-targeted metabolomics method. *Phytochem. Lett.,* **2008**, *1*(1), 18-22. [http://dx.doi.org/10.1016/j.phytol.2007.12.001]

[11] Lenz, E.M.; Wilson, I.D. Analytical strategies in metabonomics. *J. Proteome Res.,* **2007**, *6*(2), 443-458. [in English]. [http://dx.doi.org/10.1021/pr0605217] [PMID: 17269702]

[12] Lindon, J.C.; Nicholson, J.K. Spectroscopic and statistical techniques for information recovery in metabonomics and metabolomics. *Annu. Rev. Anal. Chem. (Palo Alto, Calif.),* **2008**, *1*, 45-69. [http://dx.doi.org/10.1146/annurev.anchem.1.031207.113026] [PMID: 20636074]

[13] Claudino, W.M.; Quattrone, A.; Biganzoli, L.; Pestrin, M.; Bertini, I.; Di Leo, A. Metabolomics: available results, current research projects in breast cancer, and future applications. *J. Clin. Oncol.,* **2007**, *25*(19), 2840-2846. [http://dx.doi.org/10.1200/JCO.2006.09.7550] [PMID: 17502626]

[14] Tang, D.Q.; Zou, L.; Yin, X.X.; Ong, C.N. HILIC-MS for metabolomics: An attractive and complementary approach to RPLC-MS. *Mass Spectrom. Rev.,* **2016**, *35*(5), 574-600. [http://dx.doi.org/10.1002/mas.21445] [PMID: 25284160]

[15] Ikegami, T.; Tomomatsu, K.; Takubo, H.; Horie, K.; Tanaka, N. Separation efficiencies in hydrophilic interaction chromatography. *J. Chromatogr. A,* **2008**, *1184*(1-2), 474-503. [http://dx.doi.org/10.1016/j.chroma.2008.01.075] [PMID: 18294645]

[16] Noga, S.; Bocian, S.; Buszewski, B. Hydrophilic interaction liquid chromatography columns classification by effect of solvation and chemometric methods. *J. Chromatogr. A,* **2013**, *1278*, 89-97. [http://dx.doi.org/10.1016/j.chroma.2012.12.077] [PMID: 23351397]

[17] Trivedi, D.K.; Jones, H.; Shah, A.J.; Iles, R.K. Development of Zwitterionic Hydrophilic Liquid Chromatography (ZIC-HILIC-MS) metabolomics method for Shotgun analysis of human urine. *J. Chromatogr. Sep. Tech.,* **2012**, *3*(06) [http://dx.doi.org/10.4172/2157-7064.1000144]

[18] Sun, T.; Wetzel, S.J.; Johnson, M.E.; Surlow, B.A.; Patton-Vogt, J. Development and validation of a hydrophilic interaction liquid chromatography-tandem mass spectrometry method for the quantification of lipid-related extracellular metabolites in Saccharomyces cerevisiae. *J. Chromatogr. B Analyt. Technol. Biomed. Life Sci.,* **2012**, *897*, 1-9.

[http://dx.doi.org/10.1016/j.jchromb.2012.03.034] [PMID: 22541168]

[19] Paglia, G.; Hrafnsdóttir, S.; Magnúsdóttir, M.; Fleming, R.M.; Thorlacius, S.; Palsson, B.Ø.; Thiele, I. Monitoring metabolites consumption and secretion in cultured cells using ultra-performance liquid chromatography quadrupole-time of flight mass spectrometry (UPLC-Q-ToF-MS). *Anal. Bioanal. Chem.,* **2012**, *402*(3), 1183-1198. [http://dx.doi.org/10.1007/s00216-011-5556-4] [PMID: 22159369]

[20] Gika, H.G.; Theodoridis, G.A.; Vrhovsek, U.; Mattivi, F. Quantitative profiling of polar primary metabolites using hydrophilic interaction ultrahigh performance liquid chromatography-tandem mass spectrometry. *J. Chromatogr. A,* **2012**, *1259*, 121-127. [http://dx.doi.org/10.1016/j.chroma.2012.02.010] [PMID: 22381888]

[21] Li, J.; von Pföstl, V.; Zaldivar, D.; Zhang, X.; Logothetis, N.; Rauch, A. Measuring multiple neurochemicals and related metabolites in blood and brain of the rhesus monkey by using dual microdialysis sampling and capillary hydrophilic interaction chromatography-mass spectrometry. *Anal. Bioanal. Chem.,* **2012**, *402*(8), 2545-2554. [http://dx.doi.org/10.1007/s00216-011-5427-z] [PMID: 21956265]

[22] Jaisson, S.; Gorisse, L.; Pietrement, C.; Gillery, P. Quantification of plasma homocitrulline using hydrophilic interaction liquid chromatography (HILIC) coupled to tandem mass spectrometry. *Anal. Bioanal. Chem.,* **2012**, *402*(4), 1635-1641. [http://dx.doi.org/10.1007/s00216-011-5619-6] [PMID: 22160237]

[23] Joo, K.M.; Han, J.Y.; Son, E.D.; Nam, G.W.; Chung, H.Y.; Jeong, H.J.; Cho, J.C.; Lim, K.M. Rapid, simultaneous and nanomolar determination of pyroglutamic acid and cis-/trans-urocanic acid in human stratum corneum by hydrophilic interaction liquid chromatography (HILIC)-electrospray ionization tandem mass spectrometry. *J. Chromatogr. B Analyt. Technol. Biomed. Life Sci.,* **2012**, *897*, 55-63. [http://dx.doi.org/10.1016/j.jchromb.2012.04.008] [PMID: 22560249]

[24] Xu, Y.; Yang, L.; Yang, F.; Xiong, Y.; Wang, Z.; Hu, Z. Metabolic profiling of fifteen amino acids in serum of chemical-induced liver injured rats by hydrophilic interaction liquid chromatography coupled with tandem mass spectrometry. *Metabolomics,* **2012**, *8*(3), 475-483. [http://dx.doi.org/10.1007/s11306-011-0333-0]

[25] Zhu, C.; Dane, A.; Spijksma, G.; Wang, M.; van der Greef, J.; Luo, G.; Hankemeier, T.; Vreeken, R.J. An efficient hydrophilic interaction liquid chromatography separation of 7 phospholipid classes based on a diol column. *J. Chromatogr. A,* **2012**, *1220*, 26-34. [http://dx.doi.org/10.1016/j.chroma.2011.11.034] [PMID: 22169191]

[26] Hosozumi, C.; Toriba, A.; Chuesaard, T.; Kameda, T.; Tang, N.; Hayakawa, K. Analysis of 8-hydroxy-2′-deoxyguanosine in human urine using hydrophilic interaction chromatography with tandem mass spectrometry. *J. Chromatogr. B Analyt. Technol. Biomed. Life Sci.,* **2012**, *893-894*, 173-176. [http://dx.doi.org/10.1016/j.jchromb.2012.02.043] [PMID: 22445309]

[27] Nelson, D.; Xu, N.; Carlson, J. Semi-automated quantification of methylmalonic acid in human serum by LC-MS/MS. *Scand. J. Clin. Lab. Invest.,* **2012**, *72*(6), 441-446. [http://dx.doi.org/10.3109/00365513.2012.679963] [PMID: 22935047]

[28] Huang, Y.; Tian, Y.; Zhang, Z.; Peng, C. A HILIC-MS/MS method for the simultaneous determination of seven organic acids in rat urine as biomarkers of exposure to realgar. *J. Chromatogr. B Analyt. Technol. Biomed. Life Sci.,* **2012**, *905*, 37-42. [http://dx.doi.org/10.1016/j.jchromb.2012.07.038] [PMID: 22902916]

[29] Miller, J.H., IV; Poston, P.A.; Karnes, H.T. A quantitative method for acylcarnitines and amino acids using high resolution chromatography and tandem mass spectrometry in newborn screening dried blood spot analysis. *J. Chromatogr. B Analyt. Technol. Biomed. Life Sci.,* **2012**, *903*, 142-149. [http://dx.doi.org/10.1016/j.jchromb.2012.07.008] [PMID: 22846691]

[30] Neubauer, S.; Haberhauer-Troyer, C.; Klavins, K.; Russmayer, H.; Steiger, M.G.; Gasser, B.; Sauer,

M.; Mattanovich, D.; Hann, S.; Koellensperger, G. U13C cell extract of Pichia pastoris--a powerful tool for evaluation of sample preparation in metabolomics. *J. Sep. Sci.,* **2012**, *35*(22), 3091-3105. [http://dx.doi.org/10.1002/jssc.201200447] [PMID: 23086617]

[31] Bathena, S.P.; Huang, J.; Epstein, A.A.; Gendelman, H.E.; Boska, M.D.; Alnouti, Y. Rapid and reliable quantitation of amino acids and myo-inositol in mouse brain by high performance liquid chromatography and tandem mass spectrometry. *J. Chromatogr. B Analyt. Technol. Biomed. Life Sci.,* **2012**, *893-894*, 15-20. [http://dx.doi.org/10.1016/j.jchromb.2012.01.035] [PMID: 22425384]

[32] Boudra, H.; Doreau, M.; Noziere, P.; Pujos-Guillot, E.; Morgavi, D.P. Simultaneous analysis of the main markers of nitrogen status in dairy cow's urine using hydrophilic interaction chromatography and tandem mass spectrometry detection. *J. Chromatogr. A,* **2012**, *1256*, 169-176. [http://dx.doi.org/10.1016/j.chroma.2012.07.094] [PMID: 22897860]

[33] Turner, R.; Stamp, L.K.; Kettle, A.J. Detection of allantoin in clinical samples using hydrophilic liquid chromatography with stable isotope dilution negative ion tandem mass spectrometry. *J. Chromatogr. B Analyt. Technol. Biomed. Life Sci.,* **2012**, *891-892*, 85-89. [http://dx.doi.org/10.1016/j.jchromb.2012.02.009] [PMID: 22398212]

[34] Gopu, C.L.; Hari, P.R.; George, R.; Harikrishnan, S.; Sreenivasan, K. Simultaneous determination of homocysteine and asymmetric dimethylarginine in human urine by liquid chromatography-tandem mass spectrometry. *J. Chromatogr. B Analyt. Technol. Biomed. Life Sci.,* **2013**, *939*, 32-37. [http://dx.doi.org/10.1016/j.jchromb.2013.09.010] [PMID: 24095873]

[35] Zhao, H.Q.; Wang, X.; Li, H.M.; Yang, B.; Yang, H.J.; Huang, L. Characterization of nucleosides and nucleobases in natural Cordyceps by HILIC-ESI/TOF/MS and HILIC-ESI/MS. *Molecules,* **2013**, *18*(8), 9755-9769. [http://dx.doi.org/10.3390/molecules18089755] [PMID: 23955321]

[36] Guo, S.; Duan, J.A.; Qian, D.; Wang, H.; Tang, Y.; Qian, Y.; Wu, D.; Su, S.; Shang, E. Hydrophilic interaction ultra-high performance liquid chromatography coupled with triple quadrupole mass spectrometry for determination of nucleotides, nucleosides and nucleobases in Ziziphus plants. *J. Chromatogr. A,* **2013**, *1301*, 147-155. [http://dx.doi.org/10.1016/j.chroma.2013.05.074] [PMID: 23800804]

[37] Martens-Lobenhoffer, J.; Surdacki, A.; Bode-Böger, S.M. Fast and Precise Quantification of l-Homoarginine in Human Plasma by HILIC-Isotope Dilution-MS–MS. *Chromatographia,* **2013**, *76*(23-24), 1755-1759. [http://dx.doi.org/10.1007/s10337-013-2561-0]

[38] Sriboonvorakul, N.; Leepipatpiboon, N.; Dondorp, A.M.; Pouplin, T.; White, N.J.; Tarning, J.; Lindegardh, N. Liquid chromatographic-mass spectrometric method for simultaneous determination of small organic acids potentially contributing to acidosis in severe malaria. *J. Chromatogr. B Analyt. Technol. Biomed. Life Sci.,* **2013**, *941*, 116-122. [http://dx.doi.org/10.1016/j.jchromb.2013.10.005] [PMID: 24200840]

[39] Isaguirre, A.C.; Olsina, R.A.; Martinez, L.D.; Lapierre, A.V.; Cerutti, S. Rapid and sensitive HILIC-MS/MS analysis of carnitine and acetylcarnitine in biological fluids. *Anal. Bioanal. Chem.,* **2013**, *405*(23), 7397-7404. [http://dx.doi.org/10.1007/s00216-013-7193-6] [PMID: 23877176]

[40] Peng, M.; Liu, L.; Jiang, M.; Liang, C.; Zhao, X.; Cai, Y.; Sheng, H.; Ou, Z.; Luo, H. Measurement of free carnitine and acylcarnitines in plasma by HILIC-ESI-MS/MS without derivatization. *J. Chromatogr. B Analyt. Technol. Biomed. Life Sci.,* **2013**, *932*, 12-18. [http://dx.doi.org/10.1016/j.jchromb.2013.05.028] [PMID: 23816563]

[41] Sørensen, L.K.; Rittig, N.F.; Holmquist, E.F.; Jørgensen, K.A.; Jørgensen, J.O.; Møller, N.; Johannsen, M. Simultaneous determination of β-hydroxybutyrate and β-hydroxy-β-methylbutyrate in human whole blood using hydrophilic interaction liquid chromatography electrospray tandem mass spectrometry. *Clin. Biochem.,* **2013**, *46*(18), 1877-1883.

[http://dx.doi.org/10.1016/j.clinbiochem.2013.08.011] [PMID: 23994603]

[42] Müller, D.C.; Degen, C.; Scherer, G.; Jahreis, G.; Niessner, R.; Scherer, M. Metabolomics using GC-TOF-MS followed by subsequent GC-FID and HILIC-MS/MS analysis revealed significantly altered fatty acid and phospholipid species profiles in plasma of smokers. *J. Chromatogr. B Analyt. Technol. Biomed. Life Sci.,* **2014**, *966*, 117-126.
[http://dx.doi.org/10.1016/j.jchromb.2014.02.044] [PMID: 24630914]

[43] Zhang, G.; Walker, A.D.; Lin, Z.; Han, X.; Blatnik, M.; Steenwyk, R.C.; Groeber, E.A. Strategies for quantitation of endogenous adenine nucleotides in human plasma using novel ion-pair hydrophilic interaction chromatography coupled with tandem mass spectrometry. *J. Chromatogr. A,* **2014**, *1325*, 129-136.
[http://dx.doi.org/10.1016/j.chroma.2013.12.017] [PMID: 24377733]

[44] Arora, B.; Narayanasamy, A.; Nirmal, J.; Halder, N.; Patnaik, S.; Ravi, A.K.; Velpandian, T. Development and validation of a LC-MS/MS method for homocysteine thiolactone in plasma and evaluation of its stability in plasma samples. *J. Chromatogr. B Analyt. Technol. Biomed. Life Sci.,* **2014**, *944*, 49-54.
[http://dx.doi.org/10.1016/j.jchromb.2013.11.011] [PMID: 24291720]

[45] Combes, A.; El Abdellaoui, S.; Vial, J.; Lagrange, E.; Pichon, V. Development of an analytical procedure for quantifying the underivatized neurotoxin β-N-methylamino-L-alanine in brain tissues. *Anal. Bioanal. Chem.,* **2014**, *406*(19), 4627-4636.
[http://dx.doi.org/10.1007/s00216-014-7872-y] [PMID: 24858470]

[46] Görgens, C.; Guddat, S.; Schänzer, W.; Thevis, M. Screening and confirmation of myo-inositol trispyrophosphate (ITPP) in human urine by hydrophilic interaction liquid chromatography high resolution / high accuracy mass spectrometry for doping control purposes. *Drug Test. Anal.,* **2014**, *6*(11-12), 1102-1107.
[http://dx.doi.org/10.1002/dta.1700] [PMID: 25070041]

[47] Tang, Y.B.; Sun, F.; Teng, L.; Li, W.B.; An, S.M.; Zhang, C.; Yang, X.J.; Lv, H.Y.; Ding, X.P.; Zhu, L.; Chen, H.Z. Simultaneous determination of the repertoire of classical neurotransmitters released from embryonal carcinoma stem cells using online microdialysis coupled with hydrophilic interaction chromatography-tandem mass spectrometry. *Anal. Chim. Acta,* **2014**, *849*, 70-79.
[http://dx.doi.org/10.1016/j.aca.2014.08.053] [PMID: 25300220]

[48] Liu, J.; Wang, L.; Hu, W.; Chen, X.; Zhong, D. Development of a UHPLC-MS/MS method for the determination of plasma histamine in various mammalian species. *J. Chromatogr. B Analyt. Technol. Biomed. Life Sci.,* **2014**, *971*, 35-42.
[http://dx.doi.org/10.1016/j.jchromb.2014.08.043] [PMID: 25264911]

[49] Klavins, K.; Drexler, H.; Hann, S.; Koellensperger, G. Quantitative metabolite profiling utilizing parallel column analysis for simultaneous reversed-phase and hydrophilic interaction liquid chromatography separations combined with tandem mass spectrometry. *Anal. Chem.,* **2014**, *86*(9), 4145-4150.
[http://dx.doi.org/10.1021/ac5003454] [PMID: 24678888]

[50] Van Damme, T.; Blancquaert, D.; Couturon, P.; Van Der Straeten, D.; Sandra, P.; Lynen, F. Wounding stress causes rapid increase in concentration of the naturally occurring 2′,3′-isomers of cyclic guanosine- and cyclic adenosine monophosphate (cGMP and cAMP) in plant tissues. *Phytochemistry,* **2014**, *103*, 59-66.
[http://dx.doi.org/10.1016/j.phytochem.2014.03.013] [PMID: 24735826]

[51] Virgiliou, C.; Sampsonidis, I.; Gika, H.G.; Raikos, N.; Theodoridis, G.A. Development and validation of a HILIC- MS/MS multi-targeted method for metabolomics applications. *Electrophoresis,* **2015**, (Jul), 14.
[PMID: 26180020]

[52] Tufi, S.; Lamoree, M.; de Boer, J.; Leonards, P. Simultaneous analysis of multiple neurotransmitters by hydrophilic interaction liquid chromatography coupled to tandem mass spectrometry. *J.*

Chromatogr. A, **2015**, *1395*, 79-87.
[http://dx.doi.org/10.1016/j.chroma.2015.03.056] [PMID: 25869798]

[53] Chimalakonda, K.C.; Pang, E.; Weaver, J.L.; Howard, K.E.; Patel, V.; Boyne, M.T., II Development and validation of a liquid-chromatography tandem mass spectrometry method to determine *in vitro* and *in vivo* histamine release. *J. Pharm. Biomed. Anal.,* **2015**, *102*, 494-499.
[http://dx.doi.org/10.1016/j.jpba.2014.10.016] [PMID: 25459949]

[54] Murakami, H.; Otani, E.; Iwata, T.; Esaka, Y.; Aoyama, T.; Kawasaki, M.; Tanaka, T.; Minatoguchi, S.; Uno, B. Simple Pretreatment and HILIC Separation for LC-ESI-MS/MS Determination of Adenosine in Human Plasma. *Anal. Sci.,* **2015**, *31*(11), 1189-1192.
[http://dx.doi.org/10.2116/analsci.31.1189] [PMID: 26561265]

[55] Holčapek, M.; Cífková, E.; Červená, B.; Lísa, M.; Vostálová, J.; Galuszka, J. Determination of nonpolar and polar lipid classes in human plasma, erythrocytes and plasma lipoprotein fractions using ultrahigh-performance liquid chromatography-mass spectrometry. *J. Chromatogr. A,* **2015**, *1377*, 85-91.
[http://dx.doi.org/10.1016/j.chroma.2014.12.023] [PMID: 25543301]

[56] Esch, C.; Hui, D.S.; Lee, R.; Edwards, J.L. Quantitation of thiol metabolites from mammalian cells using fluorous tagging and HILIC-MS. *Anal. Methods,* **2015**, *7*(17), 7164-7169.
[http://dx.doi.org/10.1039/C5AY00603A]

[57] Dandan Li, X.X.; Bai, Q.; Yang, W.; Zhao, R.; Zhang, A. Development and validation of an LC-MS/MS method for quantification of dipalmitoylphosphatidylcholine as a promising biomarker for renal failure in urine. *J. Chin. Pharm. Sci.,* **2015**, *24*(2), 73-79.

[58] Liu, R. Hydrophilic interaction ultra-high performance liquid chromatography coupled with triple-quadrupole mass spectrometry for determination of nucleosides and nucleobases in animal horns. *J. Liq. Chromatogr. Relat. Technol.,* **2015**, *38*(12), 1185-1193.
[http://dx.doi.org/10.1080/10826076.2015.1025142]

[59] García-Gómez, D.; Rodríguez-Gonzalo, E.; Carabias-Martínez, R. Design and development of a two-dimensional system based on hydrophilic and reversed-phase liquid chromatography with on-line sample treatment for the simultaneous separation of excreted xenobiotics and endogenous metabolites in urine. *Biomed. Chromatogr.,* **2015**, *29*(8), 1190-1196.
[http://dx.doi.org/10.1002/bmc.3407] [PMID: 25611024]

[60] Johnsen, E.; Leknes, S.; Wilson, S.R.; Lundanes, E. Liquid chromatography-mass spectrometry platform for both small neurotransmitters and neuropeptides in blood, with automatic and robust solid phase extraction. *Sci. Rep.,* **2015**, *5*, 9308.
[http://dx.doi.org/10.1038/srep09308] [PMID: 25791195]

[61] Kato, M.; Kinumi, T.; Yoshioka, M.; Goto, M.; Fujii, S.; Takatsu, A. Development of C-reactive protein certified reference material NMIJ CRM 6201-b: optimization of a hydrolysis process to improve the accuracy of amino acid analysis. *Anal. Bioanal. Chem.,* **2015**, *407*(11), 3137-3146.
[http://dx.doi.org/10.1007/s00216-014-8190-0] [PMID: 25303928]

[62] Teleki, A.; Sánchez-Kopper, A.; Takors, R. Alkaline conditions in hydrophilic interaction liquid chromatography for intracellular metabolite quantification using tandem mass spectrometry. *Anal. Biochem.,* **2015**, *475*, 4-13.
[http://dx.doi.org/10.1016/j.ab.2015.01.002] [PMID: 25600449]

[63] Shi, Y.; Xu, X.; Fang, M.; Zhang, M.; Li, Y.; Gillespie, B.; Yorke, S.; Yang, N.; McKew, J.C.; Gahl, W.A.; Huizing, M.; Carrillo-Carrasco, N.; Wang, A.Q. Quantitative hydrophilic interaction chromatography-mass spectrometry analysis of N-acetylneuraminic acid and N-acetylmannosamine in human plasma. *J. Chromatogr. B Analyt. Technol. Biomed. Life Sci.,* **2015**, *1000*, 105-111.
[http://dx.doi.org/10.1016/j.jchromb.2015.07.018] [PMID: 26218770]

[64] Pérez-Ortuño, R.; Martínez-Sánchez, J.M.; Fernández, E.; Pascual, J.A. High-throughput wide dynamic range procedure for the simultaneous quantification of nicotine and cotinine in multiple

biological matrices using hydrophilic interaction liquid chromatography-tandem mass spectrometry. *Anal. Bioanal. Chem.*, **2015**, *407*(28), 8463-8473.
[http://dx.doi.org/10.1007/s00216-015-8993-7] [PMID: 26342312]

[65] Krumpochova, P.; Bruyneel, B.; Molenaar, D.; Koukou, A.; Wuhrer, M.; Niessen, W.M.; Giera, M. Amino acid analysis using chromatography-mass spectrometry: An inter platform comparison study. *J. Pharm. Biomed. Anal.*, **2015**, *114*, 398-407.
[http://dx.doi.org/10.1016/j.jpba.2015.06.001] [PMID: 26115383]

[66] Beach, D.G.; Kerrin, E.S.; Quilliam, M.A. Selective quantitation of the neurotoxin BMAA by use of hydrophilic-interaction liquid chromatography-differential mobility spectrometry-tandem mass spectrometry (HILIC-DMS-MS/MS). *Anal. Bioanal. Chem.*, **2015**, *407*(28), 8397-8409.
[http://dx.doi.org/10.1007/s00216-015-9012-8] [PMID: 26396078]

[67] Mata, A.T.; Jorge, T.F.; Ferreira, J.; do Rosário Bronze, M.; Branco, D.; Fevereiro, P.; Araújo, S.; António, C. Analysis of low abundant trehalose-6-phosphate and related metabolites in Medicago truncatula by hydrophilic interaction liquid chromatography-triple quadrupole mass spectrometry. *J. Chromatogr. A,* **2016**, *1477*, 30-38.
[http://dx.doi.org/10.1016/j.chroma.2016.11.031] [PMID: 27908495]

[68] Steuer, C.; Schütz, P.; Bernasconi, L.; Huber, A.R. Simultaneous determination of phosphatidylcholine-derived quaternary ammonium compounds by a LC-MS/MS method in human blood plasma, serum and urine samples. *J. Chromatogr. B Analyt. Technol. Biomed. Life Sci.*, **2016**, *1008*, 206-211.
[http://dx.doi.org/10.1016/j.jchromb.2015.12.002] [PMID: 26673229]

[69] Wang, M.; Yang, R.; Dong, J.; Zhang, T.; Wang, S.; Zhou, W.; Li, H.; Zhao, H.; Zhang, L.; Wang, S.; Zhang, C.; Chen, W. Simultaneous quantification of cardiovascular disease related metabolic risk factors using liquid chromatography tandem mass spectrometry in human serum. *J. Chromatogr. B Analyt. Technol. Biomed. Life Sci.*, **2016**, *1009-1010*, 144-151.
[http://dx.doi.org/10.1016/j.jchromb.2015.12.019] [PMID: 26735710]

[70] Li, L. Analysis of phospholipids in Schizochytrium sp. S31 by using UPLC-Q-TOF-MS. *Anal. Methods,* **2016**, *8*(4), 763-770.
[http://dx.doi.org/10.1039/C5AY02795K]

[71] Konieczna, L.; Roszkowska, A.; Niedźwiecki, M.; Bączek, T. Hydrophilic interaction chromatography combined with dispersive liquid-liquid microextraction as a preconcentration tool for the simultaneous determination of the panel of underivatized neurotransmitters in human urine samples. *J. Chromatogr. A,* **2016**, *1431*, 111-121.
[http://dx.doi.org/10.1016/j.chroma.2015.12.062] [PMID: 26747692]

[72] Meiss, E. Metabolite targeting: development of a comprehensive targeted metabolomics platform for the assessment of diabetes and its complications. *Metabolomics,* **2016**, *12*(3), 1-15.
[http://dx.doi.org/10.1007/s11306-016-0958-0]

[73] Konieczna, L.; Roszkowska, A.; Synakiewicz, A.; Stachowicz-Stencel, T.; Adamkiewicz-Drożyńska, E.; Bączek, T. Analytical approach to determining human biogenic amines and their metabolites using eVol microextraction in packed syringe coupled to liquid chromatography mass spectrometry method with hydrophilic interaction chromatography column. *Talanta,* **2016**, *150*, 331-339.
[http://dx.doi.org/10.1016/j.talanta.2015.12.056] [PMID: 26838416]

[74] Zhang, L. Validation and quantification of genomic 5-carboxylcytosine (5caC) in mouse brain tissue by liquid chromatography-tandem mass spectrometry. *Anal. Methods,* **2016**, *8*(29), 5812-5817.
[http://dx.doi.org/10.1039/C6AY01445C]

[75] Prinsen, H.C.; Schiebergen-Bronkhorst, B.G.; Roeleveld, M.W.; Jans, J.J.; de Sain-van der Velden, M.G.; Visser, G.; van Hasselt, P.M.; Verhoeven-Duif, N.M. Rapid quantification of underivatized amino acids in plasma by hydrophilic interaction liquid chromatography (HILIC) coupled with tandem mass-spectrometry. *J. Inherit. Metab. Dis.*, **2016**, *39*(5), 651-660.
[http://dx.doi.org/10.1007/s10545-016-9935-z] [PMID: 27099181]

[76] Luginbühl, M.; Rutjens, S.; König, S.; Furrer, J.; Weinmann, W. N-Acetyltaurine as a novel urinary ethanol marker in a drinking study. *Anal. Bioanal. Chem.,* **2016**, *408*(26), 7529-7536. [http://dx.doi.org/10.1007/s00216-016-9855-7] [PMID: 27520321]

[77] Warth, B.; Siegwart, G.; Lemmens, M.; Krska, R.; Adam, G.; Schuhmacher, R. Hydrophilic interaction liquid chromatography coupled with tandem mass spectrometry for the quantification of uridine diphosphate-glucose, uridine diphosphate-glucuronic acid, deoxynivalenol and its glucoside: In-house validation and application to wheat. *J. Chromatogr. A,* **2015**, *1423*, 183-189. [http://dx.doi.org/10.1016/j.chroma.2015.10.070] [PMID: 26554298]

[78] Buszewski, B.; Noga, S. Hydrophilic interaction liquid chromatography (HILIC)--a powerful separation technique. *Anal. Bioanal. Chem.,* **2012**, *402*(1), 231-247. [http://dx.doi.org/10.1007/s00216-011-5308-5] [PMID: 21879300]

[79] Dinh, N.P.; Jonsson, T.; Irgum, K. Probing the interaction mode in hydrophilic interaction chromatography. *J. Chromatogr. A,* **2011**, *1218*(35), 5880-5891. [http://dx.doi.org/10.1016/j.chroma.2011.06.037] [PMID: 21803363]

[80] Cubbon, S.; Antonio, C.; Wilson, J.; Thomas-Oates, J. Metabolomic applications of HILIC-LC-MS. *Mass Spectrom. Rev.,* **2010**, *29*(5), 671-684. [http://dx.doi.org/10.1002/mas.20252] [PMID: 19557839]

[81] McCalley, D.V. Study of the selectivity, retention mechanisms and performance of alternative silica-based stationary phases for separation of ionised solutes in hydrophilic interaction chromatography. *J. Chromatogr. A,* **2010**, *1217*(20), 3408-3417. [http://dx.doi.org/10.1016/j.chroma.2010.03.011] [PMID: 20362994]

[82] McCalley, D.V.; Neue, U.D. Estimation of the extent of the water-rich layer associated with the silica surface in hydrophilic interaction chromatography. *J. Chromatogr. A,* **2008**, *1192*(2), 225-229. [http://dx.doi.org/10.1016/j.chroma.2008.03.049] [PMID: 18440008]

[83] Guo, Y.; Gaiki, S. Retention and selectivity of stationary phases for hydrophilic interaction chromatography. *J. Chromatogr. A,* **2011**, *1218*(35), 5920-5938. [http://dx.doi.org/10.1016/j.chroma.2011.06.052] [PMID: 21737083]

[84] Dragani, L.K.; Berrie, C.P.; Corda, D.; Rotilio, D. Analysis of glycerophosphoinositol by liquid chromatography-electrospray ionisation tandem mass spectrometry using a beta-cyclodextrin-bonded column. *J. Chromatogr. B Analyt. Technol. Biomed. Life Sci.,* **2004**, *802*(2), 283-289. [http://dx.doi.org/10.1016/j.jchromb.2003.12.002] [PMID: 15018789]

[85] Zhang, T.; Creek, D.J.; Barrett, M.P.; Blackburn, G.; Watson, D.G. Evaluation of coupling reversed phase, aqueous normal phase, and hydrophilic interaction liquid chromatography with Orbitrap mass spectrometry for metabolomic studies of human urine. *Anal. Chem.,* **2012**, *84*(4), 1994-2001. [http://dx.doi.org/10.1021/ac2030738] [PMID: 22409530]

[86] Guo, Y.; Gaiki, S. Retention behavior of small polar compounds on polar stationary phases in hydrophilic interaction chromatography. *J. Chromatogr. A,* **2005**, *1074*(1-2), 71-80. [http://dx.doi.org/10.1016/j.chroma.2005.03.058] [PMID: 15941041]

[87] Kawachi, Y.; Ikegami, T.; Takubo, H.; Ikegami, Y.; Miyamoto, M.; Tanaka, N. Chromatographic characterization of hydrophilic interaction liquid chromatography stationary phases: hydrophilicity, charge effects, structural selectivity, and separation efficiency. *J. Chromatogr. A,* **2011**, *1218*(35), 5903-5919. [http://dx.doi.org/10.1016/j.chroma.2011.06.048] [PMID: 21782195]

[88] Sampsonidis, I.; Witting, M.; Koch, W.; Virgiliou, C.; Gika, H.G.; Schmitt-Kopplin, P.; Theodoridis, G.A. Computational analysis and ratiometric comparison approaches aimed to assist column selection in hydrophilic interaction liquid chromatography-tandem mass spectrometry targeted metabolomics. *J. Chromatogr. A,* **2015**, *1406*, 145-155. [http://dx.doi.org/10.1016/j.chroma.2015.06.008] [PMID: 26122858]

[89] Yuan, M.; Breitkopf, S.B.; Yang, X.; Asara, J.M. A positive/negative ion-switching, targeted mass spectrometry-based metabolomics platform for bodily fluids, cells, and fresh and fixed tissue. *Nat. Protoc.,* **2012**, *7*(5), 872-881.
[http://dx.doi.org/10.1038/nprot.2012.024] [PMID: 22498707]

[90] Kloos, D.P.; Lingeman, H.; Niessen, W.M.; Deelder, A.M.; Giera, M.; Mayboroda, O.A. Evaluation of different column chemistries for fast urinary metabolic profiling. *J. Chromatogr. B Analyt. Technol. Biomed. Life Sci.,* **2013**, *927*, 90-96.
[http://dx.doi.org/10.1016/j.jchromb.2013.02.017] [PMID: 23499264]

[91] Periat, A.; Debrus, B.; Rudaz, S.; Guillarme, D. Screening of the most relevant parameters for method development in ultra-high performance hydrophilic interaction chromatography. *J. Chromatogr. A,* **2013**, *1282*, 72-83.
[http://dx.doi.org/10.1016/j.chroma.2013.01.037] [PMID: 23411147]

[92] Cífková, E.; Hájek, R.; Lísa, M.; Holčapek, M. Hydrophilic interaction liquid chromatography-mass spectrometry of (lyso)phosphatidic acids, (lyso)phosphatidylserines and other lipid classes. *J. Chromatogr. A,* **2016**, *1439*, 65-73.
[http://dx.doi.org/10.1016/j.chroma.2016.01.064] [PMID: 26858118]

[93] Kumar, A.; Heaton, J.C.; McCalley, D.V. Practical investigation of the factors that affect the selectivity in hydrophilic interaction chromatography. *J. Chromatogr. A,* **2013**, *1276*, 33-46.
[http://dx.doi.org/10.1016/j.chroma.2012.12.037] [PMID: 23332781]

[94] Spagou, K.; Tsoukali, H.; Raikos, N.; Gika, H.; Wilson, I.D.; Theodoridis, G. Hydrophilic interaction chromatography coupled to MS for metabonomic/metabolomic studies. *J. Sep. Sci.,* **2010**, *33*(6-7), 716-727.
[http://dx.doi.org/10.1002/jssc.200900803] [PMID: 20187037]

[95] Hsieh, Y. Potential of HILIC-MS in quantitative bioanalysis of drugs and drug metabolites. *J. Sep. Sci.,* **2008**, *31*(9), 1481-1491.
[http://dx.doi.org/10.1002/jssc.200700451] [PMID: 18428187]

[96] Cajka, T.; Fiehn, O. Toward Merging Untargeted and Targeted Methods in Mass Spectrometry-Based Metabolomics and Lipidomics. *Anal. Chem.,* **2016**, *88*(1), 524-545.
[http://dx.doi.org/10.1021/acs.analchem.5b04491] [PMID: 26637011]

[97] Paglia, G.; Magnúsdóttir, M.; Thorlacius, S.; Sigurjónsson, O.E.; Guðmundsson, S.; Palsson, B.Ø.; Thiele, I. Intracellular metabolite profiling of platelets: evaluation of extraction processes and chromatographic strategies. *J. Chromatogr. B Analyt. Technol. Biomed. Life Sci.,* **2012**, *898*, 111-120.
[http://dx.doi.org/10.1016/j.jchromb.2012.04.026] [PMID: 22608809]

CHAPTER 4

In Vitro Methods for the Evaluation of Oxidative Stress

Hande Gurer Orhan[1,*], Sibel Suzen[2], Tamás Bálint Csont[3], Miroslav Pohanka[4], Bożena Nejman-Faleńczyk[5], Grzegorz Węgrzyn[5] and Luciano Saso[6]

[1] *Faculty of Pharmacy, Department of Toxicology, University of Ege, 35100 Bornova, Izmir, Turkey*

[2] *Faculty of Pharmacy, Department of Pharmaceutical Chemistry, University of Ankara, 06100, Tandogan, Ankara, Turkey*

[3] *Faculty of General Medicine, Department of Biochemistry, University of Szeged, Dóm tér 9, H-6720 Szeged, Hungary*

[4] *Faculty of Military Health Sciences, University of Defence, Trebesska 1575, CZ-500 01 Hradec Kralove, Czech Republic*

[5] *Department of Molecular Biology, University of Gdansk, Wita Stwosza 59, 80-308 Gdansk, Poland*

[6] *Faculty of Pharmacy and Medicine, Department of Physiology and Pharmacology "Vittorio Erspamer", P.le Aldo Moro 5, 00185, Rome, Italy*

Abstract: The study of oxidative stress (OS) phenomena is very relevant for many disciplines including chemistry, biochemistry, physiology and pharmacology. Unfortunately, it is known that no universal test exists for the measurement of OS and different methods should be used for its evaluation *in vitro, ex vivo* and *in vivo*. This chapter reviews the recent *in vitro* methods for the evaluation OS. Firstly, the determination of lipid, protein and DNA oxidation products as biomarkers of exposure to reactive oxygen species (ROS) and how a battery of biomarkers would increase the efficiency and prevent the limitations of a single test which can end up in overestimation as well as underestimations were discussed. In detection of free radicals in biological samples by electron paramagnetic resonance (EPR) part, properties of EPR for the direct detection and identification of free radicals (paramagnetic species) in biological systems were explained. This was followed by the evaluation of fluorescent assays for the detection of ROS and reactive nitrogen species in biological systems. Description of voltammetric methods for low molecular weight antioxidants and significant advantage over the standard spectroscopic tests was discussed. Finally, measurement of OS in bacteria by stating direct and indirect probe-based measurement of the ROS and measurement of the oxidative damage to biomolecules as well as

* **Corresponding author Prof. Luciano Saso:** Faculty of Pharmacy and Medicine, Department of Physiology and Pharmacology "Vittorio Erspamer", P.le Aldo Moro 5, 00185, Rome, Italy; Tel/Fax: 905063158241; E-mail: luciano.saso@uniroma1.it

Atta-ur-Rahman, Sibel A. Ozkan & Rida Ahmed (Eds.)

assessment of the antioxidant enzymes activities. This chapter will give an overview of the processes of *in vitro* methods for the evaluation of OS and touch briefly the most favored methods and techniques.

Keywords: Bacteria, EPR, Fluorescent assays, Oxidation products, Oxidative stress, Reactive species, Voltammetry.

INTRODUCTION

The scientific interest for oxidative stress (OS) has been very high for several decades and it is still increasing (more than 160,000 articles are currently available on PubMed with more than 10,000 articles per year in the last 5 years). In fact, the study of OS phenomena is highly related to many disciplines including chemistry, biochemistry, physiology and pharmacology.

In pharmacology, the conventional paradigm of using antioxidants to treat diseases involving OS has been changed recently. In fact, unfortunately, many clinical trials of antioxidant therapeutics have failed, yielding negative or inconclusive results for various reasons which are not always clearly understood. Sometimes OS appeared as the consequence of the disease rather than its cause and its reduction in a late stage of the disease did not provide significant therapeutic advantages. OS was sometimes only one of several processes involved in the pathogenesis of the disease. In other cases, the antioxidant treatment in humans was not able to reduce OS *in vivo* especially when compounds with high first-pass metabolism and poor bioavailability such as polyphenols, were used. In neoplastic diseases, OS is often higher in cancer cells compared to normal cells and the use of pro-oxidants rather than antioxidants was recommended. Based on these evidences, in pharmacology, the term of antioxidant drugs should be replaced with "modulators of OS" because, depending on the pathological condition, reducing or increasing OS can be the most appropriate therapeutic strategy. Thus, the importance of the accurate evaluation of OS in target tissues and cells before and after the treatment and the use of appropriate biomarkers in clinical studies appears of utmost importance.

1. DETERMINATION OF LIPID, PROTEIN AND DNA OXIDATION PRODUCTS AS BIOMARKERS OF EXPOSURE TO REACTIVE OXYGEN SPECIES

The World Health Organization has defined a biomarker as any substance, structure or process that can be measured in the body or its products and influence or predict the incidence of outcome or disease [1]. Biomarkers are supposed to reflect changes in biological systems that are related to exposure to or effects of xenobiotics or other type of toxic materials/factors [2]. Biomarkers are classified into three types: biomarkers of exposure, biomarkers of effect and biomarkers of susceptibility. An ideal biomarker of an exposure or marker of an effect should have several characteristics such as specificity for the effect of interest, capacity to reflect an early effect, being unique to the disease state concerned, detectable *via* an easy, inexpensive and sensitive assay, low background level of the biomarker in the body fluid to be analysed. A well-established relationship between the biomarker and exposure or the induced damage is also needed [2 - 4].

Free radicals including reactive species like reactive oxygen (ROS) and reactive nitrogen species (RNS) are known to involve in the pathology of many diseases. Therefore, detection of reactive species (ROS/RNS) has been suggested to be important in order to investigate whether oxidative damage can cause a disease or rather be formed as a consequence of the disease [5]. However, detection of reactive species is enormously problematic since they have very short lifetimes and readily react with nucleophilic molecules that are in their vicinity. Instead, measuring stable markers in the circulation has become the preferred way of evaluating exposure to ROS and/or systemic oxidative stress [6].

ROS levels can be accumulated either by increased formation or decreased removal. Increased ROS formation can result from either endogenous sources or exposure to several xenobiotics. Endogenous sources can be leakage of superoxide anion ($O_2^{\bullet-}$) from mitochondrial electron transport chain, $O_2^{\bullet-}$ production *via* NADPH oxidase activity during phagocytosis and $O_2^{\bullet-}$, $NO^{\bullet}$ formation by xanthine oxidase and nitric oxide synthase activities in vascular endothelial cells [7, 8]. Exposure to exogenous chemicals can also be a source for exposure to ROS/RNS *via* several mechanisms: a xenobiotic itself can be a free radical (*e.g.* nitrogen dioxide); it can be metabolized to free radicals (*e.g.* carbon tetrachloride, paraquat, menadione, mitomycin C); it can stimulate endogenous generation of ROS/RNS by affecting the physiological processes we mentioned above (*e.g.* methyl chloride induced sperm damage *via* ROS/RNS formation during inflammation) [2, 8]. When ROS/RNS production increases by either endogenous or exogenous sources antioxidant defences of organisms/cells start scavenging reactive species as well as repairing and replacing the damaged biomolecules [9]. Therefore, disturbances in the components of antioxidant defence are suggested to be useful biomarkers of oxidative stress. Antioxidant defence system comprises antioxidant enzymes and non-enzymatic antioxidant molecules. Antioxidant enzymes are namely superoxide dismutase, catalase and glutathione peroxidase which are catalytically remove ROS, such as $O_2^{\bullet-}$, H_2O_2 and/or $^{\bullet}OH$. Ascorbic acid, tocopherols, uric acid and glutathione (GSH) are antioxidant molecules with reducing capacity and scavenge ROS non-enzymatically. Antioxidants have been suggested as useful biomarkers of oxidative stress since they tend to react with the excess ROS formed during oxidative stress. Human studies, where causality relation was evaluated between diseases involving ROS and antioxidant status, report contradictory data; antioxidant enzyme activities were shown to decrease in some, whereas increased in some others. However, the disturbances in antioxidant systems are suggested to be useful biomarkers of susceptibility of subjects to ROS-induced damage [2]. Pro-oxidants and antioxidants are in balance in a healthy organism. When this balance is shifted towards oxidants, "oxidative stress" occurs. As a consequence, ROS that are generated at high concentrations attack macromolecules like lipids, proteins and DNA which are then converted to their oxidized products. Molecular oxidation products generated from the reaction of ROS with lipids, proteins and DNA are generally more stable than reactive species themselves and have long been used as biomarkers of exposure to ROS [10]. One approach to overcome the difficulty of detecting instable ROS is to allow the reactive species to react with a trap molecule to give more stable products [11, 12]. This approach is called trapping. On the other hand, not measuring the ROS themselves but the damaged macromolecules or their products are highly used as an alternative to trapping and called as fingerprinting by Halliwell and Gutteridge [8].

1.1. Lipid Oxidation Products as Biomarkers

Lipid peroxidation is probably the most extensively evaluated damage induced by ROS and has been defined as "the oxidative deterioration of polyunsaturated lipids" [8]. Polyunsaturated fatty acid (PUFA) side chains of membrane lipids make them vulnerable to ROS attack and the chain reaction, termed lipid peroxidation, is initiated by abstraction of a hydrogen atom from a methylene group which separates two *cis*-double bonds of a PUFA. The overall cascade of lipid peroxidation and the products formed are already reviewed in detail [8, 13]. Peroxidation of the membrane lipids end up with an impaired membrane structure, decreased membrane fluidity and inactivation of membrane bound proteins. Lipid peroxidation results in formation of a variety of relatively stable decomposition products such as alkanes, aldehydes and isoprostanes. Reactive oxidation products of lipids are found to be toxic by themselves but biological roles such as modulation of protective cell signalling pathways and triggering antioxidant defence have been proposed lately for their low concentrations [2, 13, 14]. Regardless of their physiological or pathological roles, these products have been suggested and extensively used as biomarkers of ROS induced damage.

Aldehydes

Several aldehydic products are formed as a result of peroxidation of membrane PUFAs, namely malondialdehyde (MDA), 4-hydroxy-2-nonenal (4-HNE) and acrolein. These molecules are highly reactive and easily react with DNA and protein *via* Michael addition reactions ending with Michael adducts/advanced lipoxidation end products [15 - 17]. This further metabolism of aldehydic products of lipid peroxidation is a disadvantage for their use as biological markers *in vivo* [17]. On the other hand, aldehydes are more stable than free radicals, therefore, they can diffuse to other.

MDA is probably the most widely used biomarker of ROS-induced lipid damage. It has been measured in biological samples like plasma and urine *via* thiobarbituric acid (TBA) assay. TBA assay is known for its several disadvantages like being not specific for MDA but some other aldehydes and compounds like bilirubin can also react with TBA and absorb light at similar wavelength range. Another drawback of the assay is the artefactual formation of MDA because of the assay conditions; strong acidity and high temperatures [8, 18 - 20]. As a result, the determination of MDA with TBA assay causes overestimation of ROS effect. The specificity of the assay is increased to some extent by determining the MDA-TBA adducts with spectrofluorometer, greatly improved by measuring *via* HPLC-fluorescence detector system [21]. Several ELISA [21] and GC-MS methods [22] are also developed and performed to detect MDA. Overall TBA assay is suggested to work reasonably well when applied to isolated membrane systems *e.g.* liposomes, microsomes with tightly controlled protocol, but its application to biological samples is unreliable and its clinical use is questionable [18, 23]. There are many human studies indicating an association between increased MDA levels and diseases such as rheumatoid arthritis [24], preeclampsia [25], diabetes [26] and several cardiovascular diseases [6, 27].

4-HNE and acrolein are α, β-unsaturated aldehydes that are suggested as a biomarker of lipid peroxidation as well as cytotoxic mediators because of their high reactivity [15, 28]. HNE, arising from the peroxidation of ω6 fatty acids, is found to react with proteins, lipids and nucleic acids and cause cytotoxicity, mutagenicity, genotoxicity, inhibition of protein and DNA synthesis and inactivation of several enzymes [10, 29]. Due to its high reactivity and rapid metabolism by glutathione S-transferases, free levels of HNE is not suggested to be measured, but protein immunodetection is preferred to detect its protein bound form [30]. As a non-invasive biomarker of oxidative stress, the metabolite of HNE, 1,4-dihydroxynonane mercapturic acid (DHN-MA), was measured in human urine. Furthermore, GC-MS based sensitive methods allow the determination of etheno-deoxyguanosine adducts and the O-pentafluorobenzyl oxime derivative of HNE in urine [31]. Acrolein also reacts with protein amino acid residues of proteins mainly generating acrolein-lysine adduct [32, 33] which questions value of detecting its free levels as biomarker. Excretion of its lysine adducts has also not been widely used as a biomarker of lipid peroxidation because of several drawbacks of the assay; cross-reactivity and its further metabolism [16].

Isoprostanes

Isoprostanes, a family of stable bicyclic endoperoxides, are formed by non-enzymatic oxidation of arachidonic acid [34, 35]. Since their structure has F-type prostane rings they are called F_2-isoprostanes. Their biosynthesis starts with hydrogen abstraction at one of the bis-allylic position of arachidonic acid. Once they are formed, they are cleaved and released into the circulation by phospholipases, circulate in the plasma mostly by high density lipoproteins [36] and excreted in the urine predominantly as glucuronide conjugates [37]. F_2-isoprostanes are suggested to be the most reliable biomarker of ROS-induced lipid damage *in vivo* because they are chemically stable [4, 6] and are not influenced by intake of dietary lipids [38]. Therefore, they reflect only endogenous lipid peroxidation. Detection of high levels of F_2-isoprostanes introduce strong evidence of a role for ROS in the pathogenesis of a large number of diseases that are related to oxidative stress, *e.g.* cardiovascular diseases [39], ischemia-reperfusion injury [38], traumatic brain injury [40], Alzheimer's disease [41], diabetes [42] and smoking [43]. Mainly three methods are used to measure F_2-isoprostanes: GC-MS with electron capture negative ionization which is suggested as the "gold standard" [10, 34], LC-MS and immunoassays (ELISA) [35]. Immunoassays have two disadvantages: 1) interference possibility due to the structural similarities between F_2-isoprostanes and enzymatically generated lipid oxidation products, prostaglandins [35] and 2) poor agreement with MS data [44]. On the other hand, it is a fact that immunoassays expanded the research in this area since they are less expensive and easier to use [10]. EFSA suggested measuring changes in F_2 -isoprostanes preferentially by GC-MS in 24-h urine samples, rather than plasma, as an indicator of ROS-induced oxidative damage to lipids. Other methods *e.g.* TBA assay, lipid peroxides, conjugated dienes, exhaled alkanes are not found reliable *in vivo* markers of lipid peroxidation, but they can be used as supportive evidence beside F_2-isoprostane levels if acceptable techniques are used for analysis [45].

1.2. Protein Oxidation Products as Biomarkers

Oxidative damage to proteins can result from the direct attack of ROS/RNS upon proteins or occur as a secondary damage by the attack of lipid and carbohydrate oxidation products. Either directly or indirectly, ROS induced oxidative protein damage is important, because it can result in proteins (receptors, enzymes *etc.*) with impaired functions or it can lead to secondary damage to other molecules/systems (*i.e.* damage to DNA repair enzymes will end up with increased oxidative DNA damage or damaged proteins can be recognized as foreign by the immune system and result in immune reactions) [8]. Although proteins are important targets for ROS attack, focusing attention on investigating biomarkers of protein oxidation is delayed compared to biomarkers of lipid peroxidation. The complex structure of proteins and the lack of sensitive and specific methods are responsible for this delay [17]. Recently, several biomarkers of protein oxidation are known and measurable and their levels in biological samples are found to be in association with several diseases involving ROS.

Protein Carbonyls

Protein carbonyls are relatively stable protein oxidation products and therefore, are suggested to be the most commonly used biomarker of protein oxidation [46, 47]. Protein carbonyls can occur 1) by oxidation of amino acid side chains of proteins, 2) by the Michael addition of aldehydic products of lipid peroxidation to lysine, cysteine and histidine residues of proteins and 3) by the reaction (glycation /glycoxidation) of reducing sugars or their oxidation products with the same amino acid residues [10, 46]. The carbonyls formed by Michael addition are called "advanced lipoxidation end products" (ALEs) and the carbonyls formed by glycoxidation reactions are termed "advanced glycation end products" (AGEs) [4, 47].

Protein carbonyls are generally measured *via* derivatization of the carbonyl group with 2,4-dinitrophenylhydrazine (DNPH). The resulting dinitrophenyl (DNP) hydrazone adduct is suggested to be stable and can be analyzed either by spectrophotometer, by immunological methods (ELISA, Western blot, immunohistochemistry) or by high-performance liquid chromatography (HPLC) [4, 17, 46, 48].

AGEs also gained attention following the associations observed between their *in vivo* levels and several disease states such as cardiovascular diseases, Alzheimer's disease, cancer and diabetes [16].

There are many human studies indicating an association between increased protein carbonyls with pathologies like obesity, diabetes mellitus, hepatocellular carcinoma and aging [4, 47]. The advantages of detection of protein carbonyls as biomarkers are listed as: being easy, fast and inexpensive, high correlation with oxidative damage related diseases, giving information about several pathways [47]. Since carbonyls are formed by many different mechanisms, they are usually present at higher concentrations in biological samples compared to other biomarkers, which is also an advantage of them [49]. They are shown to be stable for three months at -80°C which makes them good biomarkers to detect after storage [17].

The limitation of protein carbonyls as a biomarker is that it detects products generated from various pathways (AGEs, ALEs *etc*.), therefore it is not capable of clarifying a specific mechanism [47]. EFSA Panel on Dietetic Products, Nutrition and Allergies reported that the methods mentioned above to detect carbonyls generally interfere with molecules other than proteins and therefore they are suggested to be used together with at least one direct marker of oxidative protein damage *in vivo* [45]. Il'yasova [16] suggested that AGEs have poor value as urinary biomarker of protein oxidation since they are not derived directly from ROS effect but their relation to oxidative damage is complex.

Oxidized Amino Acids

Among 22 amino acids, the ones which are aromatic and contain sulfhydryl, namely phenylalanine, methionine, cysteine and tryptophan, are considered to be more susceptible to oxidative attack and modification [17].

Cysteine and methionine contain free thiol groups that are prone to oxidative modification. Oxidation of the thiol group leads to disulphide formation from cysteine and can further be oxidized to cysteic acid. Oxidation of the sulphur group of methionine is also known to be vulnerable to oxidative modifications; resulting in sulfoxide and sulfone structures [17]. Free thiol groups react with Ellman's reagent (5,5'-dithiobis-(2-nitrobenzoic acid)/DTNB) resulting in formation of mixed disulphide and a chromophore which can be easily detected spectrophotometrically [50].

Aromatic amino acids are also susceptible to oxidation reactions and N-formyl kynurenine is reported to be formed by oxidation of tryptophan. Formed molecule can easily be detected by HPLC with either UV or fluorescence detector [51].

Phenylalanine is another target amino acid for hydroxyl radicals. Hydroxylation of the amino acid results in the formation of *ortho-* and *meta*-tyrosine. These molecules are suggested to be cytotoxic [52] as well as being a biomarker of the ROS-induced protein oxidation. They can easily be detected by HPLC-UV detector system.

Dityrosine is another marker formed as a result of the attack of species reactive to proteins; starting with tyrosyl radical formation, followed by radical isomerization, diradical reaction and enolization [16, 53]. Dityrosine is resistant to hydrolysis by lytic enzymes and therefore, stable which makes it a better candidate than many of other biomarkers of oxidative damage [54]. Dityrosine levels was found to be 100-fold higher in low-density lipoproteins from atherosclerotic lesions compared to controls, which persuaded scientists to consider it as a biomarker of oxidative stress-related diseases. Furthermore, urinary dityrosine levels were found to be elevated in people with systemic bacterial infections, acute myocardial infarctions, Kwashiorkor disease and who performed endurance exercise [55 - 58]. A variety of analytical methods have been developed to quantify dityrosine *in vitro* and *in vivo* [53, 59]. HPLC-based methods are used (either with UV, fluorescence or diode array detection) and have advantages of being more specific than immunochemistry and more sensitive than TLC methods [60, 61]. However, several other protein oxidation products (*e.g.* *N*-formylkynurenine, 4-HNE-lysine adducts and retinoic acid adducts with proteins) are found to have similar fluorescence characteristics as dityrosine and structurally independent moieties can co-elute with the target molecule which are the limitations of HPLC based assays [61, 62]. Triple quadrupole LC–atmospheric pressure chemical ionization (APSI)–MS/MS method developed by Orhan *et al.* [62] has high sensitivity, selectivity as well as lower detection limits. Another important advantage of the mentioned assay is the unnecessary pretreatment step of the urine samples which results in rapid analysis. Authors mentioned that this methodology reached 2.5-fold higher sensitivity compared to the HPLC–ion-trap–tandem MS method developed by the authors previously [63].

1.3. DNA Oxidation Products as Biomarkers

Large number of base and sugar modification products are formed as a result of ROS attack on DNA. DNA base lesions (especially 8-hydroxy-2-de-xy-guanosine, 8-OHdG), DNA strand breaks and DNA-protein crosslinks are among those products which have been detected by various techniques [17, 64]. While DNA degradation occurs in the late stages of cell death, oxidative DNA damage is likely to represent an early event which is an advantage for a clinical biomarker. Among all oxidative DNA damage products, 8-OHdG is the major one, formed as a result of hydroxyl radical attack on DNA and is the most commonly studied biomarker. It is considered as a reliable biomarker since it can be quantified with high sensitivity and it is present relatively at high concentrations in DNA [10, 65]. Furthermore, there are considerable amount of evidence supporting an association between 8-OHdG levels and pathologies that are known to be related to oxidative damage such as Alzheimer's disease [66], Parkinson's disease [67], cancer [68], diabetes mellitus [69] and long-term smoking [70]. Several methods for quantifying 8-OHdG have been developed: HPLC with electrochemical detection (HPLC-EC), GC/MS, LC-MS/MS and the Comet assay are the most commonly used methods [65]. The chromatography based techniques have the advantage of showing low inter-assay variability whereas they are all prone to oxidation artefacts during sample preparation which might explain big differences in values of oxidized DNA among published studies [16, 65]. The Comet assay is likely to avoid problems of false oxidation, although it has other methodological limitations [65]. Cross-reactivity is being an important problem for detection *via* ELISA method [16]. An important but neglected aspect of detection of urinary 8-OHdG levels is suggested to be inter-individual variability in DNA repair capacity, which causes false positive/negative results that are not related to ROS-induced damage [16].

In conclusion; increased levels of ROS and oxidative stress is known to be involved in many diseases, however, there are no ideal biomarkers that can be used as disease-specific biomarkers which can be of help in both diagnosis and monitoring the efficacy of therapeutic agents [71]. To be a clinically relevant and useful biomarker, a compound should either be specific for a certain disease, have a prognostic value or have a validated association with the disease status, as well as being stable, inexpensive and non-invasive [4]. As a non-invasive biological urine specimen is extensively used for detection of clinical biomarkers. As summarized above several urinary biomarkers of ROS exposure and effect of ROS are reported to be clinically valuable. Furthermore, only validated biomarkers can give reliable results. Therefore, validation of a proposed biomarker *in vivo* against a known oxidative stressor is the key requirement for applicability of the biomarker to human studies [16]. Validation of biomarkers requires several steps including analytical validation and validation of the association between the biomarker concentration and the disease status. No biomarker meets all of the "technical" requirements, but some are superior to others [10]. Another gap in the clinically validated biomarkers of oxidative stress research area is the limited number of cohort studies performed with the candidate biomarkers. Many epidemiological studies performed with oxidative stress biomarkers have been designed as cross-sectional studies [16]. However, it is a well-known fact that this type of an epidemiological study does not have a value in exhibiting causality data but case-control or cohort studies designed as prospective studies are needed to distinguish between cause and effect of a disease. Therefore, more prospective epidemiological studies are needed to test the value and clinically relevancy of the validated biomarkers of oxidative stress. Another conclusion to make based on the literature in this field is that using a battery of biomarkers would increase the efficiency and prevent the limitations of a single test which can end up in overestimation as well as underestimations.

2. DETECTION OF FREE RADICALS IN BIOLOGICAL SAMPLES BY ELECTRON PARAMAGNETIC RESONANCE SPECTROSCOPY

Biologically relevant ROS and RNS include both free radicals (*e.g.* superoxide, hydroxyl radical, nitric oxide, *etc.*) and non-radicals (*e.g.* hydrogen peroxide, hypochlorous acid, peroxynitrite, *etc.*). The unique character of free radicals is that they contain one or more unpaired electrons (*i.e.* when an atomic or molecular orbital is occupied by only one electron). Since electron paramagnetic resonance (EPR) spectroscopy (also called electron spin resonance (ESR) spectroscopy) detects the presence of unpaired electrons, EPR is considered to be the "gold standard" for the direct detection and identification of free radicals (paramagnetic species) in biological systems. The principle of EPR spectroscopy is based on the absorption of the energy of electromagnetic radiation (usually in the microwave frequency range) by paramagnetic species placed in an external magnetic field resulting in the transition of the spin states of the unpaired electron. The equation for resonance absorption is $\Delta E = h\nu = g\beta B$, in which its Planck's constant, ν is the frequency of the applied electromagnetic radiation, β is the Bohr magneton, a physical constant, g is the g-value or the Zeeman splitting factor (*i.e.* a characteristics of the paramagnetic molecule) and B is the magnetic field strength. Usually an EPR spectrum is obtained by altering the magnetic field strength while keeping the microwave frequency fixed and the magnetic field strengths at which the microwave frequency is absorbed are recorded in the EPR spectrum. A further specific property of EPR spectra originates from the coupling of the electron spin with the surrounding nuclear magnetic spins, the so called hyperfine coupling, resulting in a species-specific pattern of absorption lines in the EPR spectrum. The area under the curve on the EPR spectrum is proportional with the number of paramagnetic species in the examined sample, making the method suitable for not only qualitative but also quantitative analysis of radical species. Although the detailed theoretical background of EPR is out of the focus of the present article, we may refer to previously published excellent reviews and books [72, 73].

The advantage of EPR is that it only 'sees' paramagnetic molecules and the spectrum is a unique fingerprint of the chemical and electronic structure around the unpaired electron. However, as many radicals are short-lived and therefore the steady state concentration of most radicals is rather low in biological systems, the direct detection of radicals by EPR is limited to only a few examples of long-lived species including the ascorbyl radical, α-tocopheroxyl radical, phenoxyl radicals, nitroaromatics and semiquinone radicals for instance [74, 75]. One possibility to overcome this stability problem and obtain valuable data for free radical detection is to perform EPR measurement at low temperature on samples rapidly frozen in liquid nitrogen. This way the decay of free radicals can be slowed down resulting in longer lifetime and a possibility for longer acquisition of the EPR spectrum [74]. However, the most widely used solution for the detection of free radicals, especially ROS and RNS, is based on the application of spin traps before performing the EPR measurement [76, 77]. Spin traps are compounds that interact with the less stable radicals, producing a more stable adduct, which still has an unpaired electron and therefore, can be detected by EPR in either liquid or frozen states [78, 79].

A number of different spin traps suitable for detection of ROS and/or RNS have been developed over the past decades [80]. The spin traps differ in their chemical structure, specificity, trapping efficiency, spin-adduct stability, toxicity to living cells, cellular and subcellular distribution and price [75]. The major types of these spin traps include nitrone, nitroso, nitronyl nitroxide compounds and iron (II) dithiocarbamate complexes [74, 81].

Nitrone spin traps are commonly used for detection of various ROSs and include the cyclic 5,5-dimethyl-pyrroline-N-oxide (DMPO); 5-tert-butoxycarbony--5-methyl-1-pyrroline-N-oxide (BMPO); 5-diethoxyphosphoryl-5-meth-l-1-pyrroline-N-oxide (DEPMPO); the open-chain alpha-phenyl-N-tert-b-tyl-nitrone (PBN) and alpha (4-pyridyl-1-oxide)-N-tert-butyl-nitrone (POBN) for instance [81]. These spin traps usually react with different free radicals and EPR is able to discriminate the signals of the specific spin-adducts based on their distinctive spectral lines characterized by their g value and hyperfine coupling [75, 76]. DMPO is the most frequently applied spin trap, as it has been used for several decades leading to a significant amount of data and experience and its price is also affordable [75]. It is used for the detection of the biologically most relevant free radicals; superoxide and hydroxyl radical [82] and even for nitric oxide with less efficiency [83]. DMPO is very useful especially for investigations of simple chemical solutions or isolated enzymes, however, there are several limitations that need to be considered when using this spin trap. The rate of reaction with superoxide is relatively slow, therefore, high concentrations of DMPO (in the mid and high mM range) need to be applied for efficient scavenging, raising the possibility of potential off-target effects in biological systems [84]. Moreover, the half-life of the spin-adduct of DMPO and superoxide is moderate and may decay to DMPO-OH adducts, thereby enhancing the chance for potentially false estimation of superoxide or hydroxyl radical production [81]. The structurally related DEPMPO seems to be a better choice for radical identification as compared to DMPO [85], while BMPO forms a much more stable adduct with superoxide and therefore, can be preferred for quantitation [75, 86].

The linear nitrone spin trap PBN and its water soluble analogue POBN can also be used for the detection of various ROSs, however, identification of the trapped free radical species with these molecules is more challenging than using cyclic nitrones [81].

Nitroso spin traps including 2-methyl-2-nitrosopropane (MNP), nitrosobenzene (NOB) and its derivatives may provide more information on the trapped radical, however, these spin traps are more suitable for trapping carbon-centered radicals rather than oxygen-centered ones [81].

Nitronyl nitroxide compounds such as 2-(4-carboxyphenyl)-4,4,-,5-tetramethylimidazoline-1-oxide (carboxy-PTIO) are able to react with nitric oxide to form iminonitroxides, which have a distinguishable EPR signal from that of the parent nitroxide and therefore, this method is suitable for NO determination [87, 88]. Nevertheless, quantification of NO by this method is problematic due to altering stoichiometry of the conversion with varying NO generation and due to the occurrence of competing reactions [81, 89].

Probably the most widely used approach to measure NO is based on the use of complexes of ferrous iron with dithiocarbamates forming stable nitrosyl iron (II)-dithiocarbamate complexes that can be readily detected by EPR [90 - 92]. The EPR spectrum of iron (II)-dithiocarbamate complexes are characterized by a triplet (three-line) splitting originating from the hyperfine interaction of an unpaired electron with the ^{14}N nucleus of the NO ligand. Dithiocarbamates can be N,N-diethyl-dithiocarmabate (DETC.) or N-methyl-D-glucamine dithiocarmabate (MGD). Iron (II) forms hydrophobic complex with DETC and hydrophilic complex with MGD, therefore, these iron complexes are suitable for trapping NO in different environments [74, 90]. One of the major advantages of using these iron complexes as spin traps is that they react with NO extremely rapidly and they are suitable spin traps for NO *in vitro* in cell cultures [93 - 95], *ex vivo* in isolated organs or tissues [96, 97] or *in vivo* [98 - 100]. When used in complex biological systems, DETC penetrates into the cell wall and binds not only the free intracellular Fe^{2+} but also endogenous Cu^{2+} forming a $Cu(DETC.)_2$ complex, characterized by a four-line EPR spectrum that overlaps with the $NO\text{-}Fe^{2+}(DETC.)_2$ spectrum at low temperature [99, 101] causing difficulties for quantitation of NO at physiological concentrations [102].

Interestingly, NO can be trapped by endogenous molecules such as hemoglobin as well and the resultant hemoglobin-NO complex can be detected with EPR spectroscopy providing an alternative approach for NO detection, especially *in vivo* [103 - 105].

In addition to the application of spin traps, another approach for detection of ROS/RNS by EPR spectroscopy has been developed, *i.e.* the use of spin probes. Spin probes are EPR-silent molecules that do not trap free radicals, but rather react with certain radical and/or non-radical ROS/RNS in redox reactions giving rise to an EPR active product [106]. Typical examples for such spin probes include the cyclic hydroxylamine 1-hydroxy-3-carboxy-2,2-5,5-tetramethylpyrrolidine (CPH), 1-hydroxy-2,2,6,6-tetramethylpiperidin-4-yl-trimethylammonium (CAT1H) and 1-hydroxy-4-phosphono-oxy-2,2,6,6-tetramethylpiperidin (PPH) and other related analogues [84, 107]. The chemical properties of the different spin probes determine their solubility and permeability and therefore, the different probes may provide information on ROS/RNS formation in different cellular compartments (*e.g.* extracellular, cytosolic, mitochondrial) [108]. Hydroxylamines applied at moderate concentrations (in the μM range) may react readily with superoxide to form EPR-detectable nitroxides. These spin probes, however, are not too specific for the reactive species, they can also be oxidized by peroxynitrite (a non-radical species) and probably other oxidants as well [107]. Therefore, additional control experiments are necessary to obtain information on the nature of the reactive species responsible for the formation of the nitroxide radical signal (*e.g.* co-application of superoxide dismutase) [84].

In conclusion, EPR spectroscopy plays a substantial role in the detection, identification and quantitation of ROS and RNS due to its unique nature for being able to detect paramagnetic molecules. Superoxide, hydroxyl radical and nitric oxide can be effectively determined using a wide range of spin traps in various experimental conditions. Even detection of the non-radical peroxynitrite can be achieved using spin probes. Furthermore, commercial appearance of affordable bench-top EPR instruments may hopefully contribute to a more widespread availability of this valuable tool for detection of free radicals in biological samples.

3. FLUORESCENT ASSAYS FOR DETECTION OF REACTIVE OXYGEN AND NITROGEN SPECIES IN BIOLOGICAL SYSTEMS

Fluorescence is a form of luminescence (emittance of light at lower temperature) that is caused by the absorption of radiation at a certain wavelength (excitation wavelength) followed by nearly immediate re-radiation at a different wavelength (emission wavelength) and that ceases almost immediately when the incident radiation stops. Measurement of fluorescence is widely used for detection of various chemical species. The detection of fluorescence is very sensitive and can be carried out by different analytical instruments including for instance fluorometers, fluorescent microplate-readers, fluorescent microscopes and flow-cytometers. Many of these devices are widely available and can be operated by a general user. Fluorescence microscopy can provide 2D or even 3D imaging and high spatial-temporal resolution, therefore, seems to be a very informative detection technique for biological samples despite limited quantitation [109].

Fluorescence assays are sensitive techniques for detection of ROS and RNS. The general principle of these assays is that ROS/RNS react with a nonfluorescent compound forming a fluorescent product and the intensity of fluorescence is measured to estimate the concentration of the product which reflects ROS/RNS concentrations. The compounds converted to fluorescent products (also called frequently probes or dyes) greatly differ in their specificity for various ROSs and RNSs. Moreover, these probes vary in their applicability in biological systems. Most of the probes can be used in solutions such as isolated enzyme-substrate systems, cell lysates or tissue homogenates, while others can be used in intact cells or tissue sections as well. In the latter case, application of fluorescent probes may provide additional information on the spatial and temporal dimensions of ROS/RNS formation that can be followed by means of fluorescent microscopy. An ideal fluorescent probe should be small, cell or membrane permeable, non-toxic, preferably water soluble and photostable; should have well-separated excitation/emission wavelengths in the visible range, a large signal-to-noise ratio and linear response; and should possess a large extinction coefficient and quantum yield [110]. Several fluorescent probes have been developed and used widely over the past decades for detection of various ROS/RNS.

One of the most commonly used fluorescent probes is dichlorodihydrofluorescin diacetate (DCFH-DA). After diffusion into the cell, DCFH-DA is accumulated in the cytosol and deacetylated by intracellular esterases to a non-fluorescent compound (DCFH), which is later oxidized by various ROS (*e.g.* peroxyl, alkoxyl, nitrogen dioxide and hydroxyl radicals and peroxynitrite) into 2',7'-dichlorofluorescein (DCF) [111]. DCF is a highly fluorescent compound which can be detected by fluorescence spectroscopy with maximum excitation and emission spectra of approximately 490 nm and 525 nm respectively. However, there are many disadvantageous features of DCFH-DA. High concentrations of DCFH-DA or high light intensities may result in artefactual photochemical oxidation to fluorescent products [111, 112]. Other limitations include that DCFH does not directly react with hydrogen peroxide; several one-electron oxidizing species are able to oxidize DCFH to DCF; DCF can undergo redox cycling by producing superoxide and hydrogen peroxide thereby artificially elevating ROS levels and transition metals, cytochrome c and heme peroxidases are able to catalyse DCFH oxidation [113, 114]. Therefore, results obtained with DCFH-DA should be carefully interpreted and should be verified using other methodologies.

Dihydrorhodamine (DHR) is another frequently used probe for fluorescent detection of peroxynitrite. The principle of the assay is that DHR can be oxidized to rhodamine, which fluoresces at around 535 nm, when excited at around 500 nm. However, DHR is not specific to peroxynitrite and certain antioxidants (*e.g.* thiols, ascorbate) may interfere with the assay due to reaction with an intermediate DHR radical formed during oxidation [115]. These issues limit the use of DHR as a specific probe for peroxynitrite.

Dihydroethidium (DHE) is frequently used for detection of cellular superoxide radicals. DHE is oxidized by superoxide to a fluorescent product, 2-hydroxy-ethidium in a relatively specific reaction, however, nonspecific redox reactions also lead to formation of ethidium, another fluorescent product [116]. Since the fluorescent spectra of the two oxidation products overlap, it is challenging to detect superoxide-specific signals and excitation wavelengths should be applied carefully in order to attempt to separate these signals [117]. HPLC based separation and quantification of the oxidation products can be used for verifying the data obtained with fluorescence spectroscopy or imaging [113]. Another weakness of DHE is that it is light sensitive and spontaneously oxidized and therefore, often results in high background fluorescence [118]. DHE has been also modified by adding a triphenylphosphonium group to facilitate accumulation of the probe in the mitochondria [117]. The mitochondrium-targeted probe is referred to as mitoSOX. While DHE detects preferably cytosolic superoxide, mitoSOX was shown to react with mitochondrial superoxide [119], however, too high concentrations of mitoSOX should be avoided [117].

A highly sensitive and stable probe for hydrogen peroxide is N-acetyl-3-7-dihydroxyphenoxazine (Amplex Red). In the presence of horseradish peroxidase (HRP), the non-fluorescent Amplex Red reacts with hydrogen peroxide in a 1:1 stoichiometry to produce resorufin, a highly fluorescent compound [120]. The assay is very specific and sensitive for the detection of hydrogen peroxide [121]. A limitation of the assay is that significant levels of superoxide may alter the stoichiometry of the reaction and the quantitative measurement of hydrogen peroxide. This problem may be solved by adding superoxide dismutase to the assay [113]. Further limitations of the Amplex Red method include potential interference of reducing agents (*e.g.* NADH) with the assay, light sensitivity and auto-oxidation of Amplex Red at high concentrations [113, 122]. Amplex Red is not able to penetrate into the cells, it is therefore, only capable of detecting hydrogen peroxide diffusing out of cells or produced in cell lysates, tissue homogenates or isolated enzyme-substrate systems [113].

Another probe often used for estimation of lipid peroxidation is the C11-BODIPY$^{581/591}$ fatty acid. C11-BODIPY$^{581/591}$ penetrates into most membranes and undergoes a shift from red to green fluorescence emission upon oxidation by various ROS/RNS including hydroxyl, peroxyl, alkoxyl radicals and peroxynitrite [111, 123, 124]. This oxidation-dependent emission shift enables fluorescence ratio imaging of lipid peroxidation in live cells [125]. Other common applications of C11-BODIPY$^{581/591}$ include fluorometric assays of antioxidant efficacy in plasma and in lipid vesicles [126, 127]. The oxidation and nitroxidation products of this BODIPY fatty acid have been characterized by mass spectrometry [128].

Another promising class of ROS/RNS sensors is the boronate-based fluorescent probes. In these probes, the boronate is removed due to hydrogen peroxide leading to the formation of fluorescent products [113]. These include the cell permeable peroxyfluor-1, peroxyresorufin-1 and peroxyxanthone-1 fluorescing in green, red and blue respectively, following reaction with intracellular hydrogen peroxide [129, 130]. Similarly, boronate-based fluorescent probes such as coumarin-7-boronic acid (CBA) have been developed for detection of peroxynitrite as well [131]. These latter probes, however, are not specific to peroxynitrite as hydrogen peroxide is also capable of converting them to a fluorescent product [113]. Application of scavengers of either peroxynitrite or hydrogen peroxide may be useful to increase specificity of these probes [113].

A decade ago a novel concept has been implemented for the detection of cellular ROS and redox status. The concept is based on the expression of genetically encoded fluorescent sensors capable of detecting intracellular hydrogen peroxide. The first representative of such proteins is HyPer [132], which consists of yellow fluorescent protein inserted into the regulatory domain of OxyR, the hydrogen peroxide-sensing protein of *E. coli* [133]. Further modified variants carrying signals directed to various intracellular compartments (*e.g.* cytosol, mitochondria, nucleus, *etc.*) have also been developed and tested [134]. HyPer demonstrates submicromolar affinity to hydrogen peroxide and is insensitive to other oxidants tested such as superoxide, oxidized glutathione, nitric oxide and peroxynitrite. HyPer does not seem to cause artefactual ROS generation and can be used for detection of fast changes of hydrogen peroxide concentration in different cell compartments under various physiological and pathological conditions [134]. Additionally, another construction in which a different hydrogen peroxide sensor (*i.e.* Orp1) is connected to a redox sensitive reporter (roGFP2) has been designed recently in order to produce a reliable probe that does not depend on the cellular redox status [113, 135].

Substantially different application areas of fluorescence detection in relation to estimation of oxidative stress are also used widely. These approaches are based on specific recognition of various target molecules using primary antibodies that are detected with the help of fluorescently labelled secondary antibodies [136]. The target molecules may include some markers of ROS/RNS reactions with cellular components or the enzymes that either produce or help to decompose ROS/RNS species. The determination of ROS/RNS markers may include detection of modified and stable residues of proteins (*e.g.* carbonyls or nitrotyrosine) or nucleic acid. Moreover, unstable protein and DNA radicals trapped by certain spin trapping molecules (*e.g.* DMPO) have been also detected by the so-called immuno-spin trapping technique [137, 138]. Determination of the protein level of both ROS/RNS-producing (*e.g.* NADPH oxidases, xanthine oxidase, nitric oxide synthases, *etc.*) and antioxidant (*e.g.* superoxide dismutase, catalase, glutathione peroxidase, *etc.*) enzymes by means of immunofluorescence detection is also commonly used. Nevertheless, this latter technique does not necessarily reflect enzyme activities and thus actual ROS/RNS formation.

In conclusion, fluorescence-based detection of oxidative stress is a very promising and rapidly developing area. In general, these methods are convenient, easy to perform, suitable for subcellular imaging and even for kinetic follow up in certain cases. However, there are still several limitations including issues related to specificity, stability, intracellular distribution and determination of optimal working concentration of the probes, as well as difficulties of quantitative analysis of ROS/RNS concentrations that needs to be overcome. Nevertheless, existing methods may provide useful information on oxidative stress following careful interpretation of the obtained data and further development and optimization of new fluorescent probes and techniques hopefully will solve most of the limitations of fluorescent detection of ROS/RNS.

4. VOLTAMMETRY OF LOW MOLECULAR WEIGHT ANTIOXIDANTS.

Voltammetry or electrochemistry in a common wider terminology is a suitable technique for the determination of chemical compounds that can be oxidized or reduced by a voltage applied on working electrode. Low molecular weight antioxidants fit the condition of oxidization, hence they are ideal analytes for a voltammetric assay. The practical use of voltammetry started with discovery of polarography by Jaroslav Heyrovsky who proposed the use of dropping mercury electrode for electrochemical assays [139, 140]. In 1959, Heyrovsky won Nobel price just for his discovery of polarography because scientific community recognized importance and wide applicability of the method [141]. Since the discovery, polarography found many adaptations in assay of low molecular weight antioxidants. Ascorbic acid level in beverages [142], carotenoids and ascorbic acid content in juices [143] and assay of glutathione in mixture with ascorbic acid [144] can be mentioned as typical examples of polarography use in the field of antioxidants.

Today, polarography is frequently replaced with solid electrodes, because the mercury electrode is problematic due to high toxicity and costs connected with laboratory safety. Solid electrodes such as the one prepared by screen printing technology are a good and relatively cheap alternative [145 - 147]. The electrodes can be performed for direct determination of antioxidants [148] or further modified by giving *e.g.* structured nano-layer improving sensitivity to the most related analytes [149]. The solid electrodes can also be a part of so-called electronic tongues characterizing many parameters of food including content of antioxidants [150, 151].

Voltammetry has a significant advantage over the standard spectroscopic tests; it is the opportunity to fully or at least partially distinguish antioxidants from each other. When potential scale scanned with methods like differential pulse voltammetry, square wave voltammetry, cyclic voltammetry or linear sweep voltammetry, typical peak or wave is formed in potential typical for each antioxidant or oxidizable compound [152 - 157]. We can learn typical determination of low molecular weight antioxidants in biological samples from the quoted reference [158]. Ascorbic acid, glutathione, trolox and uric acid were analyzed by screen printed sensors having embedded platinum auxiliary and working and argent chloride reference electrode in shape of circle around the working one. Samples were applied directly over the electrodes and analysed by square wave voltammetry. The tested antioxidants formed one or to peaks when square wave voltammetry was performed. Ascorbic acid was visible at potential 310 and 699 mV, the second compound providing two peaks, uric acid, was oxidized at 452 and 770 mV. Glutathione and Trolox formed only one peak: glutathione at 764 mV and Trolox at 717 mV. The method was verified on plasma samples from accidentally poisoned vultures from Prague Zoo (Czech Republic) and the voltammetric assay well correlated with standard ferric reducing antioxidant power spectrophotometry. In a later study, the aforementioned square wave voltammetry was adopted for sensors with graphite electrode and limits of detection for ascorbic acid (0.09 mmol/l), trolox (0.03 mmol/l), N-acetylcysteine (0.07 mmol/l) and melatonin (0.04 mmol/l) were achieved [148]. Antioxidant capacity was also measured by Cata *et al.* using cyclic voltammetry [159]. For their study, the investigators have chosen cyclic voltammetry based on a three-electrode system with argent chloride reference, glassy carbon working and platinum auxiliary electrodes. They studied various fruits including strawberry, raspberry, elderberry, mulberry, blackberry and currants by cyclic voltammetry and the results were also compared with mass spectrometry with chromatography which served for specification of anthocyanins and standard spectrophotometric tests for estimation of total antioxidant capacities.

Conventional voltammetry can be further improved with better analytical parameters including bettering of specificity to the individual antioxidants. In an example, Cabaj *et al.* (2016) developed a biosensor with immobilized enzyme tyrosinase as a tool for analyzing phenolic compounds [160]. Another improvement was made in the study of Cruz *et al.* who constructed a DNA based sensor [161]. Oligonucleotide was deposited in carbon paste electrode and the electrode was placed into reactive nitrogen species generating medium commonly with argent chloride reference electrode. DNA damage followed and it was recorded by square wave voltammetry. When an antioxidant was presented in the analyzed sample, DNA became protected which was easily revealed by the square wave voltammetry. The method was verified on various commercial juices and successfully validated with the standard spectrophotometric tests like ferric reducing antioxidant power, oxygen absorbance capacity and total phenolic content assay. Chemical standards of ascorbic acid, gallic acid, caffeic acid, coumaric acid and trolox were also studied in the study. Natural antioxidants can be measured by enzyme laccase which oxidizes phenolic compounds. In their adaptation, Garcia *et al.* (2015) immobilized crude extract of *Pycnoporus sanguineus* having a huge laccase activity into graphite electrode and performed their assay by differential pulse voltammetry and cyclic voltammetry [162]. The authors investigated content of phenol antioxidants in coffee samples and expressed it as gallic acid equivalents. Further phenolic antioxidants and compounds cathecol, rutin, caffeic acid, chlorogenic acid, phenol and chlorophenol were successfully tested. The voltammetric biosensor based assay well correlated with results from DPPH ((2,2 diphenyl-picraylhydrozyl)) test. Survey of basic data about the aforementioned assays is given in Table **1**.

Table 1. Exampled voltammetry assays of low molecular weight antioxidants.

Type of assay	Electrodes	A-analyzed Compounds	Analyzed Biological Samples	Ref
SWV*	screen printed sensor, graphite working and auxiliary, argent chloride reference electrodes	ascorbic acid, trolox, N-acetylcysteine, melatonin	none	[148]
SWV*	screen printed sensor, platinum working and auxiliary, argent chloride reference electrodes	ascorbic acid, glutathione, trolox, uric acid	plasma from lead poisoned vultures	[158]
CV**	three electrode system with glassy carbon working, platinum auxiliary and argent chloride reference electrodes	anthocyanins, total antioxidant level	fruits	[159]
SWV*	carbon paste electrode with immobilized oligonucleotide	total antioxidant level, ascorbic acid, gallic acid, caffeic acid, coumaric acid	juices	[161]
DPV*** and CV**	graphite electrode containing crude extract of *Pycnoporus sanguineus* with laccase activity	phenol antioxidants and phenolic compounds, cathecol, rutin, caffeic acid, chlorogenic acid, phenol and chlorophenol	coffee	[162]

*SWV: square wave voltammetry, ** CV: cyclic voltammetry, *** DPV: differential pulse voltammetry.

5. MEASUREMENT OF OXIDATIVE STRESS IN BACTERIA

Oxidative stress plays an extremely important role in the life of bacteria, as it provides one of the most dangerous conditions in the natural habitats of these microbes. ROS are produced by unicellular eukaryotic predators hunting bacteria as well as by neutrophils responding to appearance of alien bacteria in the host organism [163]. Therefore, bacteria must be able to respond to such conditions and they had to develop various systems for detecting the oxidative stress and ameliorating effects of ROS. Modulation of the bacterial biofilm may be one of the examples of them [164]. On the other hand, the oxidative stress may be a signal for pathogenic bacteria to produce virulence factors and invade their host [165, 166]. Furthermore, it was proposed that a common mechanism for cellular death induced by bactericidal antibiotics is based on stimulation of production of highly deleterious hydroxyl radicals in bacterial cells [167 - 179]. These radicals may cause oxidation of nucleotides and subsequent DNA breaks [171]. In fact, ROS-mediated killing of bacteria upon treatment with various antibiotics has been confirmed recently [172]. In the light of the above mentioned facts, detection of the oxidative stress in bacterial cells is important, particularly in bio-medical aspects of microbiology. However, as discussed recently, identification of this stress in bacteria is not simple [173]. Oxidative stress is caused by the presence of any of a number of reactive oxygen species which the bacterial cell is unable to counterbalance. The biological targets of ROS are biomolecules, including DNA, proteins and lipids. The presence of oxidative stress in bacteria may be tested in one of three ways: (I) direct and indirect probe-based measurement of the ROS, (II) measurement of the oxidative damage to biomolecules and (III) assessment of the antioxidant enzymes activities.

5.1. Direct and Indirect Probe-based Measurement of ROS

Directly measuring ROS might seem to be the preferred method, but many reactive oxygen species are unstable, short-lived and difficult to measure [174]. An analytical technique, ESR spectroscopy can be successfully applied for direct measurement of ROS during oxidative stress in bacteria. Thomas *et al.* (2015) described the protocol for the detection of ROS generated by *Staphylococcus aureus*. This method, based on the use of 1-hydroxy-3-methoxycarbonyl-2,2-5,5-tetramethylpyrrolidine (CMH) as a cell permeable spin probe, might be easily adapted to other bacterial species [175]. Other methods employed frequently for measuring reactive oxygen species are based on the use of fluorescent probes which directly or indirectly react with ROS. The most advanced and promising tools for imaging of intracellular redox potential are genetically encoded redox probes derived from green fluorescent protein (GFP) [176]. The use of redox-sensitive green fluorescent protein (roGFP2) as a biosensor for live measurement of the intracellular redox stress generated by *Escherichia coli* [177], *Salmonella enterica* inside macrophages [178] and *Mycobacterium tuberculosis* during infection [179], have been described recently. This method is based on fluorescence patterns of excitation arising from oxidized and reduced isoforms of roGFP2 biosensor which can be expressed in a variety of different bacteria including enteropathogenic (EPEC) and enterohaemorrhagic (EHEC) *E. coli* [180]. Another probe which is used to measure intracellular generation of H_2O_2 and other oxidants in bacteria is DCFH-DA. This probe is cell-permeable but does not directly react with H_2O_2. Two-electron oxidation of DCFH results in the conversion of the non-fluorescent probe to the highly fluorescent product, dichlorofluorescein (DCF), in which fluorescence can be monitored. Investigators have routinely used DCFH-DA to measure ROS levels in *E. coli* [181, 182], *Pseudomonas aeruginosa* [183] and *Staphylococcus aureus* bacteria [184]. Other approaches for monitoring intrabacterial H_2O_2 levels include the use of Amplex Red probe which is oxidized by horseradish peroxidase (HRP) in the presence of H_2O_2 to a fluorescent product, resorufin. Choi *et al.* [185] adapted the well-established Amplex Red-based technique for use as a single-cell, intracellular, time-resolved measurement of H_2O_2, produced after treatment of *E. coli* with antimicrobial peptide CM15. The intrabacterial ROS generation can also be monitored using the luminescence analysis. As indicated previously, the lucigenin has been used as a chemiluminescent probe to demonstrate continuous intracellular production of reactive oxygen metabolites in *E. coli* [186] whereas luminophore coelenterazine was applied in response to superoxide generation in *Paracoccus denitrificans* bacteria [187]. In addition to this, two superior alternatives to lucigenin probes: MCLA [luciferin analog, 2-methyl-6-(4-methoxyphenyl) imidazo[1,2-a]pyrazin-3(7H)-one] and L012 [8-amino-5-chloro-7-phenyl-pyridopyridazine] have been used for detection of superoxide and hydroxyl radicals (respectively) generated by *Helicobacter pylori* [188]. Another kind of probes used to determine ROS levels in bacteria is chromogenic probes. Nitro blue tetrazolium salt (NBT), sulfonated tetrazolium salt (WST-1) and other tetrazolium salts are useful for superoxide determination. In the most popular version of the assay, yellow NBT is reduced by superoxide anion radicals to the blue formazan, which absorbance can be measured spectrophotometrically at 630 nm. Such a probe has been used successfully to determine superoxide anion scavenging activity in *Streptomyces* sp. MUM256 bacteria [189]. Another chromogenic substrate ortho-nitrophenyl-β-D-galactopyranoside (ONPG) has been used in SOS chromotest assay to detect oxidative *DNA damage and* measure of the SOS inducing potency in phage-infected *E. coli* bacteria in response to treatment with H_2O_2 [190]. Due to the principle of study, this test has been classified in the next category of methods and explained in details below.

5.2. Measurement of the Oxidative Damage to Biomolecules

To date, the most extensive study related to targets for ROS damage in bacteria, has been done on *E. coli*. It seems that DNA is a major target for oxidative damage in this bacterium. DNA subjected to attack by reactive oxygen species generates a wide range of base and sugar modification products and results in both single and double-strand breaks. An oxidized form of guanine, 8-OHdG, is a well-known DNA-damage product that appears in bacterial cell and causes transversion mutations. This compound can be detected by different techniques including HPLC and MS methods and was widely accepted as a sensitive marker of oxidative DNA damage and oxidative stress in *E. coli* and *S. typhimurium* bacteria [191 - 193]. Besides, measurement of 8-OHdG has been applied to monitor oxidative DNA damage in *H. pylori* bacteria by Park *et al.* [188]. They used HPLC technique to determine the amount of 8-OHdG in DNA of *H. pylori* cultured under three different O_2 tensions. Another method utilized to monitor the oxidative DNA damage in bacteria is SOS chromotest. This technique is based on the induction of the SOS response by, among others, ROS-mediated breaks of DNA and generated ssDNA, which in turn activates the RecA protein. This protein is a co-protease cleaving the LexA repressor and allowing expression of the SOS genes in bacteria. Furthermore, it plays a significant role in pathogenesis of phage-infected enterohaemorrhagic *E. coli* (EHEC) as it induces autocleavage of the phage repressor allowing for prophage induction and production of phage-encoded Shiga toxins - the main virulence factors of EHEC [190, 194]. Following this, SOS chromotest has been successfully implemented in the investigation of H_2O_2-mediated SOS response in Stx phage–infected *E. coli* bacteria [166].

Measurement of Protein Damage

Interactions of ROS with amino acid residues of proteins lead to protein changes including cross-linking, changes in conformation and loss of function. Among these modifications, carbonylation has received much greater attention because of its irreversible and irreparable nature. This process occurs by a direct metal-catalyzed oxidative attack of some amino acid chains. Contrary to the common opinions related to the more prone degradation of carbonylated proteins, Maisonneuve *et al.* [195] indicated that bacterial carbonylated proteins are not degraded *in vivo* and detectable mainly in an aggregate state. In this light, the formation of reactive carbonyl groups in proteins seems to be suitable marker to monitor oxidative stress in bacteria. The assay of protein carbonyls is based on the reaction of carbonyl groups with 2,4 dinitrophenyl hydrazine (DNPH). Determination of the dinitrophenyl (DPN)-conjugated proteins occurs by different methods, mainly by an immunoblotting analysis with specific antibody against DPN. Such an approach has been profitably applied in K_2TeO_3-treated *E. coli* [196], *H. pylori* cultured under three different oxygen tensions [188], in response to hydrogen peroxide stress in *B. subtilis* [197] and in a wide spectrum of other bacteria. Another frequently occurred protein modification promoted by oxidative stress in bacteria is S-thiolation. In this process, low molecular weight (LMW) thiols form mixed disulfides with protein thiols. The best known variant of this modification in bacteria is *S*-glutathionylation - the specific post-translational alternation of protein cysteine residues by the addition of the tripeptide glutathione (GSH) which results in protein functional changes. Interestingly, GSH occurs primarily in Gram-negative aerobic species but very rarely in anaerobes and Gram-positive strains. Instead, the *Actinobacteria* utilize mycothiol (MSH) whereas *Firmicutes* bacteria, including *Bacillus* and *Staphylococcus* species utilize bacillithiol (BSH) as thiol-redox buffer [198]. The role of the protein *S*-glutathionylation has been widely studied in Gram-negative bacteria including *E. coli* [199], *S. enterica* serovar Typhimurium [200], *Neisseria meningitides* [201], *Pseudoalteromonans haloplanktis* [202] and *Synechocystis* sp. PCC6803 [203]. Over the past few years, several proteomic strategies have been developed to identify proteins that undergo thiol modifications. Mass spectrometry-based methods applied for identification of S-glutathionylated proteins include tandem affinity purification (TAP) tag assay [204], coupled ultra–high-pressure liquid chromatography [200] or N,N-biotinyl glutathione disulfide (BioGSSG) assay [203]. Leichert and Jacob [205] invented a differential thiol-trapping technique combined with two-dimensional (2D) gel analysis to monitor the thiol-disulfide status of proteins in aerobically growing *E. coli* cells.

Measurement of Lipid Peroxidation

Lipids are major targets upon oxidative stimulation and their peroxidation is a very important response of eukaryotic organisms to oxidative stress. In this process polyunsaturated fatty acids (PUFAs) of membrane phospholipids react with oxygen to yield lipid hydroperoxides. This modification seems less likely in most bacteria because of the lack of polyunsaturated fatty acids. The ability of bacteria to produce PUFAs is characteristic for a few marine genera: *Flexibacter, Psychroflexus, Colwellia, Moritella, Shewanella, however,* not all species within genera express this ability [206]. Two other exceptions are the *H. pylori* which membranes containing up to 3% of linoleic acid [207] and *Synechocystis* cyanobacteria which synthesize and incorporate significant levels of this acid into their membranes [208]. As presented recently, also *Borrelia burgdorferi* cells are able to incorporate ROS-susceptible polyunsaturated fatty acids into their membranes. There are different methods used to detect oxidative lipids damage in bacteria. Methods from the first group are based on the measurement of the lipid content and losses of unsaturated fatty acids by lipid separation in chromatography systems [209]. Other methods allowing for detection of oxidative lipids damage in bacteria are based on lipid peroxides as well as their degradation products (*e.g.* aldehydes) and their reaction with thiobarbituric acid in colorimetric assays. Alternatively, visualization of lipid hydroperoxides in the cell membranes might be performed by fluorescence microscope. Measurement of malondialdehyde (MDA) – the stable end product of the oxidative degradation of polyunsaturated fatty acids in *Borrelia* was performed in cells which were derivatized with thiobarbituric acid and analysed by HPLC, according to the method described by Seljeskog *et al.* [210]. In addition to this method, lipid hydroperoxides in *Borrelia* cell membranes were visualized by fluorescence microscopy [209]. Similarly, *E. coli* bacteria are also able to produce thiobarbituric acid-reactive substances (TBARS) which can be used as indicators of the oxidative modification of lipids. As shown previously, the measurement of TBARS content in *E. coli* cells might be successfully applied to monitor the oxidative stress in these bacteria [211 - 214].

5.3. Measurement of Antioxidants

To this date, a lot of methods have been developed for evaluating the activity of the anti-oxidant enzymes that are considered as the two major components of the antioxidative system together with the low molecular weight antioxidants (LMWA). Reactive oxygen species function frequently as stimulators of the antioxidant enzymes activities, however, on the other hand, they are also able to directly damage these enzymes. Assays of antioxidant enzymes utilize different techniques such as spectroscopic measurements, gel-activity procedures or immunocytochemistry and are frequently based on inhibition strategy in which externally added enzyme can quench the free radical applied [174]. Amaretti *et al.* [215] have described different techniques that allow analysing antioxidant activity of 34 strains of lactic acid bacteria (including *Bifidobacterium, Lactobacillus, Lactococcus* and *Streptococcus thermophilus*). The antioxidant properties were determined by measuring inhibition of ascorbic and linolenic acid oxidation, TAA(AA) and TAA(LA), respectively, trolox-equivalent antioxidant capacity (TEAC), intracellular content of glutathione (TGSH) and superoxide dismutase (SOD) activity. Semchyshyn and colleagues [212] implemented methods which allow assessing activities of selected antioxidant enzymes in *E. coli* stressed by exposure to H_2O_2. These methods are related to the enzymes such as catalase, peroxidase and glutathione reductase of the *oxyR* regulon but also SOD and glucose-6-phosphate dehydrogenase (G6PDH) of the *soxRS* regulon [212]. A similar approach has been applied by Gonzalez-Flecha and Demple in 1997 [211]. Methods developed to assess the total antioxidant capacity in bacteria utilize an external trap compound and may be evaluated using different techniques including spectroscopical biochemical, immunohistochemical and electrochemical assays. Determination of total LMWA rather than individual antioxidants is considered as a useful tool in bacteria. Such an approach has been successfully applied recently by Pawar *et al.* [216] to screen seaweed associated pigmented bacteria for their antioxidant potentials. Free radical scavenging activities of selected species (*Pseudomonas argentinensis, Pseudomonas koreensis* and *Serratia rubidaea*) have been tested against commercial DPPH, according to the methods of Larrauri *et al.* [217] and Takao *et al.* [218]. Two other methods, the ferric reducing antioxidant power (FRAP) assay [219] and the ABTS radical cation decolourization assay [220] have been also used for this purpose [216].

CONCLUSIONS

Unfortunately, it is known that no universal test exists for the measurement of OS. Different methods should be used for its evaluation *in vitro, ex vivo* and *in vivo*. Researchers in the field should be familiar with multiple assays and select the most appropriate assays for each specific experimental condition in order to obtain adequate results.

In this chapter, we described some of the most common tests used to evaluate OS *in vitro* (and *ex vivo* in many instances), including evaluation of oxidation of lipids, proteins and nucleic acids, chemical methods such as EPR, NMR and voltammetry and fluorescent assays to evaluate OS in biological systems. Finally, since OS is a common and important condition of bacterial life and it is involved in the response to antibiotics, an entire section is dedicated to measurement of OS in bacteria.

CONSENT FOR PUBLICATION

Not applicable.

CONFLICT OF INTEREST

The authors declares no conflict of interest, financial or otherwise.

ACKNOWLEDGEMENTS

Declared none.

REFERENCES

[1] International Programme on Chemical Safety; World Health Organization. *Biomarkers and risk assessment: validity and validation*; World Health Organization: Geneva, **2001**.

[2] de Zwart, L.L.; Meerman, J.H.; Commandeur, J.N.; Vermeulen, N.P. Biomarkers of free radical damage applications in experimental animals and in humans. *Free Radic. Biol. Med.,* **1999**, *26*(1-2), 202-226.
[http://dx.doi.org/10.1016/S0891-5849(98)00196-8] [PMID: 9890655]

[3] Henderson, R.F.; Bechtold, W.E.; Bond, J.A.; Sun, J.D. The use of biological markers in toxicology. *Crit. Rev. Toxicol.,* **1989**, *20*(2), 65-82.
[http://dx.doi.org/10.3109/10408448909017904] [PMID: 2686696]

[4] Frijhoff, J.; Winyard, P.G.; Zarkovic, N.; Davies, S.S.; Stocker, R.; Cheng, D.; Knight, A.R.; Taylor, E.L.; Oettrich, J.; Ruskovska, T.; Gasparovic, A.C.; Cuadrado, A.; Weber, D.; Poulsen, H.E.; Grune, T.; Schmidt, H.H.; Ghezzi, P. Clinical Relevance of Biomarkers of Oxidative Stress. *Antioxid. Redox Signal.,* **2015**, *23*, 1144-1170.
[http://dx.doi.org/10.1089/ars.2015.6317] [PMID: 26415143]

[5] Jaeschke, H. Mechanisms of oxidant stress-induced acute tissue injury. *Proc. Soc. Exp. Biol. Med.,* **1995**, *209*(2), 104-111.
[http://dx.doi.org/10.3181/00379727-209-43885b] [PMID: 7770460]

[6] Ho, E.; Karimi Galougahi, K.; Liu, C.C.; Bhindi, R.; Figtree, G.A. Biological markers of oxidative stress: Applications to cardiovascular research and practice. *Redox Biol.,* **2013**, *1*, 483-491.
[http://dx.doi.org/10.1016/j.redox.2013.07.006] [PMID: 24251116]

[7] Mueller, C.F.; Laude, K.; McNally, J.S.; Harrison, D.G. ATVB in focus: redox mechanisms in blood vessels. *Arterioscler. Thromb. Vasc. Biol.,* **2005**, *25*(2), 274-278.
[http://dx.doi.org/10.1161/01.ATV.0000149143.04821.eb] [PMID: 15514203]

[8] Halliwell, B.; Gutteridge, J.M., Eds. *Free Radicals in Biology and Medicine,* 3rd ed; Oxford University Press: New York, **1999**.

[9] Halliwell, B. Free radicals and antioxidants: a personal view. *Nutr. Rev.,* **1994**, *52*(8 Pt 1), 253-265. [PMID: 7970288]

[10] Dalle-Donne, I.; Rossi, R.; Colombo, R.; Giustarini, D.; Milzani, A. Biomarkers of oxidative damage in human disease. *Clin. Chem.,* **2006**, *52*(4), 601-623.
[http://dx.doi.org/10.1373/clinchem.2005.061408] [PMID: 16484333]

[11] Villamena, F.A.; Zweier, J.L. Detection of reactive oxygen and nitrogen species by EPR spin trapping. *Antioxid. Redox Signal.,* **2004**, *6*(3), 619-629.
[http://dx.doi.org/10.1089/152308604773934387] [PMID: 15130289]

[12] Rosen, G.M.; Britigan, B.E.; Halpern, H.J.; Pou, S. *Free Radicals: Biology and Detection by Spin Trapping*; Oxford University Press: New York, **1999**.

[13] Niki, E. Lipid peroxidation: physiological levels and dual biological effects. *Free Radic. Biol. Med.,* **2009**, *47*(5), 469-484.
[http://dx.doi.org/10.1016/j.freeradbiomed.2009.05.032] [PMID: 19500666]

[14] Higdon, A.; Diers, A.R.; Oh, J.Y.; Landar, A.; Darley-Usmar, V.M. Cell signalling by reactive lipid species: new concepts and molecular mechanisms. *Biochem. J.,* **2012**, *442*(3), 453-464.
[http://dx.doi.org/10.1042/BJ20111752] [PMID: 22364280]

[15] Esterbauer, H.; Schaur, R.J.; Zollner, H. Chemistry and biochemistry of 4-hydroxynonenal, malonaldehyde and related aldehydes. *Free Radic. Biol. Med.,* **1991**, *11*(1), 81-128.
[http://dx.doi.org/10.1016/0891-5849(91)90192-6] [PMID: 1937131]

[16] Il'yasova, D.; Scarbrough, P.; Spasojevic, I. Urinary biomarkers of oxidative status. *Clin. Chim. Acta,* **2012**, *413*(19-20), 1446-1453.
[http://dx.doi.org/10.1016/j.cca.2012.06.012] [PMID: 22683781]

[17] Griffiths, H.R.; Møller, L.; Bartosz, G.; Bast, A.; Bertoni-Freddari, C.; Collins, A.; Cooke, M.; Coolen, S.; Haenen, G.; Hoberg, A.M.; Loft, S.; Lunec, J.; Olinski, R.; Parry, J.; Pompella, A.; Poulsen, H.; Verhagen, H.; Astley, S.B. Biomarkers. *Mol. Aspects Med.,* **2002**, *23*(1-3), 101-208.
[http://dx.doi.org/10.1016/S0098-2997(02)00017-1] [PMID: 12079771]

[18] Gutteridge, J.M. Aspects to consider when detecting and measuring lipid peroxidation. *Free Radic. Res. Commun.,* **1986**, *1*(3), 173-184.
[http://dx.doi.org/10.3109/10715768609083149] [PMID: 2577734]

[19] Draper, H.H.; Squires, E.J.; Mahmoodi, H.; Wu, J.; Agarwal, S.; Hadley, M. A comparative evaluation of thiobarbituric acid methods for the determination of malondialdehyde in biological materials. *Free Radic. Biol. Med.,* **1993**, *15*(4), 353-363.
[http://dx.doi.org/10.1016/0891-5849(93)90035-S] [PMID: 8225017]

[20] Moore, K.; Roberts, L.J., II Measurement of lipid peroxidation. *Free Radic. Res.,* **1998**, *28*(6), 659-671.
[http://dx.doi.org/10.3109/10715769809065821] [PMID: 9736317]

[21] Bevan, R.J.; Durand, M.F.; Hickenbotham, P.T.; Kitas, G.D.; Patel, P.R.; Podmore, I.D.; Griffiths, H.R.; Waller, H.L.; Lunec, J. Validation of a novel ELISA for measurement of MDA-LDL in human plasma. *Free Radic. Biol. Med.,* **2003**, *35*(5), 517-527. [http://dx.doi.org/10.1016/S0891-5849(03)00359-9] [PMID: 12927601]

[22] Yeo, H.C.; Helbock, H.J.; Chyu, D.W.; Ames, B.N. Assay of malondialdehyde in biological fluids by gas chromatography-mass spectrometry. *Anal. Biochem.,* **1994**, *220*(2), 391-396. [http://dx.doi.org/10.1006/abio.1994.1355] [PMID: 7978283]

[23] Meagher, E.A.; FitzGerald, G.A. Indices of lipid peroxidation *in vivo*: strengths and limitations. *Free Radic. Biol. Med.,* **2000**, *28*(12), 1745-1750. [http://dx.doi.org/10.1016/S0891-5849(00)00232-X] [PMID: 10946216]

[24] Ozgüneş, H.; Gürer, H.; Tuncer, S. Correlation between plasma malondialdehyde and ceruloplasmin activity values in rheumatoid arthritis. *Clin. Biochem.,* **1995**, *28*(2), 193-194. [http://dx.doi.org/10.1016/0009-9120(94)00081-6] [PMID: 7628080]

[25] Gurer-Orhan, H.; Ozguneş, H.; Beksac, S. Correlation between plasma malondialdehyde and ceruloplasmin activity values in preeclamptic pregnancies. *Clin. Biochem.,* **2000**, *34*(6), 505-506. [http://dx.doi.org/10.1016/S0009-9120(01)00238-7] [PMID: 11074244]

[26] Slatter, D.A.; Bolton, C.H.; Bailey, A.J. The importance of lipid-derived malondialdehyde in diabetes mellitus. *Diabetologia,* **2000**, *43*(5), 550-557. [http://dx.doi.org/10.1007/s001250051342] [PMID: 10855528]

[27] Heinecke, J.W. Oxidized amino acids: culprits in human atherosclerosis and indicators of oxidative stress. *Free Radic. Biol. Med.,* **2002**, *32*(11), 1090-1101. [http://dx.doi.org/10.1016/S0891-5849(02)00792-X] [PMID: 12031894]

[28] Riahi, Y.; Cohen, G.; Shamni, O.; Sasson, S. Signaling and cytotoxic functions of 4-hydroxyalkenals. *Am. J. Physiol. Endocrinol. Metab.,* **2010**, *299*(6), E879-E886. [http://dx.doi.org/10.1152/ajpendo.00508.2010] [PMID: 20858748]

[29] Guéraud, F.; Atalay, M.; Bresgen, N.; Cipak, A.; Eckl, P.M.; Huc, L.; Jouanin, I.; Siems, W.; Uchida, K. Chemistry and biochemistry of lipid peroxidation products. *Free Radic. Res.,* **2010**, *44*(10), 1098-1124. [http://dx.doi.org/10.3109/10715762.2010.498477] [PMID: 20836659]

[30] Wakita, C.; Honda, K.; Shibata, T.; Akagawa, M.; Uchida, K. A method for detection of 4-hydroxy-2-nonenal adducts in proteins. *Free Radic. Biol. Med.,* **2011**, *51*(1), 1-4.

[31] Poli, G.; Schaur, R.J.; Siems, W.G.; Leonarduzzi, G. 4-hydroxynonenal: a membrane lipid oxidation product of medicinal interest. *Med. Res. Rev.,* **2008**, *28*(4), 569-631. [http://dx.doi.org/10.1002/med.20117] [PMID: 18058921]

[32] Uchida, K.; Kanematsu, M.; Sakai, K.; Matsuda, T.; Hattori, N.; Mizuno, Y.; Suzuki, D.; Miyata, T.; Noguchi, N.; Niki, E.; Osawa, T. Protein-bound acrolein: potential markers for oxidative stress. *Proc. Natl. Acad. Sci. USA,* **1998**, *95*(9), 4882-4887. [http://dx.doi.org/10.1073/pnas.95.9.4882] [PMID: 9560197]

[33] Furuhata, A.; Ishii, T.; Kumazawa, S.; Yamada, T.; Nakayama, T.; Uchida, K. N(ε)-(--methylpyridinium)lysine, a major antigenic adduct generated in acrolein-modified protein. *J. Biol. Chem.,* **2003**, *278*(49), 48658-48665. [http://dx.doi.org/10.1074/jbc.M309401200] [PMID: 14504272]

[34] Galano, J.M.; Mas, E.; Barden, A.; Mori, T.A.; Signorini, C.; De Felice, C.; Barrett, A.; Opere, C.; Pinot, E.; Schwedhelm, E.; Benndorf, R.; Roy, J.; Le Guennec, J.Y.; Oger, C.; Durand, T. Isoprostanes and neuroprostanes: total synthesis, biological activity and biomarkers of oxidative stress in humans. *Prostaglandins Other Lipid Mediat.,* **2013**, *107*, 95-102. [http://dx.doi.org/10.1016/j.prostaglandins.2013.04.003] [PMID: 23644158]

[35] Milne, G.L.; Dai, Q.; Roberts, L.J., II The isoprostanes--25 years later. *Biochim. Biophys. Acta,* **2015**,

1851(4), 433-445.
[http://dx.doi.org/10.1016/j.bbalip.2014.10.007] [PMID: 25449649]

[36] Proudfoot, J.M.; Barden, A.E.; Loke, W.M.; Croft, K.D.; Puddey, I.B.; Mori, T.A. HDL is the major lipoprotein carrier of plasma F2-isoprostanes. *J. Lipid Res.*, **2009**, *50*(4), 716-722.
[http://dx.doi.org/10.1194/jlr.M800607-JLR200] [PMID: 19050315]

[37] Yan, Z.; Mas, E.; Mori, T.A.; Croft, K.D.; Barden, A.E. A significant proportion of F2-isoprostanes in human urine are excreted as glucuronide conjugates. *Anal. Biochem.*, **2010**, *403*(1-2), 126-128.
[http://dx.doi.org/10.1016/j.ab.2010.04.016] [PMID: 20406619]

[38] Richelle, M.; Turini, M.E.; Guidoux, R.; Tavazzi, I.; Métairon, S.; Fay, L.B. Urinary isoprostane excretion is not confounded by the lipid content of the diet. *FEBS Lett.*, **1999**, *459*(2), 259-262.
[http://dx.doi.org/10.1016/S0014-5793(99)01259-4] [PMID: 10518031]

[39] Davies, S.S.; Roberts, L.J., II F2-isoprostanes as an indicator and risk factor for coronary heart disease. *Free Radic. Biol. Med.*, **2011**, *50*(5), 559-566.
[http://dx.doi.org/10.1016/j.freeradbiomed.2010.11.023] [PMID: 21126576]

[40] Corcoran, T.B.; Mas, E.; Barden, A.E.; Durand, T.; Galano, J.M.; Roberts, L.J.; Phillips, M.; Ho, K.M.; Mori, T.A. Are isofurans and neuroprostanes increased after subarachnoid hemorrhage and traumatic brain injury? *Antioxid. Redox Signal.*, **2011**, *15*(10), 2663-2667.
[http://dx.doi.org/10.1089/ars.2011.4125] [PMID: 21702684]

[41] Praticò, D.; MY Lee, V.; Trojanowski, J.Q.; Rokach, J.; Fitzgerald, G.A. Increased F2-isoprostanes in Alzheimer's disease: evidence for enhanced lipid peroxidation *in vivo. FASEB J.*, **1998**, *12*(15), 1777-1783.
[PMID: 9837868]

[42] Davì, G.; Ciabattoni, G.; Consoli, A.; Mezzetti, A.; Falco, A.; Santarone, S.; Pennese, E.; Vitacolonna, E.; Bucciarelli, T.; Costantini, F.; Capani, F.; Patrono, C. *In vivo* formation of 8-iso-prostaglandin f2α and platelet activation in diabetes mellitus: effects of improved metabolic control and vitamin E supplementation. *Circulation,* **1999**, *99*(2), 224-229.
[http://dx.doi.org/10.1161/01.CIR.99.2.224] [PMID: 9892587]

[43] Durazzo, T.C.; Korecka, M.; Trojanowski, J.Q.; Weiner, M.W.; O' Hara, R.; Ashford, J.W.; Shaw, L.M. Alzheimer's Disease Neuroimaging Initiative. Active Cigarette Smoking in Cognitively-Normal Elders and Probable Alzheimer's Disease is Associated with Elevated Cerebrospinal Fluid Oxidative Stress Biomarkers. *J. Alzheimers Dis.*, **2016**, *54*(1), 99-107.
[http://dx.doi.org/10.3233/JAD-160413] [PMID: 27472882]

[44] Proudfoot, J.; Barden, A.; Mori, T.A.; Burke, V.; Croft, K.D.; Beilin, L.J.; Puddey, I.B. Measurement of urinary F(2)-isoprostanes as markers of *in vivo* lipid peroxidation-A comparison of enzyme immunoassay with gas chromatography/mass spectrometry. *Anal. Biochem.*, **1999**, *272*(2), 209-215.
[http://dx.doi.org/10.1006/abio.1999.4187] [PMID: 10415090]

[45] EFSA Panel on Dietetic Products. Nutrition and Allergies. *EFSA J.*, **2011**, *9*, 2474-2486.

[46] Weber, D.; Davies, M.J.; Grune, T. Determination of protein carbonyls in plasma, cell extracts, tissue homogenates, isolated proteins: Focus on sample preparation and derivatization conditions. *Redox Biol.*, **2015**, *5*, 367-380.
[http://dx.doi.org/10.1016/j.redox.2015.06.005] [PMID: 26141921]

[47] Chevion, M.; Berenshtein, E.; Stadtman, E.R. Human studies related to protein oxidation: protein carbonyl content as a marker of damage. *Free Radic. Res.*, **2000**, *33* Suppl., S99-S108.
[PMID: 11191280]

[48] Dalle-Donne, I.; Rossi, R.; Giustarini, D.; Milzani, A.; Colombo, R. Protein carbonyl groups as biomarkers of oxidative stress. *Clin. Chim. Acta,* **2003**, *329*(1-2), 23-38.
[http://dx.doi.org/10.1016/S0009-8981(03)00003-2] [PMID: 12589963]

[49] Dean, R.T.; Fu, S.; Stocker, R.; Davies, M.J. Biochemistry and pathology of radical-mediated protein

oxidation. *Biochem. J.,* **1997**, *324*(Pt 1), 1-18.
[http://dx.doi.org/10.1042/bj3240001] [PMID: 9164834]

[50] Ellman, G.L. Tissue sulfhydryl groups. *Arch. Biochem. Biophys.,* **1959**, *82*(1), 70-77.
[http://dx.doi.org/10.1016/0003-9861(59)90090-6] [PMID: 13650640]

[51] Griffiths, H.R.; Lunec, J.; Blake, D.R. Oxygen radical induced fluorescence in proteins; identification of the fluorescent tryptophan metabolite, N-formyl kynurenine, as a biological index of radical damage. *Amino Acids,* **1992**, *3*(2), 183-194.
[http://dx.doi.org/10.1007/BF00806783] [PMID: 24193030]

[52] Gurer-Orhan, H.; Ercal, N.; Mare, S.; Pennathur, S.; Orhan, H.; Heinecke, J.W. Misincorporation of free m-tyrosine into cellular proteins: a potential cytotoxic mechanism for oxidized amino acids. *Biochem. J.,* **2006**, *395*(2), 277-284.
[http://dx.doi.org/10.1042/BJ20051964] [PMID: 16363993]

[53] Giulivi, C.; Traaseth, N.J.; Davies, K.J. Tyrosine oxidation products: analysis and biological relevance. *Amino Acids,* **2003**, *25*(3-4), 227-232.
[http://dx.doi.org/10.1007/s00726-003-0013-0] [PMID: 14661086]

[54] Giulivi, C.; Davies, K.J. Mechanism of the formation and proteolytic release of H2O2-induced dityrosine and tyrosine oxidation products in hemoglobin and red blood cells. *J. Biol. Chem.,* **2001**, *276*(26), 24129-24136.
[http://dx.doi.org/10.1074/jbc.M010697200] [PMID: 11294851]

[55] Bhattacharjee, S.; Pennathur, S.; Byun, J.; Crowley, J.; Mueller, D.; Gischler, J.; Hotchkiss, R.S.; Heinecke, J.W. NADPH oxidase of neutrophils elevates o,o′-dityrosine cross-links in proteins and urine during inflammation. *Arch. Biochem. Biophys.,* **2001**, *395*(1), 69-77.
[http://dx.doi.org/10.1006/abbi.2001.2557] [PMID: 11673867]

[56] Manary, M.J.; Leeuwenburgh, C.; Heinecke, J.W. Increased oxidative stress in kwashiorkor. *J. Pediatr.,* **2000**, *137*(3), 421-424.
[http://dx.doi.org/10.1067/mpd.2000.107512] [PMID: 10969271]

[57] Mayer, F.; Pröpper, S.; Ritz-Timme, S. Dityrosine, a protein product of oxidative stress, as a possible marker of acute myocardial infarctions. *Int. J. Legal Med.,* **2014**, *128*(5), 787-794.
[http://dx.doi.org/10.1007/s00414-014-1015-z] [PMID: 24819153]

[58] Orhan, H.; van Holland, B.; Krab, B.; Moeken, J.; Vermeulen, N.P.; Hollander, P.; Meerman, J.H. Evaluation of a multi-parameter biomarker set for oxidative damage in man: increased urinary excretion of lipid, protein and DNA oxidation products after one hour of exercise. *Free Radic. Res.,* **2004**, *38*(12), 1269-1279. a
[http://dx.doi.org/10.1080/10715760400013763] [PMID: 15763951]

[59] DiMarco, T.; Giulivi, C. Current analytical methods for the detection of dityrosine, a biomarker of oxidative stress, in biological samples. *Mass Spectrom. Rev.,* **2007**, *26*(1), 108-120.
[http://dx.doi.org/10.1002/mas.20109] [PMID: 17019703]

[60] Giulivi, C.; Davies, K.J. Dityrosine: a marker for oxidatively modified proteins and selective proteolysis. *Methods Enzymol.,* **1994**, *233*, 363-371.
[http://dx.doi.org/10.1016/S0076-6879(94)33042-5] [PMID: 8015471]

[61] Heinecke, J.W.; Hsu, F.F.; Crowley, J.R.; Hazen, S.L.; Leeuwenburgh, C.; Mueller, D.M.; Rasmussen, J.E.; Turk, J. Detecting oxidative modification of biomolecules with isotope dilution mass spectrometry: sensitive and quantitative assays for oxidized amino acids in proteins and tissues. *Methods Enzymol.,* **1999**, *300*, 124-144.
[http://dx.doi.org/10.1016/S0076-6879(99)00121-4] [PMID: 9919517]

[62] Orhan, H.; Coolen, S.; Meerman, J.H. Quantification of urinary o,o′-dityrosine, a biomarker for oxidative damage to proteins, by high performance liquid chromatography with triple quadrupole tandem mass spectrometry. A comparison with ion-trap tandem mass spectrometry. *J. Chromatogr. B Analyt. Technol. Biomed. Life Sci.,* **2005**, *827*(1), 104-108.

[http://dx.doi.org/10.1016/j.jchromb.2005.03.043] [PMID: 15878312]

[63] Orhan, H.; Vermeulen, N.P.; Tump, C.; Zappey, H.; Meerman, J.H. Simultaneous determination of tyrosine, phenylalanine and deoxyguanosine oxidation products by liquid chromatography-tandem mass spectrometry as non-invasive biomarkers for oxidative damage. *J. Chromatogr. B Analyt. Technol. Biomed. Life Sci.,* **2004**, *799*(2), 245-254. b [http://dx.doi.org/10.1016/j.jchromb.2003.10.056] [PMID: 14670743]

[64] Dizdaroglu, M.; Jaruga, P.; Birincioglu, M.; Rodriguez, H. Free radical-induced damage to DNA: mechanisms and measurement. *Free Radic. Biol. Med.,* **2002**, *32*(11), 1102-1115. [http://dx.doi.org/10.1016/S0891-5849(02)00826-2] [PMID: 12031895]

[65] Valavanidis, A.; Vlachogianni, T.; Fiotakis, C. 8-hydroxy-2′ -deoxyguanosine (8-OHdG): A critical biomarker of oxidative stress and carcinogenesis. *J Environ Sci Health C Environ Carcinog Ecotoxicol Rev,* **2009**, *27*(2), 120-139. [http://dx.doi.org/10.1080/10590500902885684] [PMID: 19412858]

[66] Mecocci, P.; Polidori, M.C.; Cherubini, A.; Ingegni, T.; Mattioli, P.; Catani, M.; Rinaldi, P.; Cecchetti, R.; Stahl, W.; Senin, U.; Beal, M.F. Lymphocyte oxidative DNA damage and plasma antioxidants in Alzheimer disease. *Arch. Neurol.,* **2002**, *59*(5), 794-798. [http://dx.doi.org/10.1001/archneur.59.5.794] [PMID: 12020262]

[67] Sato, S.; Mizuno, Y.; Hattori, N. Urinary 8-hydroxydeoxyguanosine levels as a biomarker for progression of Parkinson disease. *Neurology,* **2005**, *64*(6), 1081-1083. [http://dx.doi.org/10.1212/01.WNL.0000154597.24838.6B] [PMID: 15781836]

[68] Guo, C.; Li, X.; Wang, R.; Yu, J.; Ye, M.; Mao, L.; Zhang, S.; Zheng, S. Association between Oxidative DNA Damage and Risk of Colorectal Cancer: Sensitive Determination of Urinary 8-Hydroxy-2′-deoxyguanosine by UPLC-MS/MS Analysis. *Sci. Rep.,* **2016**, *6*, 32581. [http://dx.doi.org/10.1038/srep32581] [PMID: 27585556]

[69] de M Bandeira, S.; da Fonseca, L.J.; da S Guedes, G.; Rabelo, L.A.; Goulart, M.O.; Vasconcelos, S.M. Oxidative stress as an underlying contributor in the development of chronic complications in diabetes mellitus. *Int. J. Mol. Sci.,* **2013**, *14*(2), 3265-3284. [http://dx.doi.org/10.3390/ijms14023265] [PMID: 23385234]

[70] Chen, Z.; Wang, D.; Liu, X.; Pei, W.; Li, J.; Cao, Y.; Zhang, J.; An, Y.; Nie, J.; Tong, J. Oxidative DNA damage is involved in cigarette smoke-induced lung injury in rats. *Environ. Health Prev. Med.,* **2015**, *20*(5), 318-324. [http://dx.doi.org/10.1007/s12199-015-0469-z] [PMID: 25967734]

[71] Sultana, R.; Cenini, G.; Butterfield, A. Biomarkers of oxidative stress in neurodegenerative diseases., **2013**. [http://dx.doi.org/10.1002/9781118355886.ch14]

[72] Janzen, E.G. Electron spin resonance. *Anal. Chem.,* **1972**, *44*(5), 113-121. [http://dx.doi.org/10.1021/ac60313a033] [PMID: 22401005]

[73] Weil, J.A.; Bolton, J.R. *Electron paramagnetic resonance: Elementary theory and practical applications*; Wiley and Sons: New York, **2007**.

[74] Hawkins, C.L.; Davies, M.J. Detection and characterisation of radicals in biological materials using EPR methodology. *Biochim. Biophys. Acta,* **2014**, *1840*(2), 708-721. [http://dx.doi.org/10.1016/j.bbagen.2013.03.034] [PMID: 23567797]

[75] Spasojević, I. Free radicals and antioxidants at a glance using EPR spectroscopy. *Crit. Rev. Clin. Lab. Sci.,* **2011**, *48*(3), 114-142. [http://dx.doi.org/10.3109/10408363.2011.591772] [PMID: 21875311]

[76] Villamena, F.A.; Zweier, J.L. Detection of reactive oxygen and nitrogen species by EPR spin trapping. *Antioxid. Redox Signal.,* **2004**, *6*(3), 619-629. [http://dx.doi.org/10.1089/152308604773934387] [PMID: 15130289]

[77] Janzen, E.G. Spin trapping. *Methods Enzymol.,* **1984**, *105*, 188-198. [http://dx.doi.org/10.1016/S0076-6879(84)05025-4] [PMID: 6328178]

[78] Davies, M.J. Detection and characterisation of radicals using electron paramagnetic resonance (EPR) spin trapping and related methods. *Methods,* **2016**, *109*, 21-30. [http://dx.doi.org/10.1016/j.ymeth.2016.05.013] [PMID: 27211009]

[79] Buettner, G.R. Spin trapping: ESR parameters of spin adducts. *Free Radic. Biol. Med.,* **1987**, *3*(4), 259-303. [http://dx.doi.org/10.1016/S0891-5849(87)80033-3] [PMID: 2826304]

[80] Vanin, A.; Poltorakov, A. NO spin trapping in biological systems. *Front. Biosci. (Landmark Ed.),* **2009**, *14*, 4427-4435. [http://dx.doi.org/10.2741/3538] [PMID: 19273360]

[81] He, W.; Liu, Y.; Wamer, W.G.; Yin, J.J. Electron spin resonance spectroscopy for the study of nanomaterial-mediated generation of reactive oxygen species. *J. Food Drug Anal.,* **2014**, *22*(1), 49-63. [http://dx.doi.org/10.1016/j.jfda.2014.01.004] [PMID: 24673903]

[82] Arroyo, C.M.; Kramer, J.H.; Dickens, B.F.; Weglicki, W.B. Identification of free radicals in myocardial ischemia/reperfusion by spin trapping with nitrone DMPO. *FEBS Lett.,* **1987**, *221*(1), 101-104. [http://dx.doi.org/10.1016/0014-5793(87)80360-5] [PMID: 3040465]

[83] Arroyo, C.M.; Kohno, M. Difficulties encountered in the detection of nitric oxide (NO) by spin trapping techniques. A cautionary note. *Free Radic. Res. Commun.,* **1991**, *14*(2), 145-155. [http://dx.doi.org/10.3109/10715769109094127] [PMID: 1648020]

[84] Dikalov, S.I.; Harrison, D.G. Methods for detection of mitochondrial and cellular reactive oxygen species. *Antioxid. Redox Signal.,* **2014**, *20*(2), 372-382. [http://dx.doi.org/10.1089/ars.2012.4886] [PMID: 22978713]

[85] Frejaville, C.; Karoui, H.; Tuccio, B.; Le Moigne, F.; Culcasi, M.; Pietri, S.; Lauricella, R.; Tordo, P. 5-(Diethoxyphosphoryl)-5-methyl-1-pyrroline N-oxide: a new efficient phosphorylated nitrone for the *in vitro* and *in vivo* spin trapping of oxygen-centered radicals. *J. Med. Chem.,* **1995**, *38*(2), 258-265. [http://dx.doi.org/10.1021/jm00002a007] [PMID: 7830268]

[86] Zhao, H.; Joseph, J.; Zhang, H.; Karoui, H.; Kalyanaraman, B. Synthesis and biochemical applications of a solid cyclic nitrone spin trap: a relatively superior trap for detecting superoxide anions and glutathiyl radicals. *Free Radic. Biol. Med.,* **2001**, *31*(5), 599-606. [http://dx.doi.org/10.1016/S0891-5849(01)00619-0] [PMID: 11522444]

[87] Joseph, J.; Kalyanaraman, B.; Hyde, J.S. Trapping of nitric oxide by nitronyl nitroxides: an electron spin resonance investigation. *Biochem. Biophys. Res. Commun.,* **1993**, *192*(2), 926-934. [http://dx.doi.org/10.1006/bbrc.1993.1504] [PMID: 8387295]

[88] Katayama, Y.; Soh, N.; Maeda, M. A new strategy for the design of molecular probes for investigating endogenous nitric oxide using an EPR or fluorescent technique. *ChemPhysChem,* **2001**, *2*(11), 655-661. [http://dx.doi.org/10.1002/1439-7641(20011119)2:11<655::AID-CPHC655>3.0.CO;2-S] [PMID: 23686899]

[89] Haseloff, R.F.; Zöllner, S.; Kirilyuk, I.A.; Grigor'ev, I.A.; Reszka, R.; Bernhardt, R.; Mertsch, K.; Roloff, B.; Blasig, I.E. Superoxide-mediated reduction of the nitroxide group can prevent detection of nitric oxide by nitronyl nitroxides. *Free Radic. Res.,* **1997**, *26*(1), 7-17. [http://dx.doi.org/10.3109/10715769709097780] [PMID: 9018468]

[90] Csonka, C.; Páli, T.; Bencsik, P.; Görbe, A.; Ferdinandy, P.; Csont, T. Measurement of NO in biological samples. *Br. J. Pharmacol.,* **2015**, *172*(6), 1620-1632. [http://dx.doi.org/10.1111/bph.12832] [PMID: 24990201]

[91] Mordvintcev, P.; Mülsch, A.; Busse, R.; Vanin, A. On-line detection of nitric oxide formation in liquid

aqueous phase by electron paramagnetic resonance spectroscopy. *Anal. Biochem.*, **1991**, *199*(1), 142-146.
[http://dx.doi.org/10.1016/0003-2697(91)90282-X] [PMID: 1666941]

[92] Vanin, A.F.; Liu, X.; Samouilov, A.; Stukan, R.A.; Zweier, J.L. Redox properties of iron-dithiocarbamates and their nitrosyl derivatives: implications for their use as traps of nitric oxide in biological systems. *Biochim. Biophys. Acta,* **2000**, *1474*(3), 365-377.
[http://dx.doi.org/10.1016/S0304-4165(00)00033-7] [PMID: 10779689]

[93] Ando, T.; Yonamoto, Y. In situ EPR detection of reactive oxygen species in adherent cells using polylysine-coated glass plate. *Appl. Magn. Reson.*, **2015**, *46*, 977-986.
[http://dx.doi.org/10.1007/s00723-015-0688-x]

[94] Csont, T.; Görbe, A.; Bereczki, E.; Szunyog, A.; Aypar, E.; Tóth, M.E.; Varga, Z.V.; Csonka, C.; Fülöp, F.; Sántha, M.; Ferdinandy, P. Biglycan protects cardiomyocytes against hypoxia/reoxygenation injury: role of nitric oxide. *J. Mol. Cell. Cardiol.*, **2010**, *48*(4), 649-652.
[http://dx.doi.org/10.1016/j.yjmcc.2010.01.013] [PMID: 20096286]

[95] Gáspár, R.; Pipicz, M.; Hawchar, F.; Kovács, D.; Djirackor, L.; Görbe, A.; Varga, Z.V.; Kiricsi, M.; Petrovski, G.; Gácser, A.; Csonka, C.; Csont, T. The cytoprotective effect of biglycan core protein involves Toll-like receptor 4 signaling in cardiomyocytes. *J. Mol. Cell. Cardiol.*, **2016**, *99*, 138-150.
[http://dx.doi.org/10.1016/j.yjmcc.2016.08.006] [PMID: 27515282]

[96] Csonka, C.; Szilvássy, Z.; Fülöp, F.; Páli, T.; Blasig, I.E.; Tosaki, A.; Schulz, R.; Ferdinandy, P. Classic preconditioning decreases the harmful accumulation of nitric oxide during ischemia and reperfusion in rat hearts. *Circulation,* **1999**, *100*(22), 2260-2266.
[http://dx.doi.org/10.1161/01.CIR.100.22.2260] [PMID: 10578001]

[97] Radák, Z.; Pucsok, J.; Mecseki, S.; Csont, T.; Ferdinandy, P. Muscle soreness-induced reduction in force generation is accompanied by increased nitric oxide content and DNA damage in human skeletal muscle. *Free Radic. Biol. Med.*, **1999**, *26*(7-8), 1059-1063.
[http://dx.doi.org/10.1016/S0891-5849(98)00309-8] [PMID: 10232851]

[98] Berliner, L.J.; Fujii, H. *In vivo* spin trapping of nitric oxide. *Antioxid. Redox Signal.*, **2004**, *6*(3), 649-656.
[http://dx.doi.org/10.1089/152308604773934413] [PMID: 15130292]

[99] Csont, T.; Páli, T.; Szilvássy, Z.; Ferdinandy, P. Lack of correlation between myocardial nitric oxide and cyclic guanosine monophosphate content in both nitrate-tolerant and -nontolerant rats. *Biochem. Pharmacol.*, **1998**, *56*(9), 1139-1144.
[http://dx.doi.org/10.1016/S0006-2952(98)00167-1] [PMID: 9802323]

[100] Komarov, A.; Mattson, D.; Jones, M.M.; Singh, P.K.; Lai, C.S. *In vivo* spin trapping of nitric oxide in mice. *Biochem. Biophys. Res. Commun.*, **1993**, *195*(3), 1191-1198.
[http://dx.doi.org/10.1006/bbrc.1993.2170] [PMID: 8216248]

[101] Suzuki, Y.; Fujii, S.; Tominaga, T.; Yoshimoto, T.; Yoshimura, T.; Kamada, H. The origin of an EPR signal observed in dithiocarbamate-loaded tissues. Copper(II)-dithiocarbamate complexes account for the narrow hyperfine lines. *Biochim. Biophys. Acta,* **1997**, *1335*(3), 242-245.
[http://dx.doi.org/10.1016/S0304-4165(97)00027-5] [PMID: 9202186]

[102] Mülsch, A.; Mordvintcev, P.; Vanin, A. Quantification of Nitric Oxide in Biological Samples by Electron Spin Resonance Spectroscopy. *Neuroprotocols,* **1992**, *1*(2), 165-173.
[http://dx.doi.org/10.1016/1058-6741(92)90048-3]

[103] Eriksson, L.E. Binding of nitric oxide to intact human erythrocytes as monitored by electron paramagnetic resonance. *Biochem. Biophys. Res. Commun.,* **1994**, *203*(1), 176-181.
[http://dx.doi.org/10.1006/bbrc.1994.2165] [PMID: 8074653]

[104] Henry, Y.; Ducrocq, C.; Drapier, J.C.; Servent, D.; Pellat, C.; Guissani, A. Nitric oxide, a biological effector. Electron paramagnetic resonance detection of nitrosyl-iron-protein complexes in whole cells. *Eur. Biophys. J.,* **1991**, *20*(1), 1-15.

[PMID: 1657584]

[105] Piknova, B.; Gladwin, M.T.; Schechter, A.N.; Hogg, N. Electron paramagnetic resonance analysis of nitrosylhemoglobin in humans during NO inhalation. *J. Biol. Chem.,* **2005**, *280*(49), 40583-40588. [http://dx.doi.org/10.1074/jbc.M506292200] [PMID: 16183646]

[106] Valgimigli, L.; Pedulli, G.F.; Paolini, M. Measurement of oxidative stress by EPR radical-probe technique. *Free Radic. Biol. Med.,* **2001**, *31*(6), 708-716. [http://dx.doi.org/10.1016/S0891-5849(01)00490-7] [PMID: 11557308]

[107] Dikalov, S.; Skatchkov, M.; Bassenge, E. Quantification of peroxynitrite, superoxide, and peroxyl radicals by a new spin trap hydroxylamine 1-hydroxy-2,2,6,6-tetramethyl-4-oxo-piperidine. *Biochem. Biophys. Res. Commun.,* **1997**, *230*(1), 54-57. [http://dx.doi.org/10.1006/bbrc.1996.5880] [PMID: 9020059]

[108] Dikalov, S.I.; Kirilyuk, I.A.; Voinov, M.; Grigor'ev, I.A. EPR detection of cellular and mitochondrial superoxide using cyclic hydroxylamines. *Free Radic. Res.,* **2011**, *45*(4), 417-430. [http://dx.doi.org/10.3109/10715762.2010.540242] [PMID: 21128732]

[109] Lichtman, J.W.; Conchello, J.A. Fluorescence microscopy. *Nat. Methods,* **2005**, *2*(12), 910-919. [http://dx.doi.org/10.1038/nmeth817] [PMID: 16299476]

[110] Csonka, C.; Páli, T.; Bencsik, P.; Görbe, A.; Ferdinandy, P.; Csont, T. Measurement of NO in biological samples. *Br. J. Pharmacol.,* **2015**, *172*(6), 1620-1632. [http://dx.doi.org/10.1111/bph.12832] [PMID: 24990201]

[111] Halliwell, B.; Whiteman, M. Measuring reactive species and oxidative damage *in vivo* and in cell culture: how should you do it and what do the results mean? *Br. J. Pharmacol.,* **2004**, *142*(2), 231-255. [http://dx.doi.org/10.1038/sj.bjp.0705776] [PMID: 15155533]

[112] Afzal, M.; Matsugo, S.; Sasai, M.; Xu, B.; Aoyama, K.; Takeuchi, T. Method to overcome photoreaction, a serious drawback to the use of dichlorofluorescin in evaluation of reactive oxygen species. *Biochem. Biophys. Res. Commun.,* **2003**, *304*(4), 619-624. [http://dx.doi.org/10.1016/S0006-291X(03)00641-7] [PMID: 12727198]

[113] Dikalov, S.I.; Harrison, D.G. Methods for detection of mitochondrial and cellular reactive oxygen species. *Antioxid. Redox Signal.,* **2014**, *20*(2), 372-382. [http://dx.doi.org/10.1089/ars.2012.4886] [PMID: 22978713]

[114] Kalyanaraman, B.; Darley-Usmar, V.; Davies, K.J.; Dennery, P.A.; Forman, H.J.; Grisham, M.B.; Mann, G.E.; Moore, K.; Roberts, L.J., II; Ischiropoulos, H. Measuring reactive oxygen and nitrogen species with fluorescent probes: challenges and limitations. *Free Radic. Biol. Med.,* **2012**, *52*(1), 1-6. [http://dx.doi.org/10.1016/j.freeradbiomed.2011.09.030] [PMID: 22027063]

[115] Wardman, P. Methods to measure the reactivity of peroxynitrite-derived oxidants toward reduced fluoresceins and rhodamines. *Methods Enzymol.,* **2008**, *441*, 261-282. [http://dx.doi.org/10.1016/S0076-6879(08)01214-7] [PMID: 18554539]

[116] Zhao, H.; Kalivendi, S.; Zhang, H.; Joseph, J.; Nithipatikom, K.; Vásquez-Vivar, J.; Kalyanaraman, B. Superoxide reacts with hydroethidine but forms a fluorescent product that is distinctly different from ethidium: potential implications in intracellular fluorescence detection of superoxide. *Free Radic. Biol. Med.,* **2003**, *34*(11), 1359-1368. [http://dx.doi.org/10.1016/S0891-5849(03)00142-4] [PMID: 12757846]

[117] Robinson, K.M.; Janes, M.S.; Pehar, M.; Monette, J.S.; Ross, M.F.; Hagen, T.M.; Murphy, M.P.; Beckman, J.S. Selective fluorescent imaging of superoxide *in vivo* using ethidium-based probes. *Proc. Natl. Acad. Sci. USA,* **2006**, *103*(41), 15038-15043. [http://dx.doi.org/10.1073/pnas.0601945103] [PMID: 17015830]

[118] Buxser, S.E.; Sawada, G.; Raub, T.J. Analytical and numerical techniques for evaluation of free radical damage in cultured cells using imaging cytometry and fluorescent indicators. *Methods Enzymol.,* **1999**, *300*, 256-275.

[http://dx.doi.org/10.1016/S0076-6879(99)00133-0] [PMID: 9919529]

[119] Dikalova, A.E.; Bikineyeva, A.T.; Budzyn, K.; Nazarewicz, R.R.; McCann, L.; Lewis, W.; Harrison, D.G.; Dikalov, S.I. Therapeutic targeting of mitochondrial superoxide in hypertension. *Circ. Res.*, **2010**, *107*(1), 106-116. [http://dx.doi.org/10.1161/CIRCRESAHA.109.214601] [PMID: 20448215]

[120] Zhou, M.; Diwu, Z.; Panchuk-Voloshina, N.; Haugland, R.P. A stable nonfluorescent derivative of resorufin for the fluorometric determination of trace hydrogen peroxide: applications in detecting the activity of phagocyte NADPH oxidase and other oxidases. *Anal. Biochem.*, **1997**, *253*(2), 162-168. [http://dx.doi.org/10.1006/abio.1997.2391] [PMID: 9367498]

[121] Zielonka, J.; Zielonka, M.; Sikora, A.; Adamus, J.; Joseph, J.; Hardy, M.; Ouari, O.; Dranka, B.P.; Kalyanaraman, B. Global profiling of reactive oxygen and nitrogen species in biological systems: high-throughput real-time analyses. *J. Biol. Chem.*, **2012**, *287*(5), 2984-2995. [http://dx.doi.org/10.1074/jbc.M111.309062] [PMID: 22139901]

[122] Zhao, B.; Ranguelova, K.; Jiang, J.; Mason, R.P. Studies on the photosensitized reduction of resorufin and implications for the detection of oxidative stress with Amplex Red. *Free Radic. Biol. Med.*, **2011**, *51*(1), 153-159. [http://dx.doi.org/10.1016/j.freeradbiomed.2011.03.016] [PMID: 21419845]

[123] Pap, E.H.; Drummen, G.P.; Post, J.A.; Rijken, P.J.; Wirtz, K.W. Fluorescent fatty acid to monitor reactive oxygen in single cells. *Methods Enzymol.*, **2000**, *319*, 603-612. [http://dx.doi.org/10.1016/S0076-6879(00)19056-1] [PMID: 10907547]

[124] Drummen, G.P.; van Liebergen, L.C.; Op den Kamp, J.A.; Post, J.A. C11-BODIPY(581/591), an oxidation-sensitive fluorescent lipid peroxidation probe: (micro)spectroscopic characterization and validation of methodology. *Free Radic. Biol. Med.*, **2002**, *33*(4), 473-490. [http://dx.doi.org/10.1016/S0891-5849(02)00848-1] [PMID: 12160930]

[125] Bapat, S.; Post, J.A.; Braam, B.; Goldschmeding, R.; Koomans, H.A.; Verkleij, A.J.; Joles, J.A. Visualizing tubular lipid peroxidation in intact renal tissue in hypertensive rats. *J. Am. Soc. Nephrol.*, **2002**, *13*(12), 2990-2996. [http://dx.doi.org/10.1097/01.ASN.0000036870.58561.81] [PMID: 12444219]

[126] Itoh, N.; Cao, J.; Chen, Z.H.; Yoshida, Y.; Niki, E. Advantages and limitation of BODIPY as a probe for the evaluation of lipid peroxidation and its inhibition by antioxidants in plasma. *Bioorg. Med. Chem. Lett.*, **2007**, *17*(7), 2059-2063. [http://dx.doi.org/10.1016/j.bmcl.2007.01.080] [PMID: 17300936]

[127] Zhu, M.; Qin, Z.J.; Hu, D.; Munishkina, L.A.; Fink, A.L. Alpha-synuclein can function as an antioxidant preventing oxidation of unsaturated lipid in vesicles. *Biochemistry,* **2006**, *45*(26), 8135-8142. [http://dx.doi.org/10.1021/bi052584t] [PMID: 16800638]

[128] Drummen, G.P.; Gadella, B.M.; Post, J.A.; Brouwers, J.F. Mass spectrometric characterization of the oxidation of the fluorescent lipid peroxidation reporter molecule C11-BODIPY(581/591). *Free Radic. Biol. Med.,* **2004**, *36*(12), 1635-1644. [http://dx.doi.org/10.1016/j.freeradbiomed.2004.03.014] [PMID: 15182863]

[129] Miller, E.W.; Albers, A.E.; Pralle, A.; Isacoff, E.Y.; Chang, C.J. Boronate-based fluorescent probes for imaging cellular hydrogen peroxide. *J. Am. Chem. Soc.*, **2005**, *127*(47), 16652-16659. [http://dx.doi.org/10.1021/ja054474f] [PMID: 16305254]

[130] Chang, M.C.; Pralle, A.; Isacoff, E.Y.; Chang, C.J. A selective, cell-permeable optical probe for hydrogen peroxide in living cells. *J. Am. Chem. Soc.*, **2004**, *126*(47), 15392-15393. [http://dx.doi.org/10.1021/ja0441716] [PMID: 15563161]

[131] Zielonka, J.; Sikora, A.; Joseph, J.; Kalyanaraman, B. Peroxynitrite is the major species formed from different flux ratios of co-generated nitric oxide and superoxide: direct reaction with boronate-based fluorescent probe. *J. Biol. Chem.*, **2010**, *285*(19), 14210-14216.

[http://dx.doi.org/10.1074/jbc.M110.110080] [PMID: 20194496]

[132] Belousov, V.V.; Fradkov, A.F.; Lukyanov, K.A.; Staroverov, D.B.; Shakhbazov, K.S.; Terskikh, A.V.; Lukyanov, S. Genetically encoded fluorescent indicator for intracellular hydrogen peroxide. *Nat. Methods,* **2006**, *3*(4), 281-286.
[http://dx.doi.org/10.1038/nmeth866] [PMID: 16554833]

[133] Choi, H.; Kim, S.; Mukhopadhyay, P.; Cho, S.; Woo, J.; Storz, G.; Ryu, S.E. Structural basis of the redox switch in the OxyR transcription factor. *Cell,* **2001**, *105*(1), 103-113.
[http://dx.doi.org/10.1016/S0092-8674(01)00300-2] [PMID: 11301006]

[134] Malinouski, M.; Zhou, Y.; Belousov, V.V.; Hatfield, D.L.; Gladyshev, V.N. Hydrogen peroxide probes directed to different cellular compartments. *PLoS One,* **2011**, *6*(1), e14564.
[http://dx.doi.org/10.1371/journal.pone.0014564] [PMID: 21283738]

[135] Gutscher, M.; Sobotta, M.C.; Wabnitz, G.H.; Ballikaya, S.; Meyer, A.J.; Samstag, Y.; Dick, T.P. Proximity-based protein thiol oxidation by H2O2-scavenging peroxidases. *J. Biol. Chem.,* **2009**, *284*(46), 31532-31540.
[http://dx.doi.org/10.1074/jbc.M109.059246] [PMID: 19755417]

[136] Wheatley, S.P.; Wang, Y.L. Indirect immunofluorescence microscopy in cultured cells. *Methods Cell Biol.,* **1998**, *57*, 313-332.
[http://dx.doi.org/10.1016/S0091-679X(08)61588-5] [PMID: 9648113]

[137] Gomez-Mejiba, S.E.; Zhai, Z.; Akram, H.; Deterding, L.J.; Hensley, K.; Smith, N.; Towner, R.A.; Tomer, K.B.; Mason, R.P.; Ramirez, D.C. Immuno-spin trapping of protein and DNA radicals: "tagging" free radicals to locate and understand the redox process. *Free Radic. Biol. Med.,* **2009**, *46*(7), 853-865.
[http://dx.doi.org/10.1016/j.freeradbiomed.2008.12.020] [PMID: 19159679]

[138] Mason, R.P. Using anti-5,5-dimethyl-1-pyrroline N-oxide (anti-DMPO) to detect protein radicals in time and space with immuno-spin trapping. *Free Radic. Biol. Med.,* **2004**, *36*(10), 1214-1223.
[http://dx.doi.org/10.1016/j.freeradbiomed.2004.02.077] [PMID: 15110386]

[139] Heyrovsky, J. The fundamental laws of polarography. *Analyst (Lond.),* **1947**, *72*(855), 229-234.
[http://dx.doi.org/10.1039/AN947720229B] [PMID: 20257059]

[140] Heyrovsky, J. Trends in Polarography. *Science,* **1960**, *132*(3420), 123-130.
[http://dx.doi.org/10.1126/science.132.3420.123] [PMID: 17775751]

[141] Portillo, R. [Jaroslav Heyrovsky, founder of polarography; 1959 Nobel Prize in Chemistry]. *An. Real Acad. Farm.,* **1960**, *26*, 203-218.
[PMID: 13737298]

[142] Sontag, G.; Kainz, G. Determination of ascorbic acid in fruit juices and lemonades by means of differential pulse polarography. *Mikrochim. Acta,* **1978**, *1*(1-2), 175-184.
[http://dx.doi.org/10.1007/BF01196991]

[143] Cortes, C.; Esteve, M.J.; Frigola, A.; Torregrosa, F. Changes in carotenoids including geometrical isomers and ascorbic acid content in orange-carrot juice during frozen storage. *Eur. Food Res. Technol.,* **2005**, *221*(1-2), 125-131.
[http://dx.doi.org/10.1007/s00217-004-1117-9]

[144] Coulson, D.M.; Crowell, W.R.; Friess, S.L. Polarography of reduced glutathione and glutathione ascorbic acid mixtures - amperometric method for determinattion of ascorbic acid. *Anal. Chem.,* **1950**, *22*(4), 525-529.
[http://dx.doi.org/10.1021/ac60040a005]

[145] Pohanka, M. Three-Dimensional Printing in Analytical Chemistry: Principles and Applications. *Anal. Lett.,* In press
[http://dx.doi.org/10.1080/00032719.2016.1166370]

[146] Metters, J.P.; Kadara, R.O.; Banks, C.E. New directions in screen printed electroanalytical sensors: an

overview of recent developments. *Analyst (Lond.),* **2011**, *136*(6), 1067-1076. [http://dx.doi.org/10.1039/c0an00894j] [PMID: 21283890]

[147] Cloetens, L.; Panee, J.; Åkesson, B. The antioxidant capacity of milk--the application of different methods *in vitro* and *in vivo. Cell. Mol. Biol.,* **2013**, *59*(1), 43-57. [PMID: 24200020]

[148] Kracmarova, A.; Pohanka, M. Electrochemical Determination of Low-Molecular-Weight Antioxidants in Blood Serum. *Chem. Listy,* **2014**, *108*(1), 64-69.

[149] Yang, Y.; Zhou, J.; Zhang, H.; Gai, P.; Zhang, X.; Chen, J. Electrochemical evaluation of total antioxidant capacities in fruit juice based on the guanine/graphene nanoribbon/glassy carbon electrode. *Talanta,* **2013**, *106*, 206-211. [http://dx.doi.org/10.1016/j.talanta.2012.12.030] [PMID: 23598118]

[150] Blanco, C.A.; de la Fuente, R.; Caballero, I.; Rodriguez-Mendez, M.L. Beer discrimination using a portable electronic tongue based on screen-printed electrodes. *J. Food Eng.,* **2015**, *157*, 57-62. [http://dx.doi.org/10.1016/j.jfoodeng.2015.02.018]

[151] Medina-Plaza, C.; Garcia-Hernandez, C.; de Saja, J.A.; Fernandez-Escudero, J.A.; Barajas, E.; Medrano, G.; Garcia-Cabezon, C.; Martin-Pedrosa, F.; Rodriguez-Mendez, M.L. The advantages of disposable screen-printed biosensors in a bioelectronic tongue for the analysis of grapes. *Lebensm. Wiss. Technol.,* **2015**, *62*(2), 940-947. [http://dx.doi.org/10.1016/j.lwt.2015.02.027]

[152] Ziyatdinova, G.; Budnikov, H. Electroanalysis of antioxidants in pharmaceutical dosage forms: state-of-the-art and perspectives. *Monatsh. Chem.,* **2015**, *146*(5), 741-753. [http://dx.doi.org/10.1007/s00706-014-1376-5]

[153] Jackowska, K.; Krysinski, P. New trends in the electrochemical sensing of dopamine. *Anal. Bioanal. Chem.,* **2013**, *405*(11), 3753-3771. [http://dx.doi.org/10.1007/s00216-012-6578-2] [PMID: 23241816]

[154] Kulesza, P.J.; Cox, J.A. Solid-state voltammetry - Analytical prospects. *Electroanalysis,* **1998**, *10*(2), 73-80. [http://dx.doi.org/10.1002/(SICI)1521-4109(199802)10:2<73::AID-ELAN73>3.0.CO;2-K]

[155] Ulakhovich, N.A.; Medyantseva, E.P.; Budnikov, G.K. Carbon-paste electrode as a sensor in voltammetric analysis. *J. Anal. Chem.,* **1993**, *48*(6), 682-694.

[156] Pohanka, M. Biosensors containing acetylcholinesterase and butyrylcholinesterase as recognition tools for detection of various compounds. *Chem. Pap.,* **2015**, *69*(1), 4-16. [http://dx.doi.org/10.2478/s11696-014-0542-x]

[157] Pohanka, M. Cholinesterases in biorecognition and biosensor construction, a review. *Anal. Lett.,* **2013**, *46*(12), 1849-1868. [http://dx.doi.org/10.1080/00032719.2013.780240]

[158] Pohanka, M.; Bandouchova, H.; Sobotka, J.; Sedlackova, J.; Soukupova, I.; Pikula, J. Ferric reducing antioxidant power and square wave voltammetry for assay of low molecular weight antioxidants in blood plasma: performance and comparison of methods. *Sensors (Basel),* **2009**, *9*(11), 9094-9103. [http://dx.doi.org/10.3390/s91109094] [PMID: 22291555]

[158] Cata, A.; Stefanut, M.N.; Pop, R.; Tanasie, C.; Mosoarca, C.; Zamfir, A.D. Evaluation of Antioxidant Activities of Some Small Fruits Containing Anthocyanins Using Electrochemical and Chemical Methods. *Croat. Chem. Acta,* **2016**, *89*(1), 37-48. [http://dx.doi.org/10.5562/cca2656]

[160] Cabaj, J.; Jedrychowska, A.; Swist, A.; Soloducho, J. Tyrosinase Biosensor for Antioxidants Based on Semiconducting Polymer Support. *Electroanalysis,* **2016**, *28*(6), 1383-1390. [http://dx.doi.org/10.1002/elan.201500523]

[161] Cruz, D.; Barroso, M.F.; Ramalhosa, M.J.; Coelho, A.; da Silva, H.; Duarte, A.J.; Gonzalez-Garcia,

M.B.; Carvalho, A.P.; Delerue-Matos, C. DNA-based sensor against nitrite oxide radical: Evaluation of total antioxidant capacity in beverages. *J. Electroanal. Chem.,* **2016**, *763*, 110-115. [http://dx.doi.org/10.1016/j.jelechem.2015.12.045]

[162] Garcia, L.F.; Benjamin, S.R.; Marreto, R.N.; Lopes, F.M.; Golveia, J.C.; Fernandes, N.C.; Santiago, M.F.; Gil, E.D. Laccase Carbon Paste Based Biosensors for Antioxidant Capacity. The Effect of Different Modifiers. *Int. J. Electrochem. Sci.,* **2015**, *10*(7), 5650-5660.

[163] Loś, J.M.; Loś, M.; Węgrzyn, A.; Węgrzyn, G. Altruism of Shiga toxin-producing Escherichia coli: recent hypothesis *versus* experimental results. *Front. Cell. Infect. Microbiol.,* **2013**, *2*(2), 166. [PMID: 23316482]

[164] Gambino, M.; Cappitelli, F. Mini-review: Biofilm responses to oxidative stress. *Biofouling,* **2016**, *32*(2), 167-178. [http://dx.doi.org/10.1080/08927014.2015.1134515] [PMID: 26901587]

[165] Nguyen, C.T.; Park, S.S.; Rhee, D.K. Stress responses in Streptococcus species and their effects on the host. *J. Microbiol.,* **2015**, *53*(11), 741-749. [http://dx.doi.org/10.1007/s12275-015-5432-6] [PMID: 26502957]

[166] Licznerska, K.; Nejman-Faleńczyk, B.; Bloch, S.; Dydecka, A.; Topka, G.; Gąsior, T.; Węgrzyn, A.; Węgrzyn, G. Oxidative stress in Shiga toxin production by enterohemorrhagic Escherichia coli. *Oxid. Med. Cell. Longev.,* **2016**, *2016*, 3578368. [http://dx.doi.org/10.1155/2016/3578368] [PMID: 26798420]

[167] Goswami, M.; Mangoli, S.H.; Jawali, N. Involvement of reactive oxygen species in the action of ciprofloxacin against Escherichia coli. *Antimicrob. Agents Chemother.,* **2006**, *50*(3), 949-954. [http://dx.doi.org/10.1128/AAC.50.3.949-954.2006] [PMID: 16495256]

[168] Dwyer, D.J.; Kohanski, M.A.; Hayete, B.; Collins, J.J. Gyrase inhibitors induce an oxidative damage cellular death pathway in Escherichia coli. *Mol. Syst. Biol.,* **2007**, *3*(3), 91. [PMID: 17353933]

[169] Kohanski, M.A.; Dwyer, D.J.; Hayete, B.; Lawrence, C.A.; Collins, J.J. A common mechanism of cellular death induced by bactericidal antibiotics. *Cell,* **2007**, *130*(5), 797-810. [http://dx.doi.org/10.1016/j.cell.2007.06.049] [PMID: 17803904]

[170] Kohanski, M.A.; Dwyer, D.J.; Wierzbowski, J.; Cottarel, G.; Collins, J.J. Mistranslation of membrane proteins and two-component system activation trigger antibiotic-mediated cell death. *Cell,* **2008**, *135*(4), 679-690. [http://dx.doi.org/10.1016/j.cell.2008.09.038] [PMID: 19013277]

[171] Foti, J.J.; Devadoss, B.; Winkler, J.A.; Collins, J.J.; Walker, G.C. Oxidation of the guanine nucleotide pool underlies cell death by bactericidal antibiotics. *Science,* **2012**, *336*(6079), 315-319. [http://dx.doi.org/10.1126/science.1219192] [PMID: 22517853]

[172] Yan, Z.; Li, G.; Gao, Y.; Zhai, W.; Qi, Y.; Zhai, M. The extracellular death factor (EDF) protects Escherichia coli by scavenging hydroxyl radicals induced by bactericidal antibiotics. *Springerplus,* **2015**, *4*(4), 182. [http://dx.doi.org/10.1186/s40064-015-0968-9] [PMID: 25932369]

[173] Imlay, J.A. Diagnosing oxidative stress in bacteria: not as easy as you might think. *Curr. Opin. Microbiol.,* **2015**, *24*(24), 124-131. [http://dx.doi.org/10.1016/j.mib.2015.01.004] [PMID: 25666086]

[174] Palmieri, B.; Sblendorio, V. Oxidative stress tests: overview on reliability and use. Part I. *Eur. Rev. Med. Pharmacol. Sci.,* **2007**, *11*(5), 309-342. [PMID: 18074940]

[175] Thomas, V.C.; Chaudhari, S.S.; Jones, J.; Zimmerman, M.C.; Bayles, K.W. Electron paramagnetic resonance (EPR) spectroscopy to detect reactive oxygen species in Staphylococcus aureus. *Bio Protoc.,* **2015**, *5*(17), e1586.

[http://dx.doi.org/10.21769/BioProtoc.1586] [PMID: 27182534]

[176] Meyer, A.J.; Dick, T.P. Fluorescent protein-based redox probes. *Antioxid. Redox Signal.*, **2010**, *13*(5), 621-650.
[http://dx.doi.org/10.1089/ars.2009.2948] [PMID: 20088706]

[177] Arias-Barreiro, C.R.; Okazaki, K.; Koutsaftis, A.; Inayat-Hussain, S.H.; Tani, A.; Katsuhara, M.; Kimbara, K.; Mori, I.C. A bacterial biosensor for oxidative stress using the constitutively expressed redox-sensitive protein roGFP2. *Sensors (Basel)*, **2010**, *10*(7), 6290-6306.
[http://dx.doi.org/10.3390/s100706290] [PMID: 22163550]

[178] van der Heijden, J.; Bosman, E.S.; Reynolds, L.A.; Finlay, B.B. Direct measurement of oxidative and nitrosative stress dynamics in Salmonella inside macrophages. *Proc. Natl. Acad. Sci. USA*, **2015**, *112*(2), 560-565.
[http://dx.doi.org/10.1073/pnas.1414569112] [PMID: 25548165]

[179] Bhaskar, A.; Chawla, M.; Mehta, M.; Parikh, P.; Chandra, P.; Bhave, D.; Kumar, D.; Carroll, K.S.; Singh, A. Reengineering redox sensitive GFP to measure mycothiol redox potential of Mycobacterium tuberculosis during infection. *PLoS Pathog.*, **2014**, *10*(1), e1003902.
[http://dx.doi.org/10.1371/journal.ppat.1003902] [PMID: 24497832]

[180] van der Heijden, J.; Vogt, S.L.; Reynolds, L.A.; Peña-Díaz, J.; Tupin, A.; Aussel, L.; Finlay, B.B. Exploring the redox balance inside gram-negative bacteria with redox-sensitive GFP. *Free Radic. Biol. Med.*, **2016**, *91*(91), 34-44.
[http://dx.doi.org/10.1016/j.freeradbiomed.2015.11.029] [PMID: 26627936]

[181] Herrera, G.; Martinez, A.; Blanco, M.; O'Connor, J.E. Assessment of Escherichia coli B with enhanced permeability to fluorochromes for flow cytometric assays of bacterial cell function. *Cytometry*, **2002**, *49*(2), 62-69.
[http://dx.doi.org/10.1002/cyto.10148] [PMID: 12357461]

[182] Ma, L.; Gao, Y.; Maresso, A.W. Escherichia coli free radical-based killing mechanism driven by a unique combination of iron restriction and certain antibiotics. *J. Bacteriol.*, **2015**, *197*(23), 3708-3719.
[http://dx.doi.org/10.1128/JB.00758-15] [PMID: 26391205]

[183] Dwivedi, S.; Wahab, R.; Khan, F.; Mishra, Y.K.; Musarrat, J.; Al-Khedhairy, A.A. Reactive oxygen species mediated bacterial biofilm inhibition *via* zinc oxide nanoparticles and their statistical determination. *PLoS One*, **2014**, *9*(11), e111289.
[http://dx.doi.org/10.1371/journal.pone.0111289] [PMID: 25402188]

[184] Nobre, L.S.; Todorovic, S.; Tavares, A.F.; Oldfield, E.; Hildebrandt, P.; Teixeira, M.; Saraiva, L.M. Binding of azole antibiotics to Staphylococcus aureus flavohemoglobin increases intracellular oxidative stress. *J. Bacteriol.*, **2010**, *192*(6), 1527-1533.
[http://dx.doi.org/10.1128/JB.01378-09] [PMID: 20097858]

[185] Choi, H.; Yang, Z.; Weisshaar, J.C. Single-cell, real-time detection of oxidative stress induced in Escherichia coli by the antimicrobial peptide CM15. *Proc. Natl. Acad. Sci. USA*, **2015**, *112*(3), E303-E310.
[http://dx.doi.org/10.1073/pnas.1417703112] [PMID: 25561551]

[186] Peters, T.R.; Tosk, J.M.; Goulbourne, E.A., Jr Lucigenin chemiluminescence as a probe for measuring reactive oxygen species production in Escherichia coli. *Anal. Biochem.*, **1990**, *186*(2), 316-319.
[http://dx.doi.org/10.1016/0003-2697(90)90087-P] [PMID: 2194399]

[187] Kervinen, M.; Pätsi, J.; Finel, M.; Hassinen, I.E. Lucigenin and coelenterazine as superoxide probes in mitochondrial and bacterial membranes. *J. Anal. Biochem*, **2004**, *324*(1), 45-51.
[http://dx.doi.org/https://doi.org/10.1016/j.ab.2003.09.004] [PMID: 14654044]

[188] Park, A.M.; Li, Q.; Nagata, K.; Tamura, T.; Shimono, K.; Sato, E.F.; Inoue, M. Oxygen tension regulates reactive oxygen generation and mutation of Helicobacter pylori. *Free Radic. Biol. Med.*, **2004**, *36*(9), 1126-1133.
[http://dx.doi.org/10.1016/j.freeradbiomed.2004.02.001] [PMID: 15082066]

[189] Tan, L.T.; Ser, H.L.; Yin, W.F.; Chan, K.G.; Lee, L.H.; Goh, B.H. Investigation of antioxidative and anticancer potentials of Streptomyces sp. MUM256 isolated from Malaysia mangrove soil. *Front. Microbiol.,* **2015**, *6*(6), 1316.
[PMID: 26635777]

[190] Licznerska, K.; Dydecka, A.; Bloch, S.; Topka, G.; Nejman-Faleńczyk, B.; Węgrzyn, A.; Węgrzyn, G. The role of the exo-xis region in oxidative stress-mediated induction of Shiga toxin-converting prophages. *Oxid. Med. Cell. Longev.,* **2016**, *2016*, 8453135.
[http://dx.doi.org/10.1155/2016/8453135] [PMID: 26798427]

[191] Park, J.W. Oxidative damage to bacterial DNA and evidence for its repair. *Arch. Pharm. Res.,* **1990**, (13), 252.
[http://dx.doi.org/10.1007/BF02856530]

[192] Park, E.M.; Shigenaga, M.K.; Degan, P.; Korn, T.S.; Kitzler, J.W.; Wehr, C.M.; Kolachana, P.; Ames, B.N. Assay of excised oxidative DNA lesions: isolation of 8-oxoguanine and its nucleoside derivatives from biological fluids with a monoclonal antibody column. *Proc. Natl. Acad. Sci. USA,* **1992**, *89*(8), 3375-3379.
[http://dx.doi.org/10.1073/pnas.89.8.3375] [PMID: 1565629]

[193] Yost, A.D.; Joshi, S.G. Atmospheric nonthermal plasma-treated PBS inactivates Escherichia coli by oxidative DNA damage. *PLoS One,* **2015**, *10*(10), e0139903.
[http://dx.doi.org/10.1371/journal.pone.0139903] [PMID: 26461113]

[194] Kimmitt, P.T.; Harwood, C.R.; Barer, M.R. Toxin gene expression by shiga toxin-producing Escherichia coli: the role of antibiotics and the bacterial SOS response. *Emerg. Infect. Dis.,* **2000**, *6*(5), 458-465.
[http://dx.doi.org/10.3201/eid0605.000503] [PMID: 10998375]

[195] Maisonneuve, E.; Fraysse, L.; Lignon, S.; Capron, L.; Dukan, S. Carbonylated proteins are detectable only in a degradation-resistant aggregate state in Escherichia coli. *J. Bacteriol.,* **2008**, *190*(20), 6609-6614.
[http://dx.doi.org/10.1128/JB.00588-08] [PMID: 18689474]

[196] Pérez, J.M.; Calderón, I.L.; Arenas, F.A.; Fuentes, D.E.; Pradenas, G.A.; Fuentes, E.L.; Sandoval, J.M.; Castro, M.E.; Elías, A.O.; Vásquez, C.C. Bacterial toxicity of potassium tellurite: unveiling an ancient enigma. *PLoS One,* **2007**, *2*(2), e211.
[http://dx.doi.org/10.1371/journal.pone.0000211] [PMID: 17299591]

[197] Mostertz, J.; Hecker, M. Patterns of protein carbonylation following oxidative stress in wild-type and sigB Bacillus subtilis cells. *Mol. Genet. Genomics,* **2003**, *269*(5), 640-648.
[http://dx.doi.org/10.1007/s00438-003-0877-4] [PMID: 12845527]

[198] Loi, V.V.; Rossius, M.; Antelmann, H. Redox regulation by reversible protein S-thiolation in bacteria. *Front. Microbiol.,* **2015**, *6*(6), 187.
[PMID: 25852656]

[199] Lillig, C.H.; Potamitou, A.; Schwenn, J.D.; Vlamis-Gardikas, A.; Holmgren, A. Redox regulation of 3′-phosphoadenylylsulfate reductase from Escherichia coli by glutathione and glutaredoxins. *J. Biol. Chem.,* **2003**, *278*(25), 22325-22330.
[http://dx.doi.org/10.1074/jbc.M302304200] [PMID: 12682041]

[200] Ansong, C.; Wu, S.; Meng, D.; Liu, X.; Brewer, H.M.; Deatherage Kaiser, B.L.; Nakayasu, E.S.; Cort, J.R.; Pevzner, P.; Smith, R.D.; Heffron, F.; Adkins, J.N.; Pasa-Tolic, L. Top-down proteomics reveals a unique protein S-thiolation switch in Salmonella Typhimurium in response to infection-like conditions. *Proc. Natl. Acad. Sci. USA,* **2013**, *110*(25), 10153-10158.
[http://dx.doi.org/10.1073/pnas.1221210110] [PMID: 23720318]

[201] Chen, N.H.; Couñago, R.M.; Djoko, K.Y.; Jennings, M.P.; Apicella, M.A.; Kobe, B.; McEwan, A.G. A glutathione-dependent detoxification system is required for formaldehyde resistance and optimal survival of Neisseria meningitidis in biofilms. *Antioxid. Redox Signal.,* **2013**, *18*(7), 743-755.

[http://dx.doi.org/10.1089/ars.2012.4749] [PMID: 22937752]

[202] Castellano, I.; Ruocco, M.R.; Cecere, F.; Di Maro, A.; Chambery, A.; Michniewicz, A.; Parlato, G.; Masullo, M.; De Vendittis, E. Glutathionylation of the iron superoxide dismutase from the psychrophilic eubacterium Pseudoalteromonas haloplanktis. *Biochim. Biophys. Acta,* **2008**, *1784*(5), 816-826.
[http://dx.doi.org/10.1016/j.bbapap.2008.02.003] [PMID: 18328273]

[203] Chardonnet, S.; Sakr, S.; Cassier-Chauvat, C.; Le Maréchal, P.; Chauvat, F.; Lemaire, S.D.; Decottignies, P. First proteomic study of S-glutathionylation in cyanobacteria. *J. Proteome Res.*, **2015**, *14*(1), 59-71.
[http://dx.doi.org/10.1021/pr500625a] [PMID: 25208982]

[204] Kumar, J.K.; Tabor, S.; Richardson, C.C. Proteomic analysis of thioredoxin-targeted proteins in Escherichia coli. *Proc. Natl. Acad. Sci. USA,* **2004**, *101*(11), 3759-3764.
[http://dx.doi.org/10.1073/pnas.0308701101] [PMID: 15004283]

[205] Leichert, L.I.; Jakob, U. Protein thiol modifications visualized *in vivo. PLoS Biol.,* **2004**, *2*(11), e333.
[http://dx.doi.org/10.1371/journal.pbio.0020333] [PMID: 15502869]

[206] Nichols, D.S.; McMeekin, T.A. Biomarker techniques to screen for bacteria that produce polyunsaturated fatty acids. *J. Microbiol. Methods,* **2002**, *48*(2-3), 161-170.
[http://dx.doi.org/10.1016/S0167-7012(01)00320-7] [PMID: 11777566]

[207] Ursini, F.; Bindoli, A. The role of selenium peroxidases in the protection against oxidative damage of membranes. *Chem. Phys. Lipids,* **1987**, *44*(2-4), 255-276.
[http://dx.doi.org/10.1016/0009-3084(87)90053-3] [PMID: 3311419]

[208] Tasaka, Y.; Gombos, Z.; Nishiyama, Y.; Mohanty, P.; Ohba, T.; Ohki, K.; Murata, N. Targeted mutagenesis of acyl-lipid desaturases in Synechocystis: evidence for the important roles of polyunsaturated membrane lipids in growth, respiration and photosynthesis. *EMBO J.,* **1996**, *15*(23), 6416-6425.
[PMID: 8978669]

[209] Boylan, J.A.; Lawrence, K.A.; Downey, J.S.; Gherardini, F.C. Borrelia burgdorferi membranes are the primary targets of reactive oxygen species. *Mol. Microbiol.,* **2008**, *68*(3), 786-799.
[http://dx.doi.org/10.1111/j.1365-2958.2008.06204.x] [PMID: 18373524]

[210] Seljeskog, E.; Hervig, T.; Mansoor, M.A. A novel HPLC method for the measurement of thiobarbituric acid reactive substances (TBARS). A comparison with a commercially available kit. *Clin. Biochem.,* **2006**, *39*(9), 947-954.
[http://dx.doi.org/10.1016/j.clinbiochem.2006.03.012] [PMID: 16781699]

[211] González-Flecha, B.; Demple, B. Homeostatic regulation of intracellular hydrogen peroxide concentration in aerobically growing Escherichia coli. *J. Bacteriol.,* **1997**, *179*(2), 382-388.
[http://dx.doi.org/10.1128/jb.179.2.382-388.1997] [PMID: 8990289]

[212] Semchyshyn, H.; Bagnyukova, T.; Storey, K.; Lushchak, V. Hydrogen peroxide increases the activities of soxRS regulon enzymes and the levels of oxidized proteins and lipids in Escherichia coli. *Cell Biol. Int.,* **2005**, *29*(11), 898-902.
[http://dx.doi.org/10.1016/j.cellbi.2005.08.002] [PMID: 16202627]

[213] Joshi, S.G.; Cooper, M.; Yost, A.; Paff, M.; Ercan, U.K.; Fridman, G.; Friedman, G.; Fridman, A.; Brooks, A.D. Nonthermal dielectric-barrier discharge plasma-induced inactivation involves oxidative DNA damage and membrane lipid peroxidation in Escherichia coli. *Antimicrob. Agents Chemother.,* **2011**, *55*(3), 1053-1062.
[http://dx.doi.org/10.1128/AAC.01002-10] [PMID: 21199923]

[214] Aubron, C.; Glodt, J.; Matar, C.; Huet, O.; Borderie, D.; Dobrindt, U.; Duranteau, J.; Denamur, E.; Conti, M.; Bouvet, O. Variation in endogenous oxidative stress in Escherichia coli natural isolates during growth in urine. *BMC Microbiol.,* **2012**, *12*(12), 120.
[http://dx.doi.org/10.1186/1471-2180-12-120] [PMID: 22727065]

[215] Amaretti, A.; di Nunzio, M.; Pompei, A.; Raimondi, S.; Rossi, M.; Bordoni, A. Antioxidant properties of potentially probiotic bacteria: *in vitro* and *in vivo* activities. *Appl. Microbiol. Biotechnol.,* **2013**, *97*(2), 809-817.
[http://dx.doi.org/10.1007/s00253-012-4241-7] [PMID: 22790540]

[216] Pawar, R.; Mohandass, C.; Sivaperumal, E.; Sabu, E.; Rajasabapathy, R.; Jagtap, T. Epiphytic marine pigmented bacteria: A prospective source of natural antioxidants. *Braz. J. Microbiol.,* **2015**, *46*(1), 29-39.
[http://dx.doi.org/10.1590/S1517-838246120130353] [PMID: 26221086]

[217] Larrauri, J.A.; Sanchez-Moreno, C.; Saura-Calixo, F. Effect of temperature on the free radical scavenging capacity of extracts from red and white grape pomace peels. *J. Agric. Food Chem.,* **1998**, (46), 2694-2697.
[http://dx.doi.org/10.1021/jf980017p]

[218] Takao, T.; Kitatani, F.; Watanabe, N.; Yagi, A.; Sakata, K. A simple screening method for antioxidants and isolation of several antioxidant produced by marine bacteria from fish and shellfish. *Biosci. Biotechnol. Biochem.,* **1994**, (58), 1780-1783.
[http://dx.doi.org/10.1271/bbb.58.1780]

[219] Benzie, I.F.; Strain, J.J. The ferric reducing ability of plasma (FRAP) as a measure of "antioxidant power": the FRAP assay. *Anal. Biochem.,* **1996**, *239*(1), 70-76.
[http://dx.doi.org/10.1006/abio.1996.0292] [PMID: 8660627]

[220] Re, R.; Pellegrini, N.; Proteggente, A.; Pannala, A.; Yang, M.; Rice-Evans, C. Antioxidant activity applying an improved ABTS radical cation decolorization assay. *Free Radic. Biol. Med.,* **1999**, *26*(9-10), 1231-1237.
[http://dx.doi.org/10.1016/S0891-5849(98)00315-3] [PMID: 10381194]

CHAPTER 5

The Application of Vibrational Spectroscopy in Studies of Structural Polymorphism of Drugs

Przemysław Zalewski[1], **Gabriela Wiergowska**[1,2], **Joanna Goscianska**[3], **Kornelia Lewandowska**[4] and **Judyta Cielecka-Piontek**[1,*]

[1] *Department of Pharmaceutical Chemistry, Faculty of Pharmacy, Poznan University of Medical Sciences, Grunwaldzka 6, 60-780Poznań, Poland*

[2] *PozLab sp. z o.o. (Contract Research Organization) Parkowa 2, 60-775Poznań, Poland*

[3] *Laboratory of Applied Chemistry, Faculty of Chemistry, Adam Mickiewicz University in Poznań, Umultowska 89B, 61-614Poznań, Poland*

[4] *Institute of Molecular Physics, Polish Academy of Science, Mariana Smoluchowskiego 17, 60-179Poznań, Poland*

Abstract: Structural polymorphism of active pharmaceutical ingredients (APIs) is a common phenomenon being a subject of continuous investigation. Depending on the crystalline or amorphous form, APIs may have different physicochemical properties, including solubility, which ultimately affects the various concentrations of the tested polymorphic forms of the drug in body fluids. Currently, the different scanning calorimetry and X-ray diffraction are the methods of choice in studies of structural polymorphism of drugs. Three vibrational spectroscopic techniques namely: Fourier-transform infrared, Attenuated total reflection and Raman spectroscopy may be indicated as alternative, non-destructive, fast and cheap methods. In addition, the theoretical approach based on quantum-chemical calculations and chemometrics solutions is considered an important support in identification of various polymorphic forms of drugs by determination of positions of bands and their intensities in analyzed samples. In the first part of this chapter, we will focus on description of the differences in the structure of crystalline and amorphous forms of APIs, their pharmaceutical implications and characteristic of vibrational techniques that can be used in studies on polymorphism of drugs. In the second part, we will present the most important examples of the application of the above mentioned vibrational techniques to identify polymorphic and amorphous forms of APIs with different profiles of pharmacological activity and conventional and innovative excipients. Finally, the advantages and limitations of vibrational spectroscopy to study the structural polymorphism of drugs will be indicated and discussed.

[*] **Corresponding Author Judyta Cielecka-Piontek:** Department of Pharmaceutical Chemistry, Faculty of Pharmacy, Poznan University of Medical Sciences, Grunwaldzka 6, 60-780 Poznań, Poland; Tel: +48861546649; E-mail: jpiontek@ump.edu.pl

Atta-ur-Rahman, Sibel A. Ozkan & Rida Ahmed (Eds.)

Keywords: Active Pharmaceutical Ingredients, Amorphous form, Excipients, Pharmaceutical Analysis, Pharmaceuticals, Physicochemical properties, Polymorphous forms.

INTRODUCTION

Pharmaceutical formulations consist usually of API (*active pharmaceutical ingredient*) and excipients. API can be of synthetic, biotechnology and natural origin. Pharmaceutical excipients are inorganic or organic compounds, which are necessary for the formulation of medicinal preparations suitable for direct application to a patient which lack of any pharmacological activity.

Regardless of its origin, API and excipients can appear in different polymorphic forms as well as in amorphous state. The polymorphic and amorphous forms of APIs and excipients can differ in physicochemical properties what in turn can significantly impede preformulation and formulation studies [1]. The appearance of polymorphic and amorphous forms of API and excipients can finally influence on therapeutic safety and effectiveness of the treatment [2]. At present, polymorphism of APIs and excipients is observed in approximately 80 to 90 percent of medicinal substances. According to Walter McCrone's suggestion that the number of polymorphic forms identified for a given compound is proportional to the time and money spent on researching the compound, we can suppose that the number of polymorphic and amorphous forms of APIs and excipients will be dramatically increased in the future.

Polymorphic and amorphous forms of API and excipients can change their forms during preformulation and formulation studies. At present, a tendency to develop a Combo type of preparations (containing more than one API) is observed, which is understandable considering the patient's improved comfort. However, it is a challenge for API-API systems (especially including co-amorphous APIs), where a number of additional variables exist that may result from interactions between the APIs and their changes regarding transformation of amorphous or polymorphic forms. While excipients are classified according to their role in pharmaceutical formulation and on their interactions with API influencing drug delivery. The main classic groups of excipients are: antioxidants, coating materials, diluents, disintegrants, binders, lubricants, glidants, taste and smell-improvers, conserving agents and ointment bases. Some of them may play multipurpose role in formulation, for example, methylcellulose is applied as coating material and binder in tablets. Moreover, the most modern group of excipients are innovative materials which can stabilize selected polymorphic or amorphous forms of API or excipients *e.g.* mesoporous silica materials.

An understanding of the behavior of polymorphs and amorphous forms of certain groups of drugs under exposure to such physicochemical factors as increased temperature, humidity or pressing forces may contribute to the general knowledge and form the basis for pre-formulation studies of certain active pharmaceutical ingredients [1, 2]. In order to control the above objectives, it is important to select appropriate analytical methods to assess identification of polymorphic and amorphous forms and conversion tracking between them without causing structural changes during the measurement.

Polymorphism of Active Pharmaceutical Substance

The phenomenon of structural polymorphism is defined as the possibility that a given substance exists in the same chemical composition in two or more crystalline phases and/or different molecular compositions. In addition to crystalline forms, APIs can appear as amorphous compounds which do not possess a distinguishable crystal lattice and consist of disordered arrangements of molecules. Thus polymorphism of medicinal substances, as defined by the International Conference on Harmonization Guideline Q3A, has to include the characteristic of polymorphs, solvates and amorphous forms. It was the discovery of significant differences in a range of physicochemical characteristics of polymorphs and amorphous forms and the possibility of their mutual conversion during the development of pharmaceutical formulations or storage leading to the risk of uncontrollable changes in bioavailability that stimulated research into the polymorphism of solid-state drugs [1, 2]. The polymorphism of solid-state drugs has a pivotal role in the quality, therapeutic safety and effectiveness of medicinal products as polymorphic forms of drugs which may vary in physicochemical characteristics, mainly the melting point, solubility as well as optical and mechanical properties, including the occurrence of features typical of amorphous forms [1, 2].

The Influence of Polymorphism and Amorphization on the Physicochemical Properties of Active Pharmaceutical Ingredients and Excipients

The effect of occurrence of polymorphic/amorphous forms is different apparent solubility, permeability and chemical reactivity in the solid state, being the result of differences in solid-state phase transformations, dehydration and desolvation processes. Amorphous drug substances easily interact with water. They are also more likely to undergo solid-state reactions. The most important fact is that a major part of any formulation effort is devoted to the choice of excipients and processes which minimize the chemical stability and reactivity of the drug. The potential interactions between polymorph and excipients show chemical and physical characteristics [3 - 129].

Apart from, relationship between polymorphism and chemical reactivity and stability, polymorphism can affect the mechanical properties of drug particles. The presence of polymorphs may impact the manufacturability and physical attributes of tablets, like in case of metaprolol and paracetamol [59, 93]. The scientific literature presents different effects of pharmaceutical processes on drug polymorphs, solvates and phase conversions, while mechanical properties of polymorphs are the most important factor distinguishing them during a process of development of pharmaceutical formulation. However, processing-induced transformations of polymorphs are often difficult to predict due to complex effect of polymorphism on properties such as hardness, yield pressure, elasticity, compressibility and bonding strength [6]. It should be mentioned that other process stresses such as drying, grinding, milling wet granulation can accelerate the phase conversion of polymorphic solids. Mechanical properties of amorphous drug forms may be different from those of crystalline drug due to the absence of long range packing. The mechanical properties of amorphous forms are less understood than those of crystalline polymorphs and lack of information on mechanical properties of amorphous drugs may result from physical and chemical instability of these forms.

Vibrational Methods Used for Identification of Crystalline and Amorphous Forms of Active Pharmaceutical Ingredients and Excipients

The following techniques are considered for usage in aforementioned studies of polymorphism and amorphization of drugs: crystallography, spectroscopy, microscopy and thermal analysis [130 - 133]. They permit to examine different aspects of structure, dynamics and energetics in solid state of drugs.

As the first techniques, crystal X-ray diffraction (XRD) and powder X-ray diffraction (PXRD) are recommended [127, 132, 135]. They allow to determine differences in packing and conformation of molecules as well as to monitor phase transitions. However, an important feature of these techniques is that they cannot differentiate between true polymorphs and pseudo-polymorphs. Moreover, this technique is destructive method. The most prominent methods among the vibrational spectroscopic ones for identification of polymorph are infrared, Raman, attenuated Total Reflectance and broadband dielectric spectroscopy. These techniques are especially reliable for the polymorphs studies because hydrogen-bonding patterns and other "weak" interactions distinguishing polymorphic forms and the functional groups result in display shifts of various degrees. Infrared absorption spectroscopy and Raman spectroscopy provide similar chemical information. However Raman spectroscopy has a number of important advantages such as *in situ* studies of polymorphism behind glass and in water as well as possibility of connection with microscope. The low wavenumber

Raman scattering and near infrared spectroscopy was found to be especially significant in studies of polymorphism because low wavenumber vibrations are directly related to the characteristic defining the polymorphism of a drug. Solid-state nuclear magnetic resonance (NMR) is well-suited to study amorphous forms of pharmaceuticals and solvates that are usually trivial do detect. Moreover, it can be useful technique to understand polymorphic transformations and molecular motion in the solid state of polymorphs in case of spectra collected at various temperatures [42, 47]. Identifications of vibrational spectra are more and more often supported by theoretical calculations based on density functional theory (DFT). The quantum-chemical calculations are used in order to obtain accurate normal mode descriptions in terms of the potential energy distributions (PEDs). They become also a promising and cost-effective tool for calculating the effect of hydrogen bonding on vibrational modes of drugs. Scanning electron microscopy is a rapid tool for evaluation of shapes of polymorph crystals. However, it is usually difficult to determine if differences in morphology are caused by polymorphism or are simply a result of changes in growth conditions or solvent; therefore, this method can provide differentiation among forms only in combination with other analytical methods. Differential scanning calorimetry (DSC) and thermal gravimetric analysis (TGA), separately or together, are often the first steps in a comprehensive search for polymorphs and the determination of their stability relationship [130]. However, DSC can be misleading in cases when sample decomposition occurs during heating. TGA is effective in analysis of pseudopolimorphs because it can detect decomposition that occurs with release of volatile products.

The source literature shows vast information about the application of vibrational techniques such as FT-IR, Raman and ATR for identification of polymorphic and amorphous forms of API.

Infrared Absorption Spectroscopy

Infrared spectroscopy (IR) is one of the most important analytical techniques. It is a based on the vibrations of the atoms of a molecule. An infrared spectrum is commonly obtained by passing infrared radiation through a sample and determining what fraction of the incident radiation is absorbed at a particular energy. The energy at which any peak in an absorption spectrum appears corresponds to the frequency of a vibration of a part of a sample molecule.

IR spectroscopy is one of the longest used methods for identification of APIs in the solid state [134 - 146]. Polymorphic properties remain unchanged during the analysis and very small amount of substance to be examined is used during the study. Another advantage of this method is the acquisition of high quality spectrum in a very short time. On the other hand halide exchange between potassium bromide (KBr) and analyte could take place in this method. The Fourier transform technique (FTIR) is used for adequate characterization of polymorphs [144]. The greatest advantages of this method are minimization of transmission and beam attenuation problems.

Attenuated Total Reflectance Spectroscopy

Attenuated Total Reflectance (ATR) uses a property of total internal reflection resulting in an evanescent wave. A beam of infrared light is passed through the ATR crystal in such a way that it reflects at least once off the internal surface in contact with the sample. This reflection forms the evanescent wave which extends into the sample. The penetration depth into the sample is typically between 0.5 and 2 micrometres, with the exact value being determined by the wavelength of light, the angle of incidence and the indices of refraction for the ATR crystal and the medium being probed. Attenuated Total Reflectance is today the most widely used FTIR sampling tool [144]. It is a simple technique which could be used for characterization substance in solid state [145]. Similarly to IR method a very small amount of substance to be examined is used during the analysis. On the other hand very weak penetration of IR radiation significantly reduces application of this method in characterization of polymorphs.

Raman Spectroscopy

Almost 100 years after the discovery of the Raman scattering phenomenon, related analytical techniques have emerged as important tools in pharmaceutical sciences [146 - 163]. Raman spectroscopy is frontier, non-invasive analytical technique used in molecular-based drug discovery, design of innovative drug delivery systems and quality control of finished products [151 - 153]. Raman spectrometry is a light scattering process in which the specimen under examination is irradiated with intense monochromatic light and the light scattered from the specimen is analyzed for frequency shifts. This technique is complementary to IR. However there are many advantages of Raman spectroscopy over IR. Generally band which are not intensive in IR spectrum are stronger in Raman spectra. Raman bands have very specific frequencies and finally samples could be analyzed in glass containers in a presence of water.

The above techniques have several applications about s pharmacological activity mechanisms of API such as ampicillin [164], bicalutamide [165] and furosemide [166]. For instance, changes in FT-IR spectrum were observed in the case of polymorphs of mebendazole [167] while differences in Raman dispersion spectrum were observed for carbamazepine [168] and indomethacin [169]. Spectra received in the ATR technique were registered for cimetidine [170]. The application of FT-IR and Raman techniques is particularly indicated for estimating the amorphousness of APIs form. Due to the transience of the amorphous state, it is crucial to apply diagnostic methods that do not require sample compression in the time of marking as it might limit the conversion to crystalline forms. Below the application options for other selected spectral techniques of estimating the identity of selected APIs are shown:

- Polymorphic types of I, II, III benzocaine after applying the FT-IR technique (Fig. **1**, Table **1**);
- Amorphous sample of naratriptan hydrochloride after applying the FT-IR technique (Fig. **2**, Table **2**);
- Amorphous sample of ranitidine after applying the ATR technique (Fig. **3**, Table **3**);
- Amorphous sample of vardenafil hydrochloride after applying the Raman spectroscopy (Fig. **4**, Table **4**).

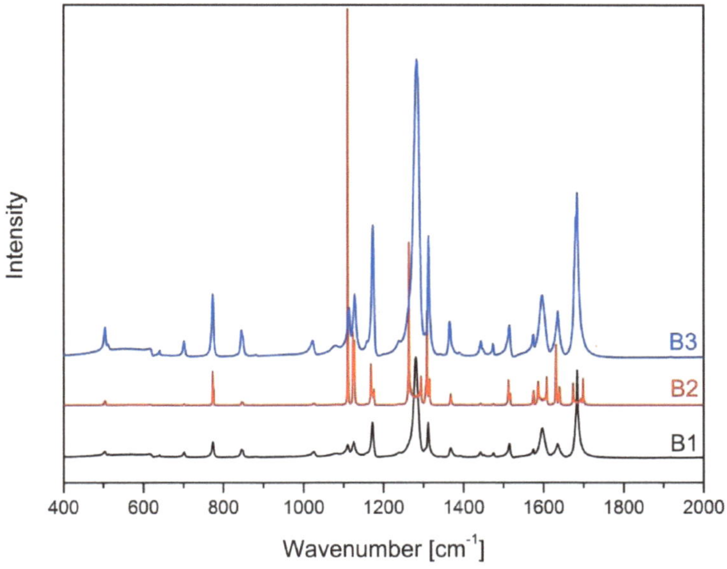

Fig. (1). IR absorption spectra first (B1), second (B2), and third (B3) form of benzocaine.

Table 1. Selected characteristic bands of benzocaine.

Theory (cm^{-1})	IR (cm^{-1})	Approximate description
703	703	O-C=O *b oop* + C-H *w*
767	773	O-C=O *boop* + C-H *w*
848	849	C-H *w* at benzene ring

(Table 1) contd.....

Theory (cm^{-1})	IR (cm^{-1})	Approximate description
1138	1110	C-O *s* + breathing benzene ring + C-H *r*
1200	1173	C-H *sc*
1306	1280	C-O *s* + C-C *s* + C-H *w*
1340	1312	C-H *r* + C-N *s*
1408	1368	C-H *r*
1431	1390	C-H *r*
1500	1444	C-H *sc*
1560	1513	C-N *s* + C-C *s* + C-C-H *b*
1610	1596	C-N *s* + C-C *s* + C-C-H *b*
1677	1596	C=C *s* + N-H *sc* + C-C *s* + C-N *s*
1785	1682	C=O s
3585	3341	N-H *s* sym.
3697	3425	N-H *s* asym.

s-stretching, b-bending, r-rocking, w-wagging, sc-scissoring, t-twisting, oop-out of the plane, sym-symetric, asym-asymetric.

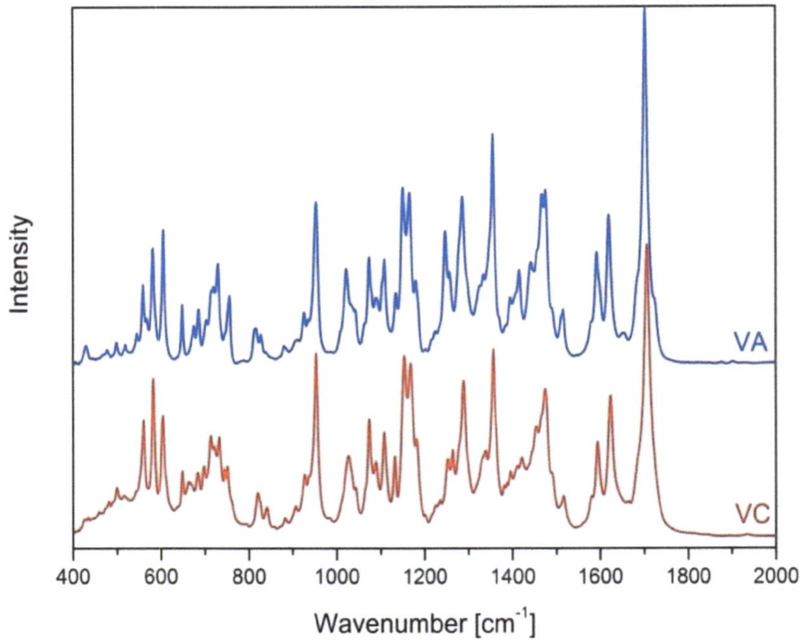

Fig. (2). IR absorption spectra crystal (VC) and amorphous (VA) form of vardenafil.

Table 2. Selected characteristic bands of vardenafil.

Theory (cm^{-1})	IR (cm^{-1})	Approximate description
465	477	def. all molecule
540	557	def. all molecule
556	581	def. all molecule
584	604	C-H *w* + O=S=O *sc*
635	649	def. phenyl ring
671	687	def. pirazolo-pirymidine ring + C-N *w* in pirazolo-pirymidine ring
695	755	N-S *s* + rocking pirazolo-pirymidine ring + def. ethylo-piperazyne group
790	815	C-H *w* + C-H *t* in ethoxy group
806	840	C-H *t* + C-H *w* in ethylo-piperazyne group
851	879	C-C *s* + C-H *w* in prophyl group
883	927	breathing piperazyne ring
903	953	C-N *s* + C-C *s* in ethylo-piperazyne group
1003	1023	C-C *s* in prophyl group + C-N *s* in pirazolo-pirymidine ring
1016	1041	C-C *s* in ethoxyl group
1058	1075	O=S *s* + C-H *r* in ethylo-piperazyne group and phenyl ring
1063	1090	O=S *s* + C-H *r* in ethylo-piperazyne group and phenyl ring
1089	1108	C-H *w* in ethoxy group and phenyl ring + O=S *s*
1106	1134	C-N *s* in ethylo-piperazyne group
1121	1148	C-C *s* in phenyl ring + C-N *s* in pirazolo-pirymidine ring + C-H *sc* in phenyl ring + def. pirazolo-pirymidine ring
1135	1164	C-H *w* in phenyl ring
1261	1248	O=S*s* + C-N *s* in pirazolo-pirymidine ring + C-H *r*
1282	1257	C-H *t* in ethylo- piperazyne group
1293	1286	C-C *s* in phenyl ring
1365	1354	C-H *sc*
1409	1393	N-N *s* in pirazolo-pirymidine ring
1427	1414	C-H *sc* + C-H *r*
1442	1442	C-H *sc* + C-H *r*
1450	1476	C-H *sc* + C-H *r*
1517	1514	C=N *s* + C-N-H *b* in pirazolo-pirymidine ring
1672	1594	C=O *s*
1672	1618	C=O *s*

(Table 2) contd.....

Theory (cm^{-1})	IR (cm^{-1})	Approximate description
1858	1702	H-Cl*s*
2921	2508	C-H *s* in prophyl ring
2934	2570	C-H *s* in CH_3 group at pirazolo-pirymidine ring
2952	2668	C-H s in ethylo- piperazyne group
2987	2927	C-H *s* in prophyl ring
3006	2964	C-H *s* in ethylo- piperazyne group
3413	3335	N-H *s* in pirazolo-pirymidine ring

s-stretching, b-bending, r-rocking, w-wagging, sc-scissoring, t-twisting, def.-deformation.

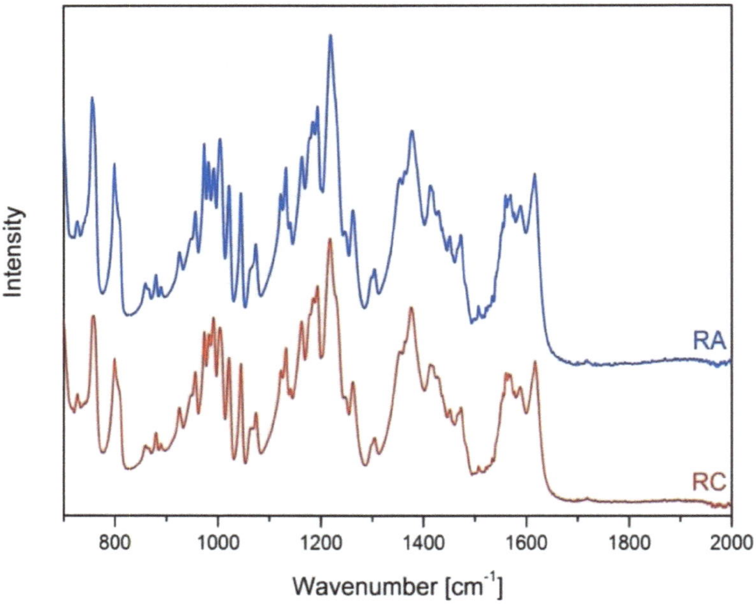

Fig. (3). ATR spectra of ranitidine; crystal (RC), and amorphous (RA) form.

Table 3. Selected characteristic bands of ranitidine.

Theory (cm^{-1})	IR (cm^{-1})	Approximate description
541	523	C-N-C *b* + C-H *w* in CH_3 in **3**-(dimethyloamine)methyl
770	761	O-N-O *sc* + N-H *w* in N-methyl and 2-nitroethene group
819	801	C-N *s* in **3**-(dimethyloamine)methyl
978	957	def. ring in **2**-furan-2-yl

(Table 3) contd.....

Theory (cm^{-1})	IR (cm^{-1})	Approximate description
1000	1006	C-O *s* + C-N *s* in (dimethyloamine)methyl group in **2**-furan-2-yl
1048	1045	C-H *w* in**2**-furan-2-yl
1098	1075	C-N *s* between amine and [methylsulfanyl] ethyl group + C-N *s* in N-methyland
1175	1134	C-H *r* in **3**-(dimethyloamine)methyl
1231	1195	C-H *t* in [methylsulfanyl] ethyl group
1263	1221	C-H *w* in [methylsulfanyl] ethyl group + C-H *w* in (dimethyloamine)methyl gropu
1298	1263	C-H *r* in 2-nitroethene group and [methylsulfanyl] ethyl group
1382	1355	C-H *w* in [methylsulfanyl] ethyl group and (dimethyloamine)methyl
1418	1381	C-H *sc* in CH_3 in **3**-(dimethyloamine)methyl
1473	1419	C-H *sc* in N-methyl and [methylsulfanyl] ethyl groups
1547	1508	N-O *s* in 2-nitroethene group + C-N-H *b* in 2-nitroethene and N-methyl groups + C-H *sc* in N-methyl group
1580	1550	C=C *s* sym. in **2**-furan-2-yl
1591	1570	C-N *s* in N-methyl group + C-N-H *b* in N-methyl group
1622	1590	C-N-H *b* in **3**-(dimethyloamine)methyl
1663	1620	C-C *s* between 2-nitroethene and N-methyl group + C-N-H *b* in N-methyl group

s-stretching, b-bending, r-rocking, w-wagging, sc-scissoring, t-twisting, sym-symetric, asym-asymetric def.-deformation.

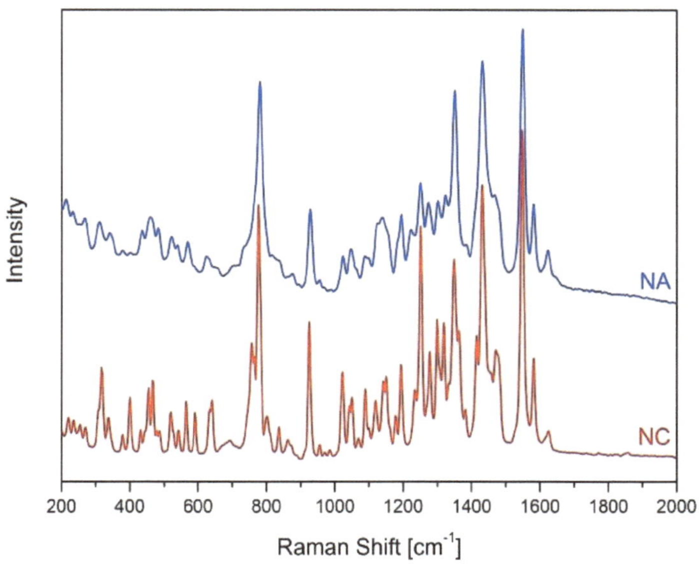

Fig. (4). Raman scattering spectra of crystal (NC) and amorphous (NA) form of naratriptan;.

Table 4. Selected characteristic bands of naratriptan.

Theory (cm^{-1})	Raman (cm^{-1})	Approximate description
391	398	N-H *w* in 1H-indol
714	691	C-S *s* + N-H w in n-methyl group + C-H *w* in all sample
772	757	Breathing 1H-indol + C-C *s* in 1H-indol and ethanesulfowanide group
795	776	Breathing methylpiperidin ring
940	926	C-C-C *b* + C-H *w* in 1H-indol
1106	1080	S=O *s*sym + CH_2t in ethanesulfowanide
1205	1193	C-C *s* in 1H-indol
1267	1251	C-N *s* + C-H *t* + N-H *r* in 1H-indol
1327	1298	C-C *s* in 1H-indol + C-H *r* in ring methylpiperidin
1390	1349	C-C *s* in 1H-indol + C-H *w* in methylpiperidin
1462	1433	C-C *s* + C-H *sc* in 1H-indol
1513	1473	C-C *s* in 1H-indol + C-S *s* in ethanesulfowanide + CH_2*sc* in N-methyl and, methylpiperidin
1588	1547	C=C *s* + C-C *s* in 1H-indol
1624	1582	C-C *s* + C-C-C *b* in 1H-indol
1667	1625	C-C *s* + C-C-C *b* in 1H-indol
3011	2863	C-H *s* in n-methyl group
3062	2868	C-H *s*sym in methylpiperidin
3109	2925	C-H *s*asym in methylpiperidin
3150	2994	C-H *s*asym in methylpiperidin
3173	3014	C-H *s*asym in 1H-indol
3266	3051	C-H *s* in 1H-indol
3589	3120	N-H *s* in n-methyl group
3684	3225	N-H *s* in 1H-indol

s-stretching, b-bending, r-rocking, w-wagging, sc-scissoring, t-twisting, sym-symetric, asym-asymetric.

In order to thoroughly analyze all the bands, the authors used theoretical analysis, which allowed them to assign precisely the changes of location and types of bands characteristic of specific forms of crystalline or amorphous states. The other approach consists in change analysis in terms of only one characteristic band in received spectra. However, it should be considered that along with an increase of matrix complexity (*e.g.* analyses of a selected medicine form) the signal alternations of changing bands may be masked. Combining experimental research results that monitor changes of selected bands with appropriate chemometric analysis tools can help solve the problem of the lack of spectrophotometric marking selectivity. It should, however, be taken into consideration that there are no procedures in terms of combining analytical methods with chemometric ones.

The following Tables (**1**, **2**, **3**, and **4**) show the most important kinds of vibration significant for identifying polymorphic and amorphous forms of examined APIs. All changes which facilitate determining differences between the examined forms based on FT-IR and Raman spectroscopy.

Polymorphism of Excipients

The polymorphism phenomenon is observed among excipients which also have crystalline forms. Polymorphism study for excipients are not as widespread as the study of active substances. However, from the pharmaceutical formulation point of view the study of polymorphism of excipients become more and more important. Various properties of active substances and excipients, such as moisture content, particle size and distribution, polymorphism, amorphism, crystal habit, hydration state, lubricant and binder level of the blend that have an influence on compaction [171]. Excipients can have an impact on stability, compressibility, wettability that are important to manufacturing process [172] and the polymorphism as the solid properties is consider more commonly. Mannitol and lactose are example of excipients with the various crystalline forms.

Mannitol is a hexahydric alcohol that has been extensively used as an excipient. Generally occurs as a crystalline excipients and was classified as alfa, beta, delta form, respectively [173]. Mannitol is a frequently used diluent in the production of tablets due to its non-hygroscopic character and low drug interaction potential. Although the δ-polymorph of mannitol has superior tablet ability in comparison to α- and β-mannitol, the latter are most commonly used because large-scale production of δ-mannitol is difficult [174].

Lactose widely used in solid form can either be in a crystalline state or in an amorphous state. Crystalline lactose can exist in a number of distinct forms. Most well-known forms are α-lactose monohydrate and β-lactose. Also, two crystalline anhydrous α-lactose types are known, a stable and an unstable (hygroscopic) form. The low hygroscopicity of crystalline lactose supports chemical inertness. The water of crystallization is bound so tightly in the crystal lattice of the lactose that it is chemically inert. α-Lactose has the peculiarity that in the crystalline state each lactose molecule is associated with 1 molecule of water. Crystals of α-lactose monohydrate possess a characteristic tomahawk-like shape. These crystals are very hard and brittle. Lactose monohydrate crystals In the pharmaceutical industry this type of lactose is used mainly as an excipient in the production of solid dosage forms, *e.g.* wet granulation processes and direct compression [175] and for capsule filling. Crystals of pure β-lactose have a characteristic kite-like form. Particles with crystalline β-lactose are more brittle than α–lactose monohydrate crystals and they do not contain crystal water. B-Lactose is often referred to as anhydrous lactose. This type of lactose is used mainly as filler-binder for tablet production *via* direct compression processes [176].

Methods of Stabilization of Polymorphic and Amorphous Forms of Active Pharmaceutical Ingredients

The stability of a given polymorphic API form reduces the necessity of using FT-IR and Raman techniques in terms of monitoring structure changes. However, an increasing interest in amorphous forms of APIs in terms of receiving pharmaceutical forms with required dispersion is becoming a tendency favoring more and more frequent using of FT-IR and Raman spectroscopy for monitoring possible structure changes during processing. One of the techniques of receiving selected pharmaceutical forms engaging amorphous API forms is hot-melt extrusion (HME). HME is a robust, novel technique to make solid dispersions in order to provide time controlled, modified, extended, and targeted drug delivery resulting in improved bioavailability as well as taste masking of bitter active pharmaceutical ingredients. The FT-IR and Raman spectroscopy techniques are also applicable in API analyses after combination with biopolymers as well as processing of given pharmaceutical forms by HME. A. Sarode used FT-IR to examine the interaction of indomethacin, itracoazole and griseofulvin with hydrophlic polymers: eudragit EPO, Eudragit L-100-55, Eudragit L-100, HPMCAS-LF, HPMCAS-MF, pharmacoat 603, Kollidon VA-64 during preparation of solid dispersions which were manufactured using HME [16, 177]. On the contrary, R. Fule applied Raman spectroscopy to identify solid dispersion of artesunate after extrution with soluplus or kollidon and additives in the form of surfactants or plasticizers [178]. It is significant that in addition to examining the changes of bands connected with the loss of APIs' amorphousness, applying of these techniques also facilitates estimating of hydrogen interactions between APIs and biopolymers.

The amorphous form of active pharmaceutical ingredients is thermodynamically unstable and usually converts back to the more stable crystalline form. To stabilize the amorphous form, molecular mobility should minimized to prevent recrystallization [179]. In the area of design of active pharmaceutical ingredients delivery systems, one of the most important problems is finding stable and selective carriers that would meet the criteria of classical auxiliary substances as to the safety of use as well as the 'intelligent' materials permitting modifications of the physicochemical properties of API towards improvement of their therapeutic effectiveness [180, 181]. The use of a proper carrier material and suitable method of API introduction onto its surface can prevent recrystallization of API and improve their bioavailability, reduce side effects and extend the time of activity of a given API. The possibility of hindering recrystallization of amorphous forms of APIs is often applied by using mesoporous structures [180, 182]. While confirming possibilities of applying mesoporous materials as stabilizers for amorphous API forms, also FT-IR and Raman techniques are frequently used.

Porous carriers of active pharmaceutical ingredients can be classified into three groups. According to IUPAC (International Union of Applied Chemistry) notation, microporous materials have pore diameters of less than 2 nm, mesoporous materials containing pores with diameters between 2-50 nm and macroporous materials have pore diameters of greater than 50 nm [183]. For stabilizing amorphous active pharmaceutical ingredients, only mesoporous carriers are suitable. Microporous materials (*e.g.* zeolites, activated carbons) have strong interaction with API molecules which causes complete filling of pores at a relatively low pressure. The uptake of the molecules is constrained by accessible pore volume only and reduces the loading capacity of the microporous carriers. Macroporous materials (*e.g.* metal oxides) have wide pores which act like a flat surface for adsorbing molecules. In the case of mesoporous carriers, the adsorption potential of the pore walls is not dominant and adsorption of active pharmaceutical ingredients depends on the interaction between pore walls and adsorbate as well as between API molecules [179].

Ordered mesoporous silicas have been recently widely used as carriers of pharmaceutically active ingredients. They are characterized by well-developed surface area, large pore volume (0.5 – 2.1 cm^3/g), wall thickness in the range 0.6 – 5 nm and uniform pores of the lengths 100 – 5000 nm, so they can be effective adsorbents. Other attractive features of silicas are their high hydrothermal resistance and universality of the methods of their syntheses [184]. The first patents on the synthesis of ordered mesoporous materials were approved in 1971, but because of the lack of their detail description they have not been widespread [185]. Only the synthesis of mesoporous silica materials of M41S type, published in 1992 by the research team from Mobil R&D Corporation has started intense studies of these materials [186, 187]. The character of synthesis permitted control and regulation of the pore size, density of active centers and composition. Traditional zeolites do not offer such opportunities. The search for new silica materials was the necessity of obtaining adsorbents for macromolecular compounds, in particular as carriers of therapeutically active ingredients. MCM-41 type silica (MCM stands for Mobil Composition of Matter) have been the first synthesized mesoporous materials of ordered structure and are the best known representatives of the family of M41S compounds. They have hexagonal structure, narrow pore size distribution and large internal surface [186, 187]. From the traditional zeolites they differ in the size of pores which makes them attractive adsorbents of large size molecules [184]. Although the walls of MCM-41 pores are amorphous, they have crystalline structure, which follows from their ordered and periodically repetitive structure. In 1998 modern compounds called SBA (Fig. **5**) were obtained when the surfactants used in the synthesis were replaced by block copolymers [188, 189]. One of the best known representatives of this group is SBA-15 of hexagonal symmetry. Although MCM-41 and SBA-15 show hexagonal ordering, they differ in the pore diameter, pore wall thickness and thus in stability. The experimentally determined surface area of SBA-15 is even five times larger than the theoretical value calculated on the basis of the pore size and pore wall thickness. This result implies that SBA-15 besides the hexagonal mesopores must also have smaller size pores distributed in the whole volume of the materials, whose contribution to the surface area reaches over 70%. It has been suggested that a large number of the micropores are localized in the SBA-15 walls between the ordered mesopores [190]. Another representative of SBA group materials is SBA-16 of *Im3m* symmetry (Fig. **5**). It is a porous solid containing ordered pores. The three-dimensional mesoporous structure is conducive to greater accessibility of pores to the adsorbate molecules. Therefore, SBA-16 is an attractive material in adsorption, catalysis and immobilization of biomolecules [191]. Besides the above mentioned silicas, there are other interesting ones for instance KIT-6 of regular structure [192], HMS of hexagonal structure (obtained with the use of nonionic surfactant as a template) [193] and FSM-16 of hexagonal distribution of one-dimensional channels [194].

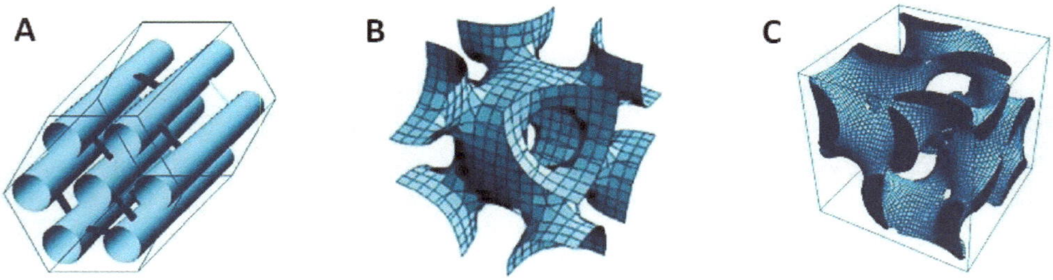

Fig. (5). Structure of typical mesoporous materials: A. SBA-15, B. SBA-16, C. KIT-6 [195].

In the context of using silicas as carriers of APIs, the character of silica surface must be thoroughly determined. First of all the nature of silanol groups present on silica surface must be understood. A very effective method for this purpose is the Fourier transform infrared spectroscopy. Silanols can occur on the silica surface as individual (isolated) groups, in geminal (double) form or in vicinal (associated) form, (Fig. **6**).

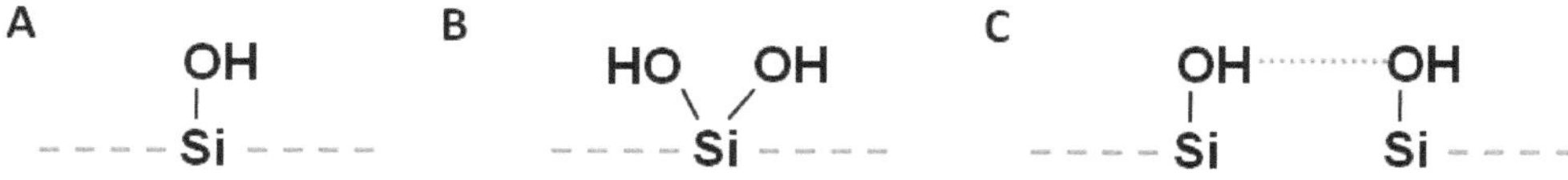

Fig. (6). Types of silanol groups occurring on the silica surface: A. isolated, B. geminal, C. vicinal.

Pairs of vicinal silanols can be connected through hydrogen bonds. It is known that silica also contains the so-called internal silanol groups hidden in the silica matrix and inaccessible to adsorbent. The bands assigned to the internal, isolated and linked silanol groups often overlap, so it is difficult to use FT-IR spectra for quantitative analysis of silanol groups of particular types. Silanol groups absorb infrared radiation in a wide range 3800 - 3000 cm^{-1}. Usually a sharp peak at about 3730-3750 cm^{-1} is accompanied by a broad absorption band in the range 3600-3000 cm^{-1}. The peak at 3730-3750 cm^{-1} is assigned to individual isolated silanol groups, the signals in the range 3580-3540 cm^{-1} are assigned to vicinal silanols and disappear after heating. The band at 3680 cm^{-1}, often overlapped by the peak corresponding to isolated silanols and the broad band coming from vicinal silanols is assigned to the internal hydroxyl groups. The absorption bands assigned to internal hydroxyl groups appear at 3660 cm^{-1} or 3650 cm^{-1} [196]. The range 1040-1020 cm^{-1} corresponds to deformation vibrations of Si–OH, while the bands at 966 cm^{-1} and 810 cm^{-1} are the characteristic stretching vibrations of unbridged oxygen atoms (Si–O-) in the Si–O–H bonds and the symmetric stretching band of $(Si–O–Si)_{sym}$ in SiO_4^- tetrahedrons in mesoporous materials of SBA-15 type [197]. The FT-IR spectra of silica can also show bands assigned to siloxane groups $–(SiO)_n–$ in the wavenumber range 1100-1000 cm^{-1}. The stretching vibrations of siloxane chains give two bands of similar intensity at about 1085 cm^{-1} and 1020 cm^{-1}, and their intensity increases with increasing chain length. The symmetric stretching vibrations of Si–O–Si are manifested as a characteristic band in the range 625-480 cm^{-1} [198]. Besides the above described bands characteristic of silica, the spectra of mesoporous silicas also show bands at 2956, 2924 and 2853 cm^{-1} assigned to the vibrations of $–CH_2$ – and $–CH_3$ bonds from the fragments of the template residues. Moreover, the bands assigned to template can appear in the range 1465-1440 cm^{-1} (asymmetric vibration of $–CH_3$), 1390-1370 cm^{-1} (symmetric vibration of $–CH_3$), and 1410-1350 cm^{-1} (bending vibration of $–CH_3$) [198].

With the help of FT-IR spectroscopy and other spectroscopic methods it is possible to identify the type and the site of interaction between the mesoporous support and the pharmaceutically active ingredient.

Ambrogi *et al.* (2008) have applied of the mesoporous silica MCM-41 as a vehicle able to improve the carbamazepine dissolution rate by preventing drug crystallization. The inclusion compound was prepared by equilibrating an almost saturated solution of the drug in an appropriate solvent with MCM-41 [199]. Carbamazepine exists in different polymorphic and solvate forms, at least four anhydrates (two monoclinic forms, one trigonal or α-polymorph and one triclinic or γ-polymorph) and one dehydrate are known [200, 201]. These forms are characterized by different thermodynamic stability and different solubility. The FT-IR spectra of crystalline carbamazepine showed the absorption peaks characteristic of polymorph III [3465 cm^{-1} (vibration of NH_2), 1677 cm^{-1} (vibration – CO–R), 1605 and 1593 cm^{-1} (vibration –C=C– and C=O and deformation –NH)] [202]. The FT-IR spectra of MCM-41 with adsorbed carbamazepine revealed many changes namely: a remarkably reduced peak at 3740 cm^{-1}, with a low shoulder at 3645 cm^{-1} and an intensive and broad band between 2800 and 3500 cm^{-1}. The physical mixture showed the presence of peaks assigned to both MCM-41 and crystalline carbamazepine. These changes suggested that the formation of hydrogen bonds occurred between the active pharmaceutical ingredients and the isolated terminal silanol groups of MCM-41. Moreover since the NH_2 peak was the most affected, it can be hypothesized that this group was involved in the interactions. Carbamazepine loading was of 14% and it did not appear in the crystalline form. *In vitro* drug release results showed a remarkable increase in carbamazepine dissolution rate in all tested media. This improvement is due both to the high specific surface area of MCM-41 and to the lack of crystalline drug; the storage stability studies under different environment conditions were conducted in order to investigate the ability of MCM-41 to prevent drug crystallization [199].

Ibuprofen (IBU) was loaded on MCM-41 and SBA-15 by spray drying method [203]. Drug molecules were entrapped inside the mesoporous channels at a high drug loading of 50:50 (w/w). NMR spectra of IBU/MCM-41 and IBU/SBA-15 exhibited very good resolution due to an efficient averaging of the homonuclear ^{1}H–^{1}H dipolar interactions coming from the reorientation of IBU within the pore structures. In comparison, the IBU crystals gave a broad line of very low intensity. The broad line of ^{1}H MAS spectroscopy is characteristic of a rigid crystalline solid resulting from the incomplete averaging of the strong homonuclear ^{1}H–^{1}H dipolar interaction [204]. The differences in the spectra of IBU/MCM-41 and IBU crystal indicated that the IBU entrapped in the pore channels of MCM-41 and SBA-15 was not in a solid crystal form. Moreover, results showed that mesoporous materials with pores of the sizes smaller than 10 nm produced amorphous state of ibuprofen due to nanospace confinement but, drug nanocrystals were obtained when ibuprofen was spray dried with mesoporous silica having large pore size above 20 nm [203]. The physical state of ibuprofen to a significant degree determines its dissolution from the solid dispersion. IBU in the amorphous state exhibited a higher dissolution rate than nanocrystalline IBU, even though the larger pore size could facilitate diffusion from the host matrix. It should be reminded that the internal surface of MCM-41 does not have as many adsorption centers as that of SBA-15 silica, which also affects the interactions with the API and the rate of its release. If SBA-15 silica is used as a support for ibuprofen, the drug molecules preferentially fill micropores and only after filling them up they go to mesopores making a thin film. In the two types of silica ibuprofen, a carboxylic acid, can undergo dimerization (Fig. **7**).

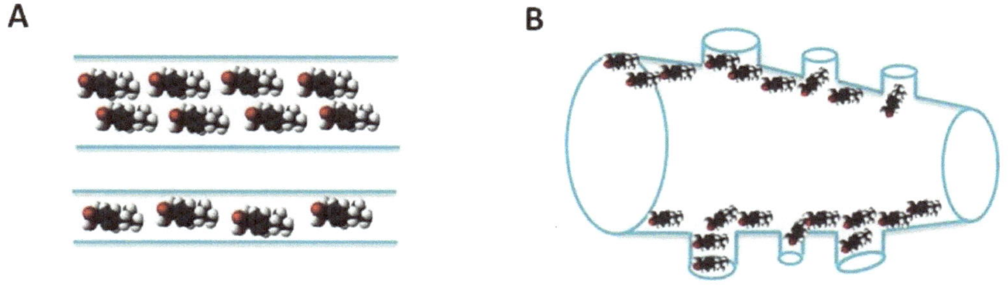

Fig. (7). Arrangement of ibuprofen molecules in pores of mesoporous silica materials: A. MCM-41, B. SBA-15.

Qu *et al.* (2006) have reported that after introduction of ibuprofen into the pores of mesoporous MCM-41 type silica, the band assigned to Si-OH groups at 3744 cm^{-1} disappeared from the FT-IR spectra [205]. Meanwhile the intensities of ν(COOH) at 1723 cm^{-1} and δ_{oop} (OH) at 950 cm^{-1} of ibuprofen significantly decreased [206, 207]. This indicates the formation of a hydrogen bond between the carboxylic group of ibuprofen and the silanol group of the mesoporous silica. Moreover, the quaternary carbon atom peaks at 1463 and 1519 cm^{-1}, and $C–H_x$ peaks at 2919 and 2962 cm^{-1} (as clearly observed in the drug-loaded samples) confirmed that the drug has been loaded in the mesoporous silica system [205].

Ayad *et al.* (2016) have applied mesoporous silica of regular structure KIT-6, modified with 3-aminopropyltriethoxysilane (APTS) as a support of ketoprofen (KP). To explore the successful loading of KP drug into the amine-functionalized KIT-6, the FT-IR measurements were conducted [208]. Firstly, the FT-IR spectrum of KP drug was recorded. The spectrum of KP revealed seven characteristic bands appearing at 2942 cm^{-1} (aromatic C-H stretching and carboxylic acid O-H stretching), 1695 cm^{-1} (C=O stretching of acid), 1630 cm^{-1} (C=O stretching of ketone), 1585 cm^{-1} (carboxylic O-H out of plane deformation), 1417 cm^{-1} ($CH-CH_3$ deformation) and 866 cm^{-1} (C-H out of plane deformation for substituted aromatic ring). After loading of KP to mesoporous silica KIT-6, the sharp band observed at 1084 cm^{-1} which is characteristic of the antisymmetric Si-O-Si stretching vibration was shifted to 1047 cm^{-1} and the intense peak at 1560 cm^{-1}, characteristic of N-H was shifted to 1552 cm^{-1}. In addition, the band that appeared at 2942 cm^{-1} in the ketoprofen spectrum characteristic of carboxylic O-H was shifted to higher wavenumber 3921 cm^{-1}, while the band at 1695 cm^{-1} characteristic of C=O was shifted to 1681 cm^{-1} and was less sharp confirming that the interaction occurred through carboxylic and carbonyl groups. The same authors studied the drug release behaviors in phosphate buffer as a function of pH (pH 2.0, 6.7 and 7.4) at room temperature [208]. It was found that at physiological pH (pH 2), the H^+ concentration increased, therefore the hydrogen bond between the carboxylic and carbonyl groups from ketoprofen and the NH_2 group from the functionalized KIT-6-NH_2 grew stronger. Consequently, small amounts of the KP drug were released. The percentage of KP released was gradually increased by decreasing the H^+ concentration (increasing pH from 6.7 to 7.4). Thus, the ketoprofen loaded-KIT-6-NH_2 was readily protonated, and the hydrogen bond interaction was weakened; then further dissociation of KP molecules occurred leading to an increase in the percentage of KP released [208].

CONCLUDING REMARKS

There is a vast potential of applying vibrational spectroscopy methods for identification of polymorphic and amorphous forms of APIs. The most significant benefits from their application are: availability of such methods in research laboratories, a simple procedure of preparing samples for measurement (especially in spectroscopy), the lack of necessity of receiving samples in a compact form as well as a possibility of carrying out measurements in selected kinds of pharmaceutical packaging (*e.g.* glass packaging). On the other hand, the biggest restraints in applying FT-IR and Raman spectroscopy are: a limited selectivity of marking (similarly, however, to other techniques recommended for examining polymorphic and amorphous forms of APIs), the lack of appropriate validation procedures for applied marking as well as possible fluorescence of selected APIs, precluding the reception of credible signals.

CONSENT FOR PUBLICATION

Not applicable.

CONFLICT OF INTEREST

The authors declares no conflict of interest, financial or otherwise.

ACKNOWLEDGEMENTS

This study was supported by SONATA grant from the National Science Centre Poland (UMO-2013/09/D/NZ7/02525)

REFERENCES

[1] Chieng, N.; Rades, T.; Aaltonen, J. An overview of recent studies on the analysis of pharmaceutical polymorphs. *J. Pharm. Biomed. Anal.,* **2011**, *55*(4), 618-644. [http://dx.doi.org/10.1016/j.jpba.2010.12.020] [PMID: 21237609]

[2] Hilfiker, R. *Polymorphism in pharmaceutical industry*; Wiley-VCH Verlag GmbH &KGaA: Weinheim, **2006**. [http://dx.doi.org/10.1002/3527607889]

[3] Guiry, K.P.; Kelleher, J.M.; Lawrence, S.E.; McAuliffe, M.T.; Moynihan, H.A.; Ryan, A.L. Crystal polymorphism of pharmaceuticals: probing crystal nucleation at the molecular level. *J. Enzyme Inhib. Med. Chem.,* **2007**, *22*(5), 550-555. [http://dx.doi.org/10.1080/14756360701425147] [PMID: 18035822]

[4] Yin, S.X.; Grosso, J.A. Selecting and controlling API crystal form for pharmaceutical development--strategies and processes. *Curr. Opin. Drug Discov. Devel.,* **2008**, *11*(6), 771-777. [PMID: 18946842]

[5] Lu, J.; Rohani, S. Polymorphism and crystallization of active pharmaceutical ingredients (APIs). *Curr. Med. Chem.,* **2009**, *16*(7), 884-905. [http://dx.doi.org/10.2174/092986709787549299] [PMID: 19275600]

[6] Desiraju, G.R. Polymorphism: the same and not quite the same. *Cryst. Growth Des.,* **2008**, *8*(1), 3-5. [http://dx.doi.org/10.1021/cg701000q]

[7] Morissette, S.L.; Almarsson, O.; Peterson, M.L.; Remenar, J.F.; Read, M.J.; Lemmo, A.V.; Ellis, S.; Cima, M.J.; Gardner, C.R. High-throughput crystallization: polymorphs, salts, co-crystals and solvates of pharmaceutical solids. *Adv. Drug Deliv. Rev.,* **2004**, *56*(3), 275-300. [http://dx.doi.org/10.1016/j.addr.2003.10.020] [PMID: 14962582]

[8] Beczkowicz, H.; Glice, M.; Korczak, K.; Kosmacińska, B.; Łaszcz, M.; Maruszak, W. Polimorfizm substancji farmaceutycznych. *Przem. Chem.,* **2000**, *85*, 354-359.

[9] Doelker, E. [Crystalline modifications and polymorphism changes during drug manufacture]. *Ann. Pharm. Fr.,* **2002**, *60*(3), 161-176. [PMID: 12050595]

[10] Blagden, N.; de Matas, M.; Gavan, P.T.; York, P. Crystal engineering of active pharmaceutical ingredients to improve solubility and dissolution rates. *Adv. Drug Deliv. Rev.,* **2007**, *59*(7), 617-630. [http://dx.doi.org/10.1016/j.addr.2007.05.011] [PMID: 17597252]

[11] Urakami, K. Characterization of pharmaceutical polymorphs by isothermal calorimetry. *Curr. Pharm. Biotechnol.,* **2005**, *6*(3), 193-203. [http://dx.doi.org/10.2174/1389201054022904] [PMID: 15974974]

[12] Newman, A.W.; Byrn, S.R. Solid-state analysis of the active pharmaceutical ingredient in drug products. *Drug Discov. Today,* **2003**, *8*(19), 898-905. [http://dx.doi.org/10.1016/S1359-6446(03)02832-0] [PMID: 14554018]

[13] Li, H.; Kiang, Y.H.; Jona, J. Multiple approaches to pharmaceutical polymorphism investigation--a case study. *Eur. J. Pharm. Sci.,* **2009**, *38*(5), 426-432. [http://dx.doi.org/10.1016/j.ejps.2009.09.003] [PMID: 19748574]

[14] Bugay, D.E. Characterization of the solid-state: spectroscopic techniques. *Adv. Drug Deliv. Rev.,* **2001**, *48*(1), 43-65. [http://dx.doi.org/10.1016/S0169-409X(01)00101-6] [PMID: 11325476]

[15] Detoisien, T.; Arnoux, M.; Taulelle, P.; Colson, D.; Klein, J.P.; Veesler, S. Thermal analysis: a further step in characterizing solid forms obtained by screening crystallization of an API. *Int. J. Pharm.,* **2011**, *403*(1-2), 29-36. [http://dx.doi.org/10.1016/j.ijpharm.2010.10.016] [PMID: 20951782]

[16] Domingos, S.; André, V.; Quaresma, S.; Martins, I.C.; Minas da Piedade, M.F.; Duarte, M.T. New forms of old drugs: improving without changing. *J. Pharm. Pharmacol.,* **2015**, *67*(6), 830-846. [http://dx.doi.org/10.1111/jphp.12384] [PMID: 25648101]

[17] Zhang, Q.; Mei, X.F. [Polymorph transformation of solid drugs]. *Yao Xue Xue Bao,* **2015**, *50*(5), 521-527. [PMID: 26234130]

[18] Ying, J.; Lü, Y.; Du, G.H. [Progress in the research of amorphous pharmaceuticals]. *Yao Xue Xue Bao,* **2009**, *44*(5), 443-448. [PMID: 19618716]

[19] Grohganz, H.; Priemel, P.A.; Löbmann, K.; Nielsen, L.H.; Laitinen, R.; Mullertz, A.; Van den Mooter, G.; Rades, T. Refining stability and dissolution rate of amorphous drug formulations. *Expert Opin. Drug Deliv.,* **2014**, *11*(6), 977-989. [http://dx.doi.org/10.1517/17425247.2014.911728] [PMID: 24754747]

[20] Antal, I.; Zelkó, R. [Pharmaceutical significance of amorphous materials]. *Acta Pharm. Hung.,* **2008**, *78*(3), 121-133. [PMID: 18986089]

[21] Rodríguez-Spong, B.; Price, C.P.; Jayasankar, A.; Matzger, A.J.; Rodríguez-Hornedo, N. General principles of pharmaceutical solid polymorphism: a supramolecular perspective. *Adv. Drug Deliv. Rev.,* **2004**, *56*(3), 241-274. [http://dx.doi.org/10.1016/j.addr.2003.10.005] [PMID: 14962581]

[22] Yu, L. Amorphous pharmaceutical solids: preparation, characterization and stabilization. *Adv. Drug Deliv. Rev.,* **2001**, *48*(1), 27-42. [http://dx.doi.org/10.1016/S0169-409X(01)00098-9] [PMID: 11325475]

[23] Katrincic, L.M.; Sun, Y.T.; Carlton, R.A.; Diederich, A.M.; Mueller, R.L.; Vogt, F.G. Characterization, selection, and development of an orally dosed drug polymorph from an enantiotropically related system. *Int. J. Pharm.,* **2009**, *366*(1-2), 1-13. [http://dx.doi.org/10.1016/j.ijpharm.2008.08.044] [PMID: 18824090]

[24] Kopp, S.; Beyer, C.; Graf, E.; Kubel, F.; Doelker, E. Methodology for a better evaluation of the relation between mechanical strength of solids and polymorphic form. *J. Pharm. Pharmacol.,* **1989**, *41*(2), 79-82. [http://dx.doi.org/10.1111/j.2042-7158.1989.tb06397.x] [PMID: 2568432]

[25] Shah, B.; Kakumanu, V.K.; Bansal, A.K. Analytical techniques for quantification of amorphous/crystalline phases in pharmaceutical solids. *J. Pharm. Sci.,* **2006**, *95*(8), 1641-1665. [http://dx.doi.org/10.1002/jps.20644] [PMID: 16802362]

[26] Sood, J.; Sapra, B.; Bhandari, S.; Jindal, M.; Tiwary, A.K. Understanding pharmaceutical polymorphic transformations I: influence of process variables and storage conditions. *Ther. Deliv.,* **2014**, *5*(10), 1123-1142. [http://dx.doi.org/10.4155/tde.14.68] [PMID: 25418270]

[27] Kuhnert-Brandstätter, M.; Geiler, M.; Wurian, I. Contribution to the Polymorphism of Organic Drugs. *Mikrochim. Acta,* **1983**, *79*(3-4), 221-235.

[28] Olímpia, S.; Martins, M. Dias, R.M.E.; Tavares, J.J.; de Sousa, L.M.E.; Savioli, S.J.; Carlos A. Polymorphism: an evaluation of the potential risk to the quality of drug products from the Farmácia Popular Rede Própria. *Braz. J. Pharm. Sci.,* **2014**, *50*(1), 1-24. [http://dx.doi.org/10.1590/S1984-82502011000100002]

[29] Airaksinen, S.; Karjalainen, M.; Räsänen, E.; Rantanen, J.; Yliruusi, J. Comparison of the effects of two drying methods on polymorphism of theophylline. *Int. J. Pharm.,* **2004**, *276*(1-2), 129-141. [http://dx.doi.org/10.1016/j.ijpharm.2004.02.017] [PMID: 15113621]

[30] Matsuo, K.; Matsuoka, M. Solid-state polymorphic transition of theophylline anhydrate and humidity

effect. *Cryst. Growth Des.*, **2007**, *7*, 411-415.
[http://dx.doi.org/10.1021/cg060299i]

[31] Hiramatsu, Y.; Suzuki, H.; Kuchiki, A.; Nakagawa, H.; Fujii, S. X-ray structural studies of lomeridine dihydrochloride polymorphs. *J. Pharm. Sci.*, **1996**, *85*(7), 761-766.
[http://dx.doi.org/10.1021/js950319v] [PMID: 8819003]

[32] Du, Y.; Zhang, H.; Xue, J.; Tang, W.; Fang, H.; Zhang, Q.; Li, Y.; Hong, Z. Vibrational spectroscopic study of polymorphism and polymorphic transformation of the anti-viral drug lamivudine. *Spectrochim. Acta A Mol. Biomol. Spectrosc.*, **2015**, *137*, 1158-1163.
[http://dx.doi.org/10.1016/j.saa.2014.08.128] [PMID: 25305607]

[33] Chemburkar, S.R.; Bauer, J.; Deming, K.; Spiwek, H.; Patel, K.; Morris, J.; Henry, R.; Spanton, S.; Dziki, W.; Porter, W.; Quick, J.; Bauer, P.; Donaubauer, J.; Narayanan, B.A.; Soldani, M.; Riley, D.; Mcfarland, K. Dealing with the impact of ritonavir polymorphs on the late stages of bulk drug process development. *Org. Process Res. Dev.*, **2000**, *4*(5), 413-417.
[http://dx.doi.org/10.1021/op000023y]

[34] Gandhi, R.B.; Bogardus, J.B.; Bugay, D.E.; Perrone, R.K.; Kaplan, M.A. Pharmaceutical relationships of three solid state forms of stavudine. *Int. J. Pharm.*, **2000**, *201*(2), 221-237.
[http://dx.doi.org/10.1016/S0378-5173(00)00419-1] [PMID: 10878328]

[35] O'Sullivan, B.; Glennon, B. Application of *in situ* FBRM and ATR-FTIR to the Monitoring of the Polymorphic Transformation of d-Mannitol. *Org. Process Res. Dev.*, **2005**, *9*(6), 884-889.
[http://dx.doi.org/10.1021/op0500887]

[36] Campbell Roberts, S.N.; Williams, A.C.; Grimsey, I.M.; Booth, S.W. Quantitative analysis of mannitol polymorphs. FT-Raman spectroscopy. *J. Pharm. Biomed. Anal.*, **2002**, *28*(6), 1135-1147.
[http://dx.doi.org/10.1016/S0731-7085(02)00059-6] [PMID: 12049978]

[37] Schöll, J.; Bonalumi, D.; Vicum, L.; Mazzotti, M. *In situ* Monitoring and Modeling of the Solvent-Mediated Polymorphic Transformation of l-Glutamic Acid. *Cryst. Growth Des.*, **2006**, *6*(4), 881-891.
[http://dx.doi.org/10.1021/cg0503402]

[38] Larkin, P.J.; Wasylyk, J.; Raglione, M. Application of Low- and Mid-Frequency Raman Spectroscopy to Characterize the Amorphous-Crystalline Transformation of Indomethacin. *Appl. Spectrosc.*, **2015**, *69*(11), 1217-1228.
[http://dx.doi.org/10.1366/15-07926] [PMID: 26647045]

[39] Ali, H.R.; Alhalaweh, A.; Velaga, S.P. Vibrational spectroscopic investigation of polymorphs and cocrystals of indomethacin. *Drug Dev. Ind. Pharm.*, **2013**, *39*(5), 625-634.
[http://dx.doi.org/10.3109/03639045.2012.671831] [PMID: 22480325]

[40] Wu, V.; Rades, T.; Saville, D.J. Stability of polymorphic forms of ranitidine hydrochloride. *Pharmazie,* **2000**, *55*(7), 508-512.
[PMID: 10944778]

[41] Chieng, N.; Zujovic, Z.; Bowmaker, G.; Rades, T.; Saville, D. Effect of milling conditions on the solid-state conversion of ranitidine hydrochloride form 1. *Int. J. Pharm.*, **2006**, *327*(1-2), 36-44.
[http://dx.doi.org/10.1016/j.ijpharm.2006.07.032] [PMID: 16949221]

[42] Mirmehrabi, M.; Rohani, S.; Murthy, K.S.; Radatus, B. Characterization of tautomeric forms of ranitidine hydrochloride: thermal analysis, solid-state NMR, X-ray. *J. Cryst. Growth,* **2004**, *260*(3–4), 517-526.
[http://dx.doi.org/10.1016/j.jcrysgro.2003.08.061]

[43] Lin, S.Y. An overview of famotidine polymorphs: solid-state characteristics, thermodynamics, polymorphic transformation and quality control. *Pharm. Res.*, **2014**, *31*(7), 1619-1631.
[http://dx.doi.org/10.1007/s11095-014-1323-5] [PMID: 24577998]

[44] Cheng, W.T.; Lin, S.Y. Famotidine polymorphic transformation in the grinding process significantly depends on environmental humidity or water content. *Int. J. Pharm.*, **2008**, *357*(1-2), 164-168.

[http://dx.doi.org/10.1016/j.ijpharm.2008.01.048] [PMID: 18403143]

[45] Koradia, V.; Chawla, G.; Bansal, A.K. Qualitative and quantitative analysis of clopidogrel bisulphate polymorphs. *Acta Pharm.,* **2004**, *54*(3), 193-204.
[PMID: 15610616]

[46] Zimper, U.; Aaltonen, J.; Krauel-Goellner, K.; Gordon, K.C.; Strachan, C.J.; Rades, T. The influence of milling on the dissolution performance of simvastatin. *Pharmaceutics,* **2010**, *2*(4), 419-431.
[http://dx.doi.org/10.3390/pharmaceutics2040419] [PMID: 27721365]

[47] Brus, J.; Urbanova, M.; Sedenkova, I.; Brusova, H. New perspectives of 19F MAS NMR in the characterization of amorphous forms of atorvastatin in dosage formulations. *Int. J. Pharm.,* **2011**, *409*(1-2), 62-74.
[http://dx.doi.org/10.1016/j.ijpharm.2011.02.030] [PMID: 21356299]

[48] Schmidt, A.C. Structural characteristics and crystal polymorphism of three local anaesthetic bases crystal polymorphism of local anaesthetic drugs: part VII. *Int. J. Pharm.,* **2005**, *298*(1), 186-197.
[http://dx.doi.org/10.1016/j.ijpharm.2005.04.025] [PMID: 15955643]

[49] Schmidt, A.C. Solid-state characterization of falicaine hydrochloride and isomorphic dyclonine hydrochloride. Part IV. Crystal polymorphism of local anaesthetic drugs. *Eur. J. Pharm. Sci.,* **2005**, *25*(4-5), 407-416.
[http://dx.doi.org/10.1016/j.ejps.2005.04.001] [PMID: 15896948]

[50] Schmidt, A.C.; Griesser, U.J.; Brehmer, T.; Harris, R.K.; King, A. Conformational polymorphism of the local anaesthetic drug oxybuprocaine hydrochloride. *Acta Crystallogr.,* **2002**, *A58*, 140-145.
[http://dx.doi.org/10.1107/S0108767302090645]

[51] Giron, D.; Draghi, M.; Goldbronn, C.; Pfeffer, S.; Piechon, P. Study of the polymorphic behaviour of some local anaesthetic drugs. *J. Therm. Anal.,* **1997**, *49*, 913-927.
[http://dx.doi.org/10.1007/BF01996777]

[52] Schmidt, A.C.; Schwarz, I. Solid-state characterization of non-stoichiometric hydrates of ester-type local anaesthetics. Part XI. Crystal polymorphism of local anaesthetic drugs. *Int. J. Pharm.,* **2006**, *320*(1-2), 4-13.
[http://dx.doi.org/10.1016/j.ijpharm.2006.04.017] [PMID: 16857329]

[53] Sarkar, A.; Rohani, S. Investigation on polymorphic behavior of progesterone and stabilization by co-crystallization: a review. *Mini Rev. Med. Chem.,* **2014**, *14*(10), 853-861.
[PMID: 25307305]

[54] Kachrimanis, K.; Braun, D.E.; Griesser, U.J. Investigation on polymorphic behavior of progesterone and stabilization by co-crystallization: a review. *J. Pharm. Biomed. Anal.,* **2007**, *43*(2), 407-412.
[http://dx.doi.org/10.1016/j.jpba.2006.07.032] [PMID: 16935450]

[55] Abu Bakar, M.R.; Nagy, Z.K.; Rielly, C.D.; Dann, S.E. Investigation of the riddle of sulfathiazole polymorphism. *Int. J. Pharm.,* **2011**, *414*(1-2), 86-103.
[http://dx.doi.org/10.1016/j.ijpharm.2011.05.004] [PMID: 21596117]

[56] Hu, Y; Erxleben, A; Ryder, AG; McArdle, P Quantitative analysis of sulfathiazole polymorphs in ternary mixtures by attenuated total reflectance infrared, near-infrared and Raman spectroscopy. *J Pharm Biomed Anal.,* **2010**, *53*(3), 412-420.

[57] Vueba, M.L.; Pina, M.E.; Batista de Carvalho, L.A. Conformational stability of ibuprofen: assessed by DFT calculations and optical vibrational spectroscopy. *J. Pharm. Sci.,* **2008**, *97*(2), 845-859.
[http://dx.doi.org/10.1002/jps.21007] [PMID: 17628493]

[58] Zimmermann, B.; Baranović, G. Thermal analysis of paracetamol polymorphs by FT-IR spectroscopies. *J. Pharm. Biomed. Anal.,* **2011**, *54*(2), 295-302.
[http://dx.doi.org/10.1016/j.jpba.2010.08.023] [PMID: 20863645]

[59] Di Martino, P.; Conflant, P.; Drache, M.; Huvenne, J.P.; Guyot-Hermann, A.M. Preparation and physical characterization of forms II and III of paracetamol. *J. Therm. Anal.,* **1997**, *48*(3), 447-458.

[http://dx.doi.org/10.1007/BF01979491]

[60] Di Martino, P.; Guyot-Hermann, A.M.; Conflant, P.; Drache, M.; Guyot, J.C. A new pure paracetamol for direct compression: the orthorhombic form. *Int. J. Pharm.*, **1996**, *128*(1/2), 1-8. [http://dx.doi.org/10.1016/0378-5173(95)04127-3]

[61] Yokoi, Y.; Yonemochi, E.; Terada, K. Changes in surface properties by granulation and physicochemical stability of granulated amorphous cefditoren pivoxil with additives. *Int. J. Pharm.*, **2004**, *280*(1-2), 67-75. [http://dx.doi.org/10.1016/j.ijpharm.2004.05.005] [PMID: 15265548]

[62] Yokoi, Y.; Yonemochi, E.; Terada, K. Effects of sugar ester and hydroxypropyl methylcellulose on the physicochemical stability of amorphous cefditoren pivoxil in aqueous suspension. *Int. J. Pharm.*, **2005**, *290*(1-2), 91-99. [http://dx.doi.org/10.1016/j.ijpharm.2004.11.020] [PMID: 15664134]

[63] Childs, S.L.; Chyall, L.J.; Dunlap, J.T.; Coates, D.A.; Stahly, B.C.; Stahly, G.P. A metastable polymorph of metformin hydrochloride: isolation and characterization using capillary crystallization and thermal microscopy techniques. *Cryst. Growth Des.*, **2004**, *4*(3), 441-449. [http://dx.doi.org/10.1021/cg034243p]

[64] Charoenlarp, P.; Waikagul, J.; Muennoo, C.; Srinophakun, S.; Kitayaporn, D. Efficacy of single-dose mebendazole, polymorphic forms A and C, in the treatment of hookworm and Trichuris infections. *Southeast Asian J. Trop. Med. Public Health,* **1993**, *24*(4), 712-716. [PMID: 7939946]

[65] Eyjolfsson, R. Enalapril maleate polymorphs: instability of form II in a tablet formulation. *Pharmazie,* **2002**, *57*(5), 347-348. [PMID: 12061263]

[66] Bharate, S.S.; Bharate, S.B.; Bajaj, A. Interactions and incompatibilities of pharmaceutical excipients with active pharmaceutical ingredients: a comprehensive review. *J. Excip. Food Chem.*, **2010**, *1*, 3-26.

[67] Aubrey-Medendorp, C.; Swadley, M.J.; Li, T. The polymorphism of indomethacin: an analysis by density functional theory calculations. *Pharm. Res.*, **2008**, *25*(4), 953-959. [http://dx.doi.org/10.1007/s11095-007-9346-9] [PMID: 17541732]

[68] Mitchell-Koch, K.R.; Matzger, A.J. Evaluating computational predictions of the relative stabilities of polymorphic pharmaceuticals. *J. Pharm. Sci.*, **2008**, *97*(6), 2121-2129. [http://dx.doi.org/10.1002/jps.21127] [PMID: 17828731]

[69] Ramkumaar, G.R.; Srinivasan, S.; Bhoopathy, T.J.; Gunasekaran, S. Quantum chemical and experimental studies on polymorphism of antiviral drug Lamivudine. *Spectrochim. Acta A Mol. Biomol. Spectrosc.*, **2012**, *98*, 265-270. [http://dx.doi.org/10.1016/j.saa.2012.08.013] [PMID: 22958980]

[70] Simon, P.; Veverka, M.; Okuliar, J. New screening method for the determination of stability of pharmaceuticals. *Int. J. Pharm.*, **2004**, *270*(1-2), 21-26. [http://dx.doi.org/10.1016/j.ijpharm.2003.10.003] [PMID: 14726118]

[71] Lin, S.Y.; Hsu, C.H.; Ke, W.T. Solid-state transformation of different gabapentin polymorphs upon milling and co-milling. *Int. J. Pharm.*, **2010**, *396*(1-2), 83-90. [http://dx.doi.org/10.1016/j.ijpharm.2010.06.014] [PMID: 20561572]

[72] Bogdanov, N.; Gorchakov, K.A.; Puchnin, V.; Stepanov, V.; Khmelevskaya, V. Structural polymorphism and solubility of medicinal drugs. *Crystallogr. Rep.*, **2011**, *56*, 608-610. [http://dx.doi.org/10.1134/S1063774511030059]

[73] Giron, D. Investigations of polymorphism and pseudo-polymorphism in pharmaceuticals by combined thermoanalytical techniques. *J. Therm. Anal. Calorim.*, **2001**, *64*, 37-60. [http://dx.doi.org/10.1023/A:1011572610005]

[74] Yu, L.; Reutzel, S.M.; Stephenson, G. Physical characterization of polymorphic drugs: an integrated

characterization strategy. *PSTT,* **1998**, *3*, 118-127.

[75] Ayala, A.P.; Siesler, H.W.; Cuffini, S.L. Polymorphism incidence in commercial tablets of medendazole: a vibrational spectroscopy investigation. *J. Raman Spectrosc.,* **2008**, *38*, 1150-1157. [http://dx.doi.org/10.1002/jrs.1952]

[76] Grant, D.J.; Byrn, S.R. A timely re-examination of drug polymorphism in pharmaceutical development and regulation. *Adv. Drug Deliv. Rev.,* **2004**, *56*(3), 237-239. [http://dx.doi.org/10.1016/j.addr.2003.12.001] [PMID: 14962580]

[77] Price, S.L. The computational prediction of pharmaceutical crystal structures and polymorphism. *Adv. Drug Deliv. Rev.,* **2004**, *56*(3), 301-319. [http://dx.doi.org/10.1016/j.addr.2003.10.006] [PMID: 14962583]

[78] Huang, L.F.; Tong, W.Q. Impact of solid state properties on developability assessment of drug candidates. *Adv. Drug Deliv. Rev.,* **2004**, *56*(3), 321-334. [http://dx.doi.org/10.1016/j.addr.2003.10.007] [PMID: 14962584]

[79] Singhal, D.; Curatolo, W. Drug polymorphism and dosage form design: a practical perspective. *Adv. Drug Deliv. Rev.,* **2004**, *56*(3), 335-347. [http://dx.doi.org/10.1016/j.addr.2003.10.008] [PMID: 14962585]

[80] Taylor, L.S.; Zografi, G. Spectroscopic characterization of interactions between PVP and indomethacin in amorphous molecular dispersions. *Pharm. Res.,* **1997**, *14*(12), 1691-1698. [http://dx.doi.org/10.1023/A:1012167410376] [PMID: 9453055]

[81] Panagopoulou-Kaplani, A.; Malamataris, S. Preparation and characterisation of a new insoluble polymorphic form of glibenclamide. *Int. J. Pharm.,* **2000**, *195*(1-2), 239-246. [http://dx.doi.org/10.1016/S0378-5173(99)00401-9] [PMID: 10675701]

[82] Liebenberg, W.; de Villiers, M.M.; Wurster, D.E.; Swanepoel, E.; Dekker, T.G.; Lötter, A.P. The effect of polymorphism on powder compaction and dissolution properties of chemically equivalent oxytetracycline hydrochloride powders. *Drug Dev. Ind. Pharm.,* **1999**, *25*(9), 1027-1033. [http://dx.doi.org/10.1081/DDC-100102265] [PMID: 10518242]

[83] Miyazaki, S.; Arita, T.; Hori, R.; Ito, K. Effect of polymorphism on the dissolution behavior and gastrointestinal absorption of chlortetracycline hydrochloride. *Chem. Pharm. Bull. (Tokyo),* **1974**, *22*(3), 638-642. [http://dx.doi.org/10.1248/cpb.22.638] [PMID: 4472578]

[84] Kokubo, H.; Morimoto, K.; Ishida, T.; Inoue, M.; Morisaka, K. Bioavailability and inhibitory effect for stress ulcer of cimetidine polymorphs in rats. *Int. J. Pharm.,* **1987**, *35*, 181-183. [http://dx.doi.org/10.1016/0378-5173(87)90088-3]

[85] Kato, Y.; Kohketsu, M. Relationship between polymorphism and bioavailability of amorbarbital in the rabbit. *Chem. Pharm. Bull. (Tokyo),* **1981**, *29*(1), 268-272. [http://dx.doi.org/10.1248/cpb.29.268] [PMID: 7261197]

[86] Maeda, T.; Takenaka, H.; Yamahira, Y.; Noguchi, T. Use of rabbits for absorption studies on polymorphs of chloramphenicol palmitate. *Chem. Pharm. Bull. (Tokyo),* **1980**, *28*(2), 431-436. [http://dx.doi.org/10.1248/cpb.28.431] [PMID: 7389018]

[87] Hancock, B.C.; Carlson, G.T.; Ladipo, D.D.; Langdon, B.A.; Mullarney, M.P. Comparison of the mechanical properties of the crystalline and amorphous forms of a drug substance. *Int. J. Pharm.,* **2002**, *241*(1), 73-85. [http://dx.doi.org/10.1016/S0378-5173(02)00133-3] [PMID: 12086723]

[88] Summers, M.P.; Enever, R.P.; Carless, J.E. Influence of crystal form on tensile strength of compacts of pharmaceutical materials. *J. Pharm. Sci.,* **1977**, *66*(8), 1172-1175. [http://dx.doi.org/10.1002/jps.2600660832] [PMID: 894506]

[89] Otsuka, M.; Hasegawa, H.; Matsuda, Y. Effect of polymorphic forms of bulk powders on pharmaceutical properties of carbamazepine granules. *Chem. Pharm. Bull. (Tokyo),* **1999**, *47*, 852-856.

[http://dx.doi.org/10.1248/cpb.47.852]

[90] Roberts, R.J.; Payne, R.S.; Rowe, R.C. Mechanical property predictions for polymorphs of sulphathiazole and carbamazepine. *Eur. J. Pharm. Sci.,* **2000**, *9*(3), 277-283. [http://dx.doi.org/10.1016/S0928-0987(99)00065-2] [PMID: 10594385]

[91] Kopp, S.; Beyer, C.; Graf, E.; Kubel, F.; Doelker, E. Methodology for a better evaluation of the relation between mechanical strength of solids and polymorphic form. *J. Pharm. Pharmacol.,* **1989**, *41*(2), 79-82. [http://dx.doi.org/10.1111/j.2042-7158.1989.tb06397.x] [PMID: 2568432]

[92] Sun, C.; Grant, D.J. Influence of crystal structure on the tableting properties of sulfamerazine polymorphs. *Pharm. Res.,* **2001**, *18*(3), 274-280. [http://dx.doi.org/10.1023/A:1011038526805] [PMID: 11442264]

[93] Beyer, T.; Day, G.M.; Price, S.L. The prediction, morphology, and mechanical properties of the polymorphs of paracetamol. *J. Am. Chem. Soc.,* **2001**, *123*(21), 5086-5094. [http://dx.doi.org/10.1021/ja0102787] [PMID: 11457339]

[94] Bunaciu, A.A.; Aboul-Enein, H.Y.; Hoang, V.D. Vibrational spectroscopy used in polymorphic analysis. *Trends Analyt. Chem.,* **2015**, *69*, 14-22. [http://dx.doi.org/10.1016/j.trac.2015.02.006]

[95] Snider, D.A.; Addicks, W.; Owens, W. Polymorphism in generic drug product development. *Adv. Drug Deliv. Rev.,* **2004**, *56*(3), 391-395. [http://dx.doi.org/10.1016/j.addr.2003.10.010] [PMID: 14962588]

[96] Raw, A.S.; Yu, L.X. Pharmaceutical solid polymorphism in drug development and regulation. *Adv. Drug Deliv. Rev.,* **2004**, *56*, 235-236. [http://dx.doi.org/10.1016/j.addr.2003.10.004]

[97] Singhal, D.; Curatolo, W. Drug polymorphism and dosage form design: a practical perspective. *Adv. Drug Deliv. Rev.,* **2004**, *56*(3), 335-347. [http://dx.doi.org/10.1016/j.addr.2003.10.008] [PMID: 14962585]

[98] Yu, L. Amorphous pharmaceutical solids: preparation, characterization and stabilization. *Adv. Drug Deliv. Rev.,* **2001**, *48*(1), 27-42. [http://dx.doi.org/10.1016/S0169-409X(01)00098-9] [PMID: 11325475]

[99] Rodríguez-Spong, B.; Price, C.P.; Jayasankar, A.; Matzger, A.J.; Rodríguez-Hornedo, N. General principles of pharmaceutical solid polymorphism: a supramolecular perspective. *Adv. Drug Deliv. Rev.,* **2004**, *56*(3), 241-274. [http://dx.doi.org/10.1016/j.addr.2003.10.005] [PMID: 14962581]

[100] Nichols, G.; Frampton, C.S. Physicochemical characterization of the orthorhombic polymorph of paracetamol crystallized from solution. *J. Pharm. Sci.,* **1998**, *87*(6), 684-693. [http://dx.doi.org/10.1021/js970483d] [PMID: 9607944]

[101] Di Martino, P.; Guyot-Hermann, A.M.; Conflant, P.; Drache, M.; Guyot, J.C. A new pure paracetamol for direct compression: the orthorhombic form. *Int. J. Pharm.,* **1996**, *128*, 1-8. [http://dx.doi.org/10.1016/0378-5173(95)04127-3]

[102] Ragnarsson, G.; Sjögren, J. Compressibility and tablet properties of two polymorphs of metoprolol tartrate. *Acta Pharm. Suec.,* **1984**, *21*(6), 321-330. [PMID: 6534070]

[103] Craig, D.Q.; Royall, P.G.; Kett, V.L.; Hopton, M.L. The relevance of the amorphous state to pharmaceutical dosage forms: glassy drugs and freeze dried systems. *Int. J. Pharm.,* **1999**, *179*(2), 179-207. [http://dx.doi.org/10.1016/S0378-5173(98)00338-X] [PMID: 10053213]

[104] Hancock, B.C.; Zografi, G. Characteristics and significance of the amorphous state in pharmaceutical systems. *J. Pharm. Sci.,* **1997**, *86*(1), 1-12.

[http://dx.doi.org/10.1021/js9601896] [PMID: 9002452]

[105] De Villiers, M.M.; van der Watt, J.G.; Lotter, A.P. Kinetic study of the solid-state photolytic degradation of two polymorphic forms of furosemide. *Int. J. Pharm.,* **1992**, *88*, 275-283. [http://dx.doi.org/10.1016/0378-5173(92)90325-V]

[106] Matsuda, Y.; Akazawa, R.; Teraoka, R.; Otsuka, M. Pharmaceutical evaluation of carbamazepine modifications: comparative study of photostability for carbamazepine polymorphs by using Fourier-transformed reflection-absorption infrared spectroscopy and calorimetric measurement. *J Pharm Phamacol,* **1993**, *46*, 162-167. [http://dx.doi.org/10.1111/j.2042-7158.1994.tb03770.x]

[107] Bauer, J.; Spanton, S.; Henry, R.; Quick, J.; Dziki, W.; Porter, W.; Morris, J. Ritonavir: an extraordinary example of conformational polymorphism. *Pharm. Res.,* **2001**, *18*(6), 859-866. [http://dx.doi.org/10.1023/A:1011052932607] [PMID: 11474792]

[108] Haleblian, J.; McCrone, W. Pharmaceutical applications of polymorphism. *J. Pharm. Sci.,* **1969**, *58*(8), 911-929. [http://dx.doi.org/10.1002/jps.2600580802] [PMID: 4899118]

[109] Yu, L.X.; Furness, M.S.; Raw, A.; Outlaw, K.P.; Nashed, N.E.; Ramos, E.; Miller, S.P.; Adams, R.C.; Fang, F.; Patel, R.M.; Holcombe, F.O., Jr; Chiu, Y.Y.; Hussain, A.S. Scientific considerations of pharmaceutical solid polymorphism in abbreviated new drug applications. *Pharm. Res.,* **2003**, *20*(4), 531-536. [http://dx.doi.org/10.1023/A:1023285627778] [PMID: 12739758]

[110] Pikal, M.J. Impact of polymorphism on the quality of lyophilized products.*Polymorphism in Pharmaceutical Sciences, Drugs and the Pharmaceutical Sciences, 95*; Brittain, H., Ed.; Marcel Dekker: New York, **1999**, pp. 395-419.

[111] Brittain, H.G.; Fiese, E.F. Effects of pharmaceutical processing on drug polymorphs and solvates.*Polymorphism in Pharmaceutical Sciences, Drugs and the Pharmaceutical Sciences, 95*; Brittain, H., Ed.; Marcel Dekker: New York, **1999**, pp. 331-361.

[112] Brittain, H.G.; Grant, D.J. Effects of polymorphism and solid-state solvation on solubility and dissolution rate.*Polymorphism in Pharmaceutical Sciences, Drugs and the Pharmaceutical Sciences, 95*; Brittain, H., Ed.; Marcel Dekker: New York, **1999**, pp. 279-330.

[113] Brittain, H.G. Methods for the characterization of polymorphs and solvatesin.*Polymorphism in Pharmaceutical Sciences, Drugs and the Pharmaceutical Sciences, 95*; Brittain, H.G., Ed.; Marcel Dekker: New York, **1999**, pp. 227-278.

[114] Guillory, J.K. Generation of polymorphs, hydrates, solvates, and amorphous solids.*Polymorphism in Pharmaceutical Sciences, Drugs and the Pharmaceutical Sciences, 95*; Brittain, H., Ed.; Marcel Dekker: New York, **1999**, pp. 183-226.

[115] Brittain, H.; Byrn, S.R. Structural aspects of polymorphism.*Polymorphism in Pharmaceutical Sciences, Drugs and the Pharmaceutical Sciences, 95*; Brittain, H., Ed.; Marcel Dekker: New York, **1999**, pp. 73-124.

[116] Brittain, H. Application of the phase rule to the characterization of polymorphic systems.*Polymorphism in Pharmaceutical Sciences, Drugs and the Pharmaceutical Sciences, 95*; Brittain, H., Ed.; Marcel Dekker: New York, **1999**, pp. 35-72.

[117] Grant, D.J. Theory and origin of polymorphism.*Polymorphism in Pharmaceutical Sciences, Drugs and the Pharmaceutical Sciences, 95*; Brittain, H., Ed.; Marcel Dekker: New York, **1999**, pp. 1-33.

[118] Aguiar, A.J.; Krc, J., Jr; Kinkel, A.W.; Samyn, J.C. Effect of polymorphism on the absorption of chloramphenicol from chloramphenicol palmitate. *J. Pharm. Sci.,* **1967**, *56*(7), 847-853. [http://dx.doi.org/10.1002/jps.2600560712] [PMID: 6034828]

[119] Aguiar, A.J.; Zelmer, J.E. Dissolution behavior of polymorphs of chloramphenicol palmitate and mefenamic acid. *J. Pharm. Sci.,* **1969**, *58*(8), 983-987.

[http://dx.doi.org/10.1002/jps.2600580817] [PMID: 5344535]

[120] Raw, A.S.; Furness, M.S.; Gill, D.S.; Adams, R.C.; Holcombe, F.O., Jr; Yu, L.X. Regulatory considerations of pharmaceutical solid polymorphism in Abbreviated New Drug Applications (ANDAs). *Adv. Drug Deliv. Rev.,* **2004**, *56*(3), 397-414. [http://dx.doi.org/10.1016/j.addr.2003.10.011] [PMID: 14962589]

[121] Yu, L.X.; Furness, M.S.; Raw, A.; Outlaw, K.P.; Nashed, N.E.; Ramos, E.; Miller, S.P.; Adams, R.C.; Fang, F.; Patel, R.M.; Holcombe, F.O., Jr; Chiu, Y.Y.; Hussain, A.S. Scientific considerations of pharmaceutical solid polymorphism in abbreviated new drug applications. *Pharm. Res.,* **2003**, *20*(4), 531-536. [http://dx.doi.org/10.1023/A:1023285627778] [PMID: 12739758]

[122] Schmidt, A.C. The role of molecular structure in the crystal polymorphism of local anesthetic drugs: crystal polymorphism of local anesthetic drugs, part X. *Pharm. Res.,* **2005**, *22*(12), 2121-2133. [http://dx.doi.org/10.1007/s11095-005-8135-6] [PMID: 16283568]

[123] Singhal, D.; Curatolo, W. Drug polymorphism and dosage form design: a practical perspective. *Adv. Drug Deliv. Rev.,* **2004**, *56*(3), 335-347. [http://dx.doi.org/10.1016/j.addr.2003.10.008] [PMID: 14962585]

[124] Łodyga-Chruścińka, E.; Zakrzewski, M.; Kuberski, S.; Paluszkiewicz, A.; Dinnebier, R.E.; Sugimoto, K. Preliminary Characterization of New Polymorphic Forms of Bupivacaine HCl. *Annals of Polish Chemical Society,* **2005**, *2*, 87-90.

[125] Pfeffer Henning, S.; Piechom, P.; Bellus, M.; Goldbronn, C.; Tedesco, E. Physico-chemical characterization of an active pharmaceutical ingredient. Crystal polymorphism and structural analysis. *J. Therm. Anal. Calorim.,* **2004**, *77*, 663-679. [http://dx.doi.org/10.1023/B:JTAN.0000039002.79914.b7]

[126] Maruyama, S.; Ooshima, H. Crystallization behavior of taltirelin polymorphs in a mixture of water and methanol. *J. Cryst. Growth,* **2000**, *212*, 239-245. [http://dx.doi.org/10.1016/S0022-0248(99)00903-3]

[127] Yamamura, S.; Momose, Y. Quantitative analysis of crystalline pharmaceuticals in powders and tablets by a pattern-fitting procedure using X-ray powder diffraction data. *Int. J. Pharm.,* **2001**, *212*(2), 203-212. [http://dx.doi.org/10.1016/S0378-5173(00)00605-0] [PMID: 11165078]

[128] Vishweshwar, P.; McMahon, J.A.; Oliveira, M.; Peterson, M.L.; Zaworotko, M.J. The predictably elusive form II of aspirin. *J. Am. Chem. Soc.,* **2005**, *127*(48), 16802-16803. [http://dx.doi.org/10.1021/ja056455b] [PMID: 16316223]

[129] Ouvrard, C.; Price, S.L. Toward Crystal Structure Prediction for Conformationally Flexible Molecules: The Headaches Illustrated by Aspirin. *Cryst. Growth Des.,* **2004**, *4*, 1119-1127. [http://dx.doi.org/10.1021/cg049922u]

[130] McGregor, C.; Bines, E. The use of high-speed differential scanning calorimetry (Hyper-DSC) in the study of pharmaceutical polymorphs. *Int. J. Pharm.,* **2008**, *350*(1-2), 48-52. [http://dx.doi.org/10.1016/j.ijpharm.2007.08.015] [PMID: 17890030]

[131] Loftsson, T.; Duchêne, D. Cyclodextrins and their pharmaceutical applications. *Int. J. Pharm.,* **2007**, *329*(1-2), 1-11. [http://dx.doi.org/10.1016/j.ijpharm.2006.10.044] [PMID: 17137734]

[132] Bates, S.; Zografi, G.; Engers, D.; Morris, K.; Crowley, K.; Newman, A. Analysis of amorphous and nanocrystalline solids from their X-ray diffraction patterns. *Pharm. Res.,* **2006**, *23*(10), 2333-2349. [http://dx.doi.org/10.1007/s11095-006-9086-2] [PMID: 17021963]

[133] Harris, R.K. Applications of solid-state NMR to pharmaceutical polymorphism and related matters. *J. Pharm. Pharmacol.,* **2007**, *59*(2), 225-239. [http://dx.doi.org/10.1211/jpp.59.2.0009] [PMID: 17270076]

[134] Blanco, M.; Villar, A. Polymorphic analysis of a pharmaceutical preparation by NIR spectroscopy. *Analyst (Lond.),* **2000**, *125*(12), 2311-2314.
[http://dx.doi.org/10.1039/b005746k] [PMID: 11219073]

[135] Otsuka, M.; Kato, F.; Matsuda, Y. Determination of indomethacin polymorphic contents by chemometric near-infrared spectroscopy and conventional powder X-ray diffractometry. *Analyst (Lond.),* **2001**, *126*(9), 1578-1582.
[http://dx.doi.org/10.1039/b103498g] [PMID: 11592653]

[136] Blanco, M.; Alcalá, M.; González, J.M.; Torras, E. Near infrared spectroscopy in the study of polymorphic transformations. *Anal. Chim. Acta,* **2006**, *567*, 262-268.
[http://dx.doi.org/10.1016/j.aca.2006.03.036]

[137] Higgins, J.P.; Arrivo, S.M.; Reed, R.A. Approach to the determination of hydrate form conversions of drug compounds and solid dosage forms by near-infrared spectroscopy. *J. Pharm. Sci.,* **2003**, *92*(11), 2303-2316.
[http://dx.doi.org/10.1002/jps.10465] [PMID: 14603515]

[138] Kalinkova, G.N. Infrared spectroscopy in pharmacy. *Vib. Spectrosc.,* **1999**, *19*, 307-320.
[http://dx.doi.org/10.1016/S0924-2031(99)00017-X]

[139] Auer, M.E.; Griesser, U.J.; Sawatzki, J. Qualitative and quantitative study of polymorphic forms in drug formulations by near infrared FT-Raman spectroscopy. *J. Mol. Struct.,* **2003**, *661/662*, 307-317.
[http://dx.doi.org/10.1016/j.molstruc.2003.09.002]

[140] Blanco, M.; Valdés, D.; Bayod, M.; Fernandez-Martí, F.; Llorente, I. Characterization and analysis of polymorphs by near-infrared spectrometry. *Anal. Chim. Acta,* **2004**, *502*, 221-227.
[http://dx.doi.org/10.1016/j.aca.2003.10.016]

[141] MacFhionnghaile, P.; Hu, Y.; McArdle, P.; Erxleben, A. A comprehensive near infrared spectroscopic study of the limits of quantitative analysis of sulfathiazole polymorphism. *J. Near Infrared Spectrosc.,* **2013**, *21*(1), 55-66.
[http://dx.doi.org/10.1255/jnirs.1036]

[142] Patel, A.D.; Luner, P.E.; Kemper, M.S. Quantitative analysis of polymorphs in binary and multi-component powder mixtures by near-infrared reflectance spectroscopy. *Int. J. Pharm.,* **2000**, *206*(1-2), 63-74.
[http://dx.doi.org/10.1016/S0378-5173(00)00530-5] [PMID: 11058811]

[143] Blanco, M.; Coello, J.; Iturriaga, H.; Maspoch, S.; Perez-Maseda, C. Determination of polymorphic purity by near infrared spectrometry. *Anal. Chim. Acta,* **2000**, *407*, 247-254.
[http://dx.doi.org/10.1016/S0003-2670(99)00828-4]

[144] Salari, A.; Young, R.E. Application of attenuated total reflectance FTIR spectroscopy to the analysis of mixtures of pharmaceutical polymorphs. *Int. J. Pharm.,* **1998**, *163*, 157-166.
[http://dx.doi.org/10.1016/S0378-5173(97)00378-5]

[145] Kendall, D.N. Identification of Polymorphic Forms of Crystals by Infrared Spectroscopy. *Anal. Chem.,* **1953**, *25*(3), 382-389.
[http://dx.doi.org/10.1021/ac60075a002]

[146] Ayala, AP Polymorphism in drugs investigated by low wavenumber Raman scatting. *Vibrational Spect,* **2007**, *452*, 112-116.

[147] Du, Y.; Xue, J. Investigation of Polymorphism and Cocrystallization of Active Pharmaceutical Ingredients Using Vibrational Spectroscopic Techniques. *Curr. Pharm. Des.,* **2016**, *22*(32), 4917-4928.
[http://dx.doi.org/10.2174/1381612822666160726104604] [PMID: 27464725]

[148] Larkin, P.J.; Dabros, M.; Sarsfield, B.; Chan, E.; Carriere, J.T.; Smith, B.C. Polymorph characterization of active pharmaceutical ingredients (APIs) using low-frequency Raman spectroscopy. *Appl. Spectrosc.,* **2014**, *68*(7), 758-776.

[http://dx.doi.org/10.1366/13-07329] [PMID: 25014842]

[149] Heinz, A.; Strachan, C.J.; Gordon, K.C.; Rades, T. Analysis of solid-state transformations of pharmaceutical compounds using vibrational spectroscopy. *J. Pharm. Pharmacol.,* **2009**, *61*(8), 971-988.
[http://dx.doi.org/10.1211/jpp.61.08.0001] [PMID: 19703341]

[150] Erxleben, A. Application of Vibrational Spectroscopy to Study Solid-state Transformations of Pharmaceuticals. *Curr. Pharm. Des.,* **2016**, *22*(32), 4883-4911.
[http://dx.doi.org/10.2174/1381612822666160726110103] [PMID: 27464723]

[151] Aaltonen, J.; Gordon, K.C.; Strachan, C.J.; Rades, T. Perspectives in the use of spectroscopy to characterise pharmaceutical solids. *Int. J. Pharm.,* **2008**, *364*(2), 159-169.
[http://dx.doi.org/10.1016/j.ijpharm.2008.04.043] [PMID: 18555625]

[152] Paudel, A.; Raijada, D.; Rantanen, J. Raman spectroscopy in pharmaceutical product design. *Adv. Drug Deliv. Rev.,* **2015**, *89*, 3-20.
[http://dx.doi.org/10.1016/j.addr.2015.04.003] [PMID: 25868453]

[153] Gala, U.; Chauhan, H. Principles and applications of Raman spectroscopy in pharmaceutical drug discovery and development. *Expert Opin. Drug Discov.,* **2015**, *10*(2), 187-206.
[http://dx.doi.org/10.1517/17460441.2015.981522] [PMID: 25399993]

[154] Smith, G.P.; McGoverin, C.M.; Fraser, S.J.; Gordon, K.C. Raman imaging of drug delivery systems. *Adv. Drug Deliv. Rev.,* **2015**, *89*, 21-41.
[http://dx.doi.org/10.1016/j.addr.2015.01.005] [PMID: 25632843]

[155] Gordon, K.C.; McGoverin, C.M. Raman mapping of pharmaceuticals. *Int. J. Pharm.,* **2011**, *417*(1-2), 151-162.
[http://dx.doi.org/10.1016/j.ijpharm.2010.12.030] [PMID: 21194560]

[156] Henson, M.J.; Zhang, L. Drug characterization in low dosage pharmaceutical tablets using Raman microscopic mapping. *Appl. Spectrosc.,* **2006**, *60*(11), 1247-1255.
[http://dx.doi.org/10.1366/000370206778998987] [PMID: 17132441]

[157] Wartewig, S.; Neubert, R.H. Pharmaceutical applications of Mid-IR and Raman spectroscopy. *Adv. Drug Deliv. Rev.,* **2005**, *57*(8), 1144-1170.
[http://dx.doi.org/10.1016/j.addr.2005.01.022] [PMID: 15885850]

[158] Mansour, H.M.; Hickey, A.J. Raman characterization and chemical imaging of biocolloidal self-assemblies, drug delivery systems, and pulmonary inhalation aerosols: a review. *AAPS PharmSciTech,* **2007**, *8*(4), E99.
[http://dx.doi.org/10.1208/pt0804099] [PMID: 18181559]

[159] Strachan, C.J.; Rades, T.; Gordon, K.C.; Rantanen, J. Raman spectroscopy for quantitative analysis of pharmaceutical solids. *J. Pharm. Pharmacol.,* **2007**, *59*(2), 179-192.
[http://dx.doi.org/10.1211/jpp.59.2.0005] [PMID: 17270072]

[160] Brown, S.C.; Claybourn, M.; Sievwright, D.; Fearnside, V.; Ashman, C. Lean Raman imaging for rapid assessment of homogeneity in pharmaceutical formulations. *Appl. Spectrosc.,* **2010**, *64*(4), 442-447.
[http://dx.doi.org/10.1366/000370210791114239] [PMID: 20412630]

[161] Hédoux, A.; Guinet, Y.; Descamps, M. The contribution of Raman spectroscopy to the analysis of phase transformations in pharmaceutical compounds. *Int. J. Pharm.,* **2011**, *417*(1-2), 17-31.
[http://dx.doi.org/10.1016/j.ijpharm.2011.01.031] [PMID: 21256937]

[162] Ikeda, Y.; Ishihara, Y.; Moriwaki, T.; Kato, E.; Terada, K. A novel analytical method for pharmaceutical polymorphs by terahertz spectroscopy and the optimization of crystal form at the discovery stage. *Chem. Pharm. Bull. (Tokyo),* **2010**, *58*(1), 76-81.
[http://dx.doi.org/10.1248/cpb.58.76] [PMID: 20045970]

[163] Nakamoto, K.; Urasaki, T.; Hondo, S.; Murahashi, N.; Yonemochi, E.; Terada, K. Evaluation of the

crystalline and amorphous states of drug products by nanothermal analysis and Raman imaging. *J. Pharm. Biomed. Anal.*, **2013**, *75*, 105-111.
[http://dx.doi.org/10.1016/j.jpba.2012.11.020] [PMID: 23246930]

[164] Baraldi, C.; Tinti, A.; Ottani, S.; Gamberini, M.C. Characterization of polymorphic ampicillin forms. *J. Pharm. Biomed. Anal.*, **2014**, *100*, 329-340.
[http://dx.doi.org/10.1016/j.jpba.2014.08.021] [PMID: 25194347]

[165] Vega, D.R.; Polla, G.; Martinez, A.; Mendioroz, E.; Reinoso, M. Conformational polymorphism in bicalutamide. *Int. J. Pharm.*, **2007**, *328*(2), 112-118.
[http://dx.doi.org/10.1016/j.ijpharm.2006.08.001] [PMID: 16978811]

[166] Schönbichler, S.A.; Bittner, L.K.; Weiss, A.K.; Griesser, U.J.; Pallua, J.D.; Huck, C.W. Comparison of NIR chemical imaging with conventional NIR, Raman and ATR-IR spectroscopy for quantification of furosemide crystal polymorphs in ternary powder mixtures. *Eur. J. Pharm. Biopharm.*, **2013**, *84*(3), 616-625.
[http://dx.doi.org/10.1016/j.ejpb.2013.01.006] [PMID: 23395969]

[167] Bunaciu, A.A.; Fleschin, S.; Aboul-Enein, H.Y. Analysis of mebendazole polymorphs by Fourier transform IR spectrometry using chemometric methods. *Spectrosc. Lett.*, **2001**, *34*(5), 527-536.
[http://dx.doi.org/10.1081/SL-100106868]

[168] Strachan, C.J.; Pratiwi, D.; Gordon, K.C.; Rades, T. Quantitative analysis of polymorphic mixtures of carbamazepine by Raman spectroscopy and principal components analysis. *J. Raman Spectrosc.*, **2004**, *35*(5), 347-352.
[http://dx.doi.org/10.1002/jrs.1140]

[169] Heinz, A.; Savolainen, M.; Rades, T.; Strachan, C.J. Quantifying ternary mixtures of different solid-state forms of indomethacin by Raman and near-infrared spectroscopy. *Eur. J. Pharm. Sci.*, **2007**, *32*(3), 182-192.
[http://dx.doi.org/10.1016/j.ejps.2007.07.003] [PMID: 17716878]

[170] Calvo, N.L.; Maggio, R.M.; Kaufman, T.S. A dynamic thermal ATR-FTIR/chemometric approach to the analysis of polymorphic interconversions. Cimetidine as a model drug. *J. Pharm. Biomed. Anal.*, **2014**, *92*, 90-97.
[http://dx.doi.org/10.1016/j.jpba.2013.12.036] [PMID: 24509287]

[171] Patel, S.; Kaushal, A.M.; Bansal, A.K. Compression physics in the formulation development of tablets. *Crit. Rev. Ther. Drug Carrier Syst.*, **2006**, *23*(1), 1-65.
[http://dx.doi.org/10.1615/CritRevTherDrugCarrierSyst.v23.i1.10] [PMID: 16749898]

[172] Bauer, J.F. Polymorphism - A Critical Consideration in Pharmaceutical Development, Manufacturing, and Stability. *Journal of Validation Technology,* **2008**, 15-23.

[173] Fronczek, F.R.; Kamel, H.N.; Slattery, M. Three polymorphs (alpha, beta, and delta) of D-mannitol at 100 K. *Acta Crystallogr. C,* **2003**, *59*(Pt 10), 567-570.
[http://dx.doi.org/10.1107/S0108270103018961] [PMID: 14532673]

[174] Vanhoorne, V.; Van Bockstal, P.J.; Van Snick, B.; Peeters, E.; Monteyne, T.; Gomes, P.; De Beer, T.; Remon, J.P.; Vervaet, C. Continuous manufacturing of delta mannitol by cospray drying with PVP. *Int J Pharm,* **2016**, *501*(1-2), 139-147.
[http://dx.doi.org/10.1016/j.ijpharm.2016.02.001]

[175] Huang, W.; Shi, Y.; Wang, C.; Yu, K.; Sun, F.; Li, Y. Using spray-dried lactose monohydrate in wet granulation method for a low-dose oral formulation of a paliperidone derivative. *Powder Technol.*, **2013**, *246*, 379-394.
[http://dx.doi.org/10.1016/j.powtec.2013.05.042]

[176] http://www.dfepharma.com/en/search-results.aspx?search=lactose+some+basic+properties

[177] Sarode, A.L.; Sandhu, H.; Shah, N.; Malick, W.; Zia, H. Hot melt extrusion (HME) for amorphous solid dispersions: predictive tools for processing and impact of drug-polymer interactions on

supersaturation. *Eur. J. Pharm. Sci.,* **2013**, *48*(3), 371-384.
[http://dx.doi.org/10.1016/j.ejps.2012.12.012] [PMID: 23267847]

[178] Fule, R.; Paithankar, V.; Amin, P. Hot melt extrusion based solid solution approach: Exploring polymer comparison, physicochemical characterization and in-vivo evaluation. *Int. J. Pharm.,* **2016**, *499*(1-2), 280-294.
[http://dx.doi.org/10.1016/j.ijpharm.2015.12.062] [PMID: 26746801]

[179] Kanaujia, P.; Poovizhi, P.; Ng, W.K.; Tan, R.B. Amorphous formulations for dissolution and bioavailability enhancement of poorly soluble APIs. *Powder Technol.,* **2015**, *285*, 2-15.
[http://dx.doi.org/10.1016/j.powtec.2015.05.012]

[180] Fahr, A.; Liu, X. Drug delivery strategies for poorly water-soluble drugs. *Expert Opin. Drug Deliv.,* **2007**, *4*(4), 403-416.
[http://dx.doi.org/10.1517/17425247.4.4.403] [PMID: 17683253]

[181] Aungst, B.J. Absorption enhancers: applications and advances. *AAPS J.,* **2012**, *14*(1), 10-18.
[http://dx.doi.org/10.1208/s12248-011-9307-4] [PMID: 22105442]

[182] Goscianska, J.; Olejnik, A.; Nowak, I.; Marciniak, M.; Pietrzak, R. Ordered mesoporous silica modified with lanthanum for ibuprofen loading and release behaviour. *Eur. J. Pharm. Biopharm.,* **2015**, *94*, 550-558.
[http://dx.doi.org/10.1016/j.ejpb.2015.07.003] [PMID: 26170160]

[183] Everett, D.H. IUPAC Manual of Symbols and Terminology for Physicochemical Quantities and Units, Appendix II: Definitions, Terminology and Symbols in Colloid and Surface Chemistry. *Pure Appl. Chem.,* **1972**, *31*, 578-38.
[http://dx.doi.org/10.1351/pac197231040577]

[184] Burleigh, M.C.; Dai, S. Functonalized nanoporous adsorbents for environmental remediation.*Nanoporous Materials: Science and Engineering*; Imperial College Press: London, **2004**.
[http://dx.doi.org/10.1142/9781860946561_0024]

[185] Chiola, V.; Ritsko, J.E.; Vanderpool, C.D. US Patent 3 556 7251971.

[186] Beck, J.S.; Vartuli, J.C.; Roth, W.J.; Leonowicz, M.E.; Kresge, C.T.; Schmitt, K.D.; Chu, C.T.; Olson, D.H.; Sheppard, E.W.; McCullen, S.B.; Higgins, J.B.; Schlenker, J.L. A New Family of Mesoporous Molecular Sieves Prepared with Liquid Crystal Templates. *J. Am. Chem. Soc.,* **1992**, *114*, 10834-10843.
[http://dx.doi.org/10.1021/ja00053a020]

[187] Kresge, C.T.; Leonowicz, M.E.; Roth, W.J.; Vartuli, J.C.; Beck, J.S. Ordered mesoporous molecular sieves synthesized by a liquid-crystal template mechanism. *Nature,* **1992**, *359*, 710-712.
[http://dx.doi.org/10.1038/359710a0]

[188] Zhao, D.; Feng, J.; Huo, Q.; Melosh, N.; Fredrickson, G.H.; Chmelka, B.F.; Stucky, G.D. Triblock copolymer syntheses of mesoporous silica with periodic 50 to 300 angstrom pores. *Science,* **1998**, *279*(5350), 548-552.
[http://dx.doi.org/10.1126/science.279.5350.548] [PMID: 9438845]

[189] Zhao, D.Y.; Huo, Q.S.; Feng, J.L.; Chmelka, B.F.; Stucky, G.D. Nonionic triblockand star diblock copolymer and oligomeric surfactant syntheses of highly ordered, hydrothermally stable, mesoporous silica structures. *J. Am. Chem. Soc.,* **1998**, *120*, 6024-6036.
[http://dx.doi.org/10.1021/ja974025i]

[190] Yu, C.; Tian, B.; Liu, X.; Fan, J.; Yang, H.; Zhao, D.Y. Advances in mesoporous materials templated by nonionic block copolymers.*Nanoporous Materials: Science and Engineering*; Imperial College Press: London, **2004**.
[http://dx.doi.org/10.1142/9781860946561_0002]

[191] Grudzień, R.M.; Grabicka, B.E.; Jaroniec, M. Adsorption studies of thermal stability of SBA-16 mesoporous silicas. *Appl. Surf. Sci.,* **2007**, *253*, 5660-5665.

[http://dx.doi.org/10.1016/j.apsusc.2006.12.033]

[192] Goscianska, J.; Olejnik, A.; Pietrzak, R. Adsorption of l-phenylalanine onto mesoporous silica. *Mater. Chem. Phys.*, **2013**, *142*, 586-593.
[http://dx.doi.org/10.1016/j.matchemphys.2013.07.057]

[193] Tanev, P.T.; Pinnavaia, T.J. A neutral templating route to mesoporous molecular sieves. *Science,* **1995**, *267*(5199), 865-867.
[http://dx.doi.org/10.1126/science.267.5199.865] [PMID: 17813916]

[194] Yanagisawa, T.; Shimizu, T.; Kuroda, K.; Kato, C. The preparation of alkyltrimethylammonium-kanemite complexes and their conversion to microporous materials. *Bull. Chem. Soc. Jpn.,* **1990**, *63*, 988-992.
[http://dx.doi.org/10.1246/bcsj.63.988]

[195] Florek, J.; Giret, S.; Juère, E.; Larivière, D.; Kleitz, F. Functionalization of mesoporous materials for lanthanide and actinide extraction. *Dalton Trans.,* **2016**, *45*(38), 14832-14854.
[http://dx.doi.org/10.1039/C6DT00474A] [PMID: 27240525]

[196] Nawrocki, J. The silanol group and its role in liquid chromatography. *J. Chromatogr. A,* **1997**, *779*, 29-71.
[http://dx.doi.org/10.1016/S0021-9673(97)00479-2]

[197] Hu, L.; Ji, S.; Jiang, Z.; Song, H.; Wu, P.; Liu, Q. Direct synthesis and structural characteristics of ordered SBA-15 mesoporous silica containing tungsten oxides and tungsten carbides. *J. Phys. Chem. C,* **2007**, *111*, 15173-15184.
[http://dx.doi.org/10.1021/jp074879h]

[198] Socrates, G. Infrared and Raman Characteristic Group Frequencies. In: *Tables and Charts,* 3rd ed; John Wiley & Sons, LTD, **2001**.

[199] Ambrogi, A.; Perioli, L.; Marmottini, F.; Accorsi, O.; Pagano, C.; Ricci, M.; Rossi, C. Role of mesoporous silicates on carbamazepine dissolution rate enhancement. *Microporous Mesoporous Mater.,* **2008**, *113*, 445-452.
[http://dx.doi.org/10.1016/j.micromeso.2007.12.003]

[200] Grzesiak, A.L.; Lang, M.; Kim, K.; Matzger, A.J. Comparison of the four anhydrous polymorphs of carbamazepine and the crystal structure of form I. *J. Pharm. Sci.,* **2003**, *92*(11), 2260-2271.
[http://dx.doi.org/10.1002/jps.10455] [PMID: 14603511]

[201] Harris, R.K.; Ghi, P.Y.; Pushmann, H.; Apperley, D.C.; Griesser, U.J.; Hammond, R.B.; Ma, C.; Roberts, K.J.; Pearce, G.J.; Yates, J.R.; Picard, C.J. Structural studies of the polymorphs of carbamazepine, its dehydrate, and two solvates. *Org. Process Res. Dev.,* **2005**, *9*, 902-910.
[http://dx.doi.org/10.1021/op0500990]

[202] Rustichelli, C.; Gamberini, G.; Ferioli, V.; Gamberini, M.C.; Ficarra, R.; Tommasini, S. Solid-state study of polymorphic drugs: carbamazepine. *J. Pharm. Biomed. Anal.,* **2000**, *23*(1), 41-54.
[http://dx.doi.org/10.1016/S0731-7085(00)00262-4] [PMID: 10898153]

[203] Shen, S.C.; Ng, W.K.; Chia, L.; Hu, J.; Tan, R.B. Physical state and dissolution of ibuprofen formulated by co-spray drying with mesoporous silica: effect of pore and particle size. *Int. J. Pharm.,* **2011**, *410*(1-2), 188-195.
[http://dx.doi.org/10.1016/j.ijpharm.2011.03.018] [PMID: 21419202]

[204] Azaïs, T.; Bonhomme-Coury, L.; Vaissermann, J.; Maquet, J.; Bonhomme, C. The first aluminophosphonate cluster analogue of the 4 = 1 SBU of zeolites: structure and multinuclear solid-state NMR study including ^{1}H NMR. *Eur. J. Inorg. Chem.,* **2002**, *2002*, 2838-2843.

[205] Qu, F.; Zhu, G.; Lin, H.; Zhang, W.; Sun, J.; Li, S.; Qiu, S. A controlled release of ibuprofen by systematically tailoring the morphology of mesoporous silica materials. *J. Solid State Chem.,* **2006**, *179*, 2027-2035.
[http://dx.doi.org/10.1016/j.jssc.2006.04.002]

[206] Rámila, A.; Muñoz, B.; Pérez-Pariente, J.; Vallet-Regí, M. Mesoporous MCM-41 as drug host system. *J. Sol-Gel Sci. Technol.,* **2003**, *26*, 1199-02. [http://dx.doi.org/10.1023/A:1020764319963]

[207] Muñoz, B.; Rámila, A.; Pérez-Pariente, J.; Díaz, I.; Vallet-Regí, M. MCM-41 organic modification as drug delivery rate regulator. *Chem. Mater.,* **2003**, *15*, 500-503. [http://dx.doi.org/10.1021/cm021217q]

[208] Ayad, M.M.; Salahuddin, N.A.; El-Nasr, A.A.; Torad, N.L. Amine-functionalized mesoporous silica KIT-6 as a controlled release drug delivery carrier. *Microporous Mesoporous Mater.,* **2016**, *229*, 166-177. [http://dx.doi.org/10.1016/j.micromeso.2016.04.029]

CHAPTER 6

Electrochemical Biosensors Based on Conductive Polymers and Their Applications in Biomedical Analysis

Sławomira Skrzypek* and **Paweł Krzyczmonik**

Department of Inorganic and Analytical Chemistry, University of Łódź, Łódź, Poland

Abstract: Biosensors are a very important subject in the contemporary analytics, mainly in biomedical area. This chapter describes biosensors built with the use of conductive polymers. Conductive polymers provide electrical conductivity of the forming layer, they may function as mediators in red-ox processes and they can enable the immobilization of a biological agent on their surface. Nowadays, in the center of attention are composite materials in which, next to conductive polymers, nanostructured materials such as graphene, carbon nanotubes and quantum dots are applied. In the chapter, the basic modes of biosensors of the first, the second and the third generation, are described. The examples of different enzymatic and affinity biosensors are presented.

Keywords: Amperometry, Antibodies, Biosensors, Cholesterol oxidase, Conductive polymers, Enzymes, Glucose oxidase, Immobilization, Laccase, PEDOT, Potentiometry, Urease.

INTRODUCTION

Biosensors are one of the most advanced tools of modern analytics. The history of their development began in 1962 when Clark and Lyons first proposed the concept of a glucose enzyme electrode, but it was not until 1977 when the name biosensor was first used. Another important step forward came with the development of research on conductive polymers and modified electrodes, as well as the development of optical fibre technology. These developments resulted in significant acceleration of the work on electrochemical and optical sensors and biosensors. The fact that, since the beginning of 1960s, there were more than 80,000 works published focusing on various biosensors, is the testament to their popularity as a salient scientific issue.

* **Corresponding authors Sławomira Skrzypek:** Department of Inorganic and Analytical Chemistry, University of Łódź, ul. Tamka 12, 91-403 Łódź, Poland; Tel/Fax: 48 42 6355785; E-mail: skrzypek@uni.lodz.pl

Atta-ur-Rahman, Sibel A. Ozkan & Rida Ahmed (Eds.)

IUPAC defines a chemical sensor as a device that transforms chemical information, ranging from the concentration of a specific sample component to total composition analysis, into an analytically useful signal. A biosensor is, on the other hand, a sensor, which contains biological material in its receptor layer. Proteins, enzymes, antibodies, tissues, or even whole organisms, such as bacteria and other biological objects, can all be examples of such biological material. Biosensor allows for utilization of high sensitivity with respect to the analyte. Enzymes, which have a wide and varied range of specificity, are classic example of this. On one hand, there are such enzymes as laccase or pancreatic lipase which have specificity towards a group of specific compounds or a specific type of reaction, and on the other hand, there is, for example, glucose oxidase which reacts only with one form of glucose, β-D-glucopyranose. The choice of appropriate biological agent determines specificity of a biosensor, which allows for determination of an analyte and simultaneously decreases or eliminates the influence of other matrix components [1, 2]. Using biological material results also in greater sensitivity of determination and a possibility of determining lower levels of concentration. This section presents an overview of biosensors built based on conductive polymers, immobilized enzymes, antibodies and DNA.

CONDUCTIVE POLYMERS

Electron-conductive polymers are macromolecular substances capable of conducting electric current in which charge carriers are either electrons or electron holes. Polyaniline (PANI), polypyrrole, polythiophene, poly (p-phenyl), poly(3,4-ethylene-1,4-dioxythiophene) (PEDOT), and their derivatives and copolymers, are the most popular and most studied representatives types of conductive polymers [3, 4].

Polyaniline is one of the conductive polymers, which have been studied for the longest time. It can be synthesized either chemically or electrochemically. Depending on the conditions of its polymerization, it is possible to obtain polyaniline with different electrical properties – from an insulator to a conductor [5]. Polyaniline capable of conducting electric current is obtained from acid solutions (pH <3), which ensures that the nitrogen atoms in a polyaniline molecule are protonated. Polyaniline also needs to be doped at the level of at least 30%. Polymers obtained under such conditions are partially oxidized, protonated, and doped, and they are characterized by high electrical conductivity. Polyaniline occurs in several redox forms and each of them may be protonated or deprotonated [6]. Polyaniline conducts current only in a protoned and partially oxidized form, often referred to as emeraldine salt. The reduced form (leucoemeraldine) does not demonstrate any conductivity. Neither does the completely oxidized form (pernigraniline), which undergoes degradation in reaction with water [5, 6]. Polyaniline has a wide variety of diverse applications. Other polymers can be obtained based on polyaniline for electrode modification. It should be noted that polymers obtained from aniline derivatives usually have lower conductivity than polyaniline, and that only some of aniline derivatives are able to polymerize. Another method is copolymerization of aniline with its derivatives, for example copolymerization of aniline with o-aminosulphonic acid [7], copolymerization of aniline derivatives such as o-toluidine and anthranilic acid [8] and copolymerization of aniline or its derivatives with other compounds, for example, copolymerization of aniline and pyrrole [9]. Polyaniline is also a component of conductive composite materials. A good example of this may be polyaniline and polystyrene sulfonic acid [10], polyaniline and polyacrylic acid [11], as well as PEDOT and polystyrene sulfonic acid [12]. In such materials, a conductive polymer provides electrical conductivity, whereas the second component may improve mechanical properties and, if it contains negatively charged groups in its structure, it may provide doping of the polymer. In this case, negative charges are permanently attached to the composite material layer. It prevents these charges from elution, and prevents the polymer from permanent loss of conductivity [13]. PANI is one of the most intensively studied conductive polymers, and it is characterized by such positive features as excellent durability, great electrical properties and strong biomolecular interactions. Polyaniline is also one of the most popular materials in the construction of biosensors with electrochemical detection. It is also used as an electrode material on which the enzymes are immobilized. Its drawback is the pH at which it is obtained ($pH<3$). As most of the biological tests are carried out in neutral or slightly acidic environment, its derivatives, which can work in a neutral or slightly acidic pH, are used more frequently than polyaniline itself [14].

Polypyrrole and polythiophene are, just like polyaniline, one of the most often studied and best known conductive polymers, whereas PEDOT is a relatively new polymer. It has been developed by a group of scientists from Bayer's laboratories in the second half of the 1980s [15 - 18]. It is one of the most durable conductive polymers, characterized by very good stability and high conductivity. The band gap of PEDOT is variable and ranges from 1.4 to 2.5 eV [12]. Its conductivity decreases in the state of complete oxidation, and the polymer behaves like a semiconductor. What is more, PEDOT demonstrates the electrochromic effect. In its reduced state, PEDOT is dark blue, and in the oxidized form, it is colorless [12]. PEDOT is usually obtained from non-aqueous solutions, for example, from acetonitrile. It is also possible to synthesize it from aqueous solutions, but because of the limited solubility of 3,4-ethylene-1,4-dioxythiophene (EDOT) in water, the obtained polymer layers are usually thin. The properties of PEDOT obtained from aqueous solutions are as good as the properties of PEDOT obtained from non-aqueous solutions. PEDOT has many applications in electrochemistry, such as batteries, fuel cells, sensors and biosensors, and is widely used in optoelectronics. A commercially available composite material, Baytron P, has been developed on the basis of PEDOT. It is a mixture of PEDOT and poly (styrenesulfonic acid sodium salt) (PSS). The properties of PEDOT were proven to be better than those of polypyrrole and polythiophene in a number of different applications. Krosa *et al.* [19] proved in their studies that polypyrrole can be used as a component of a biosensor for a very short period of time, with PEDOT, under the same conditions, allowed to obtain sensors suitable for continuous use. According to the authors, this was due because PEDOT has greater electrochemical stability than polypyrrole. In the case of systems containing biochemical objects, it is very important that PEDOT has low toxicity in comparison to, for example, polyaniline whose degradation products are carcinogenic [20]. Systems of similar structure are also used; for example, PEDOT doped with lithium polystyrenesulfonate [21].

Tuantranont *et al.* [22] presented an interesting example of the practical application of the above described materials. The authors indicated the possibility of using suspension containing graphene and PEDOT:PSS as ink for the production of printed screen electrodes. They studied both the process of obtaining electrodes and the possibility of their use in electroanalysis, and proved that these electrodes can be used to determine hydrogen peroxide, nicotinamide adenine dinucleotide and hexacyanoferrate complex. It was also demonstrated that the use of the described electrodes allowed to obtain higher current levels than the unmodified screen printed electrodes and the PEDOT-modified electrodes. The cited article suggests a very interesting direction of further research, as the development of modifying material in the form of ink results in a possibility to both immediately produce printed screen electrodes and to apply them very broadly. It seems that electrode materials based on PEDOT composites and composites of other conductive polymers have a chance to be widely used in this technology.

NANOSTRUCTURED MATERIALS

Contemporary modified electrodes have become very complex objects, which use multilayer systems and systems made of composite materials. Nanostructured materials are very popular materials used for construction of such systems, and they are very often combined with conductive polymers.

Quantum Dots

Quantum dots are one of the newest nanostructured materials used in construction of biosensors. They were discovered by Alexey Ekimov in 1981 in a glass matrix, and then in colloidal solutions by Louis E. Brus in 1985. The term "quantum dot" itself was coined by Mark Reed. Quantum dots are nanocrystals, confined in all spatial dimensions, made of semiconductor materials, which are small enough to exhibit quantum mechanical properties. They usually have a spherical shape and contain hundreds or thousands of atoms depending on their final size. CdTe, CdSe, ZnSe and ZnS are the most popular types of quantum dots. However, metals such as Ga, In, and many others, can also be used in the synthesis of quantum dots. Due to their small size, the electrons in quantum dots are confined in a small space, in a quantum box. At the quantum level, the properties of this type of system are decided by the presence of excitons. Excitons are defined as bound states of an electron excited to the conduction band, and an electron hole, which was created in the valence band. Energy of the electron band is the energy of the Coulomb interaction between the electron and the hole. When the radii of the semiconductor nanocrystals are smaller than the radius of the Bohr exciton, the quantization of energy levels occurs, in accordance with the Pauli exclusion. Imitation of the motion of a particle (quantum dot) in three dimensions means quantization in each individual direction, which means that the particle may assume only certain states, defined by the Schrödinger equation. Thus, only adequately defined energy levels, may be occupied by the particle. This is why quantum dots are sometimes called artificial atoms. What makes quantum dots attractive to physicists and chemists is the ability to control the structure of the energy levels and the number of trapped charge carriers at the stage of creation of the nanodot by means of controlling the size and shape [23 - 27].

The difference in energy between the lowest valence band and the highest increase of the conduction band is due to their respective sizes. Smaller size of the dot is equal to greater value of the band gap. Larger amount of energy is needed to excite the dots and, simultaneously, a large amount of energy is released when the crystal is returned to its basic state, causing the color to change from red to blue in the emitted light. As result of this phenomenon, quantum dots are able to emit light of any color on the basis of the same material by changing the size of the dot. Additionally, as high level of size control is possible, quantum dots can be configured already during their production. Quantum dots can be classified into different types according to their composition and structure:

Core-type Quantum Dots

Core-type quantum dots are single component materials with uniform internal compositions, such as selenides or sulfides of such metals as cadmium or zinc, for example, CdSe or CdS. The photo- and electroluminescence properties can be fine-tuned by changing their size [28 - 30].

Core-shell Quantum Dots

Quantum dots of this type are composed of two different semiconductors. We can distinguish between two different kinds of core-type quantum dots, depending on the type of materials used. The core is made of narrow band gap material and the shell of wide band gap material, or vice versa; for example, quantum dots with CdSe in the core and ZnS in the shell. Such a construction allows, for example, controlling the luminescence properties of this type of quantum dots [28 - 30].

Alloyed Quantum Dots

The ability to tune optical and electronic properties by changing the size of crystallites has become the distinguishing feature of quantum dots. However, tuning the properties by changing the crystallite size can cause problems in many applications connected with size restrictions. Multicomponent quantum dots offer an alternative method of tuning properties without changing the size of crystallites. Alloyed semiconductor quantum dots, with both homogeneous and gradient internal structures, make it possible to tune the optical and electronic properties by merely changing the composition and internal structure, without changing the crystallite size. For example, alloyed quantum dots of 6 nm diameter in the CdS_xSe_{1-x}/ZnS compositions emit light of different wavelengths in reaction to any change of the composition [28 - 30]. Quantum dots are useful in the construction of electrochemical biosensors due to their electrical conductivity, the possibility of electron exchange in the red-ox process and the possibility of immobilization on the surface of other compounds.

Carbon Nanostructured Materials

Various allotropes of carbon, such as fullerenes, carbon nanotubes and graphene, constitute another type of nanostructured materials used for the construction of biosensors. Discovery of these allotropes in the last decades resulted in a great development of various branches of science. It also had a very large impact on the electrochemistry, including also the development of sensors and biosensors. Due to their characteristic properties, they have become very attractive materials for electrochemistry. First of all, all carbon nanomaterials can be functionalized in a variety of ways. This allows to immobilize, for example, such objects as enzymes, proteins, and antibodies on their surfaces. Secondly, carbon materials have very high biocompatibility, which in the case of biosensors should positively affect their durability. Thirdly, nanotubes and graphene have high electrical conductivity, and can function as a conductive phase of an electrode. Furthermore, they can serve as mediators in red-ox processes.

Fullerenes and Nanotubes

The best known and most studies fullerene is $C_{60.}$ Isolated in its pure form it creates yellow-brown crystals with regular structure which, when dissolved in benzene, changes the color of the solution to pink-violet. C_{60} is the smallest stable fullerene whose shape resembles that of a ball. Such shape is created when combined carbon atoms create twelve five-membered rings and twenty six-membered rings. Crystalline C_{60} is dielectric and does not conduct electricity. It is, however, possible, to obtain good conductive properties by introducing an alkali atom. C_{60} and its homologues are used in many fields, such as superconductivity, photo-optics, biochemistry, materials engineering, fuel engineering, catalysis and many others [31].

Carbon nanotubes (CNT) are rolled-up graphite sheets in a cylindrical form. The change of their structure has a significant impact on their electron properties, which means that they can either function as semiconductors or have an extremely high electrical conductivity, similar to conductivity of some metals. Unique structural and electronic properties influence their wide application in many fields. The key advantage of CNTs are their small diameters and large length-t--diameter ratios, which allow them to be used as molecular wires for facilitating electron transfer between biomolecules and electrodes. Due to these unique properties, it is possible to use carbon nanotubes in electroanalysis and make them ideal candidates for the construction of high-performance sensors. There are single wall (SWNTs), double wall, and multiwall carbon nanotubes (MWNTs). According to the rolled-up direction of the graphene sheet, which constitutes the nanotube, the CNTs can be classified into three conformations: armchair, zig-zag, and chiral [31, 32]. The fact that CNTs can be functionalized is a very significant advantage. Functionalization of nanotubes by means of creating functional groups on their surfaces is the most commonly used form of functionalization. It allows attaching various types of molecules of chemical and biochemical compounds to nanotubes.

Graphene

Graphene and related materials graphene, otherwise known as graphenoids, have been thoroughly tested in recent years. Graphene and graphenoids are used also in electrochemistry.

The IUPAC definition of graphene formulated in 1995 defines it as: “a single carbon layer of the graphite structure, describing its nature by analogy to a polycyclic aromatic hydrocarbon of quasi infinite size.” However, in practice, the materials have specific sizes and are often not limited to single sheets of graphene. They are not always flat sheets. Such structures are referred to as graphenoids, which can be divided according to their structure into: sheets, ribbons, flakes and quantum dots. The term “sheet” is used when talking about graphene whose size is given in a few micrometres. Ribbons are, on the other hand, defined as narrow strips of graphene of typical width and length, of the size of tens of nanometres. Flakes can be as big as 50x50 or even 100x100 nanometres, and quantum dots can be as large as 50 nm.

Graphene oxide (GO) is a strongly oxidized form of graphene, wherein the ratio of carbon to oxygen is about 2. Preparation of graphene oxide is done by exfoliating graphene in the liquid phase by means of using ultrasound. Graphene oxide is an oxidized form of graphene, which is comprised of single graphene sheets with oxygen-containing functional groups, both on the surface and at the edges of the sheets. The structure of graphene oxide is therefore enriched with a large number of cheap groups, such as OH, C=O, COH, and COOH. A significant difference between graphite oxide and graphene oxide is the fact that graphite oxide has in its structure only atoms of sp^2 hybridization, while graphene oxide in their hexagonal structure contains both atoms with sp^2 hybridization and atoms with sp^3 hybridization, associated with the presence of functional groups. Carbon atoms in graphene oxide which are covalently bonded to functional groups (such as hydroxyl, epoxyl, and carboxyl groups) have sp^3 hybridization. Various spectroscopic techniques can be used in order to examine the internal structure of graphene oxide. Atomic force microscopy (AFM) is one of such techniques and it allows to observe the direct thickness of a single GO layer, as well as the number of layers [33].

Graphenoids, can also be obtained from the reduced form of graphene oxide by means of chemical, thermal and electrochemical processes. Reduced forms of graphene oxide with a similar heterogeneous nature are also graphenoids. There are three basic methods of obtaining reduced graphene oxide: the thermal method, the chemical method and the chemical vapour deposition (CVD). It should be noted that in electrochemical applications, the defect-free CVD graphene surface is similar to the surface of a pyrolytic graphite electrode. Choosing an appropriate substrate, which allows electrons to move at high speed, is very significant, as there is a possibility of graphite forming "islands" which can dominate the whole electrochemical process [34].

There are several known methods modifying the metallic or carbon electrodes by means of graphene oxide. The first method involves evaporating a thin layer of graphene oxide suspension on the electrode surface. The second method is spin-coating. GO sheets can also be deposed on the electrode substrate layer by layer [33].

Graphene as a material for modifying electrodes differs from other carbon materials by being a two-dimensional structure. The large surface area of graphene makes it a very convenient material for the synthesis and construction of multi-component nanomaterials. It may also have a large number of oxygen-containing functional groups, such as hydroxyl, carboxyl and epoxy groups. These properties make it possible to functionalize such electrodes by means of covalent or non-covalent bonds. As there have been numerous studies on the modification of the structure of graphene and functionalization, it can be divided into some basic categories, such as: functionalization with nanoparticles, organic compounds, as well as biomaterials and polymers [33].

Functionalization with Nanoparticles

Graphene is a substrate used for immobilization of nanoparticles, which demonstrates high conductivity and excellent thermal stability. Hydrophobic and static interactions are the main driving forces of the process of adsorbing nanoparticles on graphene surfaces. In practice, nanoparticles of the following metals were used: Ag, Au, Pt, Pd, Fe, Sn, Cu or quantum dots: Fe_2O_3, Fe_3O_4, Al_2O_3, SnO_2, MnO_2, TiO_2, ZnO, and Cu_2O [35].

Functionalization with Organic Compounds

Hybrid materials created by means of covalent or non-covalent organic compounds, which facilitate functionalization of the surface of graphene oxide, became the subject of research on the creation of hybrid nanocomposites with known functions and applications. Non-covalent compounds are characterized by strong adsorption of aromatic organic compounds onto the surface of graphene oxide, which is caused by strong hydrophobic interactions and the presence of bonds. Covalent bonds, on the other hand, concern the presence of oxidising groups on the GO surface, which results in a chemical modification using the carbon surface. As a result, the chemical compound is capable of diffusion and migration to the active area on the surface of graphene oxide, or capable of reacting with functional groups on the surface of the sheet. Porphyrins, aromatic dyes, alkyl amines, ionic liquids and cyclodextrins are examples of such compounds.

Functionalization with Polymers

GO-derived polymer nanocomposites are considered to be a promising material for electrochemistry. Nanocomposites have significantly improved properties which cannot be achieved with only the polymer or only the graphene oxide. Improvement of the properties of the nanocomposite is dependent to a large extent on the polymer used, but also on the degree of dispersion of GO nanofibers with polymer matrix. The combination of appropriate polymer with graphene oxide results in the creation of a nanocomposite, which enhances the functionality of the electrodes, due to their high electrical conductivity and chemical stability.

Polymers, such as polyaniline and PEDOT are used in order to improve the electrochemical properties and efficiency of the process. It was established that sensors created on the basis of GO with polymers have greater stability, sensitivity and selectiveness, when compared to the sensor created with only GO. There are two ways of functionalization of the graphene oxide surface with polymers: polymerization on the substrate (*in situ*) and the intercalation of the solution [33].

When it comes to the *in situ* polymerization, one of the methods of creation of a specific GO/PANI composite is polymerization of aniline monomers in the presence of graphene oxide under acid conditions. Polymerization is carried out by means of the chemical method in the presence of a particular oxidant [33].

Another method is intercalation of graphene oxide sheets in the polymer. First of all, graphene is dissolved in a polymer solution and allowed to swell. Then, the solvent is evaporated. Usually, the GO or rGO sheets are arranged appropriately in a given polymer solution by using ultrasound or mechanical agitation. One example is the functionalization of graphene oxide with the use of sulfonated polyaniline (SPANI), which improves dispersibility in water and electroconductive properties of the obtained material [33].

Functionalization with Biomaterials

GO has a large surface area and numerous functional groups, which constitute ideal place for embedding biomolecules, which are of key importance in the production of biofunctional electrodes. Studies demonstrate excellent biocompatibility of GO; it improves the electrochemical reactivity of biomolecules and the electron transfer rate [33]. The physical adsorption or creation of chemical bonds results in various biological molecules, for example: DNA, proteins, peptides, cellulose, and chitosan, being immobilized on the surface of the graphene. Such biomolecules include, for example, the ssDNA molecule, which was adsorbed on both sides of the graphene sheet by means of Π-Π bonds, due to a process known as a strong ultrasonication [35].

Another example, proposed by Zhang, is the preparation of an electrode on the basis of GO modified with the enzymes of horseradish peroxidase and lysozyme. The whole process consisted of incubation of the graphene oxide suspension. Peroxidase and lysozyme can spontaneously attach to the surface of GO, because a single sheet of graphene oxide is enriched with oxygen groups. Oxygen groups located on the graphene sheet have the ability to immobilize the enzyme without changing the surface of graphene oxide or using additional reagents [36]. It has been proven that electrostatic interactions and hydrogen bonds between the enzyme and the graphene oxide are used for this purpose [33].

TYPES OF DETECTION IN BIOSENSORS

Selecting the right transducer is a very important part of development of a biosensor. The following types of detection can be used in the case of electrochemical sensors and biosensors: potentiometric, amperometric, voltammetric, impedance, conductivity and electrogravimetric detention. Amperometric and potentiometric detections are the most common ones. This is due to the relatively simple system of measurement and a simple two-electrode measurement cell. Voltammetric, impedance and electrogravimetric detentions require a more complex three-electrode measurement cell and such advanced equipment as a potentiostat, an impedance analyzer, and a quartz crystal nanobalance. The two most popular detection systems (potentiometric and amperometric) are described below.

Potentiometric Detection

Potentiometric methods use the relationship between the logarithm of the concentration (or more precisely, activity) of the determined ion in the solution and the electrical potential of the respective electrode. The Nernst equation is the mathematical basis of potentiometry. Almost all enzymatic reactions generate or bind hydrogen ions, hydroxyl ions or other simple molecules or ions. In the case of such biosensors, linear relationships of the change in pH or the change in concentration of respective molecules are used, proportionally to the concentration of the analyte.

Well-known classical electrodes, such as glass electrodes, fluoride electrodes, gas electrodes or ion selective electrodes, are used as the basis of construction of biosensors with potentiometric detection. Surfaces of such electrodes are coated with a semipermeable layer containing a suitable enzyme. Specificity of the enzyme determines that the change in concentration of relevant ions or molecules on the surface of the electrode results from a change of the analyte concentration in the test solution. In such case, there are more complex systems, *e.g*. one containing two enzymes (Table **1**) or systems in which subsequent reactions take place. A biosensor used for the determination of penicillin can be an example of such system (Fig. **1**).

Table 1. Examples of enzymatic reactions used in the construction of potentiometric biosensors.

Analyte	Enzyme	Reaction	Product
Urea	Urease	$CO(NH_2)_2 + H_2O \xrightarrow{\text{Urease}} CO_3 + 2NH_3$	NH_3
Glucose	Glucose oxidase	$GOx + \beta\text{-D-glucose} + O_2 \xrightarrow[\text{oxidase}]{\text{Glucose}} H_2O_2 + \text{gluconic acid}$	H_2O_2

(Table 1) contd.....

Analyte	Enzyme	Reaction	Product
Glucose	Glucose oxidase Peroxidase	GOx+ β-D-glucose + O_2 $\xrightarrow[\text{oxidase}]{\text{Glucose}}$ H_2O_2 + gluconic acid H_2O_2 + 4-fluoroaniline $\xrightarrow{\text{Peroxidase}}$ F^- + polymer	F^-
Lipids	Lipase	Lipid + H_2O $\xrightarrow{\text{Lipase}}$ glycerol + fatty acid + H^+	H_3O^+
Lactates	Lactate Oxidase	Lactates + O_2 $\xrightarrow[\text{oxidase}]{\text{Lactate}}$ H_2O_2 + pyruvic acid	H_2O_2

$$\text{Penicyllin} + H_2O \xrightarrow{\text{Penicillinase}} \text{Penicillanic acid}$$

$$\text{Penicillanic acid} + H_2O \longrightarrow [\text{Penicillanic acid}]^- + H_3O^+$$

Fig. (1). The reaction used by biosensors to determine penicillin.

In this case, the sensor is built on a glass electrode, which is sensitive to pH changes caused by the hydrogen ions originating from dissociation of penicillanic acid. Penicillinase is immobilized in the membrane covering the glass electrode. The disadvantage of this type of detection in biosensors, using reactions that cause changes in pH of the medium, is that biocatalytic activity of the enzyme is pH dependent. If pH changes during the reaction are large enough, the ability of the enzyme to catalyze the reaction may vary, resulting in the change of the measured signal. In this case, the sensor response may not be reproducible. It depends on the enzyme used and on the scope of changes in pH.

The previously mentioned problems associated with biocatalytic activity were solved by means of application of gas membrane electrodes, which contributed to the development of potentiometric biosensors and increased their selectivity. If we take as an example a system for the detection of urea, in which the urease is used (Table **1**), then the detected agent will be ammonia, and the sensor must be constructed on the basis of an ammonia-sensitive gas membrane electrode. Typically, these electrodes are made of a glass electrode in combination with an internal electrolyte (for ammonia, it would be an ammonium chloride solution). Ammonia from the sample passes through the membrane in the diffusion process, up to the point at which partial pressures on both sides of the membrane were equalized. pH of the internal solution is determined by means of the concentration of the analyte in the test sample. This system was modified by the use of an ion selective electrode instead of a gas membrane electrode. It resulted in enhanced selectivity of these biosensors. Gas biosensors have one disadvantage – their low response dynamics. Sensors sensitive to ammonia have response times in the range of 0.2-2 minutes for concentrations above 0.1 mol/dm^3 [37].

Amperometric Detection

Amperometry is a measurement method based on the relationship between the energy current concentration and the concentration of electroactive substances. The measurement is carried out at a constant potential of the electrode polarization. A system of two electrode submerged in the test substance is used for this purpose. During the measurement, the solution should be stirred at a constant speed. Another form of amperometric measurement is measurement conducted under the constant electrolyte flow. This technique is widely used in the flow injection analysis (FIA) and as one of the detection methods in chromatographic techniques.

Similarly, to the potentiometric detection, amperometric detection also takes advantage of the fact that enzymatic reactions generate or bind hydrogen ions, or other simple molecules or ions. Table **2** shows examples of enzymatic reactions that can be used to construct biosensors with amperometric detection. Biosensor with this type of detection measure, for example, the production of hydrogen peroxide or nicotinamide adenine dinucleotide (NADH). The established current intensity or current variation rate are proportional to the analyte concentration level in the sample. A very large number of biosensors with amperometric detection is based on the oxidation reactions of oxygen, hydrogen dioxide, and NADH [37].

Table 2. Examples of enzymatic reactions used in the construction of amperometric biosensors.

Analyte	Enzyme	Reaction	Product
Glucose	Glucose oxidase	GOx+ β-D-glucose + O_2 $\xrightarrow[\text{oxidase}]{\text{Glucose}}$ H_2O_2 + gluconic acid	H_2O_2
Polyphenols	Polyphenol oxidase	Polyfenol + O_2 $\xrightarrow[\text{oxidase}]{\text{Polyphenol}}$ o-chinon	o-chinone
Cholesterol	Cholesterol oxidase	cholesterol + O_2 $\xrightarrow[\text{Fc}]{\text{Enzyme}}$ cholestenone + H_2O_2	H_2O_2
Ethanol	Alcohol dehydrogenase	C_2H_5OH + NAD $\xrightarrow{\text{Enzyme}}$ CH_3CHO + $NADH^+$ + H^+	$NADH^+$
Pesticide	Acetylcholin-esterase	acetylocholine + H_2O $\xrightarrow{\text{Enzyme}}$ choline + acetic acid cholina + O_2+ H_2O $\longrightarrow$ betaina + H_2O_2	H_2O_2

Detection of Oxygen

Detection of oxygen is used very often in the case of biosensor with immobilized oxidases and other enzymes that consume oxygen in reactions, which they catalyze. In the case of such biosensors, the current measured in the amperometric method is inversely proportional to the concentration of the substrate undergoing the enzymatic reaction, which consumes oxygen. In such systems, there is often a problem with high concentrations of the analyte and a small amount of oxygen; the electrode shows response to oxygen, not the substance. The classic glucose biosensor designed by Gough *et al.* constitutes one of the possible solutions to this problem [38]. It is a system composed of two Clark-type oxygen electrodes coated with layers containing glucose oxidase. In one layer the enzyme is active, in the other, inactive. Biosensors of this type measure the difference between the currents flowing through both electrodes, with the difference being proportional to the amount of consumed oxygen, and, thus, indirectly, to the glucose concentration in the sample.

Detection of Hydrogen Dioxide

This type of detection is also widely used. The enzyme is immobilized on the surface of the polarized anode. The lower limit of determination is in this case lower than the detection limit of oxygen, but selectivity is worse. The measurement is generally carried out at a potential of about 0.6V. Therefore, the components of the sample with lower oxidation potentials can interfere with the detection of H_2O_2 [37].

Detection of NADH

Amperometric detection of NADH is used in systems in which the employed enzyme is dehydrogenase. Dehydrogenase catalyzes the oxidation reaction of the determined substance with simultaneous reduction of NAD^+ to NADH. In the electrode reaction, NADH is oxidized back to NAD^+, and the anode current is measured as a result. Unfortunately, when the process is conducted on the classic solid electrodes, it is necessary to use high potentials during the oxidation process, which reduces the selectivity of the sensor. The second problem is the contamination of electrodes with emerging enzymatically inactive products, which limits the use of direct oxidation. Modified electrodes are developed which allow to increase selectivity and reduce the contamination of electrodes [37]. The development of modified electrodes allowed for the creation of systems which reduce these problems. Still, so far the issues have not been completely eliminated.

BIOSENSORS

Biosensors constitute examples of heterogeneous systems in which one of the reactants is in a solid state. Thanks to this construction, the biological agent immobilized on the surface of the sensor, an enzyme or an antibody, can be repeatedly used in subsequent reactions. Another advantage of this sensor lies in the possibility of obtaining a high analytical signal, as the signal originates from the entirety of the biological agent accumulated on the sensor surface. If a biological agent is dissolved in the solution, the analytical signal would only be generate by a small portion remaining in the vicinity of the electrode. Therefore, the process of immobilization of a biological agent has decisive importance for the good performance of a biosensor. Immobilization of a given component on the electrode must be sufficiently strong to keep the enzyme on the surface, but should not cause its destruction or blockage of its properties. Additionally, the immobilized molecule must often be appropriately positioned with respect to the electrode surface, so that its active center is in an acceptable distance, which allows electron transfer to occur. The most important methods of enzyme immobilization can be divided into methods based on binding enzymes to supports (adsorption, covalent binding, binding utilizing specific affinity, binding involving transition metals), their cross-linking and imprisonment in a matrix (Fig. **2**).

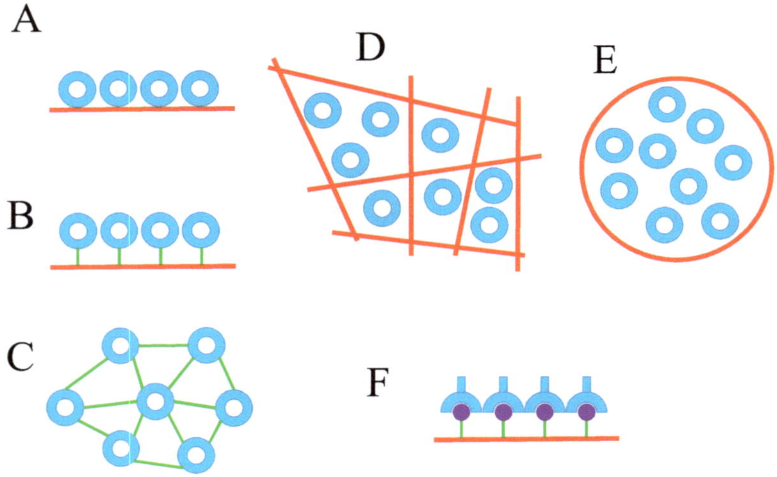

Fig. (2). Enzyme immobilization methods diagram. An adsorption, B covalent bonding, C cross-linking, D entrapment, E encapsulation, F affinity.

Glucose Biosensors

Glucose biosensors are the most popular biosensors. Out of all published works related to biosensor, as much as 10% focus on glucose biosensors, and 90% of those are devoted to the amperometric glucose biosensors. Regardless of the variety of applied techniques, materials and solutions, all glucose biosensors can be divided into three groups: biosensors of the first, second and third generation.

Biosensors with glucose oxidase (GOx) are one of the most widely studied biosensors. Biosensors of this type with electrochemical detection, which, as written above, could be biosensors of the first, second and third generations [39]. First generation biosensors are the oldest group of this type of sensors. Fig. (**3**) presents the functional diagram and reactions, which take place in the receptor layer of the sensor. Gluconolactone is obtained during the operations of a first generation biosensor as a result of the reaction of glucose oxidase in its oxidized form with a glucose molecule. Consequently, glucose oxidase is converted into its reduced form, and the obtained gluconolactone is hydrolyzed to gluconic acid by exposing it to water. The reduced form of glucose oxidase is again converted into its oxidized form under the influence of oxygen present in the solution, which functions as an electron acceptor. Hydrogen peroxide is created in the enzyme oxidization process. In such a system, current which flows through the electrode as a result of hydrogen peroxide decomposition is measured by means of amperometric detection.

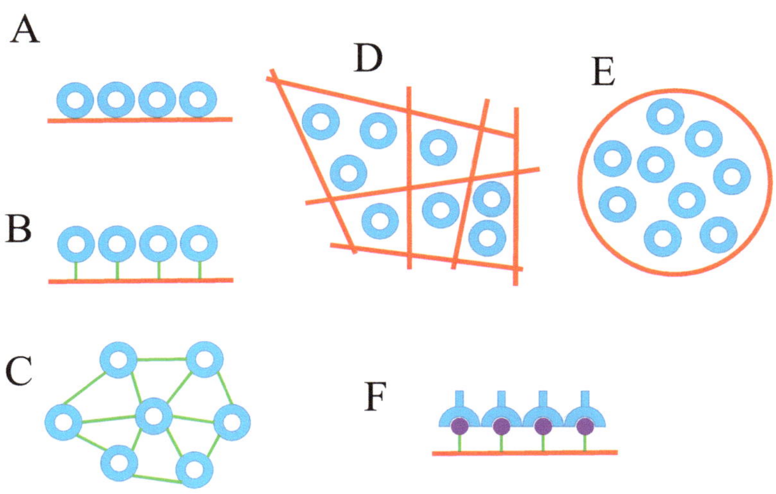

Fig. (3). Functional diagram of the first generation glucose biosensor and the equations of the reactions which take place during its operations.

$$GOx_{ox} + glucose \rightarrow GOx_{red} + gluconolactone$$

$$GOx_{red} + O_2 \rightarrow GOx_{ox} + H_2O_2$$

$$H_2O_2 \rightarrow 2H^+ + O_2 + e$$

The undoubted advantage of these biosensors is the simplicity of their operations. However, first generation biosensors have also significant drawbacks. One of them is the problem of interference of other electroactive components of blood, such as ascorbic acid and uric acid, at the operational potential (approx. 0.6V). This reduces the selectivity of a biosensor and may result in obtaining inaccurate results. Moreover, continuous supply of oxygen and fluctuations in its concentration when measured in body fluids can be another source of problems [40].

Second generation biosensors constitute an attempt to solve the above-described problems associated with first generation biosensors. These systems were created as a result of seeking alternatives to oxygen as an electron acceptor. In a second generation sensor, regeneration of glucose oxidase takes place by means of a red-ox reaction with a mediator (M), which is present in the polymer matrix. Ferrocene/ferrycynic ion system can be used as such a mediator. In such a case, the following process take place (Fig. **4**).

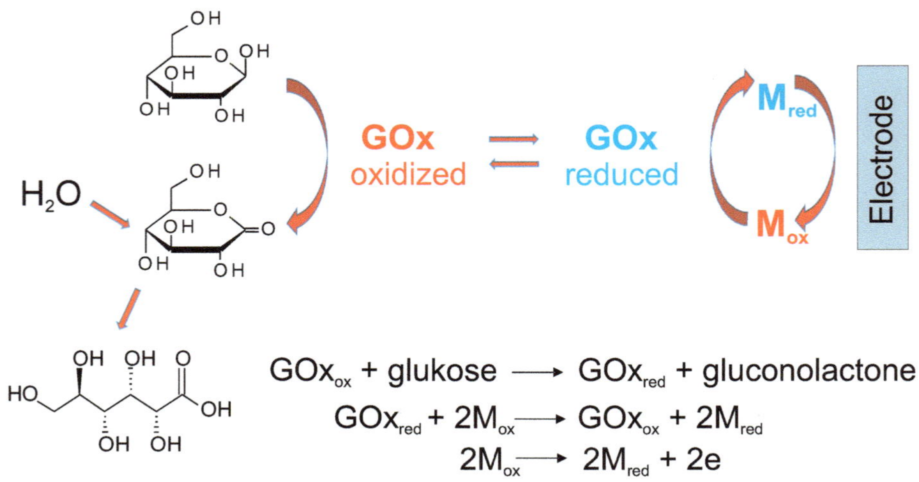

Fig. (4). The functional diagram of the second generation glucose biosensor and the equations of the reactions which take place during its operations (M – mediator).

$$GOx_{ox} + \text{glucose} \rightarrow GOx_{red} + \text{gluconolactone}$$

$$GOx_{red} + 2M_{ox} \rightarrow GOx_{ox} + 2M_{red}$$

$$2M_{red} \rightarrow 2M_{ox} + 2e$$

Gluconolactone is obtained during the operations of a second generation biosensor as a result of the reaction of glucose oxidase in its oxidized form with a glucose molecule. Consequently, glucose oxidase is converted into its reduced form, and the obtained gluconolactone is hydrolyzed to gluconic acid by exposing it to water. This part is the same as in the case of a first generation biosensor. The difference concerns oxidation of the reduced glucose oxidase. As a result of interaction with a reduced oxidase, it is oxidized again, which results in the reduction of the mediator itself. The reduced mediator is then oxidized at the electrode surface, which is the source of the analytical signal. Both amperometric and voltammetric detection can be used in the case of second generation sensors. The use of synthetic mediators allowed to eliminate the problems which plagued the first generation of sensors, which were related to the concentration of oxygen and the presence of interferents. Ferrocene and Prussian blue are some of the most popular mediators used in biosensors. A suitable mediator should have a low redox potential, low molecular weight, and low toxicity, and it should not form complexes and remain stable both in the oxidized and reduced form. A glucose sensor was created on the basis of immobilized glucose oxidase and polyaniline with tetrathiafulvalene as a mediator, using the well-known design of an electrochemical transistor [41, 42]. Systems of this type operated, as second-generation sensors, with amperometric detection. Malliara described in his work the use of PEDOT layer containing ferrocene/ferrycynic ion redox system to build the second generation sensor in the form of an organic electrochemical transistor, the operations of which was also based on the amperometric detection [43]. Unfortunately, this solution also had its drawbacks, for example, high competition between the mediator and oxygen, which results in accumulation of hydrogen peroxide at the electrode surface, reducing the enzyme activity. Another problem is the interaction between the mediator and an interferent, and the mediator being leached out of the region between the enzyme and the surface of the electrode. Works on the third generation biosensor began in order to eliminate the reoccurring problems [44].

$$\mathrm{GOx_{ox}} + \mathrm{glucose} \rightarrow \mathrm{GOx_{red}} + \mathrm{gluconolactone}$$

$$\mathrm{GOx_{red}} \rightarrow \mathrm{GOx_{ox}} + 2\mathrm{e}$$

The concept of the third generation biosensor is based on the idea of elimination of the mediator and direct electron transfer between the electrode and the active centre of enzyme (Fig. **5**).

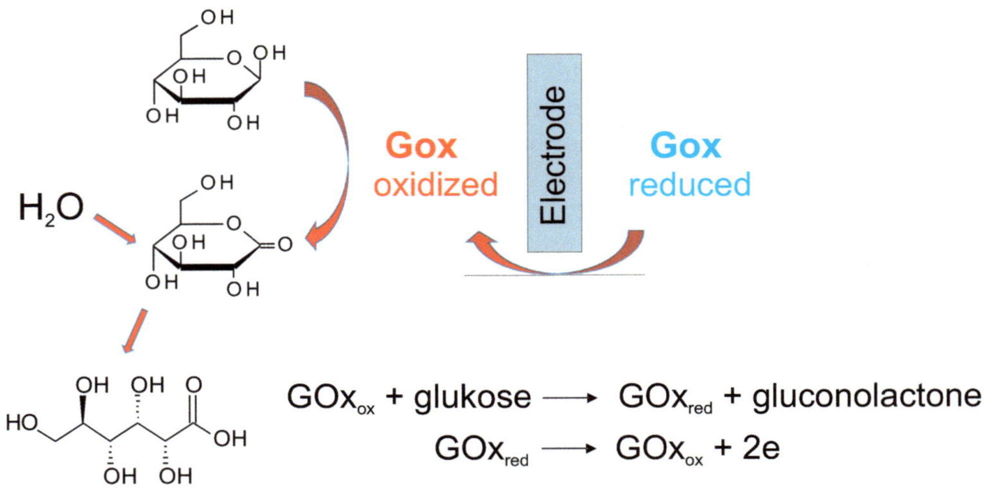

Fig. (5). The functional diagram of the third generation glucose biosensor and the equations of the reactions which take place during its operations.

Oxidation of glucose to gluconolactone and further to gluconic acid occurs during the operations of the third generation biosensor, just as in the sensors of the first and second generation. In third generation sensors, regeneration of glucose oxidase from the reduced form back to the oxidized form is carried out by means of a direct electron transfer between the electrode and the active centre of the enzyme. Both amperometric and voltammetric detection can be used in this type of sensors. Unfortunately, academic sources provide only a few well-documented examples of such transfers. In this case, the problem concerns a barrier to electron flow, that is, the globular structure of glucose oxidase, in which the active centre, containing flavin adenine dinucleotide (FAD), is hidden inside a cavity of approx 13 Å. Direct electron transfer is possible through the use of conductive organic salts, such as tetracyanoquinodimethane (TCNQ), tetrathiafulvalene (TTF), conductive polymers or carbon nanotubes [45].

A variety of substrates can be used to build sensors with glucose oxidase. For example, polyaniline and PEDOT layers, as well as layers of multicomponent materials based on these polymers (composites and nanostructure materials). Immobilization of enzymes on the surface can also be carried out in different ways. Immobilization of an enzyme, in this case glucose oxidase, by means of utilisation of weak electrostatic interactions is rarely used. This type of sensor has been built on the basis of layers of PEDOT and PEDOT – poly(N-methyl-4-pyridine) composite [46]. On the other hand, one of the most often used methods of immobilization of glucose oxidase on different surfaces is forming covalent bonds between the substrate and the enzyme. This method ensures high durability of the sensor, and at the same time, the entire enzyme is located on the surface and remains in contact with the solution, which guarantees that a large part of the enzyme remains active.

A good example of the enzyme immobilization process is shown in Fig. (**6**). It is the immobilization of glucose oxidase on the surface of a composite of PEDOT, lithium polystyrenesulfonate and polyacrylic acid (polyAAc) [47]. In this case, the substrate was obtained by the chemical polymerization of EDOT in the presence of polystyrene sulfonate and acrylic acid. Carboxyl groups are present on the surface of the created layer of the conductive composite and can be used for enzyme immobilization. In this case, glucose oxidase was immobilized by means of formation of covalent bonds with the help of diimide. Such a synthesis guarantees high stability of the sensor and allows the entirety of the enzyme to maintain contact with the solution. It results in the activity of a large part of the immobilized enzyme. The systems can operate as a second generation sensor with amperometric detection. The process of creating a biosensor is shown in the Fig. (**6**):

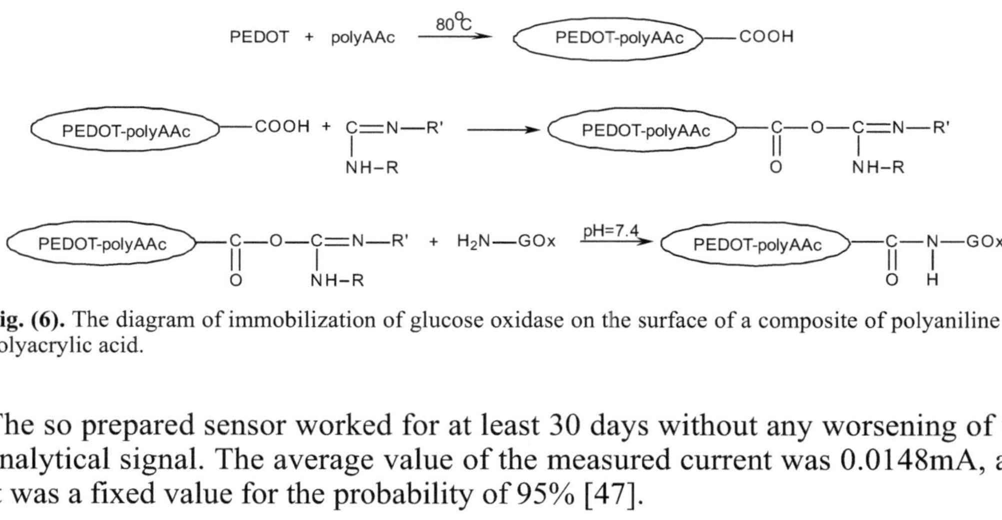

Fig. (6). The diagram of immobilization of glucose oxidase on the surface of a composite of polyaniline and polyacrylic acid.

The so prepared sensor worked for at least 30 days without any worsening of the analytical signal. The average value of the measured current was 0.0148mA, and it was a fixed value for the probability of 95% [47].

The issue of determining glucose with biosensors is still an important issue today. Many designs solutions concerning this type of sensors have been created in the course of many years of research. Table **3** shows over twenty examples of different glucose biosensors, which are meant to demonstrate the degree of diversity concerning this area of chemistry. It is also worth to pay attention to big differences between the sensors when it comes to their lifetimes, which range from a few days to several months.

Table 3. Comparison of parameters of selected amperometric glucose biosensors.

Biosensor	Linear range	Response time	Stability	Ref.
Covalent immobilized of glucose oxidase on graphite electrode with adsorbed of N-methyl-phenazinium ion.	0.5 µM – 2mM	60s	Enzyme 9 month Mediator 1 day	[48]
Cellulose acetate layer with cross-linked GOx.	0 – 28mM	100s	6 days	[49]
Amperometric biosensor with immobilized GOx into graphite paste modified with tetracyanoquinodimethane (TCNQ).	5 – 50mM	15-50s	35 days	[50]
Immobilized GOx covered with a Nafion membrane	0.5 – 40mM	30s	6 days	[51]
GOx immobilized in matrix composed of silica sol and Copolymer of poly(vinyl alcohol) with 4-vinylpyridine.	0 – 16mM	11s	5 months (storage)	[52]
Layer of Prussian blue covered with a poly(o-phenylenediamine) membrane with immobilized GOx.	8 – 14mM	4-8s	-	[53]
Poly(vinylimidazole) complex of $[Os(bpy)_2 Cl]^{+/2+}$ (PVI-Os) and glucose oxidase (GOx) on a gold electrode.	0 – 30mM	5s	64 h (50% activity)	[54]
Immobilized GOx into a sol-gel composite at the surface of a basal plane pyrolytic graphite electrode modified with multiwall carbon nanotube.	0.2 – 20mM	<5s	3 weeks	[55]
GOX adsorbed at the platinum nanoparticle-modified carbon nanotube electrode.	0.1 – 13.5mM	5s	22 days (73.5% activity)	[56]
Electrochemical entrapment of GOx into porous poly(acrylonitrile-co-acrylic acid) layer.	5µM – 3mM	<30s	5 weeks	[57]
Immobilized GOx directly onto an oxidized boron-doped diamond (BDD) electrode.	0.67µM – 2mM	<5s	1 month	[58]
A nanobiocomposite film consisted of polypyrrole, functionalized multiwalled carbon nanotubes and GOx.	0 – 4mM	8s	-	[59]
GOx immobilized on Pt nanoparticles supported on sulfonated multiwall carbon nanotubes on glassy carbon electrode.	0 – 6.4mM	-	1 h	[60]
The immobilized GOx in a titania sol-gel film, which was prepared by a vapor deposition method, on a Prussian modified electrode.	0.02 – 15mM	<10s	3 months	[61]
Graphite nanoplatelets with Pt and Pd nanoparticles in Nafion layer.	0 – 20mM	2s	1 week	[62]
The electrochemically entrapped GOx onto the inner wall of highly ordered polyaniline nanotubes.	0.01 – 5.5mM	3s	2 weeks (91% activity)	[63]

(Table 3) contd.....

Biosensor	Linear range	Response time	Stability	Ref.
GOx immobilized in thin films of chitosan containing nanocomposites of graphene and gold nanoparticles (AuNPs) at a gold electrode.	2 – 14mM	-	15 days	[64]
GOx immobilized in nanohybrid materials derived from gold nanoparticles and multi-walled carbon nanotubes.	20μM – 10mM	3s	1 week	[65]
GOx immobilized in composite film of chitosan and Pt nanoparticles.	1.2μM – 4mM	<5s	1 month (61% activity)	[66]
GOx immobilized in zinc oxide nanoparticles modified carbon paste enzyme electrodes.	9.1μM – 14.5mM	-	-	[67]
The self-assembled monolayer anchoring glucose oxidase (GOx) covered with films of Prussian blue.	0 – 25mM	-	4 days	[68]

Cholesterol Biosensors

Biosensor construction has undergone massive changes in relatively short time. Modern biosensors are systems characterised by a very high degree of complexity. Several types of material, including nanostructured materials, are used in the construction of biosensors. Layers of conductive polymers are increasingly replaced by conductive composite materials, which contain not only the polymer but also metal nanoparticles, nanodots, nanotubes, and such. The layers of polymer or composite are often created in a manner, which allows to create appropriate structures, such as, for example, nanofibers, nanospheres or porous layers with a specific pore shape and size. Subsequent examples of biosensors are characterized by exactly this type of a complex structure.

Cholesterol and its fatty acids are essential for the human body as part of the structure of biological membranes, nerve and brain cells. It is synthesized in the liver and also delivered in food. Cholesterol is one of substrates in the production processes of inter alia, hormones, bile acid, and vitamin D. Its inappropriate level in human body is associated with serious health problems. Elevated levels of cholesterol in blood can cause damage to blood vessels and lead to increased risk of heart disease, atherosclerosis, and metabolic problems. On the other hand, if its level is too low, it may cause anaemia, thyroid problems and issues with nutrient absorption. Determination of the level of cholesterol in blood is an important part of the diagnosis of the aforementioned diseases. Moreover, determination of cholesterol content in food is essential in the diets, which require consumption of its specified amounts. Thus, the creation of a rapid and sensitive method of determination is an important task for chemical and clinical analysis. Conventional methods for the qualitative cholesterol determination, such as gas chromatography and high-performance liquid chromatography, usually require complicated procedures and a lot of time, and have limited specificity. Production of highly selective and specific enzymatic method has the potential to replace the above-mentioned methods. The first enzymatic determination of cholesterol was based on the spectrophotometric detection and used three enzymes: cholesterol oxidase (ChOx), cholesterol esterase (ChEt), and horseradish peroxidase (HRP). ChEt converts esterified cholesterol, which is present in blood, into free cholesterol, and ChOx, in the presence of oxygen, catalyzes oxidation of cholesterol and produces hydrogen peroxide and cholestenon. Flavin adenine dinucleotide (FAD), which is present in ChOx, is responsible for cholesterol oxidation. Subsequent reaction of H_2O_2 with HRP results in the creation of a colored compound, which can be detected. This method, although very efficient, uses large amounts of enzymes, so the cost is relatively high [69]. Fig. (**7**) shows the principle of determination of cholesterol in blood, using cholesterol esterase and cholesterol oxidase.

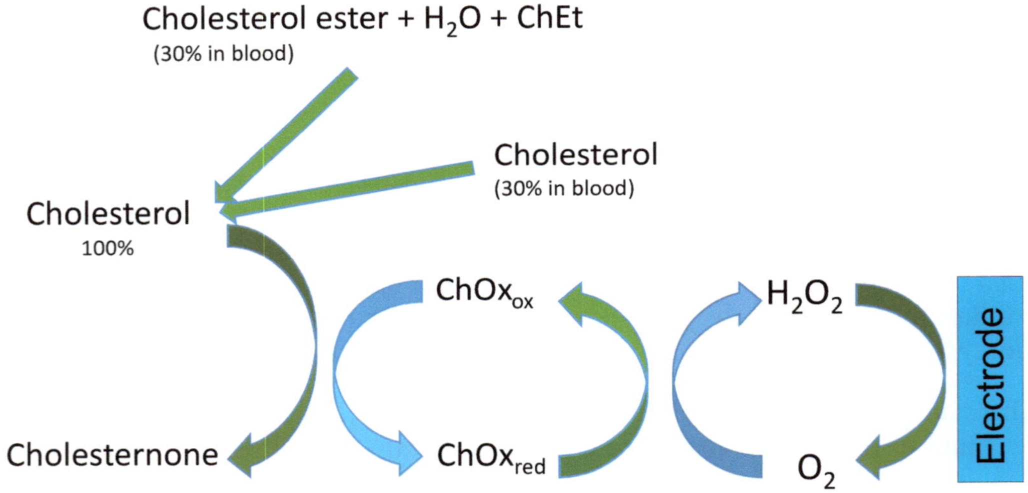

Fig. (7). The principle of determination of cholesterol in blood, using cholesterol esterase (ChEt) and cholesterol oxidase (ChOx).

Currently, most of biosensors used for cholesterol detection are based on amperometric detection because of such advantages as real-time monitoring and ease of miniaturization. Amperometric determination is most often used for determination of O_2 consumption or the amount of produced H_2O_2 (Fig. **7**). The signal is proportional to the current analyte. It is a first generation biosensor. Unfortunately, amperometric determination of the amount of H_2O_2 requires the application of the high anode potential, which can cause other substances contained in the sample to undergo oxidation, which interferes with the determination. The problem can be solved, if the signal is generated by the direct transfer of electrons through an electrochemically active centre of the enzyme (FAD), without the need for oxygen. Such direct transfer of electrons is the main feature of third generation sensors. Unfortunately, the fact that the active centre is located deep inside the protein of a given enzyme hinders the process, which results in the lower sensitivity of the sensor. Low stability is another problem. It is usually associated with immobilization of enzymes. Nanomaterials can be a solution to this problem. Their unusual chemical and physical properties mean that they can be excellent material for a substrate for immobilization. Nanotubes and nanoparticles are a good example of such materials, which due to their properties, facilitate the electron transfer between the redox centre of the enzyme and the electrode. Modification of electrodes with such material results in a porous electrode surface, which facilitates binding the enzyme to the surface. As the majority of nanoparticles have charges, immobilization of the enzyme can take place by means of electrostatic interaction. Saxena and Das compared the results of determinations made with various sensors used for cholesterol determination. They demonstrated that the best results were obtained for sensors using cholesterol oxidase and cholesterol esterase, as well as nanostructures such as carbon nanotubes, gold nanoparticles, *etc* [69].

The examples described above illustrate the principles of operations of cholesterol biosensors (Fig. **8**). A great many biosensors were based on the basis of these principles. They differ when it comes to solutions concerning the layers on which the enzyme is immobilized, possible mediators and the method of electron transport between the enzyme and the electrode. Some specific solutions used in cholesterol biosensors are presented below.

Gopalan *et al.* [70] have developed a novel amperometric biosensor based on the immobilization of cholesterol oxidase (ChOx) into a cross-linked matrix of chitosan (Chi)-room-temperature ionic liquid (IL) (1-butyl-3-methylimidazolium). The surface of a bare Indium tin oxide (ITO) electrode was modified with the electrodeposition of Au particles onto multi-walled carbon nanotubes functionalized with thiol groups. The solution of chitosan, ionic liquid and glutaraldehyde was then applied to the surface. The final step of the biosensor synthesis was dropping the solution of cholesterol oxidase onto the created layer. The authors demonstrated the synergy that occurs between nanotubes, gold particles, chitosan and ionic liquid, which results in the increased sensitivity of the sensor in question. The presence of modified nanotubes, Au nanoparticles, chitosan and ionic liquid contribute to the excellent performance of the biosensor. Scanning electron microscopy revealed the interconnected wires of chitosan and ionic liquid, covering the Au particles on the surface of the nanotubes layer. The biosensor operated in a linear range of 0.5-5mM. The response time was about 7s. The shelf-life of the sensor was estimated at 20 days with a decrease in the analytical signal of less than 5%.

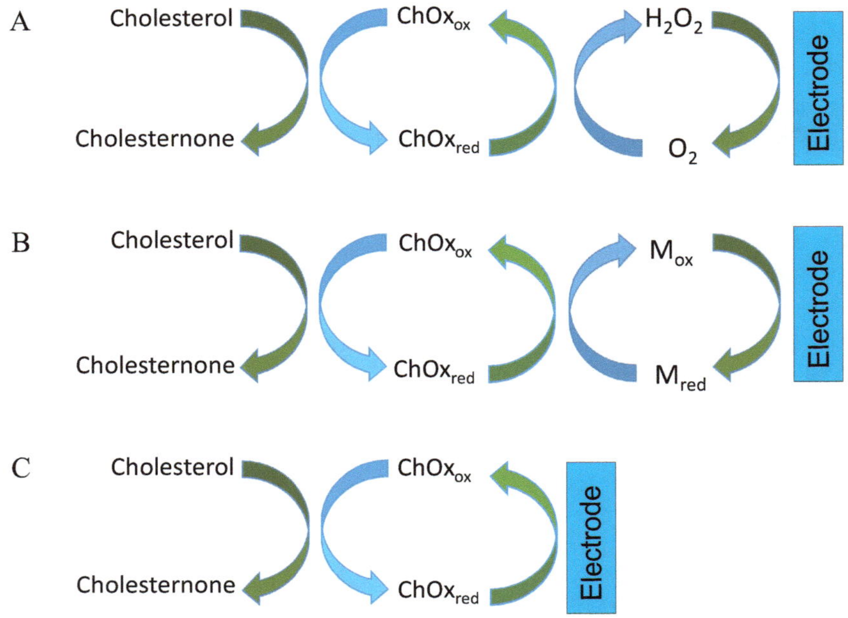

Fig. (8). Cholesterol biosensors: first generation (A), second generation (B) and third generation (C).

Malhotra *et al*. [71] described the construction of a sensor based on polymerized polyaniline nanospheres. The film was obtained by means of polymerization of the solution containing micelles composed of camphorsulfonic acid and polyaniline nanostructures. A solution of cholesterol oxidase, which was bound to the matrix *via* N-ethyl-N'-(3-dimethylaminopropyl) carbodiimide and N-hydroxysuccinimide, was then deposited on the accordingly prepared sample. The biosensor operated in the concentration range of 25 to 5500 mg/L. The response time was 10s and its shelf-life was estimated at 12 weeks.

Malhotra *et al*. [72] sonicated the chemically synthesized polyaniline and obtained solution containing polyaniline nanostructures. Subsequently, a thin film of polyaniline nanostructures was deposited onto ITO electrode using the above-mentioned solution. Cholesterol oxidase was immobilized through a covalent bonding of polyaniline amino groups and carboxyl groups of the enzyme by means of a linker N-ethyl-N'-(3-dimethylaminopropyl carbodiimide). The sensor created in this manner operated in the linear range of 25 to 400 mg/L, and the detection limit was 25mg/L.

Brahim *et al.* [73] entrapped cholesterol oxidase within a matrix composed of poly(2-hydroxyethyl methacrylate), p(HEMA)) and polypyrrole. The film was created in two stages, with the help of solution containing HEMA monomer, cross-linker and cholesterol oxidase. In the first stage, the mixture was irradiated with UV light (366 nm). Subsequently, the resulting hydrogel film was saturated with pyrrole, which was polymerized using potentiostatic electropolymerization (E=0.85V). The biosensor obtained with this method had very good properties: the linear response range from 5×10^{-4} to 1.5×10^{-2} M, the detection limit of 120μM toward cholesterol, and the response time of the biosensor at the level of 30s. The sensor was capable of analyzing 60 samples per hour. The authors claim that after 12 months of storage in a desiccated form, in the absence of a buffer, the biosensor retained 80% of its initial activity.

Yao and Takashima [74] constructed a biosensor based on a sol-gel film. 1,2-diaminobenzene was electrochemically polymerized on a platinum electrode at the potential of 0.65V, and afterwards the electrode was transferred into the solution containing tetramethoxysilane and cholesterol oxidase. After the sol-gel film had been formed, the electrode was exposed to glutaraldehyde vapour. The sensor was able to amperometrically determine glucose at the potential of 0.6 V. The biosensor operated in a linear range of 0.06-3 mM, and the response time was approximately 51 s. The lifetime of the sensor was estimated at 32 days, with a subsequent decrease in activity of the sensor of 50%.

The synergy of the whole system is of key importance in systems, which contain several components influencing the size of the final analytical signal. When using electroanalytical methods, it is expected that the total current resulting from the electrode reaction of the designated depolarizer will be equal to the sum of the currents originating from individual components of the composite. In the ideal situation, there is synergy – current for composite is higher than the sum of currents that can be achieved for the individual components separately.

Biosensor Based on Laccase

Laccases are widespread oxidoreductases occurring, inter alia, in higher plants, the majority of fungi and some bacteria. Laccase as an enzyme with low substrate specificity oxidizes various chemical compounds, from mono- and diphenols to inorganic compounds. Laccase is used in laboratories of many research centers in the process of synthesis of many pharmaceutical compounds, food additives or polymers. The broad spectrum of substrates and various types of reactions catalyzed by laccase allow this universal oxidoreductase to be used in many different types of technological processes which fulfil the white biotechnology criteria. Oxidation reactions catalyzed by laccases are accompanied by the reduction of molecular oxygen to water. The active centre of laccase contains 4 copper atoms. Only one of these atoms is the site of substrate oxidation, the remaining ones form a trinuclear centre in which molecular oxygen is bound and reduced to water. The reactions catalyzed by laccase can be divided into three groups: oxidation reactions of simple phenolic derivatives without the involvement of a mediator, oxidation of substrates, which require additional presence of a mediator and coupling reactions. The mechanisms of these processes are presented in Fig. (**9**). It demonstrates that the biosensor with laccase can operate in three different modes [75].

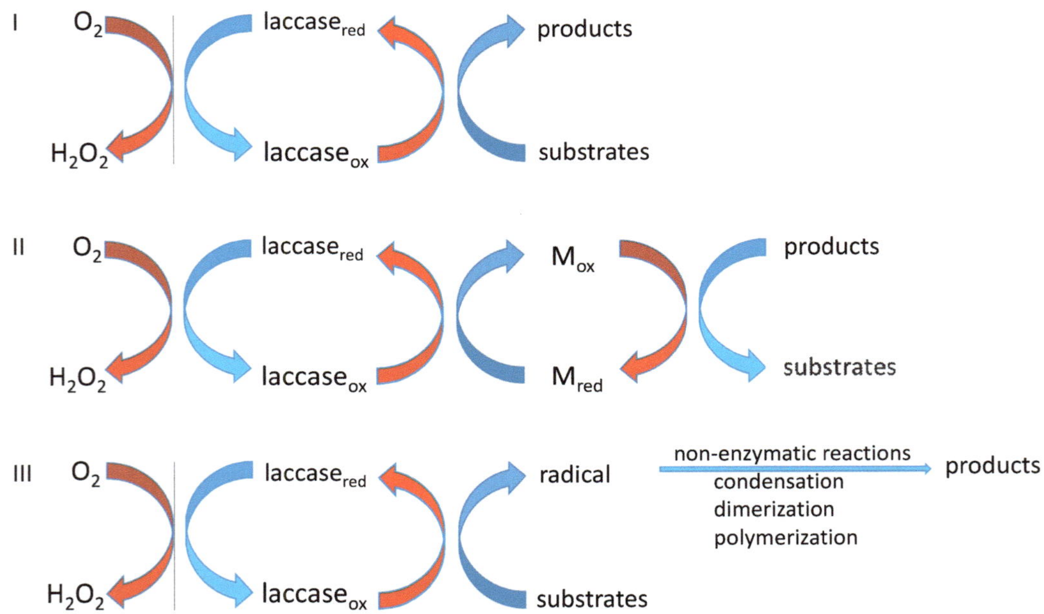

Fig. (9). Types of mechanisms of reactions catalyzed by laccase: I – oxidation of simple phenolic derivatives without the involvement of a mediator, II – oxidation of substrates which require additional presence of a mediator, III – coupling reactions.

Catehin biosensors are an example of biosensors, which use laccase. Laccase catalyses oxidation of 1,2-dihydroxybenzyl groups in flavanoids. One of many flavanoids, catechin proved to be an effective anticancer agent – one which exhibits antimutagenic, hypocholesterolemic, antibacterial and anti-inflammatory properties.

New electroanalytical methods for the detection of catechin are aimed at testing the bioavailability and pharmacokinetics of catechin in human body fluids. The following methods are nowadays frequently used for the detection of catechin:

high performance liquid chromatography, liquid chromatography at atmospheric pressure, mass spectrometry and gas chromatography (GC). Unfortunately, all these analytical methods require expensive, specialized equipment, as well as carrying out the separation process prior to detection of catechin. Therefore, there is a need for a simple, sensitive, and stable method for the catechin detection.

Rahman *et al.* [76] created a biosensor on the basis of a conductive polymer, poly (3′,4′-diamino-2,2′;5′,2″-terthiophene), which constitutes a derivative of polythiophene, which has amino groups in 3- and 4-positions in every second ring of thiopene. Dendrimers are used for the construction of the sensor in combination with nanoparticles of gold in order to increase the sensitivity and stability. Dendrimers were immobilized by means of covalent bonds to amino groups of the polymer film. Au(III) ions were first coordinated in the interior of the dendrimer with nitrogen ligands and then subsequently reduced to Au nanoparticles. Au nanoparticles encapsulated in dendrimers were afterwards used to covalently immobilize laccase by forming amide bonds between carboxyl groups of the dendrimer and amino groups of laccase. A third generation biosensor was obtained as a result, combining physical and chemical properties of Au nanoparticles and reactivity of the surface of dendrimers. This biosensor is a promising tool for the detection of catechin in beverages, food, and biological samples. Fig. (**10**) shows a scheme of the synthesis of the described sensor.

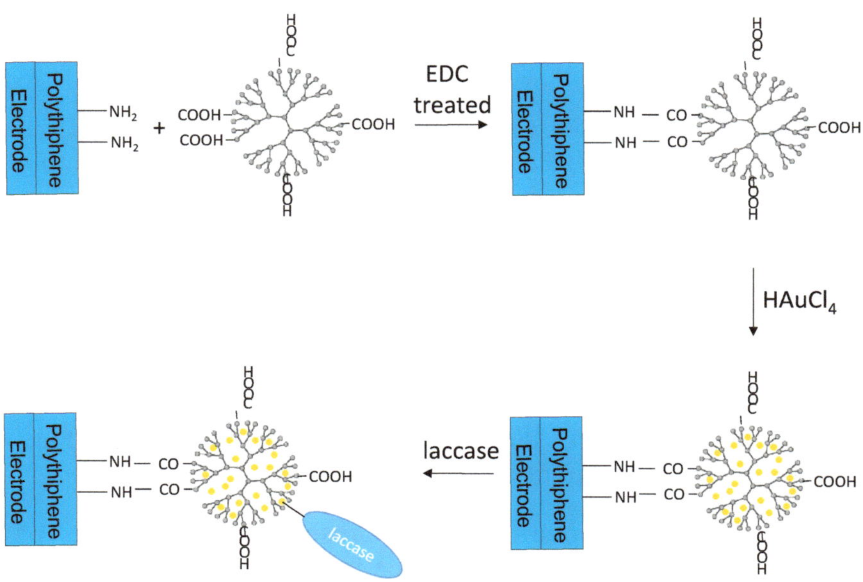

Fig. (10). A schematic representation of the processes leading to the creation of the PDATT/Den(AuNP)/laccase electrode.

On the basis of the conducted research, the authors found that the gold nanoparticles, which are present in the film, function as a kind of mediator in order to facilitate the exchange of electrons between the enzyme and the electrode surface. The described sensors have also been used successfully for the determination of catechin in the grafted samples of human urine.

Urea Biosensors

Similarly, to glucose oxidase electrodes, the electrodes with urease are also a very popular research objects. Many different electrodes of this type have been developed, utilizing various immobilization methods and detection systems. Next to glucose biosensors, biosensors with immobilized urease are among the most frequently studies sensors. Urea, as the end product of nitrogen metabolism, has great importance in clinical chemistry, because its concentration in blood is an important indicator of kidney function problems. Moreover, urea is widespread in nature and its analysis is essential in food chemistry and environmental monitoring. It is clear, therefore, that biosensors with urease have a wide variety of applications. Urease catalyses the urea hydrolysis to ammonium ion and bicarbonate ion, according to the reaction on Fig. (**11**).

$$NH_2CONH_2 + 3H_2O \xrightarrow{\text{Ureas}} 2NH_4^+ + OH^- + HCO_3^-$$

Fig. (11). The reaction catalyzed by urease.

Rajesh *et al.* developed a urea biosensor based on urease immobilized in the conductive polymer film [77, 78]. The urease was covalently immobilized to electrochemically produced copolymer poly (N-3 aminopropyl pyrrole-c--pyrrole) (PAPCP) covering an ITO electrode. In such an arrangement, the enzyme retained biological activity and allowed for the determination of urea in test samples. Potentiometric detection was used. The electrode worked in a linear range of concentration of urea from $6.3 \cdot 10^{-6}$ to $4.07 \cdot 10^{-4}$ M. The sensor showed response time of 25-50 s, for 95% of the final value of the signal. The durability of the sensor was 2 months. Control of urease activity was carried out by means of UV-Vis spectroscopy, using the reaction with Nessler reagent. This method was also used to check the response time of the sensor.

The same authors used the above-mentioned biosensor for determination of urea with amperometric detection [79]. Amperometric measurement was carried out at a fixed bias voltage of 0.0 V *vs* a silver chloride electrode (Ag/AgCl) in a phosphate buffer (pH=7.0). The electrode gave a linear response range of 0.16-5.02 mM of urea in an aqueous solution. The response time was 40 s for 95% of the final value of the signal. The durability of the sensor was 2 months.

Lakard *et al.* [80] researched the possibility of immobilization of urease in layers created from polyethylenimine on platinum microelectrodes. The layers were prepared by electropolymerization through cyclic voltammetry during 5 cycles of changes of potential in the range of 0 to + 3V, with the potential change rate of 50 mV/s. Researchers have tested various methods of synthesizing the receptor layer of the biosensor. Immobilization by physical adsorption, followed by crosslinking in dilute aqueous solutions of glutaraldehyde, was identified as the best solution. The biosensor operated on the basis of potentiometric detection. It had a short response time, ranging from 15 to 30 s, and a linear range of urea levels from $1 \cdot 10^{-2.5}$ to $1 \cdot 10^{-1.5}$ M. The durability of the sensor was estimated at 4 weeks.

Saeedfar *et al.* [81] developed a urea biosensor based on urease immobilized onto functionalized fullerene C_{60}. Such fullerene-immobilized urease (C_{60}-urease) can catalyze the hydrolysis of urea in a solution. An H^+ ion sensor was first developed by mixing adequate amounts of the n-butyl acrylate, 2,2-Dimethoxy-2-phenylacetophenone, sodium tetrakis(4-flourophenyl)borate dehydrate, HI and 1,6-Hexanediol diacrylate. The mixture (4 µL) was then drop-coated onto the Ag/AgCl SPE and photo-cured for 3 min under constant nitrogen gas flow. The biomaterial was deposited on the H^+ ion sensor by means of applying an aqueous solution of C_{60}-urease on its surface and evaporation. A schematic diagram presenting the immobilization of urease on the surface of C_{60} fullerene is shown in Fig. (**12**). The optimal pH and concentration of the phosphate buffer for the urea biosensor were 7.0 and 0.5 mM, respectively. The sensitivity of such a biosensor did not differ from the theoretical values. Common cations such as Na^+, K^+, Ca^{2+}, Mg^{2+} and showed no obvious interference with the urea biosensor's response The biosensor construction allowed also to increase its stability for up to 140 days. The linear range of the biosensor was at the level of 10^{-3}–10^{-5}M. The authors compared the results of the determination of the concentration of urea in grafted samples of urine obtained using the biosensor with the concentrations obtained by means of the reference method. UV-Vis spectroscopy was used as the reference method. The difference between the obtained values are at the level of ± 5%, which indicates that the biosensor can be successfully used for clinical measurements.

Luo and Do [82] studied urea biosensors based on a matrix of polyaniline and Nafion®. These systems operated with voltammetric detection. As the first step, a Nafion® solution was applied to a ceramic plate with sputtered layer of gold by means of spotting and allowed to dry. A layer of polyaniline was then created on the so-formed substrate by means of electropolymerization, using cyclic voltammetry. Urease was immobilized on the prepared substrate electrochemically, as well as by means of spotting the solution and evaporation. Three sensors were prepared, differing in the manner of applying the layer, which contained urease.

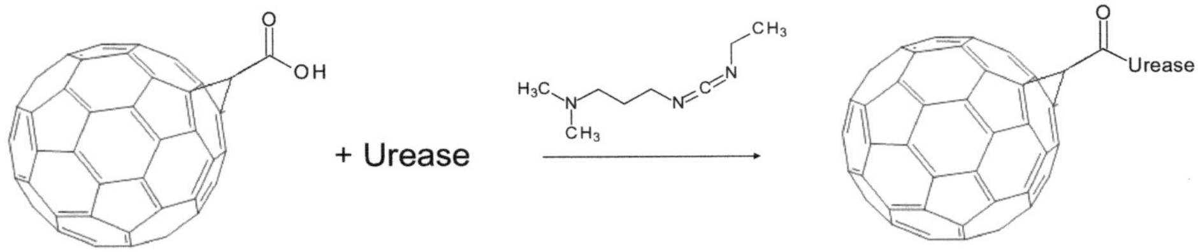

Fig. (12). A schematic of the process of immobilization of urease on the surface of fullerene C_{60}.

The enzyme was applied electrochemically from a solution containing the urease and glutaraldehyde. The process was carried out by means of cyclic voltammetry. The sensor was marked as (urease)/PANi-Nafion®.

The second sensor was created by spotting. An aliquot of urease and glutaraldehyde solution was dried at room temperature and then, in a vacuum, at 4°C. The sensor was marked as GA(urease)/PANi-Nafion®.

The next sensor was also created by spotting. An aliquot of urease and Nafion® solution was dried at room temperature and then, in a vacuum, at 4°C. The sensor was marked as Nafion® (urease)/PANi-Nafion®. According to the data in Table **4**, the sensor obtained by means of spotting urease and Nafion® solution (Nafion® (urease)/PANi-Nafion®) showed the highest level of sensitivity and the lowest LOD value of all the tested sensors.

Table 4. Comparison of characteristics of urea biosensors [82].

Modifier	Sensitivity μA $[(mg\ dl^{-1})^{-1}\ cm^{-2}]$	Linear range $[mg\ dl^{-1}]$	LOD $[mg\ dl^{-1}]$
Nafion®(urease)/PANi-Nafion®	5.27	3–30	0.3
GA(urease)/PANi-Nafion®	0.22	3–45	3
(urease)/PANi-Nafion®	0.7	6–60	6

Bertocchi *et al.* [83] have developed an amperometric urea and ammonium ion biosensor in which urease and glutamate dehydrogenase were immobilized on nylon nets coupled to a platinum electrode. The glutamate dehydrogenase along with sodium hydroxide enhanced the reproducibility and stability of the sensor. Amperometric measurements were performed at the potential of 0.4V. A comparison between the potentiometric and amperometric transducers was made, and it was discovered that that in the former case there is a problem of interference associated with the presence of potassium ions in the sample. Biosensors worked in the linear range from 10^{-5} to 3×10^{-4} M of urea, with the detection limit of 10^{-5} M and estimated 4 days of sensor stability. The biosensor was used for clinical purposes for the determination of the content of urea and ammonium ions in human saliva.

Vostiar *et al.* [84] used polytoluidine blue (PTOB) in their study as a layer allowing amperometric detection of pH changes. The PTOB film was deposited on the glassy carbon electrode. Then, a buffer solution containing urease was pipetted onto the PTOB film and left unit the water evaporated. Finally, the sensor surface was covered with a dialysis membrane. The so-constructed biosensor was able to work with amperometric detection at the potential of 0.V (*vs* SCE). The biosensor had the linear range up to 0.8 mM (Fig. **2**) and the detection limit of 0.02 mM (S/N = 3).

Academic sources provide a very large number of reports on various urea sensors constructed for different applications. Table **5** contains selected examples of urea biosensor, which were used in biomedical assays. Urea was determined in urine, blood, and serum.

Table 5. Comparison of characteristics of urea biosensors used in biomedical analysis.

Biosensor	**Linear range**	**Response time**	**Stability**	**Ref.**
Encapsulation, nylon net	4 days	10^{-5} to 3×10^{-4} M	–	[83]
Entrapment poly(carbamoylsufonate) hydrogel	4 days (buffer, 4°C) 3–4 months(dry conditions)	0.432–126mg/dl	1–2min	[85]
Physical adsorption, poly(3-cyclohexyl thiophene)	3 months (4°C)	–	–	[86]
Entrapment, gelatin	240 days (50%)	5–140mgdl^{-1}	2min	[87]
Physical adsorption, graphite and platinum composite electrode	3 months	10–250M	2min	[88]
Entrapment, poly(*N*-vinyl carbazole/stearic acid) Langmuir–Blodgett film	5 weeks (4°C)	0.5–68mM	2min	[89]
Physical adsorption, polyaniline-poly (*n*-butyl methacrylate)	–	20–120mg/dl	10min	[90]
Entrapment, polyurethane film	–	0.7–8mM	20s	[91]
Covalent binding and crosslinking, poly(vinyl alcohols) with styrylpyridinium groups	30 days (4°C)	3–300mg/dl	1–2min	[92]
Ionic binding, DEAE cellulose	2 weeks (4°C)	10–400mg/dl	–	[93]

(Table 5) contd.....

Biosensor	Linear range	Response time	Stability	Ref.
Entrapment, controlled pore glass; and entrapment, silica gel	>310h (pH=7.0, 4°C)	0.5 to 150mM	4–5mins	[94]
Adsorption, nanoporous alumina	30 days in buffer	0.5M to 3.0mM	30s	[95]
Covalent binding, PVC–COOH polymer	20 days (pH 7.5, 4°C)	10^{-4} to 10^{-2} M	1–2min	[96, 97]
Free enzyme	1 month	2–11mM	5min	[98, 99]
Physical adsorption, Whatman paper	Disposable	1.66–16.6mM	2min	[100]
Free enzyme	–	0–100mg/dl	3min	[101]

Affinity Biosensors

There are also other biological objects, other than enzymes, which make up the receptor film of biosensors and determine their properties and uses. Affinity, which is an interaction consisting of a single paratope binding an antigen, is also used in many biosensors. Antibodies also often function as such objects. Furthermore, an antibody-antigen interaction is often used in this type of sensors. An antigen immobilized in the receptor film of the biosensor binds the antigen present in the analyzed sample and allows for its reduction and oxidation.

Shim *et al.* [102] described a deposited composite material on which anti-CAP antibodies, whose affinity allowed for a selective determination of chloramphenicol, were immobilized. The construction of the sensor looks as follows: gold nanoparticles are applied on the surface of the glassy carbon (GCE), and afterwards copolymer layer (poly 5, 2′:5′, 2″-terthiophene-3′-carboxyl acid) with carboxyl groups in the 3-position. Particles of dendrimer modified with cadmium sulfide (CdS) nanodots were chemically immobilized to the carboxyl groups. Anti-CAP antibodies (anti chloramphenicol acetyl transferase), which bind chloramphenicol, are attached to the CdS particles. Thus prepared, the sensor can be used to determine the level of chloramphenicol in a sample. The authors used a system designed to measure levels of chloramphenicol in natural samples, which were different types of meat [102].

Another solution of the biosensor concept are systems containing a DNA aptamer in the receptor film, as a biological substance. Su *et al.* [103] described the synthesis and properties of polymer films obtained from 3-((2,- -dihydrothieno[3,4-b] [1, 4]dioxin-2-yl)methoxy)propane-1-thiol, which is a derivative of EDOT. This monomer has a thiol group in its structure, which allows for establishment of a monolayer on the surface of the gold electrode by means of the self-assembly. Afterwards, such a film may be electrochemically polymerized. The described method allows to obtain a nanofilm of a conductive polymer with a very regular structure on the gold surface. In another publication, the same team of researchers described the application of the layer, which they had developed, in construction of a biosensor for determination of lipopolysaccharides [104]. Gold nanoparticles (AuNPs) were electrochemically generated directly on the surface of the polymerized self-assembled monolayer (SAM) of thiol derivative 3,4-ethylenedioxythiophene (EDOT), EDTMSHA. A thiolated single-stranded DNA aptamer, with high specificity to LPS, was immobilized on the surface of golden nanoparticles. Impedimetric detection is used in the presented biosensor. In the case of the described sensor, detection consists in using the linear dependence of the resistance change transfer (R_{ct}) on the analyte concentration. The range of the determined LPS concentrations were from 0.1 ng/mL to 1 ng/mL.

CONCLUDING REMARKS

Biosensors formed on composite materials with the presence of nanostructures and conductive polymers are an important and modern tools in medical bioanalytics. An important direction in which analytics is developing are *in vivo* and *in vitro* experiments in natural matrixes without procedures of purifying and separating. The great variety and easier and easier access to biological materials, enzymes, antibodies and DNA causes that it is possible to design sensors with the determined level of selectivity or specifity used for the determination of particular analyst in the specified environment. The other important part of biosensor *i.e.* composite material in which biological compound is immobilized, also undergoes constant changes. First of all, more and more often materials with high level of bioconsistency *e.g.* carbon materials, GC, CNT, graphene and conductive polymers such as PEDOT, are being applied. Simultaneously, bigger and bigger selection of nanostructure and conductive materials causes that it is possible to design composite layers which, besides the immobilization of biological compound, improve the parameters of the determination – they increase sensitivity, lengthen the life time of biosensor and improve the stability of the analytical signal.

The investigations of biosensors are an example of interdisciplinary research. They include biology, biochemistry, chemistry and surface physics. The development in each of these fields causes the development and new solutions in biosensors. Probably the greatest progress in the development of biosensors is still ahead.

CONSENT FOR PUBLICATION

Not applicable.

CONFLICT OF INTEREST

The authors declares no conflict of interest, financial or otherwise.

ACKNOWLEDGEMENTS

Declered none.

REFERENCES

[1] Ates, M.; Sarac, A.S. Conducting polymer coated carbon surfaces and biosensor applications. *Prog. Org. Coat.,* **2009**, *66*, 337-358. [http://dx.doi.org/10.1016/j.porgcoat.2009.08.014]

[2] Nambiar, S.; Yeow, J.T. Conductive polymer-based sensors for biomedical applications. *Biosens. Bioelectron.,* **2011**, *26*(5), 1825-1832. [http://dx.doi.org/10.1016/j.bios.2010.09.046] [PMID: 21030240]

[3] Inzelt, G. *Conducting Polymers*; Springer-Verlag Berlin: Heidelberg, **2008**.

[4] Skotheim, T.A. *Handbook of conducting polymers*; Marcel Dekker: New York, **2008**.

[5] Chiang, J-C.; MacDiarmid, A.G. 'Polyaniline': Protonic acid doping of the emeraldine form to the metallic regime. *Synth. Met.,* **1986**, *13*, 193-205. [http://dx.doi.org/10.1016/0379-6779(86)90070-6]

[6] Huang, W.S.; Humphrey, B.D.; McDairmid, A.G. Polyaniline, a novel conducting polymer. Morphology and chemistry of its oxidation and reduction in aqueous electrolytes. *Journal of the Chemical Society, Faraday Transactions 1: Physical Chemistry in Condensed Phases,* **1986**, *82*(8), 2385-2400.

[7] Karyakin, A.A.; Strakhowa, A.K.; Yatsimirsky, A.K. Self-doped polianilines electrochemically active in neutral and basic aqueous solutions. Electropolymerisation of substituted anilines. *J. Electroanal. Chem.,* **1994**, *371*, 259-265. [http://dx.doi.org/10.1016/0022-0728(93)03244-J]

[8] Zhang, G-R.; Zhao, S-F.; Liu, X.; Liu, X-L.; Luo, Y-L.; Lu, J-X. Study on Electrocopolymerization of o-Toluidine and Anthranilic Acid with *in situ* UV-vis Spectroscopy. *Acta Chimi. Sin.,* **2010**, *68*(11), 1063-1069.

[9] Cakmak, G.; Küçükyavuz, Z.; Küçükyavuz, S. Conductive copolymers of polyaniline, polypyrrole and poly(dimethylsiloxane). *Synth. Met.,* **2005**, *151*(1), 10-18. [http://dx.doi.org/10.1016/j.synthmet.2005.02.019]

[10] Morita, M.; Myazaki, S.; Tanoue, H.; Ishikawa, M.; Matsuda, Y. „Electrochemical behawior of polyaniline-poly(styrene sulfonate) composite film in organic electrolyte solutions. *J. Electrochem. Soc.,* **1994**, *114*(6), 1409-1413. [http://dx.doi.org/10.1149/1.2054938]

[11] Bartlett, P.N.; Simon, E. Poly(aniline)-poly(acrylate) composite films as modified electrodes fot the oxidation of NADH. *Phys. Chem. Chem. Phys.,* **2000**, *2*, 2599-2606. [http://dx.doi.org/10.1039/b001107j]

[12] Groenendaal, L.B.; Jonas, F.; Freitag, D.; Pielartzik, H.; Reynolds, J.R. Poly(3,4-ethylenedioxythiophene) and its derivatives: past, present, and future. *Adv. Mater.,* **2000**, *12*(7), 481-494. [http://dx.doi.org/10.1002/(SICI)1521-4095(200004)12:7<481::AID-ADMA481>3.0.CO;2-C]

[13] Łapkowski, M. *Wiadomości chemiczne,* **1993**, *47*(1993), 299.

[14] Wei, D.; Ivaska, A. Electrochemical Biosensors Based on Polyaniline. *Chem. Anal. (Warsaw)*, **2006**, *51*, 839.

[15] Bayer, A.G. 1988.Eur. Patent 339 340

[16] Jonas, F.; Schrader, L. *Synth. Met.*, **1991**. 41±43, 831

[17] Heywang, G.; Jonas, F. *Adv. Mater.*, **1992**, *4*, 116. [http://dx.doi.org/10.1002/adma.19920040213]

[18] Winter, I.; Reece, C.; Hormes, J.; Heywang, G.; Jonas, F. *Chem. Phys.*, **1995**, *194*, 207. [http://dx.doi.org/10.1016/0301-0104(95)00026-K]

[19] Krosa, A.; Sommerdijkc, N.A.; Nolteb, R.J. Poly(pyrrole) *versus* poly(3,4-ethylenedioxythiophene): implications for biosensor applications. *Sens. Actuators B Chem.*, **2005**, *106*, 289-295. [http://dx.doi.org/10.1016/j.snb.2004.08.011]

[20] Fuck, F. *Farbe Lack,* **1998**, *104*, 32.

[21] Krzyczmonik, P.; Socha, E.; Andrijewski, G. Determination of Ascorbic Acid by a Composite-Modified Platinum Electrode. *Anal. Lett.*, [http://dx.doi.org/10.1080/00032719.2016.1201096]

[22] Sriprachuabwong, Ch.; Karuwan, Ch.; Wisitsorrat, A.; Phokharatkul, D.; Lomas, T.; Sritongkham, P.; Tuantranont, A. Inkjet-printed graphene-PEDOT:PSS modified screen printed carbon electrode for biochemical sensing. *J. Mater. Chem.*, **2012**, *22*, 5478. [http://dx.doi.org/10.1039/c2jm14005e]

[23] Reimann, S.M.; Manninen, M. *Rev. Mod. Phys.*, **2002**.

[24] Kastner, M.A. *Phys. Today,* **1993**.

[25] Ekimov, AI; Efros, AL; Onushchenko, AA *Quantum size effect in semiconductor microcrystals,* **1985**. [http://dx.doi.org/10.1016/S0038-1098(85)80025-9]

[26] Ekimov, A.I.; Onushchenko, A.A. *JETP Lett.*, **1981**.

[27] Couto, O.D.; Puebla, J. Charge control in InP/(Ga,In)P single quantum dots embedded in Schottky diodes. *Phys. Rev.*, **2011**. [http://dx.doi.org/10.1103/PhysRevB.84.125301]

[28] Bailey, R.E.; Nie, S. *Chemistry of Nanomaterials*; Rao, C.N.R.; Mueller, A.; Cheetham, A.K., Eds.; , **2004**.

[29] Vastola, G.; Zhang, Y-W.; Shenoy, V.B. *Curr. Opin. Solid State Mater. Sci.*, **2012**.

[30] Smith, A.M.; Nie, S. *Nat. Biotechnol.*, **2009**.

[31] Huczko, A. *Fullereny*; Wydawnictwo PWN: Warszawa, **2000**.

[32] Murat Ates, A. *Prog. Org. Coat.*, **2009**, *66*, 337-358. [http://dx.doi.org/10.1016/j.porgcoat.2009.08.014]

[33] Chen, D.; Feng, H.; Li, J. Graphene oxide: preparation, functionalization, and electrochemical applications. *Chem. Rev.*, **2012**, *112*(11), 6027-6053. [http://dx.doi.org/10.1021/cr300115g] [PMID: 22889102]

[34] Pumera, M. Electrochemistry of graphene, graphene oxide and other graphenoids. *Electrochem. Commun.*, **2013**, *36*, 14-18. [http://dx.doi.org/10.1016/j.elecom.2013.08.028]

[35] Chen, X.M.; Wu, G.H.; Jiang, Y.Q.; Wang, Y.R.; Chen, X. Graphene and graphene-based nanomaterials: the promising materials for bright future of electroanalytical chemistry. *Analyst (Lond.),* **2011**, *136*(22), 4631-4640. [http://dx.doi.org/10.1039/c1an15661f] [PMID: 21975368]

[36] Zhang, J.; Zhang, F.; Yang, H.; Huang, X.; Liu, H.; Zhang, J.; Guo, S. Graphene oxide as a matrix for enzyme immobilization. *Langmuir,* **2010**, *26*(9), 6083-6085. [http://dx.doi.org/10.1021/la904014z] [PMID: 20297789]

[37] Brzózka, Z.; Wróblewski, W. *Sensory Chemiczne*; Oficyna Wydawnicza Politechniki Warszawskiej: Warszawa, **1999**.

[38] Gough, D.A.; Lucisano, J.Y.; Tse, P.H. Two-dimensional enzyme electrode sensor for glucose. *Anal. Chem.,* **1985**, *57*(12), 2351-2357. [http://dx.doi.org/10.1021/ac00289a042] [PMID: 4061843]

[39] Toghill, K.E.; Compton, R.G. Electrochemical Non-enzymatic Glucose Sensors: A Perspective and an Evaluation. *J. Electrochem. Soc.,* **2010**, *5*, 1246-1301.

[40] Guilbault, G.G.; Lubrano, G.J. An enzyme electrode for the amperometric determination of glucose. *Anal. Chim. Acta,* **1973**, *64*(3), 439-455. [http://dx.doi.org/10.1016/S0003-2670(01)82476-4] [PMID: 4701057]

[41] Bartlett, P.; Birkin, P.R. Enzyme switch responsive to glucose. *Anal. Chem.,* **1993**, *65*, 1118-1119. [http://dx.doi.org/10.1021/ac00056a029]

[42] Bartlett, P.; Astier, Y. Microelectrochemical enzyme transistors. *Chem. Commun. (Camb.),* **2000**, 105-112. [http://dx.doi.org/10.1039/a902905b]

[43] Shim, N.Y.; Bernards, D.A.; Macaya, D.J.; Defranco, J.A.; Nikolou, M.; Owens, R.M.; Malliaras, G.G. All-plastic electrochemical transistor for glucose sensing using a ferrocene mediator. *Sensors (Basel),* **2009**, *9*(12), 9896-9902. [http://dx.doi.org/10.3390/s91209896] [PMID: 22303153]

[44] Gough, D.A.; Lucisano, J.Y.; Tse, P.H. Two-dimensional enzyme electrode sensor for glucose. *Anal. Chem.,* **1985**, *57*(12), 2351-2357. [http://dx.doi.org/10.1021/ac00289a042] [PMID: 4061843]

[45] Cass, A.E.; Davis, G.; Francis, G.D.; Hill, H.A.; Aston, W.J.; Higgins, I.J.; Plotkin, E.V.; Scott, L.D.; Turner, A.P. Ferrocene-mediated enzyme electrode for amperometric determination of glucose. *Anal. Chem.,* **1984**, *56*(4), 667-671. [http://dx.doi.org/10.1021/ac00268a018] [PMID: 6721151]

[46] Kros, A.; von Hovell, S.W.; Sommerdijk, N.A.; Nolte, R.J. Poly(3,4-ethylenedioxythiophene – based glucose biosensors. *Adv. Mater.,* **2001**, *13*(20), 1555-1557. [http://dx.doi.org/10.1002/1521-4095(200110)13:20<1555::AID-ADMA1555>3.0.CO;2-7]

[47] Krzyczmonik, P.; Socha, E.; Skrzypek, S. Immobilization of glucose oxidase on modified electrodes with composite layers based on poly(3,4-ethylenedioxythiophene). *Bioelectrochemistry,* **2015**, *101*, 8-13. [http://dx.doi.org/10.1016/j.bioelechem.2014.06.009] [PMID: 25023029]

[48] Jönsson, G.; Gorton, L. An amperometric glucose sensor made by modification of a graphite electrode surface with immobilized glucose oxidase and adsorbed mediator. *Biosensors,* **1985**, *1*(4), 355-368. [http://dx.doi.org/10.1016/0265-928X(85)80005-5] [PMID: 3842604]

[49] Kerner, W.; Zier, H.; Steinbach, G.; Brückel, J.; Pfeiffer, E.F.; Weiss, T.; Cammann, K.; Planck, H. A potentially implantable enzyme electrode for amperometric measurement of glucose. *Horm. Metab. Res. Suppl.,* **1988**, *20*, 8-13. [PMID: 3248792]

[50] Pandey, P.C.; Kayastha, A.M.; Pandey, V. Amperometric enzyme sensor for glucose based on graphite paste-modified electrodes. *Appl. Biochem. Biotechnol.,* **1992**, *33*(2), 139-144. [http://dx.doi.org/10.1007/BF02950782] [PMID: 1510468]

[51] Zhu, J.; Tian, C.; Wu, W.; Wu, J.; Zhang, H.; Lu, D.; Zhang, G. Fabrication and characterization of

glucose sensors based on a microarray H_2O_2 electrode. *Biosens. Bioelectron.,* **1994**, *9*(4-5), 295-300. [http://dx.doi.org/10.1016/0956-5663(94)80026-X] [PMID: 8068227]

[52] Wang, B.; Li, B.; Deng, Q.; Dong, S. Amperometric glucose biosensor based on sol-gel organic-inorganic hybrid material. *Anal. Chem.,* **1998**, *70*(15), 3170-3174. [http://dx.doi.org/10.1021/ac980160h] [PMID: 11013720]

[53] Garjonyte, R.; Malinauskas, A. Amperometric glucose biosensor based on glucose oxidase immobilized in poly(o-phenylenediamine) layer. *Sens. Actuators B Chem.,* **1999**, *56*, 85-92. [http://dx.doi.org/10.1016/S0925-4005(99)00163-X]

[54] Fei, J.; Wu, Y.; Ji, X.; Wang, J.; Hu, S.; Gao, Z. An amperometric biosensor for glucose based on electrodeposited redox polymer/glucose oxidase film on a gold electrode. *Anal. Sci.,* **2003**, *19*(9), 1259-1263. [http://dx.doi.org/10.2116/analsci.19.1259] [PMID: 14516076]

[55] Salimi, A.; Compton, R.G.; Hallaj, R. Glucose biosensor prepared by glucose oxidase encapsulated sol-gel and carbon-nanotube-modified basal plane pyrolytic graphite electrode. *Anal. Biochem.,* **2004**, *333*(1), 49-56. [http://dx.doi.org/10.1016/j.ab.2004.06.039] [PMID: 15351279]

[56] Tang, H.; Chen, J.; Yao, S.; Nie, L.; Deng, G.; Kuang, Y. Amperometric glucose biosensor based on adsorption of glucose oxidase at platinum nanoparticle-modified carbon nanotube electrode. *Anal. Biochem.,* **2004**, *331*(1), 89-97. [http://dx.doi.org/10.1016/S0003-2697(04)00411-7] [PMID: 15246000]

[57] Shan, D.; He, Y.; Wang, S.; Xue, H.; Zheng, H. A porous poly(acrylonitrile-co-acrylic acid) film-based glucose biosensor constructed by electrochemical entrapment. *Anal. Biochem.,* **2006**, *356*(2), 215-221. [http://dx.doi.org/10.1016/j.ab.2006.06.005] [PMID: 16842730]

[58] Wu, J.; Qu, Y. Mediator-free amperometric determination of glucose based on direct electron transfer between glucose oxidase and an oxidized boron-doped diamond electrode. *Anal. Bioanal. Chem.,* **2006**, *385*(7), 1330-1335. [http://dx.doi.org/10.1007/s00216-006-0534-y] [PMID: 16773297]

[59] Tsai, Y.C.; Li, S.C.; Liao, S.W. Electrodeposition of polypyrrole-multiwalled carbon nanotube-glucose oxidase nanobiocomposite film for the detection of glucose. *Biosens. Bioelectron.,* **2006**, *22*(4), 495-500. [http://dx.doi.org/10.1016/j.bios.2006.06.009] [PMID: 16870421]

[60] Wang, H.J.; Zhou, C.M.; Peng, H.; Yu, H. Glucose biosensor based on platinum nanoparticles supported sulfonated-carbon nanotubes modified glassy carbon electrode. *Int. J. Electrochem. Sci.,* **2007**, *2*, 508-516.

[61] Liang, R.; Jiang, J.; Qiu, J. An amperometric glucose biosensor based on titania sol-gel/Prussian Blue composite film. *Anal. Sci.,* **2008**, *24*(11), 1425-1430. [http://dx.doi.org/10.2116/analsci.24.1425] [PMID: 18997370]

[62] Lu, J.; Do, I.; Drzal, L.T.; Worden, R.M.; Lee, I. Nanometal-decorated exfoliated graphite nanoplatelet based glucose biosensors with high sensitivity and fast response. *ACS Nano,* **2008**, *2*(9), 1825-1832. [http://dx.doi.org/10.1021/nn800244k] [PMID: 19206421]

[63] Wang, Z.; Liu, S.; Wu, P.; Cai, C. Detection of glucose based on direct electron transfer reaction of glucose oxidase immobilized on highly ordered polyaniline nanotubes. *Anal. Chem.,* **2009**, *81*(4), 1638-1645. [http://dx.doi.org/10.1021/ac802421h] [PMID: 19170516]

[64] Shan, C.; Yang, H.; Han, D.; Zhang, Q.; Ivaska, A.; Niu, L. Graphene/AuNPs/chitosan nanocomposites film for glucose biosensing. *Biosens. Bioelectron.,* **2010**, *25*(5), 1070-1074. [http://dx.doi.org/10.1016/j.bios.2009.09.024] [PMID: 19883999]

[65] Si, P.; Kannan, P.; Guo, L.; Son, H.; Kim, D.H. Highly stable and sensitive glucose biosensor based on covalently assembled high density Au nanostructures. *Biosens. Bioelectron.,* **2011**, *26*(9), 3845-3851. [http://dx.doi.org/10.1016/j.bios.2011.02.044] [PMID: 21454070]

[66] Li, J.; Yuan, R.; Chai, Y.; Che, X.; Li, W. Construction of an amperometric glucose biosensor based on the immobilization of glucose oxidase onto electrodeposited Pt nanoparticles-chitosan composite film. *Bioprocess Biosyst. Eng.,* **2012**, *35*(7), 1089-1095. [http://dx.doi.org/10.1007/s00449-012-0693-5] [PMID: 22311274]

[67] Aydoğdu, G.; Zeybek, D.K.; Pekyardımcı, Ş.; Kılıç, E. A novel amperometric biosensor based on ZnO nanoparticles-modified carbon paste electrode for determination of glucose in human serum. *Artif. Cells Nanomed. Biotechnol.,* **2013**, *41*(5), 332-338. [http://dx.doi.org/10.3109/21691401.2012.744994] [PMID: 23305092]

[68] Wang, H.; Ohnuki, H.; Endo, H.; Izumi, M. Impedimetric and amperometric bifunctional glucose biosensor based on hybrid organic-inorganic thin films. *Bioelectrochemistry,* **2015**, *101*, 1-7. [http://dx.doi.org/10.1016/j.bioelechem.2014.06.007] [PMID: 25014167]

[69] Saxena, U.; Das, A.B. Nanomaterials towards fabrication of cholesterol biosensors: Key roles and design approaches. *Biosens. Bioelectron.,* **2016**, *75*, 196-205. [http://dx.doi.org/10.1016/j.bios.2015.08.042] [PMID: 26319162]

[70] Gopalan, A.I.; Lee, K.P.; Ragupathy, D. Development of a stable cholesterol biosensor based on multi-walled carbon nanotubes-gold nanoparticles composite covered with a layer of chitosan-roo--temperature ionic liquid network. *Biosens. Bioelectron.,* **2009**, *24*(7), 2211-2217. [http://dx.doi.org/10.1016/j.bios.2008.11.034] [PMID: 19167880]

[71] Dhand, C.; Singh, S.P.; Arya, S.K.; Datta, M.; Malhotra, B.D. Cholesterol biosensor based on electrophoretically deposited conducting polymer film derived from nano-structured polyaniline colloidal suspension. *Anal. Chim. Acta,* **2007**, *602*(2), 244-251. [http://dx.doi.org/10.1016/j.aca.2007.09.028] [PMID: 17933610]

[72] Dhand, C.; Das, M.; Sumana, G.; Srivastava, A.K.; Pandey, M.K.; Kim, C.G.; Datta, M.; Malhotra, B.D. Preparation, characterization and application of polyaniline nanospheres to biosensing. *Nanoscale,* **2010**, *2*(5), 747-754. [http://dx.doi.org/10.1039/b9nr00346k] [PMID: 20648320]

[73] Brahim, S.; Narinesingh, D.; Guiseppi-Elie, A. *Anal. Chim. Acta,* **2001**, *448*, 27-36. [http://dx.doi.org/10.1016/S0003-2670(01)01321-6]

[74] Yao, T.; Takashima, K. Amperometric biosensor with a composite membrane of sol-gel derived enzyme film and electrochemically generated poly(1,2-diaminobenzene) film. *Biosens. Bioelectron.,* **1998**, *13*(1), 67-73. [http://dx.doi.org/10.1016/S0956-5663(97)00076-6] [PMID: 9519451]

[75] Polak, J.; Jarosz-Wilkołazka, A. Reakcje katalizowane przez lakazę – mechanizm i zastosowanie w biotechnologii. *Biotechnologia,* **2007**, *4*, 82-94.

[76] Rahman, M.A.; Noh, H.B.; Shim, Y.B. Direct electrochemistry of laccase immobilized on au nanoparticles encapsulated-dendrimer bonded conducting polymer: application for a catechin sensor. *Anal. Chem.,* **2008**, *80*(21), 8020-8027. [http://dx.doi.org/10.1021/ac801033s] [PMID: 18841943]

[77] Rajesh, V. Bisht, W. Takashima, K. Kaneto, Development of a potentiometric urea biosensor based on copolymer poly(N-3-aminopropyl pyrrole-co-pyrrole) film. *React. Funct. Polym.,* **2005**, *62*, 51-59. [http://dx.doi.org/10.1016/j.reactfunctpolym.2004.08.009]

[78] Rajesh, V. Bisht, W. Takashima, K. Kaneto, A novel thin film urea biosensor based on copolymer poly(N-3-aminopropylpyrrole-co-pyrrole) film. *Surf. Coat. Tech.,* **2005**, *198*, 231-236. [http://dx.doi.org/10.1016/j.surfcoat.2004.10.022]

[79] Rajesh, V.; Bisht, V.; Takashima, W.; Kaneto, K. An amperometric urea biosensor based on covalent

immobilization of urease onto an electrochemically prepared copolymer poly (N-3-aminopropyl pyrrole-co-pyrrole) film. *Biomaterials,* **2005**, *26*(17), 3683-3690. [http://dx.doi.org/10.1016/j.biomaterials.2004.09.024] [PMID: 15744952]

[80] Lakard, B.; Herlem, G.; Lakard, S.; Antoniou, A.; Fahys, B. Urea potentiometric biosensor based on modified electrodes with urease immobilized on polyethylenimine films. *Biosens. Bioelectron.,* **2004**, *19*(12), 1641-1647. [http://dx.doi.org/10.1016/j.bios.2003.12.035] [PMID: 15142598]

[81] Saeedfar, K.; Heng, L.Y.; Ling, T.L.; Rezayi, M. Potentiometric urea biosensor based on an immobilised fullerene-urease bio-conjugate. *Sensors (Basel),* **2013**, *13*(12), 16851-16866. [http://dx.doi.org/10.3390/s131216851] [PMID: 24322561]

[82] Luo, Y-Ch.; Do, J-S. Urea biosensor based on PANi(urease)-Nafion/Au composite electrode. *Biosens. Bioelectron.,* **2004**, *20*(1), 15-23. [http://dx.doi.org/10.1016/j.bios.2003.11.028] [PMID: 15142572]

[83] Bertocchi, P.; Compagnone, D.; Palleschi, G. Amperometric ammonium ion and urea determination with enzyme-based probes. *Biosens. Bioelectron.,* **1996**, *11*(1-2), 1-10. [http://dx.doi.org/10.1016/0956-5663(96)83708-0] [PMID: 8600914]

[84] Vostiar, I.; Tkac, J.; Sturdik, E.; Gemeiner, P. Amperometric urea biosensor based on urease and electropolymerized toluidine blue dye as a pH-sensitive redox probe. *Bioelectrochemistry,* **2002**, *56*(1-2), 113-115. [http://dx.doi.org/10.1016/S1567-5394(02)00042-7] [PMID: 12009455]

[85] Eggenstein, C.; Borchardt, M.; Diekmann, C.; Gründig, B.; Dumschat, C.; Cammann, K.; Knoll, M.; Spener, F. A disposable biosensor for urea determination in blood based on an ammonium-sensitive transducer. *Biosens. Bioelectron.,* **1999**, *14*(1), 33-41. [http://dx.doi.org/10.1016/S0956-5663(98)00103-1] [PMID: 10028647]

[86] Pandey, P.C.; Upadhyay, S.; Singh, G.; Prakash, R.; Srivastava, R.C.; Seth, P.K. A new solid-state pH sensor and its application in the construction of solid-state urea biosensor. *Electroanalysis,* **2000**, *12*, 517-521. [http://dx.doi.org/10.1002/(SICI)1521-4109(200005)12:7<517::AID-ELAN517>3.0.CO;2-Y]

[87] Srivastava, P.K.; Kayastha, A.M.; Srinivasan, Characterization of gelatin-immobilized pigeonpea urease and preparation of a new urea biosensor. *Biotechnol. Appl. Biochem.,* **2001**, *34*(Pt 1), 55-62. [http://dx.doi.org/10.1042/BA20010016] [PMID: 11483155]

[88] Pizzariello, A.; Stredanský, M.; Stredanská, S.; Miertus, S. Urea biosensor based on amperometric pH-sensing with hematein as a pH-sensitive redox mediator. *Talanta,* **2001**, *54*(4), 763-772. [http://dx.doi.org/10.1016/S0039-9140(01)00331-9] [PMID: 18968299]

[89] Singhal, R.; Gambhir, A.; Pandey, M.K.; Annapoorni, S.; Malhotra, B.D. Immobilization of urease on poly(N-vinyl carbazole)/stearic acid Langmuir-Blodgett films for application to urea biosensor. *Biosens. Bioelectron.,* **2002**, *17*(8), 697-703. [http://dx.doi.org/10.1016/S0956-5663(02)00020-9] [PMID: 12052355]

[90] Ortega, M.M.; Rodriguez, D.E.; Ensinas, J.C.; Plascencia, M.; Velarde, F.A.; Olayo, R. Conductometric uric acid and urea biosensor prepared from electroconductive polyaniline-poly(--butyl methacrylate) composites. *Sens. Actuators B Chem.,* **2002**, *85*, 19-25. [http://dx.doi.org/10.1016/S0925-4005(02)00045-X]

[91] Kovács, B.; Nagy, G.; Dombi, R.; Tóth, K. Optical biosensor for urea with improved response time. *Biosens. Bioelectron.,* **2003**, *18*(2-3), 111-118. [http://dx.doi.org/10.1016/S0956-5663(02)00164-1] [PMID: 12485757]

[92] Wang, J.Q.; Chou, J.C.; Sun, T.P.; Hsuing, S.K.; Hsuing, G.B. pH based potentiometrical flow injection biosensor for urea. *Sens. Actuators B Chem.,* **2003**, *91*, 5-10. [http://dx.doi.org/10.1016/S0925-4005(03)00160-6]

[93] Reddy K, R.C.; Srivastava, P.K.; Dey, P.M.; Kayastha, A.M. Immobilization of pigeonpea (Cajanus cajan) urease on DEAE-cellulose paper strips for urea estimation. *Biotechnol. Appl. Biochem.*, **2004**, *39*(Pt 3), 323-327.
[http://dx.doi.org/10.1042/BA20030122] [PMID: 15154845]

[94] Limbut, W.; Thavarungkul, P.; Kanatharana, P.; Asawatreratanakul, P.; Limsakul, C.; Wongkittisuksa, B. Comparative study of controlled pore glass, silica gel and poraver for the immobilization of urease to determine urea in a flow injection conductimetric biosensor system. *Biosens. Bioelectron.*, **2004**, *19*(8), 813-821.
[http://dx.doi.org/10.1016/j.bios.2003.08.007] [PMID: 15128100]

[95] Yang, Z.; Si, S.; Dai, H.; Zhang, C. Piezoelectric urea biosensor based on immobilization of urease onto nanoporous alumina membranes. *Biosens. Bioelectron.*, **2007**, *22*(12), 3283-3287.
[http://dx.doi.org/10.1016/j.bios.2007.03.006] [PMID: 17433665]

[96] Gutiérrez, M.; Alegret, S.; del Valle, M. Potentiometric bioelectronic tongue for the analysis of urea and alkaline ions in clinical samples. *Biosens. Bioelectron.*, **2007**, *22*(9-10), 2171-2178.
[http://dx.doi.org/10.1016/j.bios.2006.10.007] [PMID: 17097867]

[97] Gutiérrez, M.; Alegret, S.; del Valle, M. Bioelectronic tongue for the simultaneous determination of urea, creatinine and alkaline ions in clinical samples. *Biosens. Bioelectron.*, **2008**, *23*(6), 795-802.
[http://dx.doi.org/10.1016/j.bios.2007.08.019] [PMID: 17931852]

[98] Jenkins, D.M.; Delwiche, M.J. Manometric biosensor for on-line measurement of milk urea. *Biosens. Bioelectron.*, **2002**, *17*(6-7), 557-563.
[http://dx.doi.org/10.1016/S0956-5663(02)00018-0] [PMID: 11959478]

[99] Jenkin, D.M.; Delwiche, M.J.; DePeters, E.J. R.H. BonDurant, Refinement of pressure assay for milk urea nitrogen. *J. Dairy Sci.*, **2002**, *83*(9), 2042-2048.
[http://dx.doi.org/10.3168/jds.S0022-0302(00)75085-5]

[100] Verma, N.; Singh, M. A disposable microbial based biosensor for quality control in milk. *Biosens. Bioelectron.*, **2003**, *18*(10), 1219-1224.
[http://dx.doi.org/10.1016/S0956-5663(03)00085-X] [PMID: 12835039]

[101] Renny, E.F.; Daniel, D.K.; Krastanov, A.I.; Zachariah, C.A.; Elizabeth, R. *Enzyme based biosensor for detection of urea in milk, Biotechnol*; Biotechnol. Eq, **2005**, pp. 198-201.

[102] Kim, D.M.; Rahman, M.A.; Do, M.H.; Ban, C.; Shim, Y.B. An amperometric chloramphenicol immunosensor based on cadmium sulfide nanoparticles modified-dendrimer bonded conducting polymer. *Biosens. Bioelectron.*, **2010**, *25*(7), 1781-1788.
[http://dx.doi.org/10.1016/j.bios.2009.12.024] [PMID: 20116233]

[103] Su, W.; Nguyen, H.T.; Cho, M.; Son, Y.; Lee, Y. Synthesis, characterization and self-assembled film of poly(3-((2,3-dihydrothieno[3,4-b][1,4]dioxin-2-yl)methoxy)propane-1-thiol) (PEDTMSHA). *Synth. Met.*, **2010**, *160*, 2471-2475.
[http://dx.doi.org/10.1016/j.synthmet.2010.09.029]

[104] Su, W.; Cho, M.; Nam, J-D.; Choe, W-S.; Lee, Y. Aptamer-Assisted Gold Nanoparticles/PEDOT Platform for Ultrasensitive Detection of LPS. *Electroanalysis,* **2013**, *25*(2), 380-386.
[http://dx.doi.org/10.1002/elan.201200453]

CHAPTER 7

New Trends in Electrochemical Sensors Modified with Carbon Nanotubes and Graphene for Pharmaceutical Analysis

Burcin Bozal-Palabiyik, Burcu Dogan-Topal, Sibel A. Ozkan and **Bengi Uslu***

Faculty of Pharmacy, Department of Analytical Chemistry, Ankara University, 06100, Tandogan, Ankara, Turkey

Abstract: Recently, carbon nanotubes (CNT) and graphene have been chosen as electrode material due to their unique mechanical, optical and electrical properties. These carbon-based materials are used to facilitate electrode reaction and produce an electrocatalytic effect. Moreover, a plethora of literature has emerged concerning CNT and graphene being employed to obtain a more selective and sensitive analysis. This chapter reviews this emerging literature focusing on drug analysis from pharmaceutical dosage forms and biological matrices. To achieve this, first, there is a summary of the basic properties, synthesis, characterization, electrode preparation and advantages/limitations of these materials, and then a selection of the literature from 2010 to 2017 is presented, which summarizes the type of electrode, electrochemical techniques, peak potential of drug compounds and the limit of detection values.

Keywords: Amperometry, Analysis, Buffer solution, Carbon nanotube, Chronoamperometry, Chronocoulometry, Composite material, Cyclic voltammetry, Determination, Drug, Electrochemistry, Electrochemical sensors, Graphene, Graphene oxide, Impedance spectroscopy, Modified electrodes, Nanosensors, Pharmaceutical formulation, Potentiometry, Stripping voltammetry.

INTRODUCTION

Electrochemical sensors, which constitute a significant category of chemical sensors, are composed of an electrode used as the transduction element and they are widely used in clinical, industrial and environmental fields [1]. One of the most important application areas of electrochemical sensors is the pharmaceutical industry, particularly for detecting drug active compounds in pharmaceutical dosage forms and biological fluids, such as serum, plasma and urine.

* **Corresponding author Bengi Uslu:** Ankara University, Faculty of Pharmacy, Department of Analytical Chemistry, 06100, Tandogan, Ankara, Turkey; Tel: +90 312 2033178; Fax: +90 312 2238243; E-mail: buslu@pharmacy.ankara.edu.tr

Atta-ur-Rahman, Sibel A. Ozkan & Rida Ahmed (Eds.)

In other words, electrochemical sensor studies connect the field of chemical sensors on the one hand and electrochemistry on the other [2].

Electrochemical sensing aims to perform more selective and sensitive detection. In order to enhance selectivity and sensitivity of electrodes, a process called modification is applied. Before modification, electrode performance is limited to and dependent on the solution in which it is immersed; however, through the modification process, the electrochemical properties of electrodes are increased significantly. Moreover, electrode modification contributes to fouling resistance and corrosion protection, and reduces interferences [3].

In recent years, modification with nanoparticles has been widely utilized for designing electrochemical sensors for pharmaceutical analysis. In modification, nanoparticles are widely used for their high surface area, biocompatibility and charge-sensitive conductance. In other words, nanoparticles are preferred for their enhanced electron-transfer rates, favorable geometry over bulk materials, and compatibility with design of nanoarrays [4].

Among the nanoparticles used for modification, carbon nanoparticles have been widely used since the 1980s. In particular, following Sumio Iijima's discovery of multi-wall carbon nanotubes (MWCNTs) in 1991, carbon nanostructures, especially CNTs and graphene (Fig. **1**), have been quite well researched in various fields, including physics, chemistry, and material science. These new nanoforms of carbon allotropes with cylindrical or planar geometry can be transformed to each other, which make them widely preferred for this purpose [5].

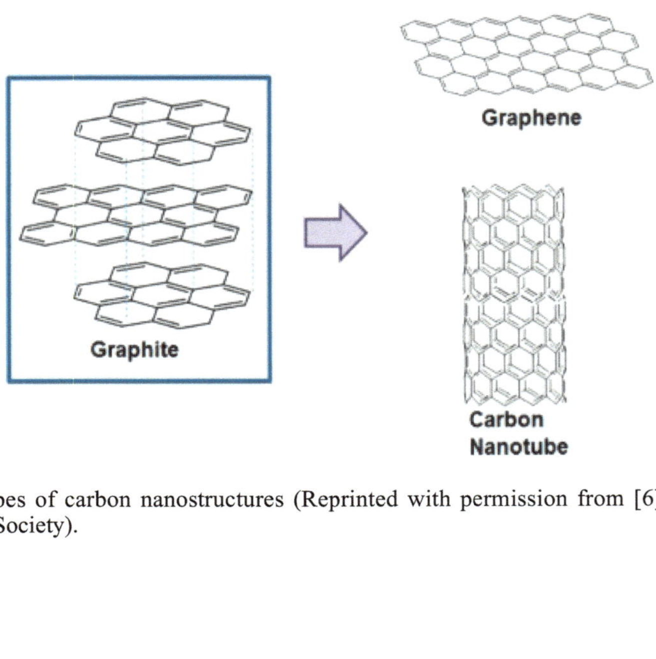

Fig. (1). Various types of carbon nanostructures (Reprinted with permission from [6]. Copyright (2017) American Chemical Society).

To start with CNTs, considering their structural and electronic uniqueness, these carbon nanoparticles have been proposed either as nanostructured bulk electrode materials for sensing and biosensing and in advanced electrochemical devices or for their significantly fast electrode kinetics [7]. CNTs behave as a metal or semiconductor depending on their structures, and CNT-modified electrodes have better conductivity than graphite [8]. Moreover, they display a superior performance compared to electrodes, such as Au and Pt and as well as other carbon-based electrodes. In addition, CNTs have a hollow core, which is suitable for storing guest molecules. The electrical conductivity of nanotubes can be improved by modifying the original CNT structure. CNTs possess fascinating mechanical strength and are the strongest and stiffest material known, which make them especially preferable in electrochemical sensor design [9]. CNTs, particularly their side walls, are relatively inert. However, the ends of nanotubes are more reactive than the cylindrical parts. CNTs are preferred in electrochemical sensing since they show an enhanced electrochemical response to some important biomolecules and promote electron transfer reactions of proteins [10].

Another significant carbon nanomaterial used for designing electrochemical sensors is graphene. Most simply defined, graphene is a single layer of carbon atoms arranged in a hexagonal lattice, and its perfect honeycomb structure has been applied as a theoretical model to understand the properties of various carbon-based materials that emerged as dimensional allotropes of graphene [11]. International Union of Pure and Applied Chemistry defines this material as "a single carbon layer of the graphite structure, describing its nature by analogy to a polycyclic aromatic hydrocarbon of quasi infinite size" [12].

Graphene is a member of carbon nanostructures and it is the essential building block for most allotropic versions of carbon-based materials, including spherical C60 buckyball, single and multi-walled carbon nanotubes, or three-dimensional graphite sheets. Originally defined as a single-layer carbon structure, graphene has been renamed as a graphene nanosheet (GNS), with strips of graphene with an ultra-thin width smaller than 50 nm being called graphene nanoribbons (GNR) and stacking of two to seven graphene sheets being generally known as graphene nanoplatelets (GNP). Recently, a new form of graphene called graphene oxide (GO) has emerged as an oxidized form of graphene chemically or electrochemically reduced before its use [13]. Compared to CNTs, graphene lacks some bandgap-induced optical properties; however, it is extremely versatile and chemically tunable, having a higher surface area and chemical/thermal stability with a broad potential window [14]. For these valuable properties, graphene-based electrochemical sensors have recently been preferred for electrochemical sensing.

This chapter aims to discuss electrochemical sensors based on CNTs and graphene for pharmaceutical analysis. The structural and electrochemical properties, synthesis, characterization, advantages and limitations of CNT-based and graphene-based electrochemical sensors are examined. Moreover, the application of these electrochemical sensors in pharmaceutical analysis from 2010 to 2017 is summarized in tables.

CARBON NANOTUBES

The Structure and Properties of CNTs

Carbon is a very special element by its nature. Its electronic configuration allows carbon atoms to be bound in three different ways (single, double or triple). Therefore, carbon-based materials with different shapes and sizes emerge; *e.g.*, including graphite derivatives, diamond, amorphous carbon materials, and fullerenes [15]. CNT, another carbon-based material, can be defined as a graphene sheet wrapped to form a hollow cylinder. Based on the number of graphene sheets used to construct CNTs, these carbon nanomaterials are divided into two main groups, single-walled carbon nanotube (SWCNT) and MWCNT [16] (Fig. **2**). SWCNT can be considered as a long wrapped graphene sheet rolled in certain directions. It consists of a side wall and an end cap with very different physical and chemical properties. The side wall comprises hexagon rings and the end-cap structure is similar to or derived from a small fullerene in which carbon atoms are in both pentagonal and hexagonal rings. MWCNT, on the other hand, can be considered as a collection of concentric SWCNTs with different diameters [17]. The diameters of CNTs range from 0.2 to 2.0 nm for SWCNTs and from 2.0 to 100.0 nm for MWCNTs, and the lengths of CNTs range from several hundred nanometers to several micrometers. Moreover, in accordance with the type of rolling of a graphene sheet, there are three types of CNTs; zigzag, armchair or chiral [18] (Fig. **3**).

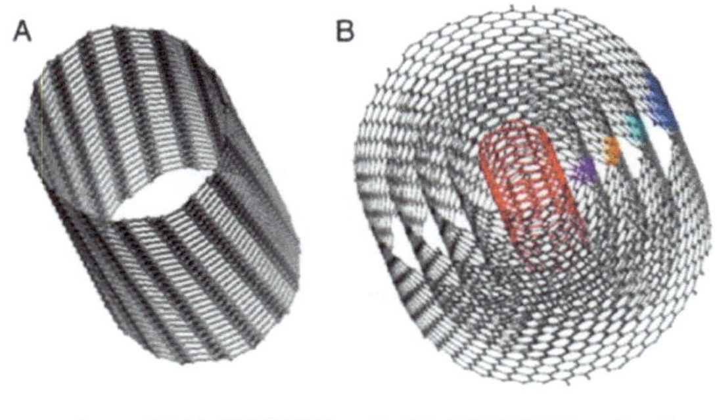

Fig. (2). Schematic representation of (A) SWCNT and (B) MWCNT; color is used to distinguish the walls (Reprinted with permission from [19]. Copyright (2017) American Chemical Society).

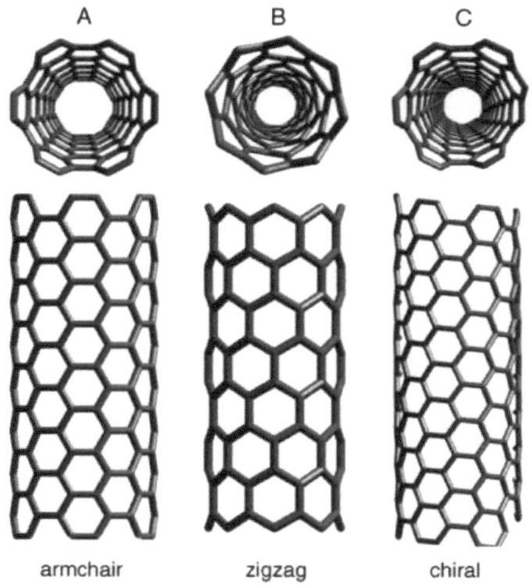

Fig. (3). Molecular models of SWCNT exhibiting different chiralities (Reprinted with permission from [19]. Copyright (2017) American Chemical Society).

CNTs have excellent electrical properties. Depending on their structure, some have metallic properties, and thus full electrical conductivity. Others are semi-conductive and have electrical behavior similar to silica. Full conductive CNTs are more conductive than copper; therefore, they provide more productive results than copper. In addition to their electrical conductivity, CNTs have high thermal conductivity, two times greater than that of diamond. Moreover, their high surface area makes them a widely-preferred electrode material. CNTs are among the strongest and stiffest fiber structures; compared to steel, they are 500 times more robust despite being six times lighter than steel [20].

All these extraordinary properties provide CNTs with an extremely wide field of application, including electronic devices (wires, transistors, electric buttons, and memory storage devices), optoelectronic technologies (LED and laser technologies), sensors, field emission devices (scan electron microscopes), batteries and power cells, composites enhanced with fibers, biomarkers for cancer recognition, and pharmaceutical carriers [21].

Synthesis and Characterization of CNTs

There are three basic methods for the synthesis of CNTs. The first and the most widely used is the arc discharge method. In this method, first, an arc discharge is created between two carbon electrodes with or without a catalyst; then, nanotubes are self-assembled from the carbon vapor that emerge out of this process. The second method is chemical vapor deposition (CVD), in which a carbon source is put in the gas phase and then a certain amount of energy produced by plasma or heated coil is transferred to a reactive gaseous carbon molecule that grows on a catalyst (mainly Ni, Fe or Co). With this method, both SWCNT and MWCNT can be produced. The last method is laser ablation, which focuses on vaporization of a mixture of carbon and transition metals by directly applying laser to the metal-graphite composite. Compared to the arc discharge method, laser ablation produces fewer but cleaner nanotubes with greater control over growth conditions [22].

In order to analyze the morphological and structural characterization of nanotubes, several methods can be used, including scanning tunneling microscopy, transmission electronic microscopy, X-ray photoelectron spectroscopy, neutron and X-ray diffraction, and infrared and Raman spectroscopy [23].

Preparation of CNT-Based Electrodes

CNTs tend to gather together in almost all kinds of organic and aqueous solutions because of the van der Waals forces between tubes. This property is inherent in CNTs and makes it difficult to produce homogenous CNT composites, greatly hampering electroanalytical applications of CNTs. Moreover, CNTs are chemically inert; therefore, in order to use them as electrodes, they should be functionalized covalently and non-covalently. Those strategies can also be used for solubilizing CNTs and further facilitating their manipulations [24].

In terms of their preparation method, there are four types of CNT-based electrodes. The first group is called vertically aligned nanotube electrode arrays which have been used for designing electrochemical sensors. This type of electrodes are prepared either by assembling CNTs vertically on the substrate surface to investigate the electron transfer mechanism or through a self-assembly strategy, in which the end tip of shortened CNTs is first modified with functional groups (such as –COOH, -SH, $-NH_2$), and then nanotubes are attached to a substrate (such as gold, polymer) through a chemical reaction [17].

An earlier example of using vertically aligned CNT-based electrodes is a glucose biosensor. Lin *et al.* (2004) attempted to covalently immobilize glucose oxidase on CNT nanoelectrode ensembles by forming amide linkages between their amine residues and carboxylic acid groups on the CNT tips and achieved selective detection of glucose [25]. More recently, Fayazfar and coworkers (2014) developed a tantalum electrode modified with vertically aligned CNTs and gold nanoparticles to detect cefazolin. The authors synthesized the vertically aligned MWCNTs on Ni-deposited Ta plates *via* the CVD technique. Then, gold nanoparticles were electrodeposited on CNTs through chronoamperometry using a 5 mM $HAuCl_4$ acidic solution. Fold nanoparticles were used for filling the holes in the sidewall of the MWCNT, which increased the surface area of the electrode [26].

The second method for a CNT-based electrode design is layer-by-layer (LBL) fabrication, which is perceived as one of the simplest and most versatile methods [27]. In this method, the substrate is repeatedly and sequentially immersed into aqueous solutions of complementarily functionalized materials, which result in the formation of ultrathin films and highly tunable surfaces [28]. The advantages of LBL assembly are manifold. Firstly, it allows uniform dispersion of CNTs into a composite by direct adsorption of CNTs from a solution to a solid state without phase segregation. Secondly, LBL-assembled electrodes have tunable multifunctional properties. Finally, the process is simple, robust, and versatile, which also makes it economical [27]. Therefore, use of LBL-assembled CNT-based electrodes electrochemical sensing is favorable. Wang *et al.* (2016) designed an LBL-assembled CNT-based electrochemical sensor for nifedipine. They successfully assembled a uniformly distributed film consisting of polyaniline and carboxylic acid-functionalized MWNTs on indium tin oxide plates using the LBL method based on electrostatic interactions. The resulting product was a highly selective sensor without interferences from glucose, urea, ascorbic acid, and trisodium citrate [29].

In another study, Kong *et al.* (2009) applied LBL-assembled CNT-based electrodes for the determination of dopamine. In this work, multiple CNTs layers with molecule recognition function were constructed using the LBL self-assembly method. The CNTs were first dispersed by β-cyclodextrin (β-CD) in a water solution with the aid of sonication. At this step, β-CD interacted with CNTs via van der Waals forces to form a stable β-CD-CNT nanocomposite. Then, the β-CD-CNTs were coated with chitosan using LBL assembly to form multiple β-CD-CNTs layers. Also dopamine was selected for testing the electrocatalytic activity of β-CD-CNT multiple layers modified glassy carbon electrode (GCE). At (β-CD-CNTs/chitosan)^6/GCE, a linear relationship was obtained over the concentration range from 0.1 to 25 μM with a detection limit of 0.06 μM [30].

The third method for preparing CNT-based electrodes is coating the electrode surface with a CNT. Although this is a simple and effective method, there is a significant problem during the coating process. As a result of strong van der Waals forces between nanotubes, CNTs are poorly soluble in water and other organic solvents, such as dimethylformamide (DMF), acetone, acetonitrile, and ethanol. In order to overcome this problem, either CNTs are functionalized or CNT solutions or suspensions are non-covalently stabilized using surfactants or polymers [31].

CNT-coated electrodes have been widely preferred in pharmaceutical studies. Salimi *et al.* (2007) used an SWCNT-coated GCE for the simultaneous determination of ranitidine and metronidazole. The electrode was prepared by dispersing SWCNTs in dimethyl sulfoxide and dropping them onto the GCE. The result was a sensitive electrochemical sensor that can be used for amperometric determination of selected analytes as low as 70 nM with good reproducibility and little fouling effects of analytes and their reduction products [32]. Jain and Sharma (2012) used an MWCNT-coated GCE for the quantification of antihistaminic drug pheniramine. In preparing this electrode, the authors dispersed MWCNTs in DMF, cast the electrode surface with this suspension and left the electrode to dry at room temperature. They were able to reach very low detection limits with this electrode showing good sensitivity and repeatability [33].

The last method for CNT-based electrode preparation is assembling CNTs with other materials. The first CNT composite electrode was developed using bromoform as the binding material to analyze the oxidation of dopamine. In another study, to obtain good electrocatalytic activity towards dopamine, uric acid and ascorbic acid, researchers mixed CNTs with mineral oil [17]. In line with these earlier examples, new materials have been utilized to design CNT composite electrodes. This type of electrodes can be categorized based on the material mixed with CNTs to form the composite electrode.

One of these categories is polymer nanocomposite electrodes. Incorporating nanotubes into polymers for protective coating is highly preferable. In order to design these composites, efficient dispersion of individual CNTs and establishment of strong chemical affinity with the surrounding polymer matrix is essential. With the addition of a whole polymer chain, nanotube properties are enhanced *via* non-covalent or covalent binding between the CNT and the polymer. While non-covalent binding involves the physical adsorption and/or wrapping of polymers to the surface of CNTs, covalent binding is based on the establishment of strong chemical bonds between nanotubes and polymers. Polymer-CNT nanocomposites have been used for transparent conductive coating, electrostatic dissipation, electrostatic painting, and electromagnetic interference shielding applications. They are also widely used in designing electrochemical sensors and biosensors due to their optical clarity, enhanced conductivity, and low cost [34].

Takeda and coworkers (2016) developed a sensor using an MWCNT and a poly(allylamine hydrochloride) polymer-modified GCE for the electrochemical determination of levodopa and carbidopa. For the preparation of the electrode, the authors first of treated the MWCNT with strong acids to remove possible metallic impurities, open the MWCNT tips and add oxygenated groups onto the MWCNT surface. Then, optimum amounts of MWCNT and poly(allylamine hydrochloride) polymer were mixed and sonicated. Finally, an optimum amount of this dispersion was dropped onto the GCE, and the solvent was evaporated. These steps are schematically presented in Fig. (**4**) [35].

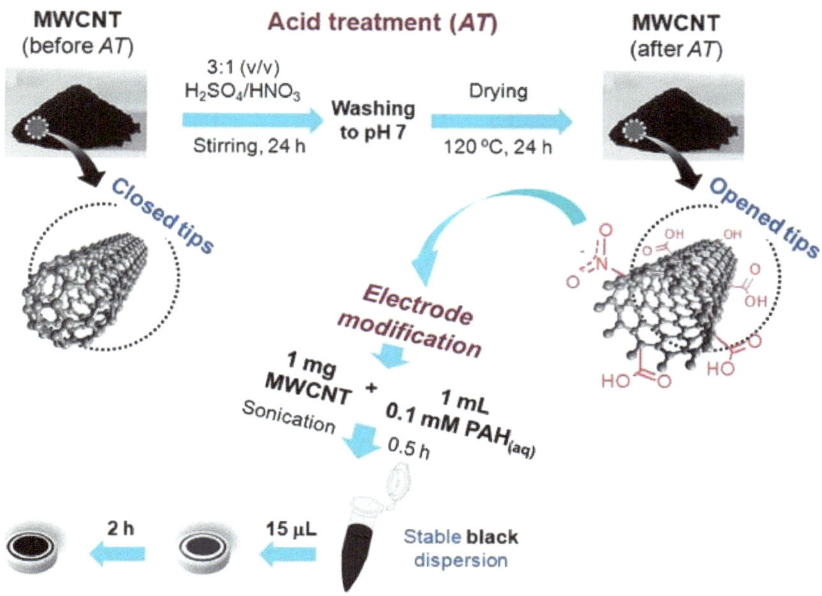

Fig. (4). Schematic representation of the steps of acid MWCNT functionalization and modification of the GCE surface with functionalized MWCNT within the poly(allylamine hydrochloride) polymer (Reprinted with permission from [35]).

Ghadimi *et al*. (2013) studied the electrochemical behaviors of paracetamol using MWCNTs-poly (4-vinylpyridine) and MWCNT composite film coated on a GCE. The results showed that the nanostructured film electrode exhibited much better electrochemical activity on the electrochemical oxidation of paracetamol compared to the bare GCE [36]. Shahrokhian *et al*. detected clozapine using a composite modified electrode composed of new coccine-doped polypyrrole and MWCNTs *via* linear sweep voltammetry (LSV). The detection limit was found to be 3.0 nM. The modified electrode demonstrated good sensitivity and stability, and further studies for pharmaceutical dosage forms and human plasma samples produced satisfactory results [37]. Nafion is another polymer used for electrode coating. A GCE modified with nafion/CNT was designed by Sanghavi *et al*. (2011) for the determination of venlafaxine and desvenlafaxine in pharmaceutical dosage forms and biological samples. Using adsorptive stripping differential pulse voltammetry, detection limits of 12.4 nM and 21.1 nM were obtained for venlafaxine and desvenlafaxine, respectively [38].

Metal nanoparticles constitute the second group of materials used for making composite electrodes by mixing with CNTs. Metal-CNT composites overcome the problem of insufficient energy density of CNTs, and electrochemical capacitance is improved by the addition of electroactive materials, such as metal oxides and metals (RuO_2, NiO, MnO_2, Ni) [39]. Compared to polymer-CNT composites, metal-CNT composites are more difficult to produce because not only do they require a high temperature and pressure, but they also need special equipment that isolate metal from the atmosphere to avoid oxidation. Moreover, CNTs might react with metals to form carbides, and hence be destroyed. Despite these difficulties, metal-CNT nanocomposites remain to be popular, particularly in automobile, aerospace and energy storage industries for their lightweight, high strength, low coefficient of thermal expansion, and high thermal conductivity [40].

Metal-CNT nanocomposites are also used in pharmaceutical studies. For example, Fayemi *et al*. (2015) determined dopamine *via* a GCE modified with MWCNTs doped with metal oxides (GCE/MWCNT/NiO, GCE/MWCNT/ZnO, GCE/MWCNT/Fe_3O_4). The authors used cyclic voltammetry (CV), electrochemical impedance spectroscopy and square wave voltammetry (SWV) techniques in a 0.1 M phosphate buffer solution (PBS) at pH 7.0. The results were promising since the MWCNTs remarkably improved the reactivity of NiO, ZnO and Fe_3O_4 for dopamine oxidation. The GC/MWCNT/NiO, GC/MWCNT/ZnO and GC/MWCNT/Fe_3O_4 electrodes exhibited good linear properties in the concentration range from 4×10^{-5} µM to 6.25 µM for the quantitative analysis of dopamine with a limit of detection of 7.99 pM, 0.3742 µM and 1.386 µM, respectively [41]. In another study by Baytak and coworkers (2016), a GCE modified with terbium oxide nanoparticles (Tb_4O_7NPs) and CNT composite were prepared for the simultaneous determination of methyldopa and paracetamol. Fig. (**5**) shows the different electrode responses of these compounds. The best response was obtained using the composite electrode [42].

The third group of materials used for designing CNT composite electrodes is ceramic materials. Although ceramic materials are more resistant to high temperatures and corrosion and have better biocompatibility compared to metal and polymer materials, their poor mechanical properties restrict their use. In order to improve their mechanical properties, ceramic materials are mixed with CNTs to form ceramic-CNT nanocomposites. There are different methods for fabricating CNTs/ceramic composites, including hot pressing, hot isostatic pressing-sintering, spark plasma sintering, microwave sintering, and high-temperature extrusion molding [43].

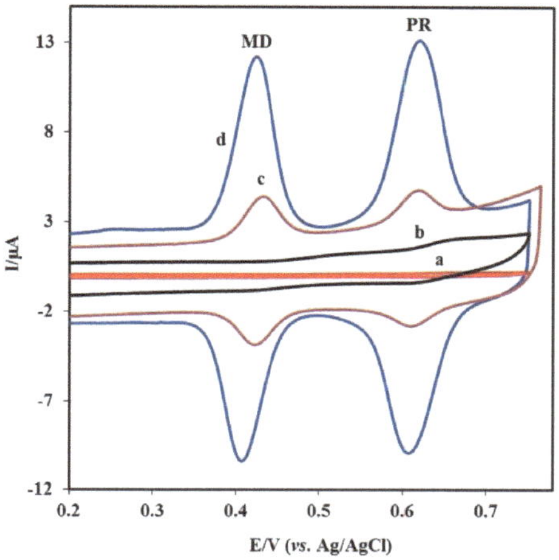

Fig. (5). Cyclic voltammograms of 8.0×10^{-7} M methydopa (MD) and 8.0×10^{-7} M paracetamol (PR) in 0.1 M PBS at pH 4.0 at bare GCE (a), Tb_4O_7NPs/GCE (b), CNTs/GCE (c), and Tb_4O_7NPs/CNTs/GCE (d). Scan rate: 50 mV s^{-1} (Reprinted with permission from [42]).

Zhu and coworkers (2007) designed ceramic-CNT composite electrodes by dispersion of MWCNT in a sol–gel solution and compared the electrochemical behavior of these electrodes to those of unmodified graphite electrodes in detecting ascorbic acid, uric acid, acetaminophenol, nicotinamide adenine dinucleotide, epinephrine, dopamine, cysteine and hydrogen peroxide using the CV technique. The results of the modified electrodes were better in terms of reversibility, peak shapes, sensitivity, and decreased over potential [44]. In another study, Salimi and coworkers (2006) designed a sol–gel carbon composite electrode by mixing a sol–gel precursor (*e.g.* methyltrimethoxysilane) and carbon powder. The authors concluded that this sensor was promising for simultaneous determination of ascorbic acid, dopamine, adrenaline, and uric acid. In addition to being highly sensitive, stable and simple, this sensor was shown to achieve simultaneous determination of uric acid, catecholamines and ascorbic acid without using any chromatography or separation systems [45].

Advantages and Limitations of CNT-Modified Electrodes

There are distinctive properties of CNTs that make them preferable as electrochemical sensor materials. One of the most important is the high surface area offered by CNTs. Although the specific surface area of CNTs seem to be lower than that of activated carbon or carbon fibers, CNT-based electrodes have a higher capacitance per unit surface area and lower internal resistance, properties that are favorable for use in pharmaceutical industry [46]. Secondly, CNTs have exceptional mechanical properties including tensile strength reaching 1000 GPa, which makes them the stiffest and strongest fibers ever produced [9]. Thirdly, the electrical conductivity of CNTs is quite high. Since CNTs have the ability to mediate electron transfer reactions with electroactive species, they are extensively used in electrochemical sensing as electrodes. Fourthly, it has been reported in the literature that CNT-based electrodes catalyze the redox reactions of analytes, including dopamine, ascorbic acid, and uric acid. Finally, CNTs can be functionalized in a way to immobilize the enzymes on them; therefore, it is possible to design different types of biosensors [17].

Despite these advantages, CNTs have certain limitations. One of the main disadvantages of CNTs is the lack of solubility in aqueous media, and to overcome this problem, the surface of CNTs is functionalized using different hydrophilic molecules, which improves their water solubility and biocompatibility. Another disadvantage of CNTs, particularly in biomedical applications, is their high toxicity. The high surface area of CNTs together with their toxicity may have harmful effects, particularly for the lungs, depending on the size of CNTs. It has been reported that particles under 100 nm generally threaten respiratory health conditions [47]. A third disadvantage is that CNTs are not homogenous in their size (both diameters and length), which could be a problem for the generation of reproducible results for the evaluation of biological activity related to specific structures [48].

Applications of CNTs in Pharmaceutical Analysis

In pharmaceutical analyses, CNT-based electrodes are developed by different techniques, including vertical alignment, LBL assembly, electrode surface coating and using different materials for designing composites, are widely-used. Table **1** summarizes recent studies from 2010 to 2017 and presents the developed electrodes, the techniques used, the supporting electrolyte studied, and the potential and limit of detection values measured.

Table 1. Selected analytical studies on drugs using CNT modified sensors.

Drugs	Matrix sample prep.	Method	Electrode	Buffer	Measured potential	Validation data (LOD)	Ref.
Sulfasalazine	Tablets and serum	CV, AdSSWV	BiNP-CNT/GCE	0.04 M BR buffer (pH 3.5)	-0.33 V (CV)	13 nM	[49]
Diazepam	Tablets, serum and urine	CV, DPV	C_{60}-CNT/IL/GCE	0.1 M PBS (pH 7.0)	-0.87 V (CV)	87 nM	[50]
Nilutamide	Tablets and human blood serum	CV, DPV	*f*MWCNT/GCE	0.05 M PBS (pH 7.0)	-0.48 V (CV)	0.2 nM	[51]
Tramadol	Tablets and human urine	CV, EIS, SWV	Polypyrrole@Sol-Gel MIP/ *f*MWCNT/GCE	pH 7.0 PBS	~0.2 V	0.03 nM	[52]
Acetaminophen	Tablets	CV, EIS, SWV	*f*MWCNT-CoPc/AuNPs/GCE	pH 5.0 McIlvaine buffer	0.48 V (SWV)	0.135 μM	[53]

(Table 1) contd.....

Drugs	Matrix sample prep.	Method	Electrode	Buffer	Measured potential	Validation data (LOD)	Ref.
Atorvastatin	Tablets, serum and urine samples	Potentiometry	ATR-CTAB-MWCNTs-BMIMPF$_6$ /Graphene paste electrode	0.1 M acetate buffer (pH 5.5)	-	1.0 nM	[54]
Diclofenac	Tablets and urine samples	CV, DPV, EIS, chronoamperometry	MWCNTs/CoHCF/ IL paste electrode	0.1 M universal buffer (pH 7.0)	0.53 V (CV)	0.3 μM	[55]
Doxorubicin	Urine samples	CV, DPV	Fe_3O_4@Pt/MWCNT/CPE	0.1 M PBS (pH 8.0)	0.37 V (DPV)	1.0 nM	[56]
Ethambutol	Human urine and blood serum samples	CV, EIS, SWV	Nafion/MWCNT-SPCE	Citrate-phosphate buffer (pH 7.4)	~ 1.0 V (SWV)	8.4×10^{-4} mg μL^{-1}	[57]
Ganciclovir	Serum and urine samples	CV, SWV	Fe_3O_4/COOH-MWCNTs/GCE	0.1 M PBS (pH 7.4)	0.93 V (CV)	20 nM	[58]
Isoproterenol	Pharmaceutical formulations and urine	chronoamperometry, CV, SWV	Pyrogallol red/MWCNTPE	0.1 M PBS (pH 7.0)	0.277 V (CV)	0.47 μM	[59]
Levodopa (LD) Carbidopa (CD)	Tablets and human serum samples	CV, DPV	MWCNT-PAH/GCE	0.1 M NaCl	-0.024 V (LD) 0.49 V (CD) (CV)	0.84 μM (LD) 0.65 μM (CD)	[35]
Linezolid	Tablets, urine and saliva	CV, DPV, EIS, chronoamperometry	MWCNT/ bromocresol green/CPE	0.04 M BR buffer (pH 7.0)	0.805 V (DPV)	7.57 nM	[60]
Mesalazine N-acetylated metabolite (simultaneous determination)	Human serum samples	CV, AdSSWV	CNTs-Nafion/GCE	0.04 M BR buffer (pH 2.0)	0.53 V 0.74 V (AdSSWV)	12 nM 26 nM	[61]
Mebendazole	Tablets	CV, DPV, AdSSWV	MWCNT/GCE	0.1 M PBS (pH 6.0)	0.960 (CV)	19 nM	[62]
Methyldopa (MD) Paracetamol (PM) (simultaneous determination)	Tablets and urine samples	CV, SWV	Tb_4O_7NPs/CNTs/GCE	0.1 M PBS (pH 4.0)	0.41 V (MD) 0.62 V (PM) (CV)	1.5 nM (MD) 0.33 nM (PM)	[42]
Morphine (MO) Codeine (CO) (simultaneous determination)	Pharmaceutical and urine samples	CV, DPV, chronoamperometry	MWCNTs/SnO_2–Zn_2SnO_4/CPE	0.1 M PBS (pH 6.0)	0.5 V (MO) 1.05 V (CO) (DPV)	9 nM (MO and CO)	[63]
Paracetamol	Human blood serum, plasma and urine	CV, EIS, SWV, chronocoulometry	CuNPs/fullerene-C_{60}/MWCNTs/CPE	0.1 M PBS (pH 6.8)	~ 0.4 V (SWV)	73 pM	[64]
Paracetamol (PM) Cetirizine (CTZ) Phenylephrine (PHE) (simultaneous determination)	Pharmaceutical formulations, blood serum and urine samples	CV, EIS, AdSDPV	MWCNT-PtNPs-CPE	0.1 M PBS (pH 5.5)	~ 0.4 V (PM) ~ 0.7 V (CTZ) ~ 0.9 V (PHE) (AdSDPV)	27.9 nM (PM) 58.6 nM (CTZ) 28.3 nM (PHE)	[65]
Pramipexole	Tablets	CV, DPV, AdSDPV	*f*MWCNT/GCE	0.2 M PBS (pH 3.0)	0.801 V (AdSDPV)	0.106 nM	[66]
Ropinirole	Tablets	CV, AdSDPV, AdSSWV	*f*MWCNT/Fe_2O_3/CPE	0.2 M PBS (pH 6.0)	0.976 V (AdSDPV) 1.049 V (AdSSWV)	1.30×10^{-3} μg mL^{-1} (AdSDPV) 4.08×10^{-4} μg mL^{-1} (AdSSWV)	[67]
Sertraline	Human urine samples	DPV	Ni (II)–Levodopa/AuNPs/MWCNT/GCE	0.1 M NaOH	0.02 V	95 nM	[68]
Sorafenib	Human serum samples	CV, AdSDPV	Chitosan/-COOH MWCNT/GCE	0.1 M PBS (pH 7.0)	0.77 V (AdSDPV)	0.44 nM	[69]
Sulfamethoxazole (SMX) Trimethoprim (TMP) (simultaneous determination)	Artificial urine samples	DPV	MWCNT/PBnc/SPE	0.1 M BR buffer (pH 7.0)	0.575 V (SMX) 0.740 V (TMP)	38 nM (SMX) 60 nM (TMP)	[70]
Sulfamethoxazole	Tablets	LSV, DPV, SWV, amperometry	Carbon nanopowder paste electrode	0.05 M PBS (pH 6.0)	0.936 V (SWV)	0.12 μM (SWV)	[71]
Terbutaline	Pharmaceutical formulations	CV, SWV	ZrO_2NPs/MWCNTs/GCE	0.1 M PBS (pH 7.0)	0.645 V (CV)	2.25 nM	[72]

(Table 1) contd.....

Drugs	Matrix sample prep.	Method	Electrode	Buffer	Measured potential	Validation data (LOD)	Ref.
Valganciclovir	Tablets and blood serum samples	CV, EIS	AuNP-MIP/MWCNT/GCE	0.01 M PBS (pH 3.0)	~ 1.05 V (CV)	0.3 nM	[73]
5-fluorouracil	Pharmaceutical dosage forms and artificial urine	CV, EIS, DPV	AuNP-MWCNT-Chitosan/GCE	0.04 M BR buffer (pH 8.0)	1.10 V (CV)	20 nM	[74]
Bicalutamide	Pharmaceutical formulations, human blood and urine samples	CV, DPV	SWCNT/CPE	0.1 M PBS (pH 7.0)	-0.84 V (DPV)	52 nM	[75]
Cisplatin	Human serum samples	CV, DPV	MWCNT-COOH/SDS/SPCEs	0.1 M NaCl (pH 7.0)	-0.25 V (CV)	4.6 μM	[76]
Hydroxychloroquine	Pharmaceutical formulations	CV, AdSDPV, chronocoulometry, chronoamperometry	MWCNTs/CPE	0.2 M PBS (pH 8.0)	~ -0.6 V (AdSDPV)	6 nM	[77]
Mefenamic acid	Capsules and urine samples	CV, EIS, DPV	Magneto MWCNT/CPE	pH 5.0 BR buffer	~ 0.7 V (DPV)	1.2 nM	[78]
Methimazole	Blood serum samples	CV, Amperometry	Rutin/MWCNTs/GCE	0.1 M PBS (pH 8.0)	0.216 V (CV)	18 nM	[79]
Nitroxoline	Tablets and human serum	CV, AdSLSV, EIS, AdSSWV	MWCNT/CPE	pH 2.0 BR buffer	~ 0.15 V (AdSSWV)	15 pM (AdSSWV) 30 pM (AdSLSV)	[80]
Norfloxacin	Urine samples	SWV	MIP/MWCNT/GCE	0.1 M H_2SO_4	1.05 V	46 nM	[81]
Paracetamol	Tablets	CV, SWV	WO_3NPs/CNTs/GCE	0.1 M PBS (pH 7.0)	0.411 V (CV)	55.4 nM	[82]
Abacavir	Pharmaceutical dosage forms and human serum	CV, AdSDPV	MWCNT/EPPGE	0.1 M H_2SO_4	0.96 V (peak 1) 1.15 V (peak 2) (AdSDPV)	23 nM (peak 1) 7.9 nM (peak 2)	[83]
Acenocoumarole	Tablets	CV, SWV	Bi_2O_3-MWCNTs/GCE	pH 3.8 BR buffer	~ -0.5 V (SWV)	0.4 ng mL^{-1}	[84]
Amitriptyline	Tablets	CV, DPV	Unmodified CNTPE	0.1 M H_2SO_4	1.35 V (CV)	1.61 μM	[85]
Atenolol	Pharmaceutical samples, human urine and blood serum	CV, DPV, chronoamperometry	AuNPs/MWCNTs/GCE	0.04 M BR buffer (pH 10.0)	0.725 V (CV)	0.5 μM (DPV) 59 nM (chronoamperometry)	[86]
Amlodipine	Pharmaceutical samples and biological fluids	CV, DPV	PLM-Au NPs/MWCNTs/GCE	pH 7.12 PBS	0.55 V (CV)	1 nM	[87]
Buspirone hydrochloride	Pharmaceutical sample	CV, DPV, amperometry	MWCNT/GCE	0.05 M PBS (pH 7.0)	0.85 V (CV)	0.22 μM	[88]
Cefazolin	Pharmaceutical sample	CV, LSV	Well aligned-MWCNTs/AuNPs/TaE	0.1 M BR buffer (pH 4.8)	-1.1 V (CV)	1 pM	[26]
Codeine (CO) Caffeine (CF)	Tablets	CV, DPV	SWCNT/CCE	0.01 M H_2SO_4 (pH 1.7)	1.05 V (CO) 1.38 V (CF) (CV)	0.11 μM (CO) 0.25 μM (CF)	[89]
Isoniazid	Tablets and artificial urine samples	CV, EIS, DPV	MWCNT–Chitosan/GCE	0.04 M BR buffer (pH 6.0)	0.15 V (DPV)	55 nM	[90]
Morphine (in the presence of diclofenac)	Pharmaceutical samples and human urine	CV, SWV	IL/NiO/CNTCPE	0.1 M PBS (pH 7.0)	0.5 V (SWV)	0.01 μM	[91]
Mycophenolate mofetil (MPM) Mycophenolic acid (MPA) (simultaneous)	Plasma and urine samples	CV, DPV	MWCNTs/GCE	0.1 M PBS (pH 5.0)	0.87 V and 1.1 V (MPM) 0.87 V (MPA) (DPV)	0.9 μM (MPM) 0.4 μM (MPA)	[92]
Pemetrexed	Pharmaceutical dosage forms	CV, AdSDPV	Over oxidized- polypyrrol /MWCNTs-COOH/GCE	0.1 M PBS (pH 2.0)	0.390 V (AdSDPV)	3.28 nM	[93]
Ropinirole (in the presence of levodopa)	Tablets and human serum samples	CV, AdSSWV	CNT-Nafion/GCE	0.1 M H_2SO_4	1.24 V (AdSSWV)	16 nM 52 nM (Levodopa)	[94]

(Table 1) contd.....

Drugs	Matrix sample prep.	Method	Electrode	Buffer	Measured potential	Validation data (LOD)	Ref.
Sulfacetamide	Tablets	CV, SWV	SWCNT/p-DAN modified EPPGE	1.0 M PBS (pH 7.2)	0.85 V (SWV)	0.11 μM	[95]
Tadalafil	Tablets and human serum	CV, AdSSWV	TiO_2-MWCNTPE	0.04 M BR buffer (pH 3.0)	1.0 V (AdSSWV)	0.27 μM	[96]
Zidovudine	Human plasma	CV, LSV, SWV, amperometry	Ag-nanofilm/M-MWCNT/GCE	0.1 M PBS (pH 7.0)	-0.65 V (CV)	0.15 μM (amperometry)	[97]
Cefexime	Human plasma	SWV	MIP/Fe@AuNPs/2-AET-*f*-MWCNTs/AP/GCE	0.1 M PBS (pH 6.0)	0.95 V	22 pM	[98]
Almotriptan maleate	Tablets	CV, DPV	MWCNT/GCE	0.2 M PBS (pH 7.0)	0.681 V (CV)	69.83 nM	[99]
Methotrexate	Tablets	CV, AdSDPV	MWCNTs-DHP/GCE	0.1 M H_2SO_4	0.75 V (CV)	33 nM	[100]
Dopamine	Dopamine hydrochloride injection and human blood serum	CV, DPV	MWCNT/poly(glycine)/CPE	0.1 M PBS (pH 7.0)	0.175 V (CV)	12 nM	[101]
Epinephrine	Serum and pharmaceutical samples	CV, DP anodic stripping voltammetry	CNT-mer dispersed MIP-modified pencil graphite electrode	0.1 M PBS (pH 6.8)	0.05 V (CV)	0.02 ng mL^{-1}	[102]
Tramadol	Human urine and tablets	CV, SWV	Nano-MIP/MWCNT/CPE	pH 7.0 BR buffer	0.82 V (CV)	4.0 nM	[103]
Ascorbic acid (AA) Adrenaline (AD) Acetaminophene Tryptophan	Pharmaceutical samples	CV, DPV	Imidazole derivative MWCNT-GCE	0.15 M PBS (pH 7.0)	0.025 V (AA) 0.181 V (AD) (CV)	0.96 μM (AA) 0.38 μM (AD)	[104]
Norepinephrine	Drug and urine	CV, SWV, chronoamperometry	MWCNTs/Carbon IL electrode	0.1 M PBS (pH 7.0)	0.36 V (CV)	0.08 μM	[105]
Metoprolol	Tablets and biological fluids	CV, DPV, AdSDPV	Nafion-CNT-GCE	pH 7.0 PBS	1.23 V (DPV)	35.1 nM	[106]
Cefixime	Tablets, urine and blood plasma samples	CV, LSV, EIS	$NiFe_2O_4$ -MWCNTs modified GCE	0.10 M PBS (pH 8.0)	0.85 V (CV)	0.02 μM	[107]
Azithromycin	Capsules, blood plasma and urine samples	DPV, EIS, amperometry	MWCNT decorated with $MgCr_2O_4$ modified GCE	0.1 M PBS (pH 8.0)	0.72 V (DPV)	0.07 μM	[108]
Zolmitriptan	Tablets and human urine	CV, DPV, AdSDPV	AgNPs/MWCNT modified GCE	0.1 M PBS (pH 7.0)	0.635 V (AdSDPV)	1.47 nM	[109]
Paracetamol	Tablets and urine samples	CV, DPV, EIS	Poly(4-vinylpyridine) MWCNT/GCE	0.1 M PBS (pH 7.0)	E_{pa}= 0.404 V E_{pc}= 0.307 V (CV)	1.69 nM	[36]
Insulin	Pharmaceutical dosage forms, blood serum samples	CV, amperometry, chronoamperometry	NiONPs/Nafion-MWCNTs/SPE	0.10 M NaOH solution	-	6.1 nM	[110]
Nifedipine	Capsule and blood serum	CV, DPV	MWCNT-β-CD modified CPE	0.05 M NaOH solution	–0.9 V (CV)	25 nM	[111]
Gemfibrozil	Tablets and urine samples	CV, AdSSWV	GCE modified with MWCNT within DHP	0.1 M PBS (pH 2.0)	1.31 V (CV)	53 nM	[112]
Clozapine	Tablets and human plasma samples	CV, LSV	MWCNTs/New Coccine Doped Polypyrrole modified GCE	0.04 M BR buffer (pH 5.44)	~ 0.5 V	3.00 nM	[37]
Methadone Acetaminophen (ACOP)	Tablets, saliva and human urine	CV, DPV	*f*MWCNT/GCE	0.1 M PBS (pH 7.0)	0.62 V (CV) 0.34 V (ACOP) (DPV)	0.28 μM 0.35 μM (ACOP)	[113]
Olanzapine	Tablets and blood serum samples	CV, SWV, chronoamperometry	Amine-functionalized TiO_2-MWCNT modified GCE	pH 5.0 PBS	0.35 V (SWV)	0.09 μM	[114]
Trifluoperazine	Tablets	CV, DPV, AdSDPV	MWCNT/GCE	0.2 M PBS (pH 3.0)	1.03 V (AdSDPV)	0.749 nM	[115]
Irinotecan	Injectable dosage form	CV, DPV	PMB/MWCNT modified GCE	0.04 M BR buffer (pH 10.0)	0.6 V (DPV)	0.214 μM	[116]

(Table 1) contd.....

Drugs	Matrix sample prep.	Method	Electrode	Buffer	Measured potential	Validation data (LOD)	Ref.
Idarubicine	Pharmaceutical dosage forms	CV, AdSDPV	MWCNT/GCE MWCNT/EPPGE	0.2 M PBS (pH 3.0)	0.636 V 0.629 V (AdSDPV)	18.7 nM 37.5 nM	[117]
Etoposide	Injectable dosage forms	CV, AdSDPV	MWCNT/GCE	0.04 M BR buffer (pH 6.0)	0.42 V (AdSDPV)	5.4 nM	[118]
Valganciclovir	Tablets	CV, DPV, AdSDPV	MWCNT/GCE	0.04 M BR buffer (pH 4.0)	1.027 V (AdSDPV)	1.52 nM	[119]
Norfloxacin	Tablets and urine	CV, SWV	MWCNT modified pyrolytic graphite electrode	1.0 M PBS (pH 2.0)	−1.285 V (SWV)	40.6 nM	[120]
Oleanic acid	Capsule	CV, EIS, SWV	MIP/CoHCNFe/ MWCNTs/PSS/Chitosan/CE	0.2 M PBS (pH 7.0)	0.28 V (SWV)	2.0 nM	[121]
Ascorbic acid (AA) Paracetamol (PM)	Pharmaceutical formulations	CV, DPV, amperometry	GCE/MWCNT–Polyhistidine	0.05 M PBS (pH 7.4)	−0.012 V (AA) 0.385 V (PM) (CV)	0.76 μM (AA) 32 nM (PM)	[122]
Furosemide	Tablets and human urine	CV, DPV	MWCNTPE	0.04 M BR buffer (pH 5.0)	i_{pa1} = 1.09 V i_{pa2} = 1.26 V (CV)	0.29 μM	[123]
Pentoxifylline	Tablets and human urine	CV, DPV	MWCNTPE	0.2 M PBS (pH 3.0)	1.36 V (CV)	0.169 μM	[124]
Theophylline	Tablets and human urine	CV, DPV	MWCNTPE	0.2 M PBS (pH 3.0)	1.24 V (CV)	19.7 nM	[125]
Tinidazole	Tablets and human plasma	CV, LSV	AuNP/CNT/GCE	0.1 M PBS (pH 7.0)	-0.565 V (LSV)	10 nM	[126]
Carbidopa	Urine samples	CV, SWV	Chloranile modified CNTPE	0.1 M PBS (pH 7.0)	0.12 V (CV)	71 nM	[127]
Amiloride	Tablets, urine and plasma	CV, EIS, AdSDPV	Nafion/CNT/GCE	pH 2.0 BR buffer	1.11 V (CV)	7.02 nM	[128]
Atropine	Ophthalmic solution	CV, DPV	MWCNTE	pH 10.5 tetramethyl ammonium hydroxide (in the presence of 0.4×10^{-4} M sodium dodecyl benzene sulfonate)	0.62 V (CV)	0.449 ng mL^{-1}	[129]
Methimazole	Tablets and human urine	CV, DPV	MWCNTs-Salen-Co(III)-MIP/GCE	0.2 M PBS (pH 6.5)	~ 0.4 V (DPV)	0.048 mg L^{-1}	[130]
6-mercaptopurine (in the presence of uric acid)	Tablets and human urine	CV, DPV, chronoamperometry, EIS	TiO_2/p-APMCNTPE	0.1 M universal buffer (pH 9.0)	0.11 V (CV)	0.065 μM	[131]
Levofloxacin	Capsules	DPV, LSV, EIS	Poly(o-aminophenol)/MWCNTs/GCE	pH 4.5 acetate buffer	~ 1.0 V (DPV)	1 μM	[132]
Amoxicillin	Capsule, urine and blood plasma samples	EIS, DPV, CV	$FeCr_2O_4$–MWCNTs modified GCE	0.1 M PBS (pH 7.5)	~ 0.6 V (DPV)	0.05 μM	[133]
Methyldopa (in the presence of uric acid)	Tablet, urine and serum samples	CV, SWV, chronoamperometry, EIS	p-chloranil/CNTPE	0.04 M BR buffer (pH 9.0)	0.045 V (SWV)	0.2 μM	[134]
Diclofenac	Ampoule, tablet, human and fish blood serum	CV, DPV, chronoamperometry	MWCNTs/$Cu(OH)_2$ NPs/IL/GCE	0.1 M PBS (pH 8.2)	0.53 V (DPV)	0.04 μM	[135]
L-Dopa	Tablets	CV, SWV	$Co(DMG)_2ClPy$/MWCNT/BPPGE	0.1 M PBS (pH 6.4)	0.18 V (SWV)	0.86 μM	[136]
Carbidopa (in the presence of folic acid and tryptophan)	Urine	CV, LSV, SWV, chronoamperometry	ferrocene dicarboxylic acid modified CNTPE	pH 5.0 PBS	0.5 V (CV)	65.0 nM	[137]
Gatifloxacin	Biological fluids	CV, DPV	MWCNTPE	pH 5.0 Acetate buffer	1.05 V (DPV)	4.5 nM	[138]
Phenobarbital Acetaminophen (ACOP)	Human urine and pharmaceutical samples	CV, DPV, chronoamperometry	PtNPs/MWCNTPE	0.1 M PBS (pH 7.0)	0.609 V 0.470 V (ACOP) (CV)	0.10 μM 0.17 μM	[139]

(Table 1) contd.....

Drugs	Matrix sample prep.	Method	Electrode	Buffer	Measured potential	Validation data (LOD)	Ref.
Cilostazol	Tablets and human plasma	CV, SWV	Bi_2O_3–MWCNTs/GCE	pH 2.5 BR buffer	~ 0.9 V (SWV)	0.76 µg mL^{-1}	[140]
Ephedrine Pseudoephedrine	Urine and pharmaceutical preparations	CV, DPV	MWCNT/GCE	pH 12.0 PBS	~ 0.45 V (DPV) 0.466 V (CV)	0.75 µM (DPV) 0.82 µM (CV)	[141]
Amlodipine besylate	Tablets	CV, SW anodic stripping voltammetry	oxidized-MWCNT/AuE	pH 11.0 PBS	0.63 V (SW anodic stripping voltammetry)	13.80 µg mL^{-1}	[142]
Cefpirome	Pharmaceutical formulations	CV, DPV, SWV	MWCNT/GCE	pH 4.51 BR buffer (presence of 1.0% CTAB)	~ -1.35 V (DPV)	0.647 µg mL^{-1}(SWV) 5.540 µg mL^{-1} (DPV)	[143]
Acetaminophen (ACOP) Ascorbic acid (AA)	Pharmaceutical and biological samples	DPV, CV	SWCNT/CCE	0.1 M PBS (pH 7.0)	0.39 V (ACOP) (CV) 0.07 V (AA) (CV)	0.12 µM (ACOP) 3.0 µM (AA)	[144]
Sumatriptan	Tablets and synthetic serum	CV, DPV	cobalt methyl-salophen complex and MWCNT modified CPE	0.1 M acetate buffer (pH 5.0)	0.94 V (CV)	0.3 µM	[145]
Sulfamethoxazole	Tablets	CV, DPV, chronoamperometry, amperometry	CNTPE	0.1 M KCl and 0.1 M acetate buffer (pH 6.0)	0.9 V (DPV)	0.1 µg mL^{-1}	[146]
Dopamine (in the presence of ascorbic acid)	Human blood serum	CV, SWV	MWCNTs/CILE	0.1 M PBS (pH 7.0)	0.2 V (SWV)	0.01 µM	[147]
Paracetamol	Tablets	CV	MWCNTs/GCE	0.1 M PBS (pH 6.5)	0.41 V	20 nM	[148]
Levodopa (LD) Carbidopa Tryptophan	Pharmaceutical, water and urine samples	CV, LSV, DPV, chronoamperometry	EBNBHCNPE	0.1 M PBS (pH 7.0)	0.46 V (LD) (DPV)	94 nM	[149]
Captopril	Tablets and human patient urine	CV, chronoamperometry, EIS, DPV	DHPBMCNTPE	0.1 M PBS (pH 8.0)	~ 0.3 V (CV)	34 nM	[150]
Methocarbamol	Human plasma samples	CV, DPV	MWCNT/GCE	0.1 M PBS (pH 3.0)	~ 1.22 V (CV)	0.025 µg mL^{-1}	[151]
Midecamycin	Tablets	CV	MWCNT/AuE	0.1 M PBS (pH 7.0)	0.64 V	0.5 µM	[152]
Benzylpenicillin	Human blood plasma	CV, DPV	MIP/Fe_3O_4@SiO_2–MWCNTs–CTS/CE	0.2 M PBS (pH 7.0)	0.28 V (DPV)	1.5 nM	[153]
Dopamine	Drug samples	CV, DPV	SWCNT/GCE	0.1 M PBS (pH 7.5)	~ 0.1 V (DPV)	0.021 ppm	[154]
Paracetamol Epinephrine	Human urine	CV, SWV	AuNP/MWNT/ITOE	1 M PBS (pH 7.2)	0.508 V (CV) 0.3 V (CV)	4.6 nM 4.2 nM	[155]
Isoproterenol Uric acid	Ampoule and urine samples	CV, EIS, DPV, chronoamperometry	p-chloranil/CNTPE	0.04 M universal buffer (pH 10.5)	0.1 V (Isoproterenol) (CV)	0.009 µM 2.3 µM	[156]
Acetaminophen	Pharmaceutical samples	CV, EIS, SWV, chronocouloumetry	MWCNT/ alumina-coated silica/GCE	0.1 M BR buffer (pH 9.0)	0.32 V (CV)	0.05 µM	[157]
Venlafaxine (VEN) Desvenlafaxine (DES)	Pharmaceutical formulations, urine and blood serum samples	CV, AdSDPV, EIS, chronocouloumetry	Nafion/CNT/GCE	0.05 M BR buffer (pH 7.0) for VEN 0.1 M acetate buffer (pH 5.0) for DES	0.78 V (VEN) 0.72 V (DES) (AdSDPV)	12.4 nM (VEN) 21.1 nM (DES)	[38]
Gemifloxacin	Tablets	CV, SWV, DPV, coulometry	MWCNT/GCE	pH 2.0 PBS (presence of CTAB)	1.29 V (CV)	0.90 ng mL^{-1} (SWV) 3.0 ng mL^{-1} (DPV)	[158]
Cetirizine	Tablets and urine samples	CV	MWCNT/GCE	0.2 M PBS (pH 3.0)	1.08 V	70.7 nM	[159]
Roxithromycin	Capsules	CV, coulometry	SWCNT/GCE	0.1 M PBS (pH 7.0)	0.86 V (CV)	0.5 µM	[160]
Furazolidone	Pharmaceutical formulations	CV, chronoamperometry	MWCNT/GCE	0.04 M BR buffer (pH 6.0) and 30% DMF solution	-0.45 V (CV)	2.30 µM	[161]

(Table 1) contd.....

Drugs	Matrix sample prep.	Method	Electrode	Buffer	Measured potential	Validation data (LOD)	Ref.
Norepinephrine (in the presence of acetaminophen and tryptophan)	Ampoule	CV, DPV, chronoamperometry	DDP/CNTPE	0.1 M PBS (pH 7.0)	0.195 V (DPV)	77 nM	[162]
Amlodipine besylate	Tablets and human urine sample	CV, SWV	SWCNT/EPPGE MWCNT/EPPGE	1.0 M PBS (pH 7.2)	0.524 V 0.605 V (SWV)	1.0 nM 5.0 nM	[163]
Dopamine (DA) Acetaminophen (ACOP)	Pharmaceutical formulations and human serum	CV, DPV	*f*MWCNTs/GCE	0.1 M PBS (pH 8.0)	0.125 V (DA) 0.307 V (ACOP) (DPV)	0.8 μM (DA) 0.6 μM (ACOP)	[164]
Carbamazepine	Tablets and wastewater samples	CV, LSV	MWCNTs/GCE	0.1 M PBS (pH 6.89)	1.09 V (CV)	40 nM	[165]
Acetaminophen (ACOP) Aspirin (ASA) Caffeine (CF)	Pharmaceutical formulations, urine and blood serum samples	CV, EIS, AdSDPV, chronocouloumetry	ISSM-MWCNTPE	0.1 M PBS (pH 7.0) (presence of TX 100)	0.30 V (ACOP) 0.85 V (ASA) 1.27 V (CF) (AdSDPV)	25.8 nM (ACOP) 84.7 nM (ASA) 88.3 nM (CF)	[166]
Captopril	Tablets and patient human urine	CV, DPV	FMCAMCNTPE	0.1 M PBS (pH 7.0)	0.47 V (CV)	0.3 μM	[167]
Dopamine	Dopamine hydrochloride injection	CV, EIS, SWV	EPPGE/MWCNT/ Fe_2O_3	0.1 M PBS (pH 7.0)	0.2 V (SWV)	0.33 μM	[168]
Penicillamine (PA) Uric acid (UA)	Tablets and urine	CV, DPV, EIS	p-APMCNTPE/TiO_2/PE	0.1 M PBS (pH 6.0)	0.18 V (PA) (CV)	0.1 μM (PA) 1.1 μM (UA)	[169]
Diclofenac	Pharmaceutical, urine and blood samples	CV, SWV	SWCNT/EPPGE	1.0 M PBS (pH 7.2)	~0.439 V and ~0.854 V (SWV)	0.82 nM (peak 1) 22.5 nM (peak 2)	[170]
Ciprofloxacin	Urine and serum samples	CV, CA	MWCNT/GCE	pH 7.0 PBS	0.97 V (CV)	6.0 μM	[171]
Isoproterenol	Ampoule and urine	CV, EIS, DPV, chronoamperometry	FMAMCNTPE	0.1 M PBS (pH 5.0)	0.52 V (CV)	0.2 μM	[172]
Dexamethasone	Pharmaceutical preparations and human urine	CV, SWV	SWCNT/EPPGE	0.1 M PBS (pH 7.2) (presence of CTAB)	-1.195 V (SWV)	0.91 nM	[173]
Ascorbic acid (AA) Dopamine (DA)	Dopamine injections and human urine	CV, LSV, amperometry	Polymerized acid yellow 9/ MWCNTs/GCE	0.1 M PBS (pH 7.2)	-0.012 V (AA) 0.20 V (DA) (CV)	0.2 μM (AA) 0.1 μM (DA)	[174]
Thiamazole	Pharmaceutical formulations, human serum and urine	CV, DPV, LSV	MWCNT/GCE	0.1 M PBS (pH 6.86)	0.36 V (DPV)	30 nM	[175]
Acetylsalicylic acid	Pharmaceutical formulations	CV, SWV	MWCNT/ alumina-coated silica/GCE	0.1 M H_2SO_4 (pH 1.0)	1.090 V (CV)	3.77 μM	[176]
Dopamine	Dopamine hydrochloride injection	CV, SWV, EIS	Fe_2O_3/SWCNT/EPPGE	0.1 M PBS (pH 7.0)	0.2 V (SWV)	0.36 μM	[177]
Salbutamol	Pharmaceutical formulations	CV, SWV	MWCNT/GCE	pH 7.0 PBS	0.6 V (SWV)	0.2 μM	[178]
Betamethasone sodium phosphate	Tablets and urine samples of pregnant women	CV, SWV	SWNTs/CTAB/EPPGE	1.0 M PBS (pH 7.2)	-1.062 V (SWV)	0.25 nM	[179]
Testosterone Epitestosterone	Urine samples	SWV	SWCNT/EPPGE	1 M PBS (pH 7.2)	-1.152 V -1.338 V	2.8 nM 4.1 nM	[180]
Ascorbic acid (AA) Acetaminophen (ACOP) Isoniazid (INZ)	Tablets and human blood serum	CV, DPV	Thionine immobilized MWCNT/CPE	0.1 M acetate buffer (pH 4.0)	0.413 V (AA) 0.686 V (ACOP) 1.028 V (INZ) (CV)	0.3 μM (AA) 0.05 μM (ACOP) 0.5 μM (INZ)	[181]

GRAPHENE

The Structure and Properties of Graphene

Graphene is a two-dimensional carbon allotrope, composed of carbon atoms forming six-membered rings. It has a thickness of one carbon atom. Although it is theoretically a single-layer material, bi-layer or multi-layer forms can also be produced. As other fullerenes, the graphene structure starts with six carbon atoms chemically bound together tightly in a way to form a regular hexagonal lattice [13, 182, 183].

The properties of graphene make it a fascinating material for various fields of science. The two dimensionality of graphene means that this material has the highest surface-to-volume ratio. Although it is expected to be perfectly flat, there are ripples on its surface because of thermal fluctuations increasing the surface area of this material [182]. Another significant property of graphene is its extreme mechanical strength that is higher than that of any known material in the world. Moreover, graphene has significant thermal and electrical conductivity, which makes it preferable for use in the design of electrochemical sensors. Its tremendous transparency and sheet resistivity has resulted in its use in transparent electrode applications [183].

These valuable physical and electrochemical properties contributed to the use of graphene extensively in various fields. In industrial applications, graphene is used in the design of hydrogen storage devices and development of lithium-ion batteries. Moreover, graphene is used in the manufacturing of touch screens, transparent conducting electrodes, flexible electronic devices, organic light-emitting diodes, solar cells, organic photovoltaic cells, fuel cells, ultra-capacitors, transistors, sound transducers, *etc* [184]. Graphene is also used as field-effect transistors (FETs) which behave as metals and show strong ambipolar field effects. Graphene-based materials are also widely preferred by those developing memory devices and used as a replacement for indium tin oxide in the field of flexible electronics [185].

The second field for which graphene is a promising material is chemical and biosensing applications. In this area, three types of graphene-based sensors are developed. The first type is chemoresistive chemical sensors used for the detection of gas-phase chemical or vapor, such as H_2, NO_2, NH_3, H_2O, CO, dinitrotoluene, iodine, and ethanol and hydrazine hydrate. The second is electrochemical sensors, particularly due to the wide electrochemical potential window of graphene with lower charge-transfer resistance. Several studies demonstrate that graphene-based electrochemical sensors are quite promising for sensing many electroactive analytes including drugs. Such electrochemical sensors are also developed for environmental analysis of heavy metal ions. The third type is graphene-based biosensors which employ graphene/biological receptors as an active sensing material and can be classified under the three categories of electrochemical, FET and electrochemiluminescence or photochemiluminescence biosensors [186].

Synthesis and Characterization of Graphene

There are various procedures for producing graphene sheets, of which the most widely used, is the mechanical or chemical exfoliation of graphite, also known as the "scotch tape method" that follows the cleaving of a sample of graphite with a cellophane-based adhesive tape [13]. In this method, contrary to the stacking of graphene sheets, weak bonding and large lattice spacing is used in the perpendicular direction to obtain graphene sheets with different thicknesses [187]. The chemical exfoliation procedure is relatively cheaper and easier since it is simply based on ultrasonication with a water-surfactant solution. As a result, graphite flakes are dispersed in this solution, forming monolayer graphene sheets [13]. This process has two steps: first, the interlayer space is increased to reduce van der Waals forces in order to intercalate graphene, and the second step involves heating and sonication to obtain graphene [187].

The second procedure for graphene production is reduction of GO. In this method, graphite oxide is stirred and sonicated to obtain single layers of GO, which is later reduced to form graphene *via* chemical, thermal or electrochemical reduction. While this method is rapid and cost-effective, the product may have structural defects and graphitic impurities [13].

The third procedure for graphene production is the CVD method, which is widely used in electrochemical applications to produce uniform graphene sheets with high crystal quality and large surface areas. This method employs decomposition of a carbon feedstock *via* heating to provide a carbon source which can be rearranged to form sp^2 carbon species [13].

Finally, the arc-discharge method is employed to synthesize graphene by arc evaporation of graphite in the presence of hydrogen *via* rapid heating. The resulting product is two-layer graphene sheets with a width of 100-200 nm [187].

Characterization of graphene is undertaken with several techniques, including atomic force microscopy, scanning tunneling microscopy, transmission electronic microscopy, and scanning electron microscopy (Fig. **6**), X-ray diffraction, X-ray photoelectron spectroscopy, and Raman spectroscopy. All these techniques provide an understanding of the morphology, structures, crystal structure, chemical compositions and intrinsic properties of graphene-based materials [188].

Fig. (6). Scanning electron microscopy images of graphene nanosheets (I & II), microparticles (III & IV) and nanoparticles (V & VI) of molecularly imprinted polymer (Reprinted with permission from [189]).

Preparation of Graphene-Based Electrodes

In the field of electrosensing, graphene was first used in a study on the simultaneous determination of dopamine, ascorbic acid and uric acid. The electrode designed with graphene nanoplatelets was found to have fast electron-transfer kinetics and better performance compared to bare GCE and edge-plane pyrolytic graphite electrode. This was followed by another study on a graphene-chitosan modified GCE developed for sensing ascorbic acid [190]. Furthermore, using a graphene-modified GCE, Kim and coworkers (2010) achieved lower detection limits for dopamine in presence of ascorbic acid [191]. In addition to dopamine, ascorbic acid and uric acid, graphene-based electrodes have been utilized to determine drug ingredients. For example, Kang and coworkers (2010) employed a graphene-modified GCE for the direct electrochemical determination of paracetamol and reported very low detection limits [192].

Graphene paste electrodes have also been employed for electroanalytical detection. One of the earliest studies on these electrodes was conducted by Parvin (2011), who used a functionalized graphene paste electrode and obtained more sensitive and selective results in the detection of chlorpromazine compared to carbon electrodes [193].

Another type of graphene-based electrodes is screen-printed electrodes made of graphene ink. One recent study aimed to develop a screen-printed electrode with graphene ink comprising chemically reduced GO and a cellulose acetate binder solution. This electrode showed a significant performance for the simultaneous detection of ascorbic acid, dopamine and uric acid. Similarly, in another significant study on graphene-based screen printed electrodes, a graphene-based screen printed ink was designed by electrolytically mixing exfoliated graphene with carbon paste, which was later used as working and counter electrodes when printed onto a poly-vinyl chloride substrate [194].

In order to increase the effectiveness of graphene-based electrodes, certain modifications have been undertaken, including polymer composites. Polymer-modified electrodes have the advantages of diffusion control, reduction of interference, and mechanical protection of the electrode surface, which ultimately lead to better sensitivity, selectivity, stability and reproducibility of the electrode response. Polymer-graphene composites demonstrate superior electrical conductive properties [195]. Motaharian and coworkers (2016) developed a carbon paste electrode (CPE) composed of graphene nanosheets and molecularly imprinted polymers (MIP) for the determination of chlordiazepoxide (Fig. **7**). In designing this electrode, the authors mixed graphite, graphene and molecularly imprinted polymer at optimum amounts and packed this paste tightly to a glass holder, which included a copper wire as the electrical conductor [189].

In addition to polymers, metal nanoparticles have also been combined with graphene for the preparation of electrodes. Wong and coworkers (2016) designed a GCE modified with GO and gold nanoparticles within a crosslinked chitosan film. This electrochemical sensor is used for the determination of clindamycin *via* SWV [196]. Similarly, Fan and coworkers (2011) designed a Nafion/TiO_2-Graphene modified GCE for the determination of paracetamol and observed that addition of TiO_2 enhanced the electrochemical performance of graphene [197].

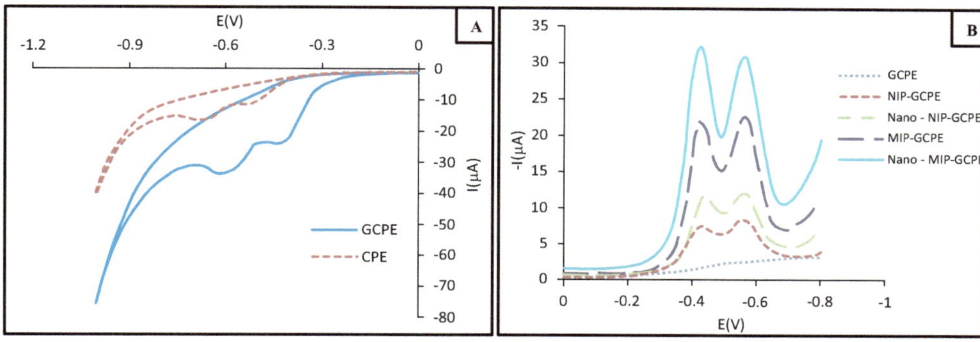

Fig. (7). (A) Cyclic voltammetry behaviour of a 5.0×10^{-5} M chlordiazepoxide solution on the CPE and graphene modified CPE (GCPE). (B) SWV signals for the CP and GCP electrodes after a 10 min extraction in 1.0×10^{-6} M chlordiazepoxide solutions (pH 4.5) (the SWV conditions: $\Delta E = 20$ mV, $V_{step} = 5$ mV, $t_{step} = 0.3$ s, frequency = 20 Hz) (Reprinted with permission from [189]).

Another method for increasing the efficiency of graphene sensing is doping graphene sheets with foreign atoms, most notably with nitrogen and boron. Sheng and coworkers (2012) used a nitrogen-doped graphene electrode for the simultaneous determination of dopamine, ascorbic acid and uric acid and achieved very low limits of detection. Doping with nitrogen enhanced the charge transfer kinetics of these molecules by activating the hydroxyl and amine groups [198]. The substitution of carbon atoms in the honeycomb lattice of graphene by heteroatoms disrupts the hybridization of carbon atoms and successfully tailors the electronic properties of the electrode [195].

Advantages and Limitations of Graphene-Modified Electrodes

The electrochemical advantages of graphene sheets are significant. Mechanically, graphene is a very stiff material, closely adhered to any substrate through van der Waals forces. Some studies have even shown that it is totally impermeable to helium atoms [199]. Robustness of graphene is so evident that it has received the title of "strongest material ever" [11]. Secondly, due to the one-atom thickness of graphene, it has a very large surface area, 2600 m^2/g, which surpasses that of both graphite and CNTs. In other words, graphene is entirely composed of a surface, without any bulk material [199]. Thirdly, its electrical conductivity is very high, approximately sixty times higher than that of SWCNTs. Moreover, the conductivity of graphene is not much affected by temperature, which is quite useful in electrochemical measurements. Furthermore, the electron mobility of graphene is 200 times higher than that of silica [13]. Other advantages of graphene include high charge carrier mobility at room temperature, good optical transparency, excellent thermal conductivity, and the quantum hall effect [200]. Moreover, graphene has a tunable band gap and high elasticity [182].

Graphene is an excellent electrode material because of its high electrical conductivity, low optical absorption, good surface smoothness, chemical stability, and flexibility. However, there are still several significant problems to be overcome in the design of graphene-based electrodes. Firstly, some graphene films are reported to have low conductivity due to the non-optimal fabrication technique of graphene-based electrodes. Secondly, hydrophobicity of graphene results in difficulties with coating the hole extraction layer [201]. Thirdly, although graphene is a great conductor of electricity, it does not have a band gap; in other words, it cannot be switched off [202]. Fourthly, graphene is very susceptible to oxidative environments; thus, local oxidation of graphene can occur in air at room temperature, particularly more in homogeneous atmospheres [203]. Finally, like CNTs, graphene has a degree of toxicity. Studies have shown that graphene can be accumulated on the membrane of kidney and lung epithelial cells and induce apoptosis [204].

Applications of Graphene in Pharmaceutical Analysis

Graphene has been used as an electrode material combined with many other modification agents, such as polymers, metal or metal oxide nanoparticles and even CNTs. Studies on this topic have been intensified in recent years. Table **2** summarizes recent studies from 2010 to 2017 including the electrodes developed, the techniques used, the supporting electrolyte studied, and the potential and limit of detection values measured.

Table 2. Selected analytical studies on drugs using graphene modified sensors.

Drugs	Matrix sample prep.	Method	Electrode	Buffer	Measured potential	Validation data (LOD)	Ref.
Baicalin	Capsules	CV, SWV	CoNPs doped-amino-graphene/GCE	0.1 M PBS (pH 4.0)	~ 0.35 V (SWV)	5.0 nM	[205]
Dopamine (in the presence of ascorbic acid and serotonin)	Urine samples	CV, DPV, LSV, chronoamperometry	GO/1,4-bis(aminomethyl)benzene/Co(OH)/GCE	0.1 M PBS (pH 7.4)	0.215 V (CV)	0.4 μM	[206]
Lomefloxacin (LMF) Amoxicillin (AMX) (simultaneous determination)	Tablets and human urine samples	SWV	AuNP-PdNP-ERGO/GCE	1.0 M PBS (pH 7.2)	~ 0.88 V (LMF) ~ 0.70 V (AMX)	81 nM (LMF) 9 μM (AMX)	[207]
Droxidopa (DDP) (in presence of phenobarbital-PB)	Human blood serum and urine samples	CV, DPV	RGO/PtNPs/GCE	0.1 M PBS (pH 7.0)	0.364 V (DDP) 0.65 V (PB) (CV)	31 nM (DDP) 23.6 nM (PB)	[208]
Methotrexate	Tablets and blood serum	CV, DPV	3D porous graphene-CNT network/GCE	0.1 M PBS (pH 6.0)	0.632 V (DPV)	70 nM	[209]
Metoprolol	Tablets	CV, AdSDPV	Nafion/PtNPs/ graphene/GCE	0.1 M BR buffer (pH 7.0)	1.325 V (AdSDPV)	4.3 nM	[210]
Rutin Isoquercitrin	Pharmaceutical dosage forms	DPV	thio-β-CD/ graphene/ PdNPs/GCE	0.04 M BR buffer (pH 3.0) for rutin 0.1 M PBS (pH 6.0) for isoquercitrin	0.55 V (Rutin) 0.45 V (Isoquercitrin)	0.3 nM (Rutin) 1.6 pM (Isoquercitrin)	[211]

(Table 2) contd.....

Drugs	Matrix sample prep.	Method	Electrode	Buffer	Measured potential	Validation data (LOD)	Ref.
Tetracycline	Human urine samples	CV, DPV	Polymelamine/ERGO/GCE	0.2 M PBS (pH 3.0)	0.85 V (DPV)	5.0 μM	[212]
Tizanidine	Tablet	CV, SWV	Graphene-ZnO nanorods/GCE	pH 6.5 BR buffer (presence of 0.1% CTAB)	-1.0 V (SWV)	0.10 ng mL^{-1}	[213]
Sulfamethoxazole	Tablet and urine	CV, EIS, chronoamperometry	NiO/GO/1M3BIB/ CPE	pH 7.0 PBS	0.81 V (CV)	0.04 μM	[214]
Chlordiazepoxide	Tablet, human serum and urine samples	CV, SWV	Nano MIP/Graphene/CPE	pH 4.5 BR buffer	~ -0.4 and -0.6 V (SWV)	0.26 nM	[189]
Pyridoxine	Tablet and human serum	Amperometry, DPV, LSV	ERGO/GCE	0.2 M PBS (pH 6.0)	0.61 V (DPV)	56 nM (amperometry)	[215]
Ofloxacin	Ophtalmic sample and human urine	CV, SWAdASV, EIS	GO-IL/CPE	0.1 M PBS (pH 6.0)	0.83 V (SWAdASV)	0.28 nM	[216]
Fluoxetine	Capsules and spiked plasma	CV, DPV, EIS	Graphene/MIP/CPE	0.1 M KCl (pH 8.0)	0.5 V (DPV)	2.8 nM	[217]
Gatifloxacin	Tablets and human urine	EIS, CV, DPV	β-CD/RGO/GCE	0.1 M PBS (pH 6.0)	0.87 V	0.02 μM	[218]
Gemcitabine	Injection samples and spiked urine sample	LSV, CV, DPV, EIS	ERGO/BiNPs/GCE	pH 3.0 PBS	1.144 V	0.05 μM	[219]
Nebivolol	Tablet and human plasma	CV, AdSDPV	Graphene/Nafion/GCE	0.1 M BR buffer (pH 8.0)	1.325 V (AdSDPV)	46 nM	[220]
Levamlodipine	Tablet	EIS, CV, SWV	Polyaniline/GO/GCE	pH 7.0 PBS	0.8 V (SWV)	1.07 ng mL^{-1}	[221]
Clindamycin	Tablet, synthetic urine, river water samples	CV, SWV	AuNPs/GO/chitosan/epichlorohydrin/GCE	0.1 M PBS (pH 7.0)	0.8 V (SWV)	0.29 μM	[196]
Paracetamol (PM) Aspirin (ASA) Caffeine (CAF)	Tablet	CV, AdASSWV	Graphene-Nafion/GCE	pH 2.5 PBS	0.64 V (PM) 1.04 V (ASA) 1.44 V (CAF) (AdASSWV)	1.2 nM (PM) 65 nM (ASA) 38 nM (CAF)	[222]
Folic acid	Human serum samples	CV, DPV	MoS_2-RGO/GCE	0.1 M PBS (pH 7.0)	0.7 V (DPV)	10 nM	[223]
Doxorubicin	Human plasma, urine, cerebrospinal fluid	CV, DPV	GQD/GCE	pH 4.0 PBS	-0.56 V (DPV)	0.016 μM	[224]
Dopamine	Injection sample, rat brain	CV, DPV	GO-C_{60}/GCE	pH 7.0 PBS	0.168 V (DPV)	0.008 μM	[225]
Phenazopyridine	Tablet, blood and urine samples	CV, DPV, EIS	AuNPs on layered double hydroxide/RGO/GCE	0.1 M PBS (pH 3.0)	0.81 V (DPV)	0.009 μM	[226]
Ascorbic acid	-	CV, DPV	[5-(4-Aminophenyl)- 10,15-20-triphenylporphyrin]Mn(III)-graphene/GCE	0.1 M PBS (pH 7.0)	0.15 V (DPV)	8.5 μM	[227]
Azathioprine	Tablet	CV	GNS/AgNPs/pyrolytic graphite electrode	0.1 M PBS (pH 6.0)	0.55 V	68 nM	[228]
Ascorbic acid (AA) Acetaminophen (ACOP) Tyrosine (TYR)	Urine, serum, pharmaceutical samples	CV, DPV	Au-Pd/RGO/CPE	pH 7.0 BR buffer	0.28 V (AA) 0.50 V (ACOP) 0.65 V (TYR) (DPV)	15.7 nM (AA) 7.6 nM (ACOP) 11.1 nM (TYR)	[229]
Chloramphenicol	Eye drop	CV	Tetrapods-like TiN-RGO/GCE	0.1 M PBS (pH 6.0)	-0.65 V	0.02 μM	[230]
Dopamine (DA) Uric acid (UA) Ascorbic acid (AA)	Human serum, urine and tablet samples	CV, DPV	Fe_3O_4-SnO_2-Graphene/GCE	pH 5.5 BR buffer	0.22 V (AA) 0.40 V (DA) 0.58 V (UA) (DPV)	62.0 nM (AA) 7.1 nM (DA) 5.0 nM (UA)	[231]

(Table 2) contd.....

Drugs	Matrix sample prep.	Method	Electrode	Buffer	Measured potential	Validation data (LOD)	Ref.
Vitamin K_3	Animal blood and feedstuff samples	CV, LSV	PEDOT- poly(styrene sulfonate/ carboxymethyl cellulose/RGO/Pd	0.1 M PBS (pH 7.0)	-0.12V (LSV)	14 nM	[232]
Pyrazinamide	Pharmaceutical, urine and blood serum samples	CV, DPV, EIS	GNS/ZnO/CPE	0.04 M BR buffer (pH 7.0)	-0.97 V (DPV)	43.1 nM	[233]
Citalopram	Tablet, human plasma samples	CV, SWV	Au–PdNPs-Graphene/AuE	0.1 M PBS (pH 8.0)	0.90 V (SWV)	0.049 μM	[234]
Acetaminophen (ACOP) Valacyclovir (VAL)	Tablets, spiked human serum	CV, DPV	RGO/GCE	0.1 M PBS (pH 7.2)	0.381 V (ACOP) 0.953 V (VAL) (DPV)	4.65 nM (ACOP) 1.34 nM (VAL)	[235]
Ascorbic acid (AA) Dopamin (DA) Acetaminophen (ACOP)	Human serum and urine sample	CV, DPV	Au/ZnO/N-doped graphene/GCE	0.1 M PBS (pH 7.4)	0.022 V (AA) 0.23 V (DA) 0.42 V (ACOP) (CV)	5.00 μM (AA) 0.40 μM (DA) 0.80 μM (ACOP)	[236]
Carbamazepin (CBZ) Paracetamol (PM)	Tablet, urine samples	CV, DPV	Graphene/MWCNT/CILPE	0.04 M BR buffer (pH 2.0)	1.25 V (CBZ) 0.65 V (PA) (DPV)	0.233 μM (CBZ) 0.262 μM (PM)	[237]
Doxorubicin	Human plasma	CV, DPV	GQD/GCE	0.1 M PBS (pH 4.0)	-0.53 V (DPV)	0.016 μM	[238]
Chloramphenicol	Milk samples	CV, LSV	RGO/Pt-Pd nanocubes/GCE	0.1 M PBS (pH 8.0)	-0.68 V (LSV)	0.1 μM	[239]
Acetaminophen	Tablet	CV, SWV	Cu_2O/Graphene/GCE	0.1 M PBS (pH 7.0)	0.35 V (SWV)	6.67 nM	[240]
Paracetamol (PM) Domperidone (DM)	Pharmaceutical formulations, human urine, and blood serum samples	AdSSWV	Nafion/PtNPs/Graphene/GCE	0.1 M PBS (pH 6.0)	0.4 V (PM) 0.9 V (DM)	0.106 nM (PM) 0.106 nM (DM)	[241]
Metronidazole (MN) Chloramphenicol (CP)	Shrimp	AdSDPV, CV, chronocoulometry	AgNPs/Sulfonated functionalized graphene/GCE	0.10 M citric acid-sodium citrate buffer (pH 4.0)	-0.38 V (MN) -0.49 V (CP) (CV)	0.01 μM (CP) 0.05 μM (MN)	[242]
Levodopa (LD) Uric acid (UA) Folic acid (FA)	Tablet, urine and human blood serum	SWV, CV, DPV, EIS	RGO/AuNPs/2-(3,4-dihydroxy phenyl) benzothiazole/GCE	0.1 M PBS (pH 7.0)	0.195 V (LD) 0.400 V (UA) 0.720 V (FA) (SWV)	0.018 μM (LD)	[243]
Isoniazid	Tablet	CV, DPV, EIS	RGO/AuNPs/GCE	0.2 M PBS (pH 7.0)	0.15 V (DPV)	10 nM	[244]
Isoniazid	Tablet	CV, LSV	RGO/GCE	0.1 M PBS (pH 7.0)	0.25 V (CV)	15 nM	[245]
Isoniazid (INZ) Uric acid (UA)	Urine samples	CV, DPV, EIS	poly(sulfosalicylic acid)/electroreduced carboxylated graphene/GCE	NH_3 – NH_4Cl buffer (pH 9.0)	-0.096V (IS) 0.016V (UA) (DPV)	12 nM (INZ and UA)	[246]
Hydrochlorothiazide (HCT) Folic acid (FA)	Tablet, human blood serum and urine	SWV, CV, chronoamperometry	GO sheets/ferrocene derivative/CPE	0.1 M PBS (pH 7.0)	0.610 V (HCT) 0.800 V (FA) (SWV)	20 nM (HCT)	[247]
Folic acid	Tablet	CV, DPV	AuNPs/Graphene/ CILE	pH 5.5 PBS	1.05 V (DPV)	2.7 nM	[248]
Diazepam	Tablet, injection and human blood plasma	CV, DPV	Ag nanodendrimers/ GNs/GCE	0.1 M PBS (pH 7.0)	-1.07 V (DPV)	85.6 nM	[249]
Clenbuterol	Liver samples	CV, DPV	sulfonated graphene sheets/MWCNTs-COOH/GCE	0.1 M PBS (pH 7.0)	0.7 V (DPV)	4.6 nM	[250]
Acetaminophen	Human serum	CV, DPV, chronoamperometry	ERGO/GCE	0.1 M PBS (pH 7.4)	0.37 V (DPV)	1.2 μM	[251]

(Table 2) contd.....

Drugs	Matrix sample prep.	Method	Electrode	Buffer	Measured potential	Validation data (LOD)	Ref.
Tizanidine	Tablet	CV, SWV	Graphene/silicon dioxide/GCE	pH 6.5 BR buffer (presence of 0.1% CTAB)	-0.9 V (SWV)	40.05 ng mL^{-1}	[252]
Rosuvastatin	Tablet, human urine and human serum	CV, SWAdASV	vertically aligned CNTs-GO electrode	pH 2.0 PBS	1.26 V	0.06 μM	[253]
Tetracycline	Tablet, artificial urine and river water	CV, AdSDPV	MWCNT-COOH-GO/CPE	0.1 M PBS (pH 7.0)	0.25 V (AdSDPV)	0.36 μM	[254]
Silodosin	Capsule	CV, AdSDPV	GNS/AuNPs/Nafion/GCE	pH 4.0 BR buffer	0.65 V (AdSDPV)	3.8 nM	[255]
Rutin	Tablet	CV, DPV	YHCFNPs/CRGO/CPE	0.1 M PBS (pH 1.5)	0.73 V (DPV)	0.82 nM	[256]
Rutin	Tablet	EIS, CV, DPV	PEDOT nanorods/GO nanosheets/GCE	0.1 M PBS (pH 3.0)	0.48 V (DPV)	1.25 nM	[257]
Rivastigmine	Tablet, urine and blood serum	CV, EIS, chronocoulometry, AdSDPV	GNS/AuNP/CPE	0.1 M PBS (pH 7.0)	0.88 V (AdSDPV)	53 nM	[258]
Nimodipine	Tablet	EIS, CV	Nitrogen-doped graphene/GCE	pH 6.5 PBS	-0.78 V	9.1 nM	[259]
Nifedipine	Tablet	LSV, DPV	hybrid PdAgNPs/IL/GNR/GCE	pH 7.0 PBS buffer	-0.66 V (DPV)	4.0 nM	[260]
Buspirone (BPH) Isoniazid (INZ) Pyrazinamide (PZ) (simultaneous determination)	Tablets and human blood serum	CV, DPV	GO/poly-L-arginine/GCE	0.05 M PBS (pH 7.0)	0.9 V (BHP) 0.3 V (INZ) -0.78 V (PZ) (CV)	3.54 μM (BPH) 2.59 μM (INZ) 3.28 μM (PZ)	[261]
Diclofenac	Tablets, human urine and blood serum samples	CV, LSV	GO-COOH/GCE	0.05 M PBS (pH 7.0)	0.58 V (CV)	0.09 μM	[262]
Acetaminophen	Tablets and urine samples	CV, DPV	ERG/Ni_2O_3–NiO NPs/GCE	0.1 M PBS (pH 7.0)	~ 0.3 V (DPV)	0.02 μM	[263]
Acetyl salicylic acid	Tablet and human oral fluid	CV	Graphene/EPPGE	pH 4.0 PBS	0.72 V	3.0 nM	[264]
Atorvastatin calcium	Tablets, synthetic urine and human serum samples	CV, AdSDPV	vertically aligned CNT/GO electrode (immobilized on a copper plate using conductive silver paste)	0.2 M PBS (pH 2.0)	1.08 V (CV)	9.4 nM	[265]
Ciprofloxacin (in the presence of ascorbic acid, uric acid and dopamine)	Tablets and human serum samples	CV, DPV	poly(alizarin red)/EGR)/GCE	0.1 M PBS (pH 5.5)	~ 1.0 V (DPV)	0.01 μM	[266]
Dopamine Paracetamol (in the presence of ascorbic acid)	Pharmaceutical formulations in human urine samples	CV, LSV, EIS	ERGO/ZrO_2 nanocomposite modified GCE	0.05 M PBS (pH 7)	0.34 V (DA) 0.53 V (PA) (LSV)	9.0 μM	[267]
Etodolac	Tablets	CV, SWV	Graphene-Polyaniline-Bi_2O_3/GCE	pH 3.8 BR buffer	~ 0.6 V (SWV)	10.03 ng mL^{-1}	[268]
Linezolid	Tablets, human urine and plasma samples	DPV, AdSLSV, AdSDPV	ERGO/Bentonite/GCE	pH 7.0 PBS	~ 1.0 V (AdSLSV)	0.0338 μM (AdSDPV) 0.1002 μM (DPV) 0.5461 μM (AdSLSV)	[269]
Pindolol	Tablets and urine samples	CV, AdSSWV, chronoamperometry	RGO–GCE	pH 5.0 BR buffer	~ 1.0 V (CV)	26 nM	[270]
Sumatriptan	Tablets, urine and blood serum	CV, EIS, chronocoulometry, AdSDPV	Graphene/AuNP/ Nafion/GCE	0.1 M PBS (pH 7.0)	~ 0.75 V (AdSDPV)	0.703 nM	[271]
Topotecan	Injection dosage form	CV, SWV	TiO_2/Graphene/ Chitosan/GCE	pH 5.6 BR buffer	~ 0.7 V (SWV)	122.07 ng mL^{-1}	[272]

(Table 2) contd.....

Drugs	Matrix sample prep.	Method	Electrode	Buffer	Measured potential	Validation data (LOD)	Ref.
Indomethacin	Tablet, capsule, human blood serum and urine	CV, SWV	Au nanorods-GO-CNT paste/GCE	pH 8.0 PBS	0.5 V (SWV)	17 nM	[273]
Acetaminophen	Pharmaceuticals	CV, DPV chronocoulometry,	Graphene/Chitosan/GCE	40 mM BR buffer (pH 7.0)	0.32 V (DPV)	0.13 μM	[274]
Quercetin	Tablet	CV, SWV	AuNPs/p-aminothiophenol functionalized GO sheets/GCE	0.1 M acetate buffer (pH 5.5)	0.2 V (SWV)	0.3 pM	[275]
Quinidine	Tablet	LSV, CV	nano graphene-MWCNT/GCE	pH 7.9 BR buffer	0.9 V (LSV)	0.186 ng mL^{-1}	[276]
Mebendazole	Tablet	CV, ASDPV	GNS/carbon nanospheres/ Chitosan/GCE	0.1 M PBS (pH 7.0)	0.82 V (ASDPV)	10.5 nM	[277]
Tyrosine (Try) Paracetamol (PM)	Pharmaceuticals and human blood serum	CV, DPV	MWCNTs-GNS/GCE	pH 7.5 PBS	0.64 V (DPV) (Tyr) 0.37 V (DPV) (PM)	0.19 μM (Tyr) 0.10 μM (PM)	[278]
Dopamin (DA) Paracetamol (PM)	Dopamine hydrochloride injection and Paracetamol tablet	CV, DPV, EIS, amperometry	MWCNTs/GO/GCE	pH 7.0 PBS	0.18 V (DA) 0.4 V (PM)	22 nM (DA) 47 nM (PM)	[279]
Codeine	Syrup and urine sample	CV, SWV	Graphene/Nafion/ GCE	pH 4.0 PBS	1.13 V (SWV)	15 nM	[280]
Ascorbic acid (AA) Dopamine (DA) Uric acid (UA) Tryptophan (Trp)	Vitamin C tablets, dopamine injections and human urine samples	CV, LSV, EIS	AgNPs/RGO/GCE	pH 3.5 PBS	0.23 V (AA) 0.49 V (DA) 0.62 V (UA) 0.93 V (Trp) (LSV)	9.6 μM (AA) 5.4 μM (DA) 8.2 μM (UA) 7.5 μM (Trp)	[281]
Dopamine	Dopamine hydrochloride injection	CV, DPV	PTCA-GS/MCNT/IL modified electrode	pH 5.0 PBS	0.309 V (CV)	1.2 nM	[282]
Picoplatin	Picoplatin injection samples	DPV	Graphene MWCNT/GCE	0.05 M KCl solution (pH 7.4)	E_{pa} = −0.15 V E_{pc} = −0.32 V	0.21 μM	[283]
Terbutaline	Tablet	CV, LSV, chronocoulometry	Poly (L-arginine)/graphene–Nafion/GCE	0.1 M BR buffer (pH 7.0)	0.7 V (LSV)	12 nM	[284]
Metronidazole	Pharmaceutical dosage forms	CV, DPV	Graphene-IL/GCE	pH 7.0 PBS	−0.69 V (DPV)	47 nM	[285]
Theophylline	Tablets, tea	CV, DPV	AuNPs/L-cysteine/Graphene/Nafion/GCE	0.1 M H_2SO_4	1.3 V (DPV)	0.4 nM	[286]
Ascorbic acid (AA) Dopamine (DA) Uric acid (UA)	-	CV, DPV	Nitrogen doped graphene/GCE	0.1 M PBS (pH 6.0)	0.0 V (AA) −0.22 V (DA) −0.36 V (UA) (DPV)	2.2 μM (AA) 0.25 μM (DA) 0.045 μM (UA)	[198]
Acetaminophen	Pharmaceutical dosage forms	CV, DPV	poly(methyl red)/TiO_2/Graphene/GCE	0.1 M PBS (pH 7.0)	0.41 V (DPV)	25 nM	[287]
Rutin	Tablet, human serum and urine	CV, DPV, chronocoulometry	Reduced graphene/Carbon IL electrode	pH 2.5 PBS	0.49 V (DPV)	24.0 nM	[288]
Azithromycin	Capsules and tablets	DPV, EIS, chronocoulometry	Graphene/ $BMIMPF_6$/GCE	pH 5.8 PBS	0.82 V (DPV)	0.19 μg mL^{-1}	[289]
Paracetamol	Tablet	CV, DPV	Nafion/TiO_2-Graphene/GCE	0.1 M PBS (pH 7.0)	0.46 V (DPV)	0.21 μM	[197]
Rutin	Tablet	CV, DPV	β-CD/CRG/Nafion/ GCE	0.1 M PBS (pH 3.0)	0.48 V (DPV)	2.0 nM	[290]
Chlorpromazine	Tablet	CV, DPV	Graphene paste electrode	0.5 M phosphate solution (pH 4.0)	0.72 V (DPV)	6.0 nM	[193]

(Table 2) contd.....

Drugs	Matrix sample prep.	Method	Electrode	Buffer	Measured potential	Validation data (LOD)	Ref.
Paracetamol	Tablet and spiked human plasma	CV, SWV	Graphene/GCE	0.1 M $NH_3.H_2O–NH_4Cl$ (pH 9.3)	0.27 V (SWV)	32 nM	[192]

CONCLUSION

Carbon has been preferred as an electrode material for decades because of high surface activity, relatively inert electrochemistry, rich surface electrochemistry, low cost and wide potential window of carbon-based electrodes [291]. With their unique characteristics, carbon materials have become a basic component of nanotechnology, particularly in the field of pharmaceutical analysis. Since the early 1990s, CNTs (SWCNTs and MWCNTs) have been extensively used as an electrode material and modifier. Their mechanical, optical and electrical properties have allowed the researchers to design various forms of electrodes for different purposes. Graphene is a relatively novel material introduced in the mid-2000s. Contrary to three-dimensional cylindrical CNTs, graphene nanosheets are two-dimensional planar materials having almost an infinite surface area. Similar to CNTs, graphene has significant mechanical and electrical advantages, which make it an important source of electrochemical sensing.

This chapter focuses on CNT- and graphene-modified electrodes. It reviews basic properties of these materials, their synthesis and characterization, as well as the production of these electrodes. Moreover, it examines the use of these electrodes in pharmaceutical analyses and presents recent studies based on the type of electrode, the electrochemical technique, buffer solution, peak potential, detection limit and the application on real samples, such as urine, serum, and blood.

As a result of this examination, it was observed that CNT- or graphene-modified electrodes tend to be preferred to establish an electrocatalytical effect. These electrodes provide a remarkable improvement in the analytical sensitivity in terms of pharmaceuticals with voltammetric sensing resulting in a low limit of detection.

ABBRAVIATIONS

1M3BIB	Ionic liquids (1-methyl-3-butylimidazolium bromide)
AdASSWV	Adsorptive anodic stripping square wave voltammetry
AdSDPV	Adsorptive stripping differential pulse voltammetry
AdSLSV	Adsorptive stripping linear sweep voltammetry
AdSSWV	Adsorptive stripping square wave voltammetry

ATR	Atorvastatine
$BMIMPF_6$	1-butyl-3-methyl-imidazolium hexafluorophosphate
BPPGE	basal plane pyrolytic graphite electrode
BR	Britton Robinson
CCE	Carbon ceramic electrode
CE	carbon electrode
CNT	Carbon nanotube
CNTPE	Carbon nanotube paste electrode
$Co(DMG)_2ClPy$	chloro(pyridine)bis(dimethylglyoximato)cobalt(III)
CoHCNFe	cobalt hexacyanoferrates nanoparticles
CoPc	Cobalt (II) phthalocyanine
CPE	Carbon paste electrode
CRG	Chemically reduced graphene
CRGO	Chemically reduced graphene oxide
CTAB	cetyltrimethylammonium bromide
CV	Cyclic voltammetry
DDP	3,4-dihydroxybenzaldehyde-2,4-dinitrophenylhydrazone
DHP	dihexadecylhydrogenphosphate film
DHPBMCNTPE	N-(3,4-dihydroxyphenethyl)-3,5-dinitrobenzamide modified carbon nanotubes paste electrode
DPV	Differential pulse voltammetry
EBNBHCNTPE	2, 2'-[1,2-ethanediylbis(nitriloethylidyne)]-bis-hydroquinone modified carbon nanotube paste electrode
EIS	Electrochemical impedance spectroscopy
EPPGE	Edge plane pyrolytic graphite electrodes
ERG	Electrochemically reduced graphene
ERGO	Electrochemically reduced graphene oxide
FMCAMCNTPE	ferrocenemonocarboxylic acid modified carbon nanotubes paste electrode
***f*MWCNT**	Functionalized multi walled carbon nanotube
GCE	Glassy carbon electrode
GNR	Graphene nanoribbons
GNS	Graphene nanosheets
GO	Graphene oxide
GQD	Graphene quantum dots
HCF	Hexacyanoferrate
IL	Ionic liquid

ISSM	*in situ* surfactant modified
LSV	Lineer sweep voltammetry
MIP	Molecularly imprinted polymer
MWCNTPE	Multi walled carbon nanotubes paste electrode
NP	nanoparticle
PAH	poly(allylamine hydrochloride)
p-APMCNTPE	p-aminophenol modified carbon nanotubes paste electrode
PBS	Phosphate buffer solution
p-DAN	poly 1,5-diaminonapthalene
PEDOT	poly (3,4-ethylenedioxythiophene)
PLM	poly-L-methionine
RGO	Reduced graphene oxide
SPE	Screen printed electrode
SWCNT	single walled carbon nanotubes
SWV	Square wave voltammetry
YHCF	Yttrium hexacyanoferrate
β-CD	β-cyclodextrine

CONSENT FOR PUBLICATION

Not applicable.

CONFLICT OF INTEREST

The authors declares no conflict of interest, financial or otherwise.

ACKNOWLEDGEMENTS

Declared none.

REFERENCES

[1] Zhang, X.; Ju, H.; Wang, J. Preface.*Electrochemical sensors, biosensors and their biomedical applications*; Zhang, X.; Ju, H.; Wang, J., Eds.; Elsevier: San Diego, CA, **2008**, p. xxi. [http://dx.doi.org/10.1016/B978-012373738-0.50002-7]

[2] Gründler, P. *Chemical sensors: An introduction for scientists and engineers*; Springer: Heidelberg, **2007**.

[3] Gooding, J.J.; Lai, L.M.; Goon, I.Y. Nanostructured electrodes with unique properties for biological and other applications.*Advances in Electrochemical Science and Engineering: Chemically Modified Electrodes*; Alkire, R.C.; Kolb, D.M.; Lipkowski, J.; Ross, P.N., Eds.; Wiley-VCH: Weinheim, **2009**, Vol. 11, pp. 1-56. [http://dx.doi.org/10.1002/9783527627059.ch1]

[4] Cox, J.A.; Zou, S. Electroanalytical measurements at electrodes modified with metal nanoparticles.*Trace analysis with nanomaterials*; Pierce, D.T.; Zhao, J.X., Eds.; Wiley-VCH: Weinheim, **2011**, pp. 283-317.

[5] Guldi, D.M.; Martin, N. Preface.*Carbon nanotubes and related structures: synthesis, characterization, functionalization, and applications*; Guldi, D.M.; Martin, N., Eds.; Wiley-VCH: Weinheim, **2010**, pp. xv-xvii.
[http://dx.doi.org/10.1002/9783527629930]

[6] Cha, C.; Shin, S.R.; Annabi, N.; Dokmeci, M.R.; Khademhosseini, A. Carbon-based nanomaterials: multifunctional materials for biomedical engineering. *ACS Nano,* **2013**, *7*(4), 2891-2897.
[http://dx.doi.org/10.1021/nn401196a] [PMID: 23560817]

[7] Iurlo, M.; Marcaccio, M.; Paolucci, F. Electrochemistry.*Carbon nanotubes and related structures: Synthesis, characterization, functionalization, and applications*; Guldi, D.M.; Martin, N., Eds.; Wiley-VCH: Weinheim, **2010**, pp. 53-75.

[8] Rahman, M.M.; Leon, I.C. Electrochemical behavior of SWCNT-TPLF electrode compared to SWCNT-coated GCE and GE electrodes. *Croat. Chem. Acta,* **2007**, *80*, 67-74.

[9] Harris, P.J. *Carbon nanotube science: Synthesis, properties and applications*; Cambridge University Press: Cambridge, **2009**.
[http://dx.doi.org/10.1017/CBO9780511609701]

[10] Calo, S.; Curulli, A.; Zane, D.; Caschera, D.; Ingo, G.M.; Padeletti, G. Single walled carbon nanotubes (SWCNTs)/Gold nanoparticles (AuNps) nanocomposites for enhancing electrochemical response to detect neurotransmitters. *ECS Trans.,* **2010**, *25*, 33-41.

[11] Rocha, C.G.; Rümmeli, M.H.; İbrahim, I. Tailoring the physical properties of graphene.*Graphene: Synthesis and applications*; Choi, W.; Li, J-w., Eds.; CRC Press: Boca Raton, FL, **2012**, pp. 1-26.

[12] Brownson, D.A.; Kampouris, D.K.; Banks, C.E. Graphene electrochemistry: fundamental concepts through to prominent applications. *Chem. Soc. Rev.,* **2012**, *41*(21), 6944-6976.
[http://dx.doi.org/10.1039/c2cs35105f] [PMID: 22850696]

[13] Brownson, D.A.; Banks, C.E. *The handbook of graphene electrochemistry*; Springer: Heidelberg, **2014**.
[http://dx.doi.org/10.1007/978-1-4471-6428-9]

[14] Liu, M.; Chen, W. Graphene-supported metal nanostructures with controllable size and shape as advanced electrocatalysts for fuel cells.*Graphene-based energy devices*; Yusoff, R.M., Ed.; Wiley-VCH: Weinheim, **2015**, pp. 307-338.
[http://dx.doi.org/10.1002/9783527690312.ch11]

[15] Suarez-Martinez, I.; Grobert, N.; Ewels, C. Encyclopedia of carbon nanoforms.*Advances in carbon nanomaterials: Science and applications*; Tagmatarchis, N., Ed.; Pan Stanford Publishing Pte Ltd: Singapore, **2012**, pp. 1-65.
[http://dx.doi.org/10.1201/b11990-2]

[16] Kazaoui, S.; Minami, N. Optical and electrical properties of C60, C70, nanotubes and endohedral fullerenes.*Macromolecular science and engineering: New aspects*; Tanabe, Y., Ed.; Springer: Heidelberg, **1999**, pp. 363-392.
[http://dx.doi.org/10.1007/978-3-642-58559-3_23]

[17] Feng, M.; Han, H.; Zhang, J.; Tachikawa, H. Electrochemical sensors based on carbon nanotubes.*Electrochemical sensors, biosensors and their biomedical applications*; Zhang, X.; Ju, H.; Wang, J., Eds.; Elsevier: Amsterdam, **2008**, pp. 460-503.
[http://dx.doi.org/10.1016/B978-012373738-0.50017-9]

[18] Sun, G.; Nicklaus, M.; Kertesz, M. Single-walled carbon nanotubes: Geometries, electronic properties, and actuation.*Dekker encyclopedia of nanoscience and nanotechnology*; Schwarz, J.A.; Contescu, C.I.; Putyera, K., Eds.; CRC Press: Boca Raton, FL, **2004**, Vol. 5, pp. 3605-3615.

[http://dx.doi.org/10.1201/9781439834398.ch301]

[19] Basu-Dutt, S.; Minus, M.L.; Jain, R.; Nepal, D.; Kumar, S. Chemistry of carbon nanotubes for everyone. *J. Chem. Educ.*, **2012**, *89*, 221-229.
[http://dx.doi.org/10.1021/ed1005163]

[20] Mongillo, J.F. *Nanotechnology 101*; Greenwood Press: Westport, CT, **2007**.

[21] Hennrich, F.; Chan, C.; Moore, V.; Rolandi, M.; O'Connell, M. The Element Carbon.*Carbon nanotubes: Properties and applications*; O'Connell, M.J., Ed.; CRC Press: Boca Raton, FL, **2006**, pp. 1-18.
[http://dx.doi.org/10.1201/9781420004212.ch1]

[22] Chen, C.; Zhang, Y. *Nanowelded carbon nanotubes: from field-effect transistors to solar microcells*; Springer: Heidelberg, **2009**.
[http://dx.doi.org/10.1007/978-3-642-01499-4]

[23] Belin, T.; Epron, F. Characterization methods of carbon nanotubes: a review. *Mater. Sci. Eng. B*, **2005**, *119*, 105-118.
[http://dx.doi.org/10.1016/j.mseb.2005.02.046]

[24] Gong, K.; Yan, Y.; Zhang, M.; Su, L.; Xiong, S.; Mao, L. Electrochemistry and electroanalytical applications of carbon nanotubes: a review. *Anal. Sci.*, **2005**, *21*(12), 1383-1393.
[http://dx.doi.org/10.2116/analsci.21.1383] [PMID: 16379375]

[25] Lin, Y.; Lu, F.; Tu, Y.; Ren, Z.F. Glucose biosensors based on carbon nanotube nanoelectrode ensembles. *Nano Lett.*, **2004**, *4*, 191-195.
[http://dx.doi.org/10.1021/nl0347233]

[26] Fayazfar, H.; Afshar, A.; Dolati, A. Tantalum electrodes modified with well-aligned carbon nanotube-Au nanoparticles: application to the highly sensitive electrochemical determination of cefazolin. *Appl. Biochem. Biotechnol.*, **2014**, *173*(6), 1511-1528.
[http://dx.doi.org/10.1007/s12010-014-0944-9] [PMID: 24817553]

[27] Shim, B.S.; Kotov, N.A. Layer-by-layer assembly of multifunctional carbon nanotube thin films.*Carbon nanotubes and related structures: Synthesis, characterization, functionalization, and applications*; Guldi, D.M.; Martin, N., Eds.; Wiley-VCH: Weinheim, **2010**, pp. 305-320.
[http://dx.doi.org/10.1002/9783527629930.ch10]

[28] Ju, H.; Zhang, X.; Wang, J. *NanoBiosensing: Principles, development and application*; Springer: Heidelberg, **2011**.
[http://dx.doi.org/10.1007/978-1-4419-9622-0]

[29] Wang, Q; Zhao, R; Wang, S A highly selective electrochemical sensor for nifedipine based on layer-by-layer assembly films from polyaniline and multiwalled carbon nanotube. *J Appl Polym Sci*, **2016**, *133*(21), 43452/1-43452/9.

[30] Kong, B.; Zeng, J.; Luo, G.; Luo, S.; Wei, W.; Li, J. Layer-by-layer assembled carbon nanotube films with molecule recognition function and lower capacitive background current. *Bioelectrochemistry*, **2009**, *74*(2), 289-294.
[http://dx.doi.org/10.1016/j.bioelechem.2008.10.005] [PMID: 19058766]

[31] Srinivasan, S.; Ajayaghosh, A. Interaction of carbon nanotubes and small molecules.*Supramolecular soft matter: Applications in materials and organic electronics*; Nakanishi, T., Ed.; John Wiley & Sons: Hoboken, NJ, **2011**, pp. 381-406.
[http://dx.doi.org/10.1002/9781118095331.ch19]

[32] Salimi, A.; Izadi, M.; Hallaj, R.; Rashidi, M. Simultaneous determination of ranitidine and metronidazole at glassy carbon electrode modified with single wall carbon nanotubes. *Electroanalysis*, **2007**, *19*, 1668-1676.
[http://dx.doi.org/10.1002/elan.200703911]

[33] Jain, R.; Sharma, S. Glassy carbon electrode modified with multi-walled carbon nanotubes sensor for

the quantification of antihistamine drug pheniramine in solubilized systems. *J. Pharm. Anal.*, **2012**, *2*, 56-61.
[http://dx.doi.org/10.1016/j.jpha.2011.09.013]

[34] Spitalskya, Z.; Tasis, D.; Papagelis, K.; Galiotis, C. Carbon nanotube–polymer composites: Chemistry, processing, mechanical and electrical properties. *Prog. Polym. Sci.*, **2010**, *35*, 357-01.
[http://dx.doi.org/10.1016/j.progpolymsci.2009.09.003]

[35] Takeda, H.H.; Silva, T.A.; Janegitz, B.C.; Vicentini, F.C.; Mattoso, L.H.; Fatibello-Filho, O. Electrochemical sensing of levodopa or carbidopa using a glassy carbon electrode modified with carbon nanotubes within a poly(allylamine hydrochloride) film. *Anal. Methods*, **2016**, *8*, 1274-1280.
[http://dx.doi.org/10.1039/C5AY03041B]

[36] Ghadimi, H.; Tehrani, R.M.; Ali, A.S.; Mohamed, N.; Ab Ghani, S. Sensitive voltammetric determination of paracetamol by poly (4-vinylpyridine)/multiwalled carbon nanotubes modified glassy carbon electrode. *Anal. Chim. Acta*, **2013**, *765*, 70-76.
[http://dx.doi.org/10.1016/j.aca.2012.12.039] [PMID: 23410628]

[37] Shahrokhian, S.; Kamalzadeh, Z.; Hamzehloei, A. Electrochemical determination of clozapine on MWCNTs/new coccine doped PPY modified GCE: an experimental design approach. *Bioelectrochemistry*, **2013**, *90*, 36-43.
[http://dx.doi.org/10.1016/j.bioelechem.2012.10.002] [PMID: 23178559]

[38] Sanghavi, B.J.; Srivastava, A.K. Adsorptive stripping differential pulse voltammetric determination of venlafaxine and desvenlafaxine employing Nafion–carbon nanotube composite glassy carbon electrode. *Electrochim. Acta*, **2011**, *56*, 4188-4196.
[http://dx.doi.org/10.1016/j.electacta.2011.01.097]

[39] Hsieh, C.T.; Lin, J.Y. Electrochemical capacitance from carbon nanotubes decorated with cobalt oxide nanoparticles in alkali electrolyte. *ECS Trans.*, **2008**, *13*, 73-89.

[40] Agarwal, A.; Bakshi, S.R.; Lahiri, D. *Carbon nanotubes: Reinforced metal matrix composites*; CRC Press: Boca Raton, FL, **2011**.

[41] Fayemi, O.E.; Adekunle, A.S.; Ebenso, E.E. Metal oxide nanoparticles/multi-walled carbon nanotube nanocomposite modified electrode for the detection of dopamine: Comparative electrochemical study. *J Biosens Bioelectron*, **2015**, *6*, 190.
[http://dx.doi.org/10.4172/2155-6210.1000190]

[42] Kutluay Baytak, A.; Duzmen, S.; Teker, T.; Aslanoglu, M. A novel modified electrode based on terbium oxide and carbon nanotubes for the simultaneous determination of methyldopa and paracetamol. *Anal. Methods*, **2016**, *8*, 4711-4719.
[http://dx.doi.org/10.1039/C6AY00969G]

[43] Wang, W.; Zhu, Y.; Liao, S.; Li, J. Carbon nanotubes reinforced composites for biomedical applications. *BioMed Res. Int.*, **2014**, *2014*, 518609.
[http://dx.doi.org/10.1155/2014/518609] [PMID: 24707488]

[44] Zhu, L.; Tian, C.; Zhai, J.; Yang, R. Sol–gel derived carbon nanotubes ceramic composite electrodes for electrochemical sensing. *Sens. Actuators B Chem.*, **2007**, *125*, 254-261.
[http://dx.doi.org/10.1016/j.snb.2007.02.012]

[45] Salimi, A.; Mamkhezri, H.; Hallaj, R. Simultaneous determination of ascorbic acid, uric acid and neurotransmitters with a carbon ceramic electrode prepared by sol-gel technique. *Talanta*, **2006**, *70*(4), 823-832.
[http://dx.doi.org/10.1016/j.talanta.2006.02.015] [PMID: 18970846]

[46] Acton, A. *Fullerenes – Advances in research and application*; **2013** Atlanta GE: Scholarly Editions

[47] Eatemadi, A.; Daraee, H.; Karimkhanloo, H.; Kouhi, M.; Zarghami, N.; Akbarzadeh, A.; Abasi, M.; Hanifehpour, Y.; Joo, S.W. Carbon nanotubes: properties, synthesis, purification, and medical applications. *Nanoscale Res. Lett.*, **2014**, *9*(1), 393-06.

[http://dx.doi.org/10.1186/1556-276X-9-393] [PMID: 25170330]

[48] Shao, W.; Arghya, P.; Yiyong, M.; Rodes, L.; Satya, P. Carbon nanotubes for use in medicine: Potentials and limitations. *Syntheses and Applications of carbon nanotubes and their composites, InTech,* Suzuki, S. **2013**. http://www.intechopen.com/books/syntheses-and-applications-of- carbon-nanotubes-and-their-composites/carbon-nanotubes-for-use-in-medicine-potentials-and-limitations

[49] Nigović, B.; Jurić, S.; Mitrović, I. Bismuth nanoparticles-carbon nanotubes modified sensor for sulfasalazine analysis. *Talanta,* **2017**, *164*, 201-208. [http://dx.doi.org/10.1016/j.talanta.2016.11.059] [PMID: 28107918]

[50] Rahimi-Nasrabadi, M.; Khoshroo, A.; Mazloum-Ardakani, M. Electrochemical determination of diazepam in real samples based on fullerene-functionalized carbon nanotubes/ionic liquid nanocomposite. *Sens. Actuators B Chem.,* **2017**, *240*, 125-131. [http://dx.doi.org/10.1016/j.snb.2016.08.144]

[51] Karthik, R.; Sasikumar, R.; Chen, S-M.; Vinoth Kumar, J.; Elangovan, A.; Muthuraj, V.; Muthukrishnan, P.; Al-Hemaid, F.M.; Ajmal Ali, M.; Elshikh, M.S. A highly sensitive and selective electrochemical determination of non-steroidal prostate anti-cancer drug nilutamide based on f-MWCNT in tablet and human blood serum sample. *J. Colloid Interface Sci.,* **2017**, *487*, 289-296. [http://dx.doi.org/10.1016/j.jcis.2016.10.047] [PMID: 27792936]

[52] Deiminiat, B.; Rounaghi, G.H.; Arbab-Zavar, M.H. Development of a new electrochemical imprinted sensor based on poly-pyrrole, sol–gel and multiwall carbon nanotubes for determination of tramadol. *Sens. Actuators B Chem.,* **2017**, *238*, 651-659. [http://dx.doi.org/10.1016/j.snb.2016.07.110]

[53] de Holanda, L.F.; Ribeiro, F.W.; Sousa, C.P. Paulo Naftali da Silva Casciano, Pedro de Lima-Neto, Adriana Nunes Correia, Multi-walled carbon nanotubes–cobalt phthalocyanine modified electrode for electroanalytical determination of acetaminophen. *J. Electroanal. Chem.,* **2016**, *772*, 9-16. [http://dx.doi.org/10.1016/j.jelechem.2016.04.021]

[54] Jalali, F.; Ardeshiri, M. Application of carbon nanotubes-ionic liquid hybrid in a sensitive atorvastatin ion-selective electrode. *Mater. Sci. Eng. C,* **2016**, *69*, 276-282. [http://dx.doi.org/10.1016/j.msec.2016.06.099] [PMID: 27612714]

[55] Damiri, S.; Oskoei, Y.M.; Fouladgar, M. Highly sensitive voltammetric and impedimetric sensor based on an ionic liquid/cobalt hexacyanoferrate nanoparticle modified multiwalled carbon nanotubes electrode for diclofenac analysis. *J. Exp. Nanosci.,* **2016**, *11*, 1384-01. [http://dx.doi.org/10.1080/17458080.2016.1233581]

[56] Madrakian, T.; Asl, K.D.; Ahmadi, M.; Afkhami, A. Fe_3O_4@Pt/MWCNT/carbon paste electrode for determination of a doxorubicin anticancer drug in a human urine sample. *RSC Advances,* **2016**, *6*, 72803-72809. [http://dx.doi.org/10.1039/C6RA13935C]

[57] Couto, R.A.; Quinaz, M.B. Development of a Nafion/MWCNT-SPCE-based portable sensor for the voltammetric analysis of the anti-tuberculosis drug ethambutol. *Sensors (Basel),* **2016**, *16*(7), 1015-1028. [http://dx.doi.org/10.3390/s16071015] [PMID: 27376291]

[58] Paimard, G.; Gholivand, M.B.; Shamsipur, M. Determination of ganciclovir as an antiviral drug and its interaction with DNA at Fe_3O_4/carboxylated multi-walled carbon nanotubes modified glassy carbon electrode. *Measurement,* **2016**, *77*, 269-277. [http://dx.doi.org/10.1016/j.measurement.2015.09.019]

[59] Keyvanfard, M.; Alizad, K. Determination of isoproterenol in pharmaceutical and biological samples using a pyrogallol red multiwalled carbon nanotube paste electrode as a sensor. *Chin. J. Catal.,* **2016**, *37*, 579-583. [http://dx.doi.org/10.1016/S1872-2067(15)61036-1]

[60] Mohamed, M.A.; Abdelwahab, N.S.; Banks, C.E. Electroanalytical sensing of the antimicrobial drug

linezolid utilising an electrochemical sensing platform based upon a multiwalled carbon nanotubes/bromocresol green modified carbon paste electrode. *Anal. Methods,* **2016**, *8*, 4345-4353. [http://dx.doi.org/10.1039/C6AY00454G]

[61] Nigović, B.; Sadiković, M.; Jurić, S. Electrochemical sensing of mesalazine and its N-acetylated metabolite in biological samples using functionalized carbon nanotubes. *Talanta,* **2016**, *147*, 50-58. [http://dx.doi.org/10.1016/j.talanta.2015.09.036] [PMID: 26592575]

[62] Ghalkhani, M.; Beheshtian, J.; Salehi, M. Electrochemical and DFT study of an anticancer and active anthelmintic drug at carbon nanostructured modified electrode. *Mater. Sci. Eng. C,* **2016**, *69*, 1345-1353. [http://dx.doi.org/10.1016/j.msec.2016.08.031] [PMID: 27612835]

[63] Taei, M.; Hasanpour, F.; Hajhashemi, V.; Movahedi, M.; Baghlani, H. Simultaneous detection of morphine and codeine in urine samples of heroin addicts using multi-walled carbon nanotubes modified SnO_2–Zn_2SnO_4 nanocomposites paste electrode. *Appl. Surf. Sci.,* **2016**, *363*, 490-498. [http://dx.doi.org/10.1016/j.apsusc.2015.12.074]

[64] Brahman, P.K.; Suresh, L.; Lokesh, V.; Nizamuddin, S. Fabrication of highly sensitive and selective nanocomposite film based on CuNPs/fullerene-C_{60}/MWCNTs: An electrochemical nanosensor for trace recognition of paracetamol. *Anal. Chim. Acta,* **2016**, *917*, 107-116. [http://dx.doi.org/10.1016/j.aca.2016.02.044] [PMID: 27026607]

[65] Kalambate, P.K.; Srivastava, A.K. Simultaneous voltammetric determination of paracetamol, cetirizine and phenylephrine using a multiwalled carbon nanotube-platinum nanoparticles nanocomposite modified carbon paste electrode. *Sens. Actuators B Chem.,* **2016**, *233*, 237-248. [http://dx.doi.org/10.1016/j.snb.2016.04.063]

[66] Bozal-Palabiyik, B.; Uslu, B. Comparative study for voltammetric investigation and trace determination of pramipexole at bare and carbon nanotube-modified glassy carbon electrodes. *Ionics,* **2016**, *22*, 2519-2528. [http://dx.doi.org/10.1007/s11581-016-1774-2]

[67] Bozal-Palabiyik, B.; Uslu, B. A novel electroanalytical nanosensor based on MWCNT/Fe_2O_3 nanoparticles for the determination of antiparkinson drug ropinirole. *Ionics,* **2016**, *22*, 115-123. [http://dx.doi.org/10.1007/s11581-015-1525-9]

[68] Shoja, Y.; Rafati, A.A.; Ghodsi, J. Electropolymerization of Ni–LD metallopolymers on gold nanoparticles enriched multi-walled carbon nanotubes as nano-structure electrocatalyst for efficient voltammetric sertraline detection in human serum. *Electrochim. Acta,* **2016**, *203*, 281-291. [http://dx.doi.org/10.1016/j.electacta.2016.03.117]

[69] Karadas-Bakirhan, N.; Patris, S.; Ozkan, S.A.; Can, A.; Kauffmann, J.M. Determination of the anticancer drug sorafenib in serum by adsorptivestripping differential pulse voltammetry using a chitosan/multiwall carbon nanotube modified glassy carbon electrode. *Electroanalysis,* **2016**, *28*, 358-365. [http://dx.doi.org/10.1002/elan.201500384]

[70] Sgobbi, L.F.; Razzino, C.A.; Machado, S.A. A disposable electrochemical sensor for simultaneous detection of sulfamethoxazole and trimethoprim antibiotics in urine based on multiwalled nanotubes decorated with Prussian blue nanocubes modified screen-printed electrode. *Electrochim. Acta,* **2016**, *191*, 1010-1017. [http://dx.doi.org/10.1016/j.electacta.2015.11.151]

[71] Lahcen, A.A.; Errayess, S.A.; Amine, A. Voltammetric determination of sulfonamides using paste electrodes based on various carbon nanomaterials. *Mikrochim. Acta,* **2016**, *183*, 2169-2176. [http://dx.doi.org/10.1007/s00604-016-1850-3]

[72] Baytak, A.K.; Teker, T.; Duzmen, S.; Aslanoglu, M. A sensitive determination of terbutaline in pharmaceuticals and urine samples using a composite electrode based on zirconium oxide nanoparticles. *Mater. Sci. Eng. C,* **2016**, *67*, 125-131.

[http://dx.doi.org/10.1016/j.msec.2016.05.008] [PMID: 27287106]

[73] Gholivand, M.B.; Torkashvand, M. The fabrication of a new electrochemical sensor based on electropolymerization of nanocomposite gold nanoparticle-molecularly imprinted polymer for determination of valganciclovir. *Mater. Sci. Eng. C,* **2016**, *59*, 594-603. [http://dx.doi.org/10.1016/j.msec.2015.09.016] [PMID: 26652412]

[74] Satyanarayana, M.; Goud, K.Y.; Reddy, K.K.; Gobi, K.V. Biopolymer stabilized nanogold particles on carbon nanotube support as sensing platform for electrochemical detection of 5-Fluorouracil invitro. *Electrochim. Acta,* **2015**, *178*, 608-616. [http://dx.doi.org/10.1016/j.electacta.2015.08.036]

[75] Pandit, U.J.; Khan, I.; Wankar, S.; Raj, K.K.; Limaye, S.N. Development of an electrochemical method for the determination of bicalutamide at the SWCNT/CPE in pharmaceutical preparations and human biological fluids. *Anal. Methods,* **2015**, *7*, 10192-10198. [http://dx.doi.org/10.1039/C5AY02025E]

[76] Materon, E.M.; Wong, A.; Klein, S.I.; Liu, J.; Sotomayor, M.D. Multi-walled carbon nanotubes modified screen-printed electrodes for cisplatin detection. *Electrochim. Acta,* **2015**, *158*, 271-276. [http://dx.doi.org/10.1016/j.electacta.2015.01.184]

[77] Ghoreishi, S.M.; Attaran, A.M.; Amin, A.M.; Khoobi, A. Multiwall carbon nanotube-modified electrode as a nanosensor for electrochemical studies and stripping voltammetric determination of an antimalarial drug. *RSC Advances,* **2015**, *5*, 14407-14415. [http://dx.doi.org/10.1039/C4RA16357E]

[78] Madrakian, T.; Haghshenas, E.; Ahmadi, M.; Afkhami, A. Construction a magneto carbon paste electrode using synthesized molecularly imprinted magnetic nanospheres for selective and sensitive determination of mefenamic acid in some real samples. *Biosens. Bioelectron.,* **2015**, *68*, 712-718. [http://dx.doi.org/10.1016/j.bios.2015.02.001] [PMID: 25679118]

[79] Dorraji, P.S.; Jalali, F. Sensitive amperometric determination of methimazole based on the electrocatalytic effect of rutin/multi-walled carbon nanotube film. *Bioelectrochemistry,* **2015**, *101*, 66-74. [http://dx.doi.org/10.1016/j.bioelechem.2014.07.009] [PMID: 25128894]

[80] Ghoneim, MM; El-Desoky, HS; Matsuda, A; Hattori, T; Abdel-Galeil, MM Voltammetric analysis of nitroxoline in tablets and human serum using modified carbon paste electrodes incorporating mesoporous carbon or multiwalled carbon nanotubes. *RSC Adv,* **2015**, *5*(69), 56086-56097. [http://dx.doi.org/10.1039/C5RA05086C]

[81] da Silva, H.; Pacheco, J.; Silva, J.; Viswanathan, S.; Delerue-Matos, C. Molecularly imprinted sensor for voltammetric detection of norfloxacin. *Sens. Actuators B Chem.,* **2015**, *219*, 301-307. [http://dx.doi.org/10.1016/j.snb.2015.04.125]

[82] Baytak, A.K.; Duzmen, S.; Teker, T.; Aslanoglu, M. A novel composite electrode based on tungsten oxide nanoparticles and carbon nanotubes for the electrochemical determination of paracetamol. *Mater. Sci. Eng. C,* **2015**, *57*, 164-170. [http://dx.doi.org/10.1016/j.msec.2015.07.060] [PMID: 26354251]

[83] Ashrafi, A.M.; Kurbanoglu, S.; Vytras, K.; Uslu, B.; Ozkan, S.A. Electrochemical mechanism and sensitive assay of antiretroviral drug Abacavir in biological sample using multiwalled carbon nanotube modified pyrolytic graphite electrode. *J. Electroanal. Chem.,* **2014**, *712*, 178-184. [http://dx.doi.org/10.1016/j.jelechem.2013.11.012]

[84] Jain, R. Dhanjai. A highly sensitive and selective bismuth oxide-multiwalled carbon nanotubes hybrid film sensor for sensing of acenocoumarole. *J. Electrochem. Soc.,* **2014**, *161*, H29-H35. [http://dx.doi.org/10.1149/2.078401jes]

[85] Duarte, E.H.; dos Santos, W.P.; Hudari, F.F.; Bott Neto, J.L.; Sartori, E.R.; Dall'Antonia, L.H.; Pereira, A.C.; Tarley, C.R. A highly improved method for sensitive determination of amitriptyline in pharmaceutical formulations using an unmodified carbon nanotube electrode in the presence of

sulfuric acid. *Talanta,* **2014**, *127*, 26-32.
[http://dx.doi.org/10.1016/j.talanta.2014.03.068] [PMID: 24913853]

[86] Shamsipur, M.; Saber, R.; Emami, M. A highly sensitive electrochemical sensor based on gold nanoparticles/multiwall carbon nanotubesmodified glassy carbon electrode for selective determination of traces of atenolol. *Anal. Methods,* **2014**, *6*, 7038-7045.
[http://dx.doi.org/10.1039/C4AY01538J]

[87] Emami, M.; Shamsipur, M.; Saber, R. Design of poly-L-methionine–gold nanocomposit/multi-walled carbon nanotube modified glassy carbon electrode for determination of amlodipine in human biological fluids. *J. Solid State Electrochem.,* **2014**, *18*, 985-992.
[http://dx.doi.org/10.1007/s10008-013-2345-3]

[88] Cheemalapati, S.; Palanisamy, S.; Chen, S-M. A simple and sensitive electroanalytical determination of anxiolytic buspirone hydrochloride drug based on multiwalled carbon nanotubes modified electrode. *J. Appl. Electrochem.,* **2014**, *44*, 317-323.
[http://dx.doi.org/10.1007/s10800-013-0637-z]

[89] Habibi, B.; Abazari, M.; Pournaghi-Azar, M.H. Simultaneous determination of codeine and caffeine using single-walled carbon nanotubes modified carbon-ceramic electrode. *Colloids Surf. B Biointerfaces,* **2014**, *114*, 89-95.
[http://dx.doi.org/10.1016/j.colsurfb.2013.09.026] [PMID: 24176887]

[90] Satyanarayana, M.; Reddy, K.K.; Gobi, K.V. Multiwall carbon nanotube ensemble biopolymer electrode for selective determination of isoniazid *in vitro. Anal. Methods,* **2014**, *6*, 3772-3778.
[http://dx.doi.org/10.1039/C4AY00154K]

[91] Sanati, A.L.; Karimi-Maleh, H.; Badiei, A.; Biparva, P.; Ensafi, A.A. A voltammetric sensor based on NiO/CNTs ionic liquid carbon paste electrode for determination of morphine in the presence of diclofenac. *Mater. Sci. Eng. C,* **2014**, *35*, 379-385.
[http://dx.doi.org/10.1016/j.msec.2013.11.031] [PMID: 24411391]

[92] Madrakian, T.; Soleimani, M.; Afkhami, A. Simultaneous determination of mycophenolate mofetil and its active metabolite, mycophenolic acid, by differential pulse voltammetry using multi-walled carbon nanotubes modified glassy carbon electrode. *Mater. Sci. Eng. C,* **2014**, *42*, 38-45.
[http://dx.doi.org/10.1016/j.msec.2014.05.012] [PMID: 25063089]

[93] Karadas, N.; Ozkan, S.A. Electrochemical preparation of sodium dodecylsulfate doped over-oxidized polypyrrole/multi-walled carbon nanotube composite on glassy carbon electrode and its application on sensitive and selective determination of anticancer drug: pemetrexed. *Talanta,* **2014**, *119*, 248-254.
[http://dx.doi.org/10.1016/j.talanta.2013.10.065] [PMID: 24401411]

[94] Sadikovic, M.; Nigovic, B.; Juric, S.; Mornar, A. Voltammetric determination of ropinirole in the presence of levodopa at the surface of a carbon nanotubes based electrochemical sensor in pharmaceuticals and human serum. *J. Electroanal. Chem.,* **2014**, *733*, 60-68.
[http://dx.doi.org/10.1016/j.jelechem.2014.09.020]

[95] Yadav, S.K.; Choubey, P.K.; Agrawal, B.; Goyal, R.N. Carbon nanotube embedded poly 1,5-diaminonapthalene modified pyrolytic graphite sensor for the determination of sulfacetamide in pharmaceutical formulations. *Talanta,* **2014**, *118*, 96-103.
[http://dx.doi.org/10.1016/j.talanta.2013.09.061] [PMID: 24274275]

[96] Demir, E.; Inam, R.; Ozkan, S.A.; Uslu, B. Electrochemical behavior of tadalafil on TiO_2 nanoparticles–MWCNT composite paste electrode and its determination in pharmaceutical dosage forms and human serum samples using adsorptive stripping square wave voltammetry. *J. Solid State Electrochem.,* **2014**, *18*, 2709-2720.
[http://dx.doi.org/10.1007/s10008-014-2529-5]

[97] Rafati, A.A.; Afraz, A. Amperometric sensing of anti-HIV drug zidovudine on Ag nanofilm-multiwalled carbon nanotubes modified glassy carbon electrode. *Mater. Sci. Eng. C,* **2014**, *39*, 105-112.

[http://dx.doi.org/10.1016/j.msec.2014.02.037] [PMID: 24863205]

[98] Yola, M.L.; Eren, T.; Atar, N. Molecularly imprinted electrochemical biosensor based on Fe@Au nanoparticles involved in 2-aminoethanethiol functionalized multi-walled carbon nanotubes for sensitive determination of cefexime in human plasma. *Biosens. Bioelectron.*, **2014**, *60*, 277-285. [http://dx.doi.org/10.1016/j.bios.2014.04.045] [PMID: 24832202]

[99] Narayan, P.S.; Teradal, N.L.; Kalanur, S.S.; Seetharamappa, J. Fabrication of an electrochemical sensor based on multiwalled carbon nanotubes for almotriptan. *Electroanalysis*, **2013**, *25*, 2684-2690. [http://dx.doi.org/10.1002/elan.201300335]

[100] Oliveira, G.G.; Janegitz, B.C.; Zucolotto, V.; Fatibello-Filho, O. Differential pulse adsorptive stripping voltammetric determination of methotrexate using a functionalized carbon nanotubes-modified glassy carbon electrode. *Cent. Eur. J. Chem.*, **2013**, *11*, 1837-1843.

[101] Thomas, T.; Mascarenhas, R.J.; Swamy, B.E.; Martis, P.; Mekhalif, Z.; Sherigara, B.S. Multi-walled carbon nanotube/poly(glycine) modified carbon paste electrode for the determination of dopamine in biological fluids and pharmaceuticals. *Colloids Surf. B Biointerfaces*, **2013**, *110*, 458-465. [http://dx.doi.org/10.1016/j.colsurfb.2013.03.056] [PMID: 23770784]

[102] Prasad, B.B.; Prasad, A.; Tiwari, M.P.; Madhuri, R. Multiwalled carbon nanotubes bearing 'terminal monomeric unit' for the fabrication of epinephrine imprinted polymer-based electrochemical sensor. *Biosens. Bioelectron.*, **2013**, *45*, 114-122. [http://dx.doi.org/10.1016/j.bios.2013.01.042] [PMID: 23455050]

[103] Afkhami, A.; Ghaedi, H.; Madrakian, T.; Ahmadi, M.; Mahmood-Kashani, H. Fabrication of a new electrochemical sensor based on a new nano-molecularly imprinted polymer for highly selective and sensitive determination of tramadol in human urine samples. *Biosens. Bioelectron.*, **2013**, *44*, 34-40. [http://dx.doi.org/10.1016/j.bios.2012.11.030] [PMID: 23391704]

[104] Nasirizadeh, N.; Shekari, Z.; Zare, H.R.; Shishehbore, M.R.; Fakhari, A.R.; Ahmar, H. Electrosynthesis of an imidazole derivative and its application as a bifunctional electrocatalyst for simultaneous determination of ascorbic acid, adrenaline, acetaminophen, and tryptophan at a multi-wall carbon nanotubes modified electrode surface. *Biosens. Bioelectron.*, **2013**, *41*, 608-614. [http://dx.doi.org/10.1016/j.bios.2012.09.028] [PMID: 23122703]

[105] Ansari, M.; Kazemi, S.; Khalilzadeh, M.A.; Karimi-Maleh, H.; Zanousi, M.B. Sensitive and stable voltammetric measurements of norepinephrine at ionic liquid–carbon nanotubes paste electrodes. *Int. J. Electrochem. Sci.*, **2013**, *8*, 1938-1948.

[106] Desai, P.B.; Srivastava, A.K. Adsorptive stripping differential pulse voltammetric determination of metoprolol at Nafion-CNT-nano-composite film sensor. *Sens. Actuators B Chem.*, **2013**, *176*, 632-638. [http://dx.doi.org/10.1016/j.snb.2012.10.073]

[107] Ensafi, A.A.; Allafchian, A.R. Multiwall carbon nanotubes decorated with $NiFe_2O_4$ magnetic nanoparticles, a new catalyst for voltammetric determination of cefixime. *Colloids Surf. B Biointerfaces*, **2013**, *102*, 687-693. [http://dx.doi.org/10.1016/j.colsurfb.2012.09.037] [PMID: 23107947]

[108] Ensafi, A.A.; Allafchian, A.R.; Rezaei, B. A sensitive and selective voltammetric sensor based on multiwall carbon nanotubes decorated with $MgCr_2O_4$ for the determination of azithromycin. *Colloids Surf. B Biointerfaces*, **2013**, *103*, 468-474. [http://dx.doi.org/10.1016/j.colsurfb.2012.11.021] [PMID: 23261568]

[109] Karadas, N.; Bozal-Palabiyik, B.; Uslu, B.; Ozkan, S.A. Functionalized carbon nanotubes—With silver nanoparticles to fabricate a sensor for the determination of zolmitriptan in its dosage forms and biological samples. *Sens. Actuators B Chem.*, **2013**, *186*, 486-494. [http://dx.doi.org/10.1016/j.snb.2013.06.055]

[110] Rafiee, B.; Fakhari, A.R. Electrocatalytic oxidation and determination of insulin at nickel oxide nanoparticles-multiwalled carbon nanotube modified screen printed electrode. *Biosens. Bioelectron.*, **2013**, *46*, 130-135.

[http://dx.doi.org/10.1016/j.bios.2013.01.037] [PMID: 23531859]

[111] Kor, K.; Zarei, K. β-Cyclodextrin incorporated carbon nanotube paste electrode as electrochemical sensor for nifedipine. *Electroanalysis,* **2013**, *25*, 1497-04. [http://dx.doi.org/10.1002/elan.201200652]

[112] Ardila, J.A.; Oliveira, G.G.; Medeiros, R.A.; Fatibello-Filho, O. Determination of gemfibrozil in pharmaceutical and urine samples by square-wave adsorptive stripping voltammetry using a glassy carbon electrode modified with multi-walled carbon nanotubes within a dihexadecyl hydrogen phosphate film. *J. Electroanal. Chem.,* **2013**, *690*, 32-37. [http://dx.doi.org/10.1016/j.jelechem.2012.11.038]

[113] Amiri-Aref, M.; Raoof, J.B.; Ojani, R. Electrocatalytic oxidation and selective determination of an opioid analgesic methadone in the presence of acetaminophen at a glassy carbon electrode modified with functionalized multi-walled carbon nanotubes: application for human urine, saliva and pharmaceutical samples analysis. *Colloids Surf. B Biointerfaces,* **2013**, *109*, 287-293. [http://dx.doi.org/10.1016/j.colsurfb.2013.03.055] [PMID: 23680846]

[114] Arvand, M.; Palizkar, B. Development of a modified electrode with amine-functionalized TiO2/multi-walled carbon nanotubes nanocomposite for electrochemical sensing of the atypical neuroleptic drug olanzapine. *Mater. Sci. Eng. C,* **2013**, *33*(8), 4876-4883. [http://dx.doi.org/10.1016/j.msec.2013.08.002] [PMID: 24094200]

[115] Dogan-Topal, B. Electrooxidative behavior and determination of trifluoperazine at multiwalled carbon nanotube-modified glassy carbon electrode. *J. Solid State Electrochem.,* **2013**, *17*, 1059-1066. [http://dx.doi.org/10.1007/s10008-012-1967-1]

[116] Karadas, N.; Sanli, S.; Akmese, B.; Dogan-Topal, B.; Can, A.; Ozkan, S.A. Analytical application of polymethylene blue-multiwalled carbon nanotubes modified glassy carbon electrode on anticancer drug irinotecan and determination of its ionization constant value. *Talanta,* **2013**, *115*, 911-919. [http://dx.doi.org/10.1016/j.talanta.2013.07.006] [PMID: 24054682]

[117] Kurbanoglu, S.; Dogan-Topal, B.; Uslu, B.; Can, A.; Ozkan, S.A. Electrochemical investigations of the anticancer drug idarubicin using multiwalled carbon nanotubes modified glassy carbon and pyrolytic graphite electrodes. *Electroanalysis,* **2013**, *25*, 1473-1482. [http://dx.doi.org/10.1002/elan.201300048]

[118] Bozal-Palabiyik, B.; Dogan-Topal, B.; Uslu, B.; Can, A.; Ozkan, S.A. Sensitive voltammetric assay of etoposide using modified glassy carbon electrode with a dispersion of multi-walled carbon nanotube. *J. Solid State Electrochem.,* **2013**, *17*, 2815-2822. [http://dx.doi.org/10.1007/s10008-013-2184-2]

[119] Dogan-Topal, B.; Bozal-Palabıyık, B.; Uslu, B.; Ozkan, S.A. Multi-walled carbon nanotube modified glassy carbon electrode as a voltammetric nanosensor for the sensitive determination of anti-viral drug valganciclovir in pharmaceuticals. *Sens. Actuators B Chem.,* **2013**, *177*, 841-847. [http://dx.doi.org/10.1016/j.snb.2012.11.111]

[120] Agrawal, B.; Chandra, P.; Goyal, R.N.; Shim, Y-B. Detection of norfloxacin and monitoring its effect on caffeine catabolism in urine samples. *Biosens. Bioelectron.,* **2013**, *47*, 307-312. [http://dx.doi.org/10.1016/j.bios.2013.03.025] [PMID: 23587793]

[121] Hu, Y.; Zhang, Z.; Li, J.; Zhang, H.; Luo, L.; Yao, S. Electrochemical imprinted sensor for determination of oleanic acid based on poly (sodium 4-styrenesulfonate-co-acrylic acid)-grafted multi-walled carbon nanotubes-chitosan and cobalt hexacyanoferrate nanoparticles. *Biosens. Bioelectron.,* **2012**, *31*(1), 190-196. [http://dx.doi.org/10.1016/j.bios.2011.10.016] [PMID: 22099956]

[122] Dalmasso, P.R.; Pedano, M.L.; Rivas, G.A. Electrochemical determination of ascorbic acid and paracetamol in pharmaceutical formulations using a glassy carbon electrode modified with multi-wall carbon nanotubes dispersed in polyhistidine. *Sens. Actuators B Chem.,* **2012**, *173*, 732-736. [http://dx.doi.org/10.1016/j.snb.2012.07.087]

[123] Malode, S.J.; Abbar, J.C.; Shetti, N.P.; Nandibewoor, S.T. Voltammetric oxidation and determination of loop diuretic furosemide at a multi-walled carbon nanotubes paste electrode. *Electrochim. Acta,* **2012**, *60*, 95-01.
[http://dx.doi.org/10.1016/j.electacta.2011.11.011]

[124] Abbar, J.C.; Malode, S.J.; Nandibewoor, S.T. Electrochemical determination of a hemorheologic drug, pentoxifylline at a multi-walled carbon nanotube paste electrode. *Bioelectrochemistry,* **2012**, *83*, 1-7.
[http://dx.doi.org/10.1016/j.bioelechem.2011.06.008] [PMID: 21784715]

[125] Malode, S.J.; Shetti, N.P.; Nandibewoor, S.T. Voltammetric behavior of theophylline and its determination at multi-wall carbon nanotube paste electrode. *Colloids Surf. B Biointerfaces,* **2012**, *97*, 1-6.
[http://dx.doi.org/10.1016/j.colsurfb.2012.04.010] [PMID: 22584261]

[126] Shahrokhian, S.; Rastgar, S. Electrochemical deposition of gold nanoparticles on carbon nanotube coated glassy carbon electrode for the improved sensing of tinidazole. *Electrochim. Acta,* **2012**, *78*, 422-429.
[http://dx.doi.org/10.1016/j.electacta.2012.06.035]

[127] Rastakhiz, N.; Beitollahi, H.; Kariminik, A.; Karimi, F. Voltammetric determination of carbidopa in the presence of uric acid and folic acid using a modified carbon nanotube paste electrode. *J. Mol. Liq.,* **2012**, *172*, 66-70.
[http://dx.doi.org/10.1016/j.molliq.2012.04.013]

[128] Desai, P.B.; Srivastava, A.K. Determination of amiloride at Nafion–CNT-nano-composite film sensor employing adsorptive stripping differential pulse voltammetry. *Sens. Actuators B Chem.,* **2012**, *169*, 341-348.
[http://dx.doi.org/10.1016/j.snb.2012.05.007]

[129] Dar, R.A.; Brahman, P.K.; Tiwari, S.; Pitre, K.S. Electrochemical determination of atropine at multi-wall carbon nanotube electrode based on the enhancement effect of sodium dodecyl benzene sulfonate. *Colloids Surf. B Biointerfaces,* **2012**, *91*, 10-17.
[http://dx.doi.org/10.1016/j.colsurfb.2011.10.020] [PMID: 22118893]

[130] Pan, M.; Fang, G.; Duan, Z.; Kong, L.; Wang, S. Electrochemical sensor using methimazole imprinted polymer sensitized with MWCNTs and Salen-Co(III) as recognition element. *Biosens. Bioelectron.,* **2012**, *31*(1), 11-16.
[http://dx.doi.org/10.1016/j.bios.2011.09.008] [PMID: 22169817]

[131] Ensafi, A.A.; Karimi-Maleh, H. Determination of 6-mercaptopurine in the presence of uric acid using modified multiwall carbon nanotubes-TiO_2 as a voltammetric sensor. *Drug Test. Anal.,* **2012**, *4*(12), 970-977.
[http://dx.doi.org/10.1002/dta.286] [PMID: 21538998]

[132] Wen, W.; Zhao, D-M.; Zhang, X-H. One-step fabrication of poly(o-aminophenol)/multi-walled carbon nanotubes composite film modified electrode and its application for levofloxacin determination in pharmaceuticals. *Sens. Actuators B Chem.,* **2012**, *174*, 202-209.
[http://dx.doi.org/10.1016/j.snb.2012.08.010]

[133] Ensafi, A.A.; Allafchian, A.R.; Rezaei, B. Multiwall carbon nanotubes decorated with $FeCr_2O_4$, a new selective electrochemical sensor for amoxicillin determination. *J. Nanopart. Res.,* **2012**, *14*, 1244-1255.
[http://dx.doi.org/10.1007/s11051-012-1244-3]

[134] Karimi-Maleh, H.; Khalilzadeh, M.A.; Ranjbarha, Z.; Beitollahi, H.; Ensafi, A.A.; Zareyee, D. p-Chloranil modified carbon nanotubes paste electrode as a voltammetric sensor for the simultaneous determination of methyldopa and uric acid. *Anal. Methods,* **2012**, *4*, 2088-2094.
[http://dx.doi.org/10.1039/c2ay05865k]

[135] Arvand, M.; Gholizadeh, T.M.; Zanjanchi, M.A. MWCNTs/$Cu(OH)_2$ nanoparticles/IL nanocomposite modified glassy carbon electrode as a voltammetric sensor for determination of the non-steroidal anti-

inflammatory drug diclofenac. *Mater. Sci. Eng. C,* **2012**, *32*(6), 1682-1689.
[http://dx.doi.org/10.1016/j.msec.2012.04.066] [PMID: 24364977]

[136] Leite, F.R.; Maroneze, C.M.; de Oliveira, A.B.; dos Santos, W.T.; Damos, F.S.; Silva Luz, Rde.C. Development of a sensor for L-Dopa based on Co(DMG)$(_2)$ClPy/multi-walled carbon nanotubes composite immobilized on basal plane pyrolytic graphite electrode. *Bioelectrochemistry,* **2012**, *86*, 22-29.
[http://dx.doi.org/10.1016/j.bioelechem.2012.01.001] [PMID: 22284852]

[137] Esfandiari Baghbamidi, S.; Beitollahi, H.; Karimi-Maleh, H.; Soltani-Nejad, S.; Soltani-Nejad, V.; Roodsaz, S. Modified carbon nanotube paste electrode for voltammetric determination of carbidopa, folic Acid, and tryptophan. *J. Anal. Methods Chem.,* **2012**, *2012*, 305872.
[http://dx.doi.org/10.1155/2012/305872] [PMID: 22666634]

[138] Brahman, P.K.; Dar, R.A.; Tiwari, S.; Pitre, K.S. Electrochemical behavior of gatifloxacin at multi-walled carbon nanotube paste electrode and its interaction with DNA. *Rev. Anal. Chem.,* **2012**, *31*, 83-92.
[http://dx.doi.org/10.1515/revac-2011-0032]

[139] Raoof, J.B.; Baghayeri, M.; Ojani, R. A high sensitive voltammetric sensor for qualitative and quantitative determination of phenobarbital as an antiepileptic drug in presence of acetaminophen. *Colloids Surf. B Biointerfaces,* **2012**, *95*, 121-128.
[http://dx.doi.org/10.1016/j.colsurfb.2012.02.038] [PMID: 22465049]

[140] Jain, R.; Sharma, R. Novel bismuth/multi-walled carbon nanotubes-based electrochemical sensor for the determination of neuroprotective drug cilostazol. *J. Appl. Electrochem.,* **2012**, *42*, 341-348.
[http://dx.doi.org/10.1007/s10800-012-0402-8]

[141] Ahmar, H.; Fakhari, A.R. Electro-oxidation and adsorptive stripping voltammetric determination of ephedrine and pseudoephedrine at carboxylated multi-walled carbon nanotube-modified electrode. *Anal. Methods,* **2012**, *4*, 812-818.
[http://dx.doi.org/10.1039/c2ay05680a]

[142] Stoiljković, Z.Ž.; Ivić, M.L.; Petrović, S.D. Voltammetric and square-wave anodic stripping determination of amlodipine besylate on gold electrode. *Int. J. Electrochem. Sci.,* **2012**, *7*, 2288-2303.

[143] Jain, R.; Vikas, Voltammetric determination of cefpirome at multiwalled carbon nanotube modified glassy carbon sensor based electrode in bulk form and pharmaceutical formulation. *Colloids Surf. B Biointerfaces,* **2011**, *87*(2), 423-426.
[http://dx.doi.org/10.1016/j.colsurfb.2011.06.001] [PMID: 21724377]

[144] Habibi, B.; Jahanbakhshi, M.; Pournaghi-Azar, M.H. Differential pulse voltammetric simultaneous determination of acetaminophen and ascorbic acid using single-walled carbon nanotube-modified carbon-ceramic electrode. *Anal. Biochem.,* **2011**, *411*(2), 167-175.
[http://dx.doi.org/10.1016/j.ab.2011.01.005] [PMID: 21236238]

[145] Amiri, M.; Pakdel, Z.; Bezaatpour, A.; Shahrokhian, S. Electrocatalytic determination of sumatriptan on the surface of carbon-paste electrode modified with a composite of cobalt/Schiff-base complex and carbon nanotube. *Bioelectrochemistry,* **2011**, *81*(2), 81-85.
[http://dx.doi.org/10.1016/j.bioelechem.2011.03.003] [PMID: 21470920]

[146] Arvand, M.; Ansari, R.; Heydari, L. Electrocatalytic oxidation and differential pulse voltammetric determination of sulfamethoxazole using carbon nanotube paste electrode. *Mater. Sci. Eng. C,* **2011**, *31*, 1819-1825.
[http://dx.doi.org/10.1016/j.msec.2011.08.014]

[147] Li, Y.; Liu, X.; Liu, X.; Mai, N.; Li, Y.; Wei, W.; Cai, Q. Application of multi-walled carbon nanotubes modified carbon ionic liquid electrode for electrocatalytic oxidation of dopamine. *Colloids Surf. B Biointerfaces,* **2011**, *88*(1), 402-406.
[http://dx.doi.org/10.1016/j.colsurfb.2011.07.021] [PMID: 21831611]

[148] Hou, X.; Shen, G.; Meng, L.; Zhu, L.; Guo, M. Multi-walled carbon nanotubes modified glass carbon

electrode and its electrocatalytic activity towards oxidation of paracetamol. *Russ. J. Electrochem.*, **2011**, *47*, 1262-1267.
[http://dx.doi.org/10.1134/S1023193511110103]

[149] Mazloum-Ardakani, M.; Ganjipour, B.; Beitollahi, H. Simultaneous determination of levodopa, carbidopa and tryptophan using nanostructured electrochemical sensor based on novel hydroquinone and carbon nanotubes: Application to the analysis of some real samples. *Electrochim. Acta,* **2011**, *56*, 9113-9120.
[http://dx.doi.org/10.1016/j.electacta.2011.07.021]

[150] Ensafi, A.A.; Karimi-Maleh, H.; Mallakpour, S.; Rezaei, B. Highly sensitive voltammetric sensor based on catechol-derivative-multiwall carbon nanotubes for the catalytic determination of captopril in patient human urine samples. *Colloids Surf. B Biointerfaces,* **2011**, *87*(2), 480-488.
[http://dx.doi.org/10.1016/j.colsurfb.2011.06.013] [PMID: 21726986]

[151] Gholivand, M.B.; Khodadadian, M. Rationally designed molecularly imprinted polymers for selective extraction of methocarbamol from human plasma. *Talanta,* **2011**, *85*(3), 1680-1688.
[http://dx.doi.org/10.1016/j.talanta.2011.06.066] [PMID: 21807239]

[152] Wan, H.; Zhao, F.; Zeng, B. Direct electrochemistry and voltammetric determination of midecamycin at a multi-walled carbon nanotube coated gold electrode. *Colloids Surf. B Biointerfaces,* **2011**, *86*(1), 247-250.
[http://dx.doi.org/10.1016/j.colsurfb.2011.03.037] [PMID: 21543194]

[153] Hu, Y.; Li, J.; Zhang, Z.; Zhang, H.; Luo, L.; Yao, S. Imprinted sol-gel electrochemical sensor for the determination of benzylpenicillin based on Fe_3O_4@SiO_2/multi-walled carbon nanotubes-chitosans nanocomposite film modified carbon electrode. *Anal. Chim. Acta,* **2011**, *698*(1-2), 61-68.
[http://dx.doi.org/10.1016/j.aca.2011.04.054] [PMID: 21645660]

[154] Chuekachang, S.; Phanichphant, S.; Chaiyasit, S. Fabrication of modified SWNTs/glassy carbon electrode for the determination of dopamine. *Mol. Cryst. Liq. Cryst. (Phila. Pa.),* **2011**, *538*, 292-297.
[http://dx.doi.org/10.1080/15421406.2011.564472]

[155] Goyal, R.N.; Rana, A.R.; Aziz, M.A.; Oyama, M. Effect of gold nanoparticle attached multi-walled carbon nanotube-layered indium tin oxide in monitoring the effect of paracetamol on the release of epinephrine. *Anal. Chim. Acta,* **2011**, *693*(1-2), 35-40.
[http://dx.doi.org/10.1016/j.aca.2011.03.026] [PMID: 21504808]

[156] Ensafi, A.A.; Dadkhah, M.; Karimi-Maleh, H. Determination of isoproterenol and uric acid by voltammetric method using carbon nanotubes paste electrode and p-chloranil. *Colloids Surf. B Biointerfaces,* **2011**, *84*(1), 148-154.
[http://dx.doi.org/10.1016/j.colsurfb.2010.12.028] [PMID: 21256720]

[157] Lu, T-L.; Tsai, Y-C. Sensitive electrochemical determination of acetaminophen in pharmaceutical formulations at multiwalled carbon nanotube-alumina-coated silica nanocomposite modified electrode. *Sens. Actuators B Chem.,* **2011**, *153*, 439-444.
[http://dx.doi.org/10.1016/j.snb.2010.11.013]

[158] Jain, R.; Rather, J.A. Voltammetric determination of antibacterial drug gemifloxacin in solubilized systems at multi-walled carbon nanotubes modified glassy carbon electrode. *Colloids Surf. B Biointerfaces,* **2011**, *83*(2), 340-346.
[http://dx.doi.org/10.1016/j.colsurfb.2010.12.003] [PMID: 21177081]

[159] Patil, R.H.; Hegde, R.N.; Nandibewoor, S.T. Electro-oxidation and determination of antihistamine drug, cetirizine dihydrochloride at glassy carbon electrode modified with multi-walled carbon nanotubes. *Colloids Surf. B Biointerfaces,* **2011**, *83*(1), 133-138.
[http://dx.doi.org/10.1016/j.colsurfb.2010.11.008] [PMID: 21145217]

[160] Wan, H.; Zhao, F.; Wu, W.; Zeng, B. Direct electron transfer and voltammetric determination of roxithromycin at a single-wall carbon nanotube coated glassy carbon electrode. *Colloids Surf. B Biointerfaces,* **2011**, *82*(2), 427-431.

[http://dx.doi.org/10.1016/j.colsurfb.2010.09.014] [PMID: 20965702]

[161] Fotouhi, L.; Nemati, M.; Heravi, M.M. Electrochemistry and voltammetric determination of furazolidone with a multi-walled nanotube composite film-glassy carbon electrode. *J. Appl. Electrochem.,* **2011**, *41*, 137-142.
[http://dx.doi.org/10.1007/s10800-010-0216-5]

[162] Mazloum-Ardakani, M.; Beitollahi, H.; Amini, M.K.; Mirkhalaf, F.; Mirjalili, B-F. A highly sensitive nanostructure-based electrochemical sensor for electrocatalytic determination of norepinephrine in the presence of acetaminophen and tryptophan. *Biosens. Bioelectron.,* **2011**, *26*(5), 2102-2106.
[http://dx.doi.org/10.1016/j.bios.2010.09.014] [PMID: 20947326]

[163] Goyal, R.N.; Bishnoi, S. Voltammetric determination of amlodipine besylate in human urine and pharmaceuticals. *Bioelectrochemistry,* **2010**, *79*(2), 234-240.
[http://dx.doi.org/10.1016/j.bioelechem.2010.06.004] [PMID: 20615763]

[164] Alothman, Z.A.; Bukhari, N.; Wabaidur, S.M.; Haider, S. Simultaneous electrochemical determination of dopamine and acetaminophen using multiwall carbon nanotubes modified glassy carbon electrode. *Sens. Actuators B Chem.,* **2010**, *146*, 314-320.
[http://dx.doi.org/10.1016/j.snb.2010.02.024]

[165] Veiga, A.; Dordio, A.; Carvalho, A.J.; Teixeira, D.M.; Teixeira, J.G. Ultra-sensitive voltammetric sensor for trace analysis of carbamazepine. *Anal. Chim. Acta,* **2010**, *674*(2), 182-189.
[http://dx.doi.org/10.1016/j.aca.2010.06.031] [PMID: 20678628]

[166] Sanghavi, B.J.; Srivastava, A.K. Simultaneous voltammetric determination of acetaminophen, aspirin and caffeine using an *in situ* surfactant-modified multiwalled carbon nanotube paste electrode. *Electrochim. Acta,* **2010**, *55*, 8638-8648.
[http://dx.doi.org/10.1016/j.electacta.2010.07.093]

[167] Khalilzadeh, M.A.; Karimi-Maleh, H.; Amiri, A.; Gholami, F.; Mazhabi, R.M. Determination of captopril in patient human urine using ferrocenemonocarboxylic acid modified carbon nanotubes paste electrode. *Chin. Chem. Lett.,* **2010**, *21*, 1467-1470.
[http://dx.doi.org/10.1016/j.cclet.2010.06.020]

[168] Adekunle, A.S.; Ozoemena, K.I. Voltammetric and impedimetric properties of nano-scaled γ-Fe_2O_3 catalysts supported on multi-walled carbon nanotubes: Catalytic detection of dopamine. *Int. J. Electrochem. Sci.,* **2010**, *5*, 1726-1742.

[169] Ensafi, A.A.; Khoddami, E.; Rezaei, B.; Karimi-Maleh, H. p-Aminophenol-multiwall carbon nanotubes-TiO_2 electrode as a sensor for simultaneous determination of penicillamine and uric acid. *Colloids Surf. B Biointerfaces,* **2010**, *81*(1), 42-49.
[http://dx.doi.org/10.1016/j.colsurfb.2010.06.020] [PMID: 20655185]

[170] Goyal, R.N.; Chatterjee, S.; Rana, A.R. The effect of modifying an edge-plane pyrolytic graphite electrode with single-wall carbon nanotubes on its use for sensing diclofenac. *Carbon,* **2010**, *48*, 4136-4144.
[http://dx.doi.org/10.1016/j.carbon.2010.07.024]

[171] Fotouhi, L.; Alahyari, M. Electrochemical behavior and analytical application of ciprofloxacin using a multi-walled nanotube composite film-glassy carbon electrode. *Colloids Surf. B Biointerfaces,* **2010**, *81*(1), 110-114.
[http://dx.doi.org/10.1016/j.colsurfb.2010.06.030] [PMID: 20655184]

[172] Ensafi, A.A.; Karimi-Maleh, H. A Multiwall carbon nanotubes paste electrode as a sensor and ferrocenemonocarboxylic acid as a mediator for electrocatalytic determination of isoproterenol. *Int. J. Electrochem. Sci.,* **2010**, *5*, 1484-1495.

[173] Goyal, R.N.; Chatterjee, S.; Rana, A.R. Effect of cetyltrimethyl ammonium bromide on electrochemical determination of dexamethasone. *Electroanalysis,* **2010**, *22*, 2330-2338.
[http://dx.doi.org/10.1002/elan.201000227]

[174] Kumar, S.A.; Wang, S-F.; Yang, T.C.; Yeh, C.T. Acid yellow 9 as a dispersing agent for carbon nanotubes: preparation of redox polymer-carbon nanotube composite film and its sensing application towards ascorbic acid and dopamine. *Biosens. Bioelectron.*, **2010**, *25*(12), 2592-2597. [http://dx.doi.org/10.1016/j.bios.2010.04.024] [PMID: 20462750]

[175] Xi, X.; Ming, L.; Liu, J. Electrochemical determination of thiamazole at a multi-wall carbon nanotube modified glassy carbon electrode. *J. Appl. Electrochem.*, **2010**, *40*, 1449-1454. [http://dx.doi.org/10.1007/s10800-010-0122-x]

[176] Lu, T-L.; Tsai, Y-C. Electrocatalytic oxidation of acetylsalicylic acid at multiwalled carbon nanotube-alumina-coated silica nanocomposite modified glassy carbon electrodes. *Sens. Actuators B Chem.*, **2010**, *148*, 590-594. [http://dx.doi.org/10.1016/j.snb.2010.05.037]

[177] Adekunle, A.S.; Agboola, B.O.; Pillay, J.; Ozoemena, K.I. Electrocatalytic detection of dopamine at single-walled carbon nanotubes–iron (III) oxide nanoparticles platform. *Sens. Actuators B Chem.*, **2010**, *148*, 93-02. [http://dx.doi.org/10.1016/j.snb.2010.03.088]

[178] Wei, Y.; Zhang, Q.; Shao, C.; Li, C.; Zhang, L.; Li, X. Voltammetric determination of salbutamol on a glassy carbon electrode coated with a nanomaterial thin film. *J. Anal. Chem.*, **2010**, *65*, 398-03. [http://dx.doi.org/10.1134/S1061934810040106]

[179] Goyal, R.N.; Bishnoi, S. Effect of single walled carbon nanotube-cetyltrimethyl ammonium bromide nanocomposite film modified pyrolytic graphite on the determination of betamethasone in human urine. *Colloids Surf. B Biointerfaces*, **2010**, *77*(2), 200-205. [http://dx.doi.org/10.1016/j.colsurfb.2010.01.024] [PMID: 20172696]

[180] Goyal, R.N.; Gupta, V.K.; Chatterjee, S. Electrochemical investigations of corticosteroid isomers--testosterone and epitestosterone and their simultaneous determination in human urine. *Anal. Chim. Acta*, **2010**, *657*(2), 147-153. [http://dx.doi.org/10.1016/j.aca.2009.10.035] [PMID: 20005326]

[181] Shahrokhian, S.; Asadian, E. Simultaneous voltammetric determination of ascorbic acid, acetaminophen and isoniazid using thionine immobilized multi-walled carbon nanotube modified carbon paste electrode. *Electrochim. Acta*, **2010**, *55*, 666-672. [http://dx.doi.org/10.1016/j.electacta.2009.08.065]

[182] Rao, C.N.; Subrahmanyam, K.S.; Matte, H.S.; Govindaraj, A. Graphene: synthesis, functionalization and properties.*Graphene and its fascinating attributes*; Pati, S.K.; Enoki, T.; Rao, C.N., Eds.; World Scientific Publishing: Singapore, **2011**, pp. 1-32. [http://dx.doi.org/10.1142/9789814329361_0001]

[183] Choi, W.; Li, J-w., Eds. *Graphene: Synthesis and applications*; CRC Press: Boca Raton, FL, **2012**.

[184] Ray, S.C. *Applications of graphene and graphene-oxide based nanomaterials*; Elsevier: Amsterdam, **2015**.

[185] Bose, S.; Kuila, T.; Kim, N.H.; Lee, J.H. Graphene produced by electrochemical exfoliation.*Graphene: properties, preparation, characterisation and devices*; Skakalova, V.; Kaiser, A.B., Eds.; Woodhead Publishing: Cambridge, **2014**, pp. 81-98. [http://dx.doi.org/10.1533/9780857099334.1.81]

[186] Wisitsoraat, A.; Tuantranont, A. Graphene-based chemical and biosensors.*Applications of nanomaterials in sensors and diagnostics*; Tuantranont, A., Ed.; Springer: Heidelberg, **2013**, pp. 103-142. [http://dx.doi.org/10.1007/5346_2012_47]

[187] Rao, C.N.; Maitra, U.; Matte, H.S. Synthesis, characterization, and selected properties of graphene.*Graphene: synthesis, properties and phenomena*; Rao, C.N.; Sood, A.K., Eds.; Wiley-VCH Verlag: Weinheim, **2013**, pp. 1-48.

[188] Chen, D.; Tang, L.; Li, J. Graphene-based materials in electrochemistry. *Chem. Soc. Rev.,* **2010**, *39*(8), 3157-3180.
[http://dx.doi.org/10.1039/b923596e] [PMID: 20589275]

[189] Motaharian, A.; Hosseini, M.R. Electrochemical sensor based on a carbon paste electrode modified by graphene nanosheets and molecularly imprinted polymer nanoparticles for determination of a chlordiazepoxide drug. *Anal. Methods,* **2016**, *8*, 6305-6312.
[http://dx.doi.org/10.1039/C6AY01594H]

[190] Galal, A.; Atta, N.F.; Hassan, H.K. Graphene as electrochemical sensor and biosensor: Synthesis, characterization and applications.*Nanosensors: Materials and technologies*; Atta, N.F., Ed.; IFSA Publishing: Barcelona, **2013**, pp. 47-104.

[191] Kim, Y-R.; Bong, S.; Kang, Y-J.; Yang, Y.; Mahajan, R.K.; Kim, J.S.; Kim, H. Electrochemical detection of dopamine in the presence of ascorbic acid using graphene modified electrodes. *Biosens. Bioelectron.,* **2010**, *25*(10), 2366-2369.
[http://dx.doi.org/10.1016/j.bios.2010.02.031] [PMID: 20307965]

[192] Kang, X.; Wang, J.; Wu, H.; Liu, J.; Aksay, I.A.; Lin, Y. A graphene-based electrochemical sensor for sensitive detection of paracetamol. *Talanta,* **2010**, *81*(3), 754-759.
[http://dx.doi.org/10.1016/j.talanta.2010.01.009] [PMID: 20298849]

[193] Parvin, M.H. Graphene paste electrode for detection of chlorpromazine. *Electrochem. Commun.,* **2011**, *13*, 366-369.
[http://dx.doi.org/10.1016/j.elecom.2011.01.027]

[194] Iniesta, J.; Garcia-Cruz, L.; Gomis-Berenguer, A.; Ania, C.O. Carbon materials based on screen-printing electrochemical platforms in biosensing applications.*Electrochemistry*; Bangs, C.; Mortimer, R.; McIntosh, S., Eds.; Royal society of Chemistry, **2015**, Vol. 13, pp. 133-169.
[http://dx.doi.org/10.1039/9781782620273-00133]

[195] Zhang, Q.; Zhang, L.; Liu, Y.; Li, J. Graphene-based electrochemical biosensor.*Handbook of carbon nano materials: Graphene – energy and sensor applications*; D'Souza, F.; Kadish, K.M.; McIntosh, S., Eds.; World Scientific Publishing: Singapore, **2014**, Vol. 6, pp. 147-188.
[http://dx.doi.org/10.1142/9789814566704_0008]

[196] Wong, A.; Razzino, C.A.; Silva, T.A.; Fatibello-Filho, O. Square-wave voltammetric determination of clindamycin using a glassy carbon electrode modified with graphene oxide and gold nanoparticles within a crosslinked chitosan film. *Sens. Actuators B Chem.,* **2016**, *231*, 183-193.
[http://dx.doi.org/10.1016/j.snb.2016.03.014]

[197] Fan, Y.; Liu, J-H.; Lu, H-T.; Zhang, Q. Electrochemical behavior and voltammetric determination of paracetamol on Nafion/TiO_2-graphene modified glassy carbon electrode. *Colloids Surf. B Biointerfaces,* **2011**, *85*(2), 289-292.
[http://dx.doi.org/10.1016/j.colsurfb.2011.02.041] [PMID: 21435844]

[198] Sheng, Z-H.; Zheng, X-Q.; Xu, J-Y.; Bao, W-J.; Wang, F-B.; Xia, X-H. Electrochemical sensor based on nitrogen doped graphene: simultaneous determination of ascorbic acid, dopamine and uric acid. *Biosens. Bioelectron.,* **2012**, *34*(1), 125-131.
[http://dx.doi.org/10.1016/j.bios.2012.01.030] [PMID: 22342696]

[199] Wolf, E.L. *Applications of graphene*; Springer: Heidelberg, **2014**.
[http://dx.doi.org/10.1007/978-3-319-03946-6]

[200] Hu, Y.; Li, F.; Han, D.; Niu, L. *Biocompatible graphene for bioanalytical applications*; Springer: Heidelberg, **2015**.
[http://dx.doi.org/10.1007/978-3-662-45695-8]

[201] Liu, J.; Dai, L. High-performance polymer solar cells containin carbon nanomaterials.*Carbon nanomaterials for advanced energy systems: Advances in materials synthesis and device applications*; Lu, W.; Baek, J-B.; Dai, L., Eds.; John Wiley & Sons: Hoboken, NJ, **2015**, pp. 165-190.

[http://dx.doi.org/10.1002/9781118980989.ch5]

[202] Scalise, E. *Vibrational properties of defective oxides and 2D nanolattices*; Springer: Heidelberg, **2014**. [http://dx.doi.org/10.1007/978-3-319-07182-4]

[203] Lanza, M.; Wang, Y.; Gao, T. Electrical and mechanical performance of graphene sheets exposed to oxidative environments. *Nano Res.*, **2013**, *6*, 485-495. [http://dx.doi.org/10.1007/s12274-013-0326-6]

[204] Jabbari, E. Nanoparticles for stem-cell engineering.*Stem-cell nanoengineering*; Baharvand, H.; Aghdami, N., Eds.; John Wiley & Sons: Hoboken, NJ, **2015**, pp. 143-170.

[205] Sheng, K.; Wang, L.; Li, H.; Zou, L.; Ye, B. Green synthesized Co nanoparticles doped amino-graphene modified electrode and its application towards determination of baicalin. *Talanta,* **2017**, *164*, 249-256. [http://dx.doi.org/10.1016/j.talanta.2016.11.050] [PMID: 28107925]

[206] Ejaz, A.; Joo, Y.; Jeon, S. Fabrication of 1,4-bis(aminomethyl)benzene and cobalt hydroxide @graphene oxide for selective detection of dopamine in the presence ofascorbic acid and serotonin. *Sens. Actuators B Chem.*, **2017**, *240*, 297-07. [http://dx.doi.org/10.1016/j.snb.2016.08.171]

[207] Kumar, N.; Sharma, R.; Goyal, R.N. Gold-palladium nanoparticles aided electrochemically reduced graphene oxide sensor for the simultaneous estimation of lomefloxacin and amoxicillin. *Sens. Actuators B Chem.*, **2017**, *243*, 658-668. [http://dx.doi.org/10.1016/j.snb.2016.12.025]

[208] Baghayeri, M. Pt nanoparticles/reduced graphene oxide nanosheets as a sensing platform: Application to determination of droxidopa in presence of phenobarbital. *Sens. Actuators B Chem.*, **2017**, *240*, 255-263. [http://dx.doi.org/10.1016/j.snb.2016.08.161]

[209] Asadian, E.; Shahrokhian, S.; Zad, A.I.; Ghorbani-Bidkorbeh, F. Glassy carbon electrode modified with 3D graphene–carbon nanotube network for sensitive electrochemical determination of methotrexate. *Sens. Actuators B Chem.*, **2017**, *239*, 617-627. [http://dx.doi.org/10.1016/j.snb.2016.08.064]

[210] Er, E.; Çelikkan, H.; Erk, N. A novel electrochemical nano-platform based on graphene/platinum nanoparticles/nafion composites for the electrochemical sensing of metoprolol. *Sens. Actuators B Chem.*, **2017**, *238*, 779-787. [http://dx.doi.org/10.1016/j.snb.2016.07.108]

[211] Liu, Z.; Xue, Q.; Guo, Y. Sensitive electrochemical detection of rutin and isoquercitrin based on SH---cyclodextrin functionalized graphene-palladium nanoparticles. *Biosens. Bioelectron.*, **2017**, *89*(Pt 1), 444-452. [http://dx.doi.org/10.1016/j.bios.2016.04.056] [PMID: 27133027]

[212] Kesavana, S.; Kumar, D.R.; Lee, Y.R.; Shim, J-J. Determination of tetracycline in the presence of major interference inhuman urine samples using polymelamine/electrochemically reducedgraphene oxide modified electrode. *Sens. Actuators B Chem.*, **2017**, *241*, 455-465. [http://dx.doi.org/10.1016/j.snb.2016.10.091]

[213] Jain, R. Dhanjai, Sinha A. Graphene-zinc oxide nanorods nanocomposite based sensor forvoltammetric quantification of tizanidine in solubilized system. *Appl. Surf. Sci.*, **2016**, *369*, 151-158. [http://dx.doi.org/10.1016/j.apsusc.2016.02.077]

[214] Shabani-Nooshabadi, M.; Roostaee, M. Modification of carbon paste electrode with NiO/graphene oxide nanocomposite and ionic liquids for fabrication of high sensitive voltammetric sensor on sulfamethoxazole analysis. *J. Mol. Liq.*, **2016**, *220*, 329-333. [http://dx.doi.org/10.1016/j.molliq.2016.05.001]

[215] Raj, M.A.; Gowthaman, N.S.; John, S.A. Highly sensitive interference-free electrochemical determination of pyridoxine at graphene modified electrode: Importance in Parkinson and Asthma treatments. *J. Colloid Interface Sci.,* **2016**, *474*, 171-178. [http://dx.doi.org/10.1016/j.jcis.2016.04.025] [PMID: 27124811]

[216] Wong, A.; Silva, T.A.; Vicentini, F.C.; Fatibello-Filho, O. Electrochemical sensor based on graphene oxide and ionic liquid for ofloxacin determination at nanomolar levels. *Talanta,* **2016**, *161*, 333-341. [http://dx.doi.org/10.1016/j.talanta.2016.08.035] [PMID: 27769415]

[217] Alizadeh, T.; Azizi, S. Graphene/graphite paste electrode incorporated with molecularly imprinted polymer nanoparticles as a novel sensor for differential pulse voltammetry determination of fluoxetine. *Biosens. Bioelectron.,* **2016**, *81*, 198-206. [http://dx.doi.org/10.1016/j.bios.2016.02.052] [PMID: 26946258]

[218] Jiang, Z.; Li, G.; Zhang, M. Electrochemical sensor based on electro-polymerization of β-cyclodextrin and reduced-graphene oxide on glassy carbon electrode for determination of gatifloxacin. *Sens. Actuators B Chem.,* **2016**, *228*, 59-65. [http://dx.doi.org/10.1016/j.snb.2016.01.013]

[219] Tandel, R.; Teradal, N.; Satpati, A.; Jaladappagari, S. Fabrication of the electrochemically reduced graphene oxide-bismuth 3 nanoparticles composite and its analytical application for an anticancer. *Chin. Chem. Lett.,* **2016**, *28*(7), 1429-1437. [http://dx.doi.org/10.1016/j.cclet.2016.11.028]

[220] Er, E.; Çelikkan, H.; Erk, N. Highly sensitive and selective electrochemical sensor based onhigh-quality graphene/nafion nanocomposite for voltammetric determination of nebivolol. *Sens. Actuators B Chem.,* **2016**, *224*, 170-177. [http://dx.doi.org/10.1016/j.snb.2015.10.028]

[221] Jain, R.; Sinha, A.; Khan, A.L. Polyaniline-graphene oxide nanocomposite sensor for quantification of calcium channel blocker levamlodipine. *Mater. Sci. Eng. C,* **2016**, *65*, 205-214. [http://dx.doi.org/10.1016/j.msec.2016.03.115] [PMID: 27157745]

[222] Yiğit, A.; Yardım, Y.; Çelebi, M.; Levent, A.; Şentürk, Z. Graphene/Nafion composite film modified glassy carbon electrode for simultaneous determination of paracetamol, aspirin and caffeine in pharmaceutical formulations. *Talanta,* **2016**, *158*, 21-29. [http://dx.doi.org/10.1016/j.talanta.2016.05.046] [PMID: 27343573]

[223] Chekin, F.; Teodorescu, F.; Coffinier, Y.; Pan, G.H.; Barras, A.; Boukherroub, R.; Szunerits, S. MoS_2/reduced graphene oxide as active hybrid material for the electrochemical detection of folic acid in human serum. *Biosens. Bioelectron.,* **2016**, *85*, 807-813. [http://dx.doi.org/10.1016/j.bios.2016.05.095] [PMID: 27288713]

[224] Hashemzadeh, N.; Hasanzadeh, M.; Shadjou, N.; Eivazi-Ziaei, J.; Khoubnasabjafari, M.; Jouyban, A. Graphene quantum dot modified glassy carbon electrode for the determination of doxorubicin hydrochloride in human plasma. *J. Pharm. Anal.,* **2016**, *6*, 235-241. [http://dx.doi.org/10.1016/j.jpha.2016.03.003]

[225] Thirumalraj, B.; Palanisamy, S.; Chen, S-M.; Lou, B-S. Preparation of highly stable fullerene C_{60} decorated graphene oxide nanocomposite and its sensitive electrochemical detection of dopamine in rat brain and pharmaceutical samples. *J. Colloid Interface Sci.,* **2016**, *462*, 375-381. [http://dx.doi.org/10.1016/j.jcis.2015.10.009] [PMID: 26484603]

[226] Taei, M.; Hasanpour, F.; Dinari, M.; Dehghani, E. Au nanoparticles decorated reduced graphene oxide/layered double hydroxide modified glassy carbon as a sensitive sensor for electrocatalytic determination of phenazopyridine. *Measurement,* **2016**, *99*, 90-97. [http://dx.doi.org/https://doi.org/10.1016/j.measurement.2016.12.012]

[227] Chen, L.; Guo, X.; Guo, B.; Cheng, S.K.; Wang, F. Electrochemical investigation of a metalloporphyrin–graphene composite modified electrode and its electrocatalysis on ascorbic acid. *J. Electroanal. Chem.,* **2016**, *760*, 105-112.

[http://dx.doi.org/10.1016/j.jelechem.2015.11.018]

[228] Asadian, E.; Iraji Zad, A.; Shahrokhian, S. Voltammetric studies of Azathioprine on the surface of graphite electrode modified with graphene nanosheets decorated with Ag nanoparticles. *Mater. Sci. Eng. C,* **2016**, *58*, 1098-1104. [http://dx.doi.org/10.1016/j.msec.2015.09.022] [PMID: 26478409]

[229] Tadayon, F.; Vahed, S.; Bagheri, H. Au-Pd/reduced graphene oxide composite as a new sensing layer for electrochemical determination of ascorbic acid, acetaminophen and tyrosine. *Mater. Sci. Eng. C,* **2016**, *68*, 805-813. [http://dx.doi.org/10.1016/j.msec.2016.07.039] [PMID: 27524083]

[230] Kong, F-Y.; Chen, T-T.; Wang, J-Y.; Fang, H-L.; Fan, D-H.; Wang, W. UV-assisted synthesis of tetrapods-like titanium nitride-reduced graphene oxide nanohybrids for electrochemical determination of chloramphenicol. *Sens. Actuators B Chem.,* **2016**, *225*, 298-04. [http://dx.doi.org/10.1016/j.snb.2015.11.041]

[231] Bagheri, H.; Pajooheshpour, N.; Jamali, B.; Amidi, S.; Hajian, A.; Khoshsafar, H. A novel electrochemical platform for sensitive and simultaneous determination of dopamine, uric acid and ascorbic acid based on Fe_3O_4-SnO_2-Gr ternary nanocomposite. *Microchem. J.,* **2016**, *131*, 120-129. [http://dx.doi.org/10.1016/j.microc.2016.12.006]

[232] Zhang, Z.; Zhang, J.; Zhang, H.; Xu, J.; Wen, Y.; Ding, W. Characterization of PEDOT:PSS-reduced graphene oxide@Pd composite electrode and its application in voltammetric determination of vitamin K3. *J. Electroanal. Chem.,* **2016**, *775*, 258-266. [http://dx.doi.org/10.1016/j.jelechem.2016.06.005]

[233] Kalambate, P.K.; Rawool, C.R.; Srivastava, A.K. Voltammetric determination of pyrazinamide at graphene-zinc oxide nanocomposite modified carbon paste electrode employing differential pulse voltammetry. *Sens. Actuators B Chem.,* **2016**, *237*, 196-05. [http://dx.doi.org/10.1016/j.snb.2016.06.019]

[234] Daneshvar, L.; Rounaghi, G.H.; Es'haghi, Z.; Chamsaz, M.; Tarahomi, S. Fabrication a new modified electrochemical sensor based on Au-Pd bimetallic nanoparticle decorated graphene for citalopram determination. *Mater. Sci. Eng. C,* **2016**, *69*, 653-660. [http://dx.doi.org/10.1016/j.msec.2016.07.025] [PMID: 27612758]

[235] Adhikari, B-R.; Govindhan, M.; Schraft, H.; Chen, A. Simultaneous and sensitive detection of acetaminophen and valacyclovir based on two dimensional graphene nanosheets. *J. Electroanal. Chem.,* **2016**, *780*, 241-248. [http://dx.doi.org/10.1016/j.jelechem.2016.09.023]

[236] Zhang G, Shi L, Pan S, Liu W, Pan H. Au/ZnO hybrid nanocatalysts impregnated in N-doped graphene for simultaneous determination of ascorbic acid, acetaminophen and dopamine. *Mater. Sci. Eng. C,* **2016**, *65*, 80-89.

[237] Daneshvar, L.; Rounaghi, G.; E'shaghi, Z.; Chamsaz, M.; Tarahomi, S. Electrochemical determination of carbamazepin in the presence of paracetamol using a carbon ionic liquid paste electrode modified with a three-dimensional graphene/MWCNT hybrid composite film. *J. Mol. Liq.,* **2016**, *215*, 316-322. [http://dx.doi.org/10.1016/j.molliq.2015.12.059]

[238] Hasanzadeh, M.; Hashemzadeh, N.; Shadjoue, N.; Eivazi-Ziaei, J.; Khoubnasabjafari, M.; Jouyban, A. Sensing of doxorubicin hydrochloride using graphene quantum dot modified glassy carbon electrode. *J. Mol. Liq.,* **2016**, *221*, 354-357. [http://dx.doi.org/10.1016/j.molliq.2016.05.082]

[239] Kong, F-Y.; Luo, Y.; Zhang, J-W.; Wang, J-Y.; Li, W-W.; Wang, W. Facile synthesis of reduced graphene oxide supported Pt-Pd nanocubes with enhanced electrocatalytic activity for chloramphenicol determination. *J. Electroanal. Chem.,* **2016**, *781*, 389-394. [http://dx.doi.org/10.1016/j.jelechem.2016.06.018]

[240] Yang, H.; Liu, B.; Ding, Y.; Li, L.; Ouyang, X. Fabrication of cuprous oxide nanoparticles-graphene

nanocomposite for determination of acetaminophen. *J. Electroanal. Chem.,* **2015**, *757*, 88-93. [http://dx.doi.org/10.1016/j.jelechem.2015.09.014]

[241] Kalambate, P.K.; Sanghavi, B.J.; Karna, S.P.; Srivastava, A.K. Simultaneous voltammetric determination of paracetamol and domperidone based on a graphene/platinum nanoparticles/nafion composite modified glassy carbon electrode. *Sens. Actuators B Chem.,* **2015**, *213*, 285-294. [http://dx.doi.org/10.1016/j.snb.2015.02.090]

[242] Zhai, H.; Liang, Z.; Chen, Z.; Wang, H.; Liu, Z.; Su, Z.; Zhou, Q. Simultaneous detection of metronidazole and chloramphenicol by differential pulse stripping voltammetry using a silver nanoparticles/sulfonate functionalized graphene modified glassy carbon electrode. *Electrochim. Acta,* **2015**, *171*, 105-113. [http://dx.doi.org/10.1016/j.electacta.2015.03.140]

[243] Benvidi, A.; Dehghani-Firouzabadi, A.; Mazloum-Ardakani, M.; Mirjalili, B-B.; Zare, R. Electrochemical deposition of gold nanoparticles on reduced graphene oxide modified glassy carbon electrode for simultaneous determination of levodopa, uric acid and folic acid. *J. Electroanal. Chem.,* **2015**, *736*, 22-29. [http://dx.doi.org/10.1016/j.jelechem.2014.10.020]

[244] Guo, Z.; Wang, Z.Y.; Wang, H.H.; Huang, G.Q.; Li, M.M. Electrochemical sensor for Isoniazid based on the glassy carbon electrode modified with reduced graphene oxide-Au nanomaterials. *Mater. Sci. Eng. C,* **2015**, *57*, 197-204. [http://dx.doi.org/10.1016/j.msec.2015.07.045] [PMID: 26354255]

[245] Zhu, X.; Xu, J.; Duan, X. Controlled synthesis of partially reduced graphene oxide: Enhance electrochemical determination of isoniazid with high sensitivity and stability. *J. Electroanal. Chem.,* **2015**, *757*, 183-191. [http://dx.doi.org/10.1016/j.jelechem.2015.09.038]

[246] Yan, H.; Xiao, H.; Xiea, Q. Simultaneous electroanalysis of isoniazid and uric acid at poly(sulfosalicylic acid)/electroreduced carboxylated graphene modified glassy carbon electrode. *Sens. Actuators B Chem.,* **2015**, *207*, 167-176. [http://dx.doi.org/10.1016/j.snb.2014.10.002]

[247] Beitollahi, H.; Hamzavi, M.; Torkzadeh-Mahani, M. Electrochemical determination of hydrochlorothiazide and folic acid in real samples using a modified graphene oxide sheet paste electrode. *Mater. Sci. Eng. C,* **2015**, *52*, 297-305. [http://dx.doi.org/10.1016/j.msec.2015.03.031] [PMID: 25953571]

[248] Wang, X.; You, Z.; Cheng, Y. Application of nanosized gold and graphene modified carbon ionic liquid electrode for the sensitive electrochemical determination of folic acid. *J. Mol. Liq.,* **2015**, *204*, 112-117. [http://dx.doi.org/10.1016/j.molliq.2015.01.036]

[249] Majidi, M.R.; Ghaderi, S.; Asadpour-Zeynali, K.; Dastangoo, H. Synthesis of dendritic silver nanostructures supported by graphene nanosheets and its application for highly sensitive detection of diazepam. *Mater. Sci. Eng. C,* **2015**, *57*, 257-264. [http://dx.doi.org/10.1016/j.msec.2015.07.037] [PMID: 26354262]

[250] Zhai, H.; Liu, Z.; Chen, Z.; Liang, Z.; Su, Z.; Wang, S. A sensitive electrochemical sensor with sulfonated graphene sheets/oxygen-functionalized multi-walled carbon nanotubes modified electrode for the detection of clenbuterol. *Sens. Actuators B Chem.,* **2015**, *210*, 483-490. [http://dx.doi.org/10.1016/j.snb.2014.12.121]

[251] Adhikari, B-R.; Govindhan, M.; Chen, A. Sensitive detection of acetaminophen with graphene-based electrochemical sensor. *Electrochim. Acta,* **2015**, *162*, 198-04. [http://dx.doi.org/10.1016/j.electacta.2014.10.028]

[252] Sinha, A. Dhanjai, Jain R. Electrocatalytic determination of α 2-adrenergic agonist tizanidine at graphene–silicon dioxide nanocomposite sensor. *Mater. Res. Bull.,* **2015**, *65*, 307-314.

[http://dx.doi.org/10.1016/j.materresbull.2015.02.001]

[253] Silva, T.A.; Zanin, H.; Vicentinia, F.C.; Corat, E.J.; Fatibello-Filho, O. Electrochemical determination of rosuvastatin calcium in pharmaceutical and human body fluid samples using a composite of vertically aligned carbon nanotubes and graphene oxide as the electrode material. *Sens. Actuators B Chem.*, **2015**, *218*, 51-59. [http://dx.doi.org/10.1016/j.snb.2015.04.085]

[254] Wong, A.; Scontri, M.; Materon, E.M.; Lanza, M.R.; Sotomayor, M.D. Development and application of an electrochemical sensor modified with multi-walled carbon nanotubes and graphene oxide for the sensitive and selective detection of tetracycline. *J. Electroanal. Chem.*, **2015**, *757*, 250-257. [http://dx.doi.org/10.1016/j.jelechem.2015.10.001]

[255] Er, E.; Çelikkan, H.; Erk, N.; Aksu, M.L. A new generation electrochemical sensor based on Graphene nanosheets/Gold nanoparticles/Nafion nanocomposite for determination of Silodosin. *Electrochim. Acta,* **2015**, *157*, 252-257. [http://dx.doi.org/10.1016/j.electacta.2015.01.020]

[256] Yang, S.; Li, G.; Wang, G.; Zhao, J.; Qiao, Z.; Qu, L. Decoration of chemically reduced graphene oxide modified carbon paste electrode with yttrium hexacyanoferrate nanoparticles for nanomolar detection of rutin. *Sens. Actuators B Chem.*, **2015**, *206*, 126-132. [http://dx.doi.org/10.1016/j.snb.2014.09.027]

[257] Zhang, K.; Xu, J.; Zhu, X. Poly(3,4-ethylenedioxythiophene) nanorods grown on graphene oxide sheets as electrochemical sensing platform for rutin. *J. Electroanal. Chem.*, **2015**, *739*, 66-72. [http://dx.doi.org/10.1016/j.jelechem.2014.12.013]

[258] Kalambate, P.K.; Biradar, M.R.; Karna, S.P.; Srivastava, A.K. Adsorptive stripping differential pulse voltammetry determination of rivastigmine at graphene nanosheet-gold nanoparticle/carbon paste electrode. *J. Electroanal. Chem.*, **2015**, *757*, 150-158. [http://dx.doi.org/10.1016/j.jelechem.2015.09.027]

[259] Lei, W.; Si, W.; Hao, Q.; Han, Z.; Zhang, Y.; Xia, M. Nitrogen-doped graphene modified electrode for nimodipine sensing. *Sens. Actuators B Chem.*, **2015**, *212*, 207-213. [http://dx.doi.org/10.1016/j.snb.2015.01.126]

[260] Shang, L.; Zhao, F.; Zeng, B. Highly dispersive hollow PdAg alloy nanoparticles modified ionic liquid functionalized graphene nanoribbons for electrochemical sensing of nifedipine. *Electrochim. Acta,* **2015**, *168*, 330-336. [http://dx.doi.org/10.1016/j.electacta.2015.04.024]

[261] Devadas, B.; Cheemalapati, S.; Chen, S-M.; Ali, M.A.; Al-Hemaid, F.M. Highly sensing graphene oxide/poly-arginine-modified electrode for the simultaneous electrochemical determination of buspirone, isoniazid and pyrazinamide drugs. *Ionics,* **2015**, *21*, 547-555. [http://dx.doi.org/10.1007/s11581-014-1179-z]

[262] Karuppiah, C.; Cheemalapati, S.; Chen, S-M.; Palanisamy, S. Carboxyl-functionalized graphene oxide-modified electrode for the electrochemical determination of nonsteroidal anti-inflammatory drug diclofenac. *Ionics,* **2015**, *21*, 231-238. [http://dx.doi.org/10.1007/s11581-014-1161-9]

[263] Liu, G-T.; Chen, H-F.; Lin, G-M.; Ye, P.P.; Wang, X.P.; Jiao, Y.Z.; Guo, X.Y.; Wen, Y.; Yang, H.F. One-step electrodeposition of graphene loaded nickel oxides nanoparticles for acetaminophen detection. *Biosens. Bioelectron.*, **2014**, *56*, 26-32. [http://dx.doi.org/10.1016/j.bios.2014.01.005] [PMID: 24462827]

[264] Kruanetr, S.; Pollard, P.; Fernandez, C.; Prabhu, R. Electrochemical oxidation of acetyl salicylic acid and its voltammetric sensing in real samples at a sensitive edge plane pyrolytic graphite electrode modified with graphene. *Int. J. Electrochem. Sci.*, **2014**, *9*, 5699-5711.

[265] Silva, T.A.; Zanin, H.; Vicentini, F.C.; Corat, E.J.; Fatibello-Filho, O. Differential pulse adsorptive stripping voltammetric determination of nanomolar levels of atorvastatin calcium in pharmaceutical

and biological samples using a vertically aligned carbon nanotube/graphene oxide electrode. *Analyst (Lond.),* **2014**, *139*(11), 2832-2841.
[http://dx.doi.org/10.1039/c4an00111g] [PMID: 24752746]

[266] Zhang, X.; Wei, Y.; Ding, Y. Electrocatalytic oxidation and voltammetric determination of ciprofloxacin employing poly(alizarin red)/graphene composite film in the presence of ascorbic acid, uric acid and dopamine. *Anal. Chim. Acta,* **2014**, *835*, 29-36.
[http://dx.doi.org/10.1016/j.aca.2014.05.020] [PMID: 24952626]

[267] Ezhil Vilian, A.T.; Rajkumar, M.; Chen, S-M. *In situ* electrochemical synthesis of highly loaded zirconium nanoparticles decorated reduced graphene oxide for the selective determination of dopamine and paracetamol in presence of ascorbic acid. *Colloids Surf. B Biointerfaces,* **2014**, *115*, 295-301.
[http://dx.doi.org/10.1016/j.colsurfb.2013.12.014] [PMID: 24384145]

[268] Jain, R.; Shrivastava, S. A graphene-polyaniline-Bi_2O_3 hybrid film sensor for voltammetric quantification of anti-inflammatory drug etodolac. *J. Electrochem. Soc.,* **2014**, *161*, H189-H194.
[http://dx.doi.org/10.1149/2.043404jes]

[269] Prashanth, S.N.; Teradal, N.L.; Seetharamappa, J.; Satpati, A.K.; Reddy, A.V. Fabrification of electroreduced graphene oxide–bentonite sodium composite modified electrode and its sensing application for linezolid. *Electrochim. Acta,* **2014**, *133*, 49-56.
[http://dx.doi.org/10.1016/j.electacta.2014.04.022]

[270] Smarzewska, S.; Ciesielski, W. Electroanalysis of pindolol on a GCE modified with reduced graphene oxide. *Anal. Methods,* **2014**, *6*, 5038-5046.
[http://dx.doi.org/10.1039/c4ay00648h]

[271] Sanghavi, B.J.; Kalambate, P.K.; Karna, S.P.; Srivastava, A.K. Voltammetric determination of sumatriptan based on a graphene/gold nanoparticles/Nafion composite modified glassy carbon electrode. *Talanta,* **2014**, *120*, 1-9.
[http://dx.doi.org/10.1016/j.talanta.2013.11.077] [PMID: 24468334]

[272] Saxena, S.; Shrivastava, R.; Satsangee, S.P.; Srivastava, S. TiO_2/Graphene/Chitosan-nanocomposit--based electrochemical sensor for the sensing of anti-HIV drug topotecan. *J. Electrochem. Soc.,* **2014**, *161*, H934-H940.
[http://dx.doi.org/10.1149/2.0891414jes]

[273] Arvand, M.; Gholizadeh, T.M. Gold nanorods–graphene oxide nanocomposite incorporated carbon nanotube paste modified glassy carbon electrode for voltammetric determination of indomethacin. *Sens. Actuators B Chem.,* **2013**, *186*, 622-632.
[http://dx.doi.org/10.1016/j.snb.2013.06.059]

[274] Zheng, M.; Gao, F.; Wang, Q.; Cai, X.; Jiang, S.; Huang, L.; Gao, F. Electrocatalytical oxidation and sensitive determination of acetaminophen on glassy carbon electrode modified with graphene-chitosan composite. *Mater. Sci. Eng. C,* **2013**, *33*(3), 1514-1520.
[http://dx.doi.org/10.1016/j.msec.2012.12.055] [PMID: 23827603]

[275] Yola, M.L.; Atar, N.; Üstündağ, Z.; Solak, A.O. A novel voltammetric sensor based on p-aminothiophenol functionalized graphene oxide/gold nanoparticles for determining quercetin in the presence of ascorbic acid. *J. Electroanal. Chem.,* **2013**, *698*, 9-16.
[http://dx.doi.org/10.1016/j.jelechem.2013.03.016]

[276] Jain, R.; Dhanjai, Nano graphene based sensor for antiarrhythmic agent quinidine in solubilized system. *Colloids Surf. B Biointerfaces,* **2013**, *105*, 278-283.
[http://dx.doi.org/10.1016/j.colsurfb.2013.01.015] [PMID: 23384690]

[277] Ghalkhani, M.; Shahrokhian, S. Adsorptive stripping differential pulse voltammetric determination of mebendazole at a graphene nanosheets and carbon nanospheres/chitosan modified glassy carbon electrode. *Sens. Actuators B Chem.,* **2013**, *185*, 669-674.
[http://dx.doi.org/10.1016/j.snb.2013.05.054]

[278] Arvand, M.; Gholizadeh, T.M. Simultaneous voltammetric determination of tyrosine and paracetamol using a carbon nanotube-graphene nanosheet nanocomposite modified electrode in human blood serum and pharmaceuticals. *Colloids Surf. B Biointerfaces,* **2013**, *103*, 84-93.

[279] Cheemalapati, S.; Palanisamy, S.; Mani, V.; Chen, S-M. Simultaneous electrochemical determination of dopamine and paracetamol on multiwalled carbon nanotubes/graphene oxide nanocomposite-modified glassy carbon electrode. *Talanta,* **2013**, *117*, 297-304. [http://dx.doi.org/10.1016/j.talanta.2013.08.041] [PMID: 24209344]

[280] Li, Y.; Li, K.; Song, G.; Liu, J.; Zhang, K.; Ye, B. Electrochemical behavior of codeine and its sensitive determination on graphene-based modified electrode. *Sens. Actuators B Chem.,* **2013**, *182*, 401-407. [http://dx.doi.org/10.1016/j.snb.2013.03.023]

[281] Kaur, B.; Pandiyan, T.; Satpati, B.; Srivastava, R. Simultaneous and sensitive determination of ascorbic acid, dopamine, uric acid, and tryptophan with silver nanoparticles-decorated reduced graphene oxide modified electrode. *Colloids Surf. B Biointerfaces,* **2013**, *111*, 97-106.

[282] Niu, X.; Yang, W.; Guo, H.; Ren, J.; Gao, J. Highly sensitive and selective dopamine biosensor based on 3,4,9,10-perylene tetracarboxylic acid functionalized graphene sheets/multi-wall carbon nanotubes/ionic liquid composite film modified electrode. *Biosens. Bioelectron.,* **2013**, *41*, 225-231. [http://dx.doi.org/10.1016/j.bios.2012.08.025] [PMID: 22951031]

[283] Ye, L.; Xiang, M.; Zhang, Y. A novel electrochemical method for sensitive detection of anticancer drug picoplatin with graphene multi-walled carbon nanotubes modified glassy carbon electrode. *Int. J. Electrochem. Sci.,* **2013**, *8*, 12726-12734.

[284] Li, Y.; Ye, Z.; Zhou, J. A new voltammetric sensor based on poly(L-arginine)/graphene–Nafion composite film modified electrode for sensitive determination of Terbutaline sulfate. *J. Electroanal. Chem.,* **2012**, *687*, 51-57. [http://dx.doi.org/10.1016/j.jelechem.2012.09.045]

[285] Peng, J.; Hou, C.; Hu, X. Determination of metronidazole in pharmaceutical dosage forms based on reduction at graphene and ionic liquid composite film modified electrode. *Sens. Actuators B Chem.,* **2012**, *169*, 81-87. [http://dx.doi.org/10.1016/j.snb.2012.03.040]

[286] Zi, L.; Li, J.; Mao, Y.; Yang, R.; Qu, L. High sensitive determination of theophylline based on gold nanoparticles/l-cysteine/Graphene/Nafion modified electrode. *Electrochim. Acta,* **2012**, *78*, 434-439. [http://dx.doi.org/10.1016/j.electacta.2012.06.018]

[287] Xu, C-X.; Huang, K-J.; Fan, Y.; Wu, Z-W.; Li, J. Electrochemical determination of acetaminophen based on TiO_2–graphene/poly(methyl red) composite film modified electrode. *J. Mol. Liq.,* **2012**, *165*, 32-37. [http://dx.doi.org/10.1016/j.molliq.2011.10.006]

[288] Gao, F.; Qi, X.; Cai, X.; Wang, Q.; Gao, F.; Sun, W. Electrochemically reduced graphene modified carbon ionic liquid electrode for the sensitive sensing of rutin. *Thin Solid Films,* **2012**, *520*, 5064-5069. [http://dx.doi.org/10.1016/j.tsf.2012.03.002]

[289] Peng, J.Y.; Hou, C.T.; Liu, X.X.; Li, H.B.; Hu, X.Y. Electrochemical behavior of azithromycin at graphene and ionic liquid composite film modified electrode. *Talanta,* **2011**, *86*, 227-232.

[290] Liu, K.; Wei, J.; Wang, C. Sensitive detection of rutin based on β-cyclodextrin@chemically reduced graphene/Nafion composite film. *Electrochim. Acta,* **2011**, *56*, 5189-5194. [http://dx.doi.org/10.1016/j.electacta.2011.03.042]

[291] Pujado, M.P. *Carbon nanotubes as platforms for biosensors with electrochemical and electronic transduction*; Springer: Heidelberg, **2012**. [http://dx.doi.org/10.1007/978-3-642-31421-6]

CHAPTER 8

Electrochemical Nanobiosensors in Pharmaceutical Analysis

Sevinc Kurbanoglu[1,2], **Sibel A. Ozkan**[1] and **Arben Merkoçi**[2,3,*]

[1] *Faculty of Pharmacy, Department of Analytical Chemistry, Ankara University, Tandogan, Ankara 06100, Turkey*

[2] *Nanobioelectronics and Biosensors Group, Catalan Institute of Nanoscience and Nanotechnology (ICN2), CSIC and The Barcelona Institute of Science and Technology, Campus UAB, Bellaterra, Barcelona 08193, Spain*

[3] *ICREA, Pg. Lluís Companys, 23, Barcelona 08010, Spain*

Abstract: Biosensors can combine both modern electronic techniques and selectivity properties of biological or synthetic molecules for the applications in pharmaceutical analyses. Electrochemical biosensors can be defined as analytical devices that are combining receptors such as cells, cell receptors, antibodies, enzymes, microorganism tissues, nucleic acids and DNA or other synthetic compounds with a physicochemical transducer especially electrical/electrochemical-based ones. Electrochemical biosensors are the most used biosensors for analysis of drugs or other analytes in pharmaceutical area being those based on nanomaterials (called also nanobiosensors), the emerging ones in the last decade. Electrochemical biosensors have found an important place due to their high sensitivity, selectivity, and characteristics such as simplicity, the ease of mass production, the low cost and availability of instrumentation. Up to now, various electrochemical technique such as voltammetry; cyclic voltammetry, step and pulse voltammetry, stripping voltammetry, conductometry, potentiometry and mainly amperometry have been used in electrochemical biosensors. In this chapter, electrochemical nanobiosensors employed in pharmaceutical analyses of various pharmacological groups are discussed.

Keywords: Amperometry, Analytical Chemistry, Analytical devices, Biorecognition, Conductometry, DNA, DNA biosensors, Electroanalytical Chemistry, Electrochemical Biosensors, Electrochemistry, Enzyme, Enzyme-based biosensor, Immunosensors, Nanobiosensors, Pharmaceutical Analysis, Pharmaceuticals, Potentiometry, Recognition, Transducer, Voltammetry.

* **Corresponding author Arben Merkoçi:** Nanobioelectronics and Biosensors Group, Catalan Institute of Nanoscience and Nanotechnology (ICN2), CSIC and The Barcelona Institute of Science and Technology, Campus UAB, Bellaterra, Barcelona 08193, Spain; Tel: +34937374604; E-mail: arben.merkoci@icn2.cat

Atta-ur-Rahman, Sibel A. Ozkan & Rida Ahmed (Eds.)

INTRODUCTION

All living beings immediately perceive the changes in their environment and try to follow the changes in order to survive. This sensing mechanism is the basis for the *in vitro* use of biosensors. Living beings' models have given birth to biosensors that combine these stimuli with analysis systems of biologically feasible systems. History of biosensors started in the mid-1950s, in a hospital in America (Cincinnati Hospital) by monitoring the amount of blood with an electrode during surgery and then biosensor technology developed rapidly [1 - 5]. Biosensors are devices that are developed by combining multiple science disciplines within the framework of multi-disciplinary understanding, combining the selectivity properties of biological or synthetic molecules with the processing capabilities of modern electronic techniques. Biosensors can be classified according to the biomaterials they use, enzymes, microbial, immuno, organs and tissues. Biosensors are small sensing devices used to control target analytes through their involvement in biological or chemical reactions. Receptor and transducer are the main parts of a biosensor. The task of the receptor (biochemical part *etc.*) is to recognize the analyte (substance to be analyzed) through specific interaction with it. The transducer part, is responsible for translating this recognition phenomenon into a readable (measurable) numerical value [1 - 10] (Fig. **1**).

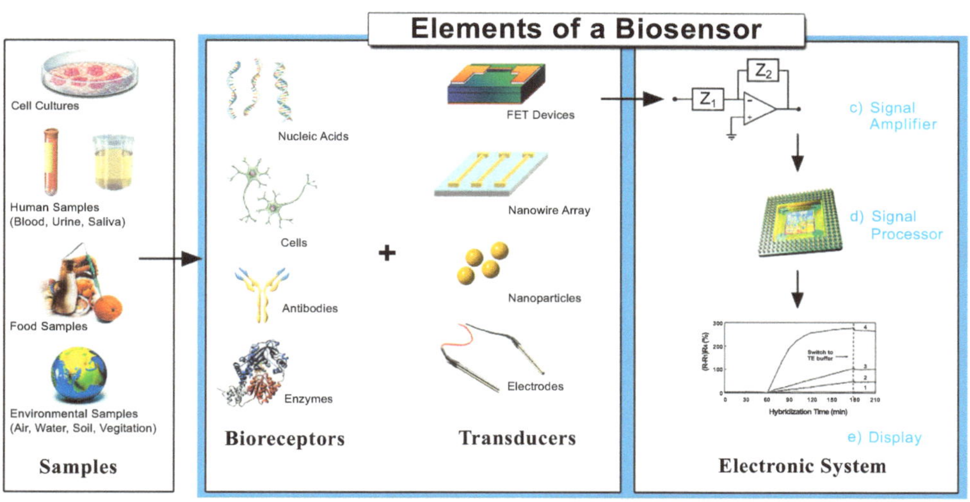

Fig. (1). Main components of a typical biosensor. (adapted with permission Ref [35]).

Biosensors have gained great interest over recent years due to their simplicity and fast response. The main task of electrochemical biosensors is to convert a biological/chemical event (interaction of the receptor with the analyte) into an electrical signal. Indeed, biosensors can be used in many different areas such as food, environmental, biomedical and pharmaceutical.

The first biosensor, enzymatic one for glucose, was made in 1962 by Clark and Lyons [16]. Glucose biosensor is the best example of biosensors that patients can use at home [17 - 19]. Biosensors also can be used in monitoring / control of drugs used in cancer therapy and other diseases, in the diagnosis of infectious diseases, analysis of hormone levels, in soil and water analysis, identification of pesticides and other toxic substances controlling the impact of the industry on the environment, monitoring plants diseases, food freshness control, identification of disease-causing microorganisms such as *E. coli*, examination of the fermentation process, in bacterial and virus analysis and many other applications [20 - 30]. Biosensors are commonly used in areas such as clinical diagnosis, biomedical sector, process / bioreactor control, food production and analysis, agriculture and veterinary medicine, diagnosis of bacteria and viruses, drug analysis, industrial waste water control, environmental protection and pollution control, analysis of toxic gas in mining enterprises, military applications and so on [20 - 35].

(Bio)Receptors

(Bio)Receptors are bioactive components that interact specifically with the substance to be analyzed. Considering the historical development of biosensor technology, it appears that the first studies in this area started with enzymatic sensors that use enzyme as receptor for glucose. Enzyme used as receptor can be pure enzyme, tissue or microorganism and can interact specifically with the analyte to form a chemical reaction product. A physical signal is produced and accordingly measured. However other (bio)receptors such as immuno agents, nucleic acids and receptor molecules have a biospecific effect and do not form a product by a chemical reaction with the substance to be analyzed. They bind specifically to the substance to be analyzed as the result of the interaction and the physical signal is afterwards produced [11 - 15, 36 - 39].

Biosensors can be grouped under two headings according to the biological recognition element; Biocatalytic biosensors and Affinity biosensors [1 - 10, 40 - 44].

Biosensors in which enzymes, tissues, bacteria and cells are used as recognition element are called biocatalytic biosensors. Enzymes are the most widely used bioreceptors for biosensors. They are substances that catalyze biological reactions. Enzymes accelerate the formation of product by catalyzing the reaction of the substrate. Enzymes are very specific to the substrate of the reaction they catalyze. This specificity allows the biosensors to be selective [11 - 15]. For example, the biosensor using the urease enzyme only gives results on the amount of urea. Even if it is chemically related to other substances, it does not recognize other substances [45, 46]. Other used receptors are glucose oxidase for glucose sensor, alcohol oxidase for ethanol or lactate oxidase for lactate sensor. The limited stability of the isolated enzymes, the inability of certain enzymes to be in pure state and the costly formation of some enzymes have increased the use of cellular materials such as plant tissues, bacterial cells, *etc*. The function of these biocatalytic electrodes is similar to that of conventional enzyme electrodes. The enzyme present in the tissue or cell is the receptor used for the biosensing operation. For example, banana tissues are rich in polyphenol oxidase enzyme. For this reason, carbon paste matrix can be mixed with banana tissues and immobilized on the electrode surface building in this way as an enzymatic biosensor [47 - 51].

Affinity biosensors are based on the use of specific biomolecules (antibody, receptor, polynucleotides) to specifically bind target analytes. These sensors are based on measuring the signals resulting from the bonding process. The affinity and high specificity of biochemical binding reactions lead to the production of highly sensitive and selective devices [40 - 44]. Affinity biosensors can be divided in three groups: immunosensors, receptor-based biosensors, and nucleic acid-based biosensors. Antibodies are molecules that are used as affinity bioreceptor. These are produced by an animal defense system against foreign substances entering the body. Biosensors that work by a mechanism similar to such defense system are called immunosensors. Antibodies do not take part in catalysis or chemical modification of chemical reactions. Antibodies bind to foreign substances in a highly selective way. Antibodies are proteins produced by the organism to bind foreign molecules such as antigen. Immunosensors are based on the recording of direct changes in antibody or antigen concentration at the end of binding or displacement events. Since the reaction of a large majority of these sensors is irreversible, they cannot be reused [52 - 60].

Nucleic acids, RNA and DNA based biosensors, are combined with a transducer as a recognition strand to form a new and important pathway of affinity biosensors. Two approaches have been developed to identify target biosensors and target analytes. The first is the introduction of analytes between the double strand structure of DNA and the second is the direct measurement of guanine, the most electroactive species of DNA structure. Compared to traditional DNA detection methods, they are faster and cheaper ensuring right genetic information. DNA is the largest known molecule of biological importance for all kinds of life. It may interact with smaller molecules electrostatically or not. Biosensors using DNA as a biomolecular recognition element are very effective and have some advantages such as high sensitivity, selectivity, fast response, low cost, minimal power requirement of the related instrument/reader, simple design, suitability for microfabrication and portability [13, 60, 61].

A newly proposed sensitive pathway is the use of chemoreceptors as biological recognition element. Receptors are protein molecules, which are placed on cellular membranes that specifically bind the target analyte. The selective binding between the membrane receptor and the target analyte is measured by measuring the change in conductivity in the membrane. A simple and selective binding phenomenon between the membrane receptor and the target analyte is the result of the increase in the conductivity of the transducing membrane, which contains thousands of ions [62 - 65].

Immobilization of the Biological Materials

Once the appropriate (bio) receptor and physical component (transducer) have been selected, it is the most important question how to connect them together. This attachment process is called bio-immobilization. A variety of methods can be used for this purpose. Which method is used depends on the selected biocompatibility and physical composition [66 - 68]. Some of the methods used for immobilization are physical adsorption, entrapment, or chemical bonding [66 - 71] (Fig. **2**).

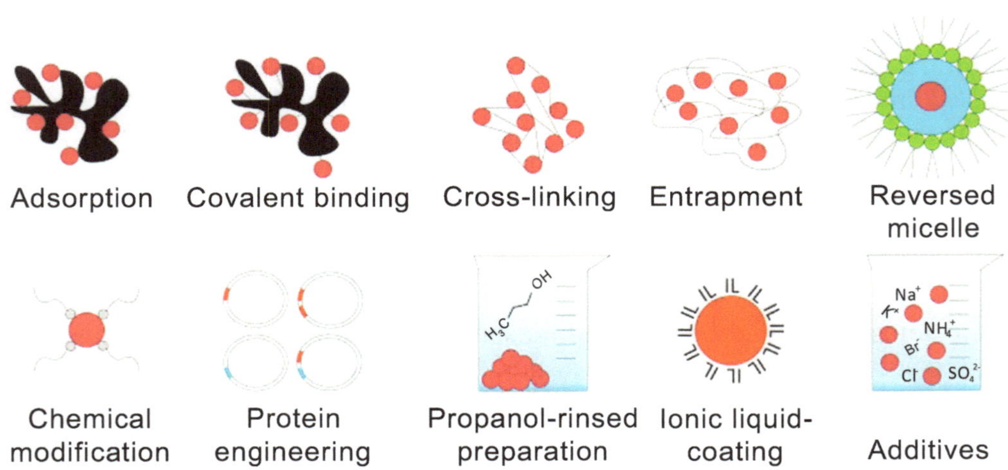

Fig. (2). Immobilization strategies used to connect (bio)receptors with transducer surface (reprinted with permission Ref 95).

Adsorption

Adsorption is the oldest and simplest method used in immobilization. The biocomponent is attached onto the transduser surface with non-covalent interactions (hydrogen bonds, multiple salt bridges, electron transfer complexes and Van der Waals forces). Immobilization of many biological material takes place according to the method of adsorption [72 - 76].

The main advantage of the adsorption method is that it does not require the use of chemicals and can be carried out in few simple steps. As a result, adsorption is cheap, easy to apply and does not affect enzymes. On the other hand, desorption of the biological material (due to changes in pH, temperature and ionic charge) is seen as a problem due to poor binding. Another disadvantage is that since it is not a specific adsorption, it can interfere with other biological substances in the structure of biological material [66 - 76].

Encapsulation

This method is performed by encapsulating the biological material in the membrane of the electrode surface. From the appropriate cavities formed in the membrane, diffusion of the substrate occurs and the reaction takes place. A wide variety of membranes such as nylon, cellulose nitrate, cellulose acetate, epoxy resin, collagen, polysulfones, polyacrylates and polycarbonates has been used. The disadvantages of this method relate to the movement of substrates, formed products and inhibitors within the membrane. These substances may cause erroneous measurement by inhibition of the biological material layer. Encapsulating can also be done in polymer gel matrices or more simply on dialysis membranes [75 - 81].

Entrapment

During the polymerization, the biological material is immobilized within the formed polymer network. Numerous materials such as polyacrylamide gels, polyvinyl alcohols, charged polymers, cationic and anionic groups are used in the entrapment method [82 - 84].

Usually this immobilization method is used for all types of biological material especially enzymes (dehydrogenases, alcohol oxidases, cholinesterases, choline oxidases, glucose oxidases and tyrosinase) and it is easy to implement and cheap compared to other systems. The disadvantages of this method are that they can lose biological material activity when free radicals are present in the medium or when ultraviolet radiation is applied to the electrode to complete the polymerization after immobilization has been achieved. These problems affect the biosensor's response time and stability. Recently, gel matrix based immobilization methods have been extensively used to build biosensors more robust, resistant to heat, not affected by solvent environments, and able to work in organic and aqueous environments [66 - 71].

Biological materials can also be entrapped in conducting polymer matrices such as polypyrrole, polyphenol, polyaniline and polythiophene. Polymerization of the monomers using electrochemical potentiostatic and galvanostatic methods provides immobilization of sensitive and thin films incorporating enzymes. The advantages of this technique are that the immobilization process can be controlled, that many electrode materials are available (platinum, gold and carbon) and are suitable for miniaturization. Although the preparation process is simple, the disadvantage of this method is that it is difficult to exactly determine the amount of biological material that is immobilized [85 - 88].

Covalent Binding

It is based on a covalent attachment of the (bio)receptor directly onto the transducer surface. For covalent attachment to occur, it is necessary to have reactive groups on the bonding surface and / or on the biomolecule such as thiol (-SH), hydroxyl (-OH), amine ($-NH_2$), carbonyl (-C = O), carboxyl (-COOH). In order to increase the stability of the biomolecules, which is methylated, the covalent binding method is carried out in the presence of inert proteins. The advantages of this immobilization method when used for enzymes, are enzyme-enzyme and enzyme-protein binding. As the covalent binding method can form structures that may prevent diffusion, the biosensor response time prepared by this method can be high [66 - 71, 88 - 90].

Crosslinking

Cross-linking is based on the principle of forming insoluble complexes between biocomponent and adsorbent using chemical reagents containing small molecules with bi or multifunctional groups. Glutaraldehyde enmeshed the biomolecules in the form of agglutination, making them insoluble and resistant. Besides glutaraldehyde, hexamethylene diisocyanate can be used in cross-linking chemistry such as 2-isocyanato -4-isothiocyanato-toluene, 1,5-difluoro-2-4-dinitrobenzene, bisdiazo benzidine-2,2-disulfonic acid [90 - 95].

Transducer Part of the Biosensors

Transducer is the biosensor component that turns biological response into a measurable physical signal. The substance to be analyzed is recognized by the biosensor and the produced physical signal translated by transducer and monitored by using the amplifier and bringing it together with the computer [1 - 10].

The biochemical reactions occurring during analyte recognition must be converted to an electrical signal by means of a suitable transducer. Electrochemical biosensors usually use amperometric, potentiometric, conductometric *etc*.

Amperometry is generally based on the measurement of a certain current magnitude. The current density is defined as a function of the concentration of the electroactive species that is oxidized or reduced on the electrode. After calibration, the current densities are used to determine the concentrations of the respective species. Potentiometric measurement methods are simple and inexpensive. A reference, working electrode (ion selective electrode) and the potential measuring device are required for the measurements. When the reference electrode and the ion selective electrode are in contact with the sample to be measured, the ion-selective electrode exhibits a different potential depending on the analyte ion being analyzed, while the reference electrode exhibits a constant potential. With this difference, the activities of the ions are determined. The properties sought for reference electrodes must have unchanging potential over time, conform to the Nernst equation and be reversible [96 - 102].

Conductometric biosensors are capacitors composed for example of planar microelectrodes with gold interspersed with an insulating chip such as alumina or silicon dioxide. In conductometric biosensors, sensing is accomplished by modification with a biomolecule that provides chemical selectivity to the gap between two microelectrodes that are very close together. The conductivity of the immobilized film varies when the analyte binds to the selective agent. The greatest drawbacks of such sensors are the complex equipment requirement for low response time, signal generation and evaluation between others [96 - 102].

Nanomaterials in Nanobiosensing

In biosensors nanomaterials are mainly used for the modification of the surface of the transducer due to nanomaterials' unique physical and chemical properties. Nanomaterials can be classified as carbon-based and metal-based nanomaterials [100, 114 - 117] (Fig. **3**).

Carbon has exceptional hybridization properties such as sp, sp^2 or sp^3 due to its ground-state orbital configuration, leading itself to develop carbonaceous nanostructures. Carbon-based nanobiosensors can be prepared by various carbon nanomaterials, such as fullerene, carbon nanotube, graphene, carbon dot, and nanodiamond. These carbon based nanomaterials are the most used nanomaterials among all nanomaterials, receiving huge attention in the past decade due to their unique mechanical, thermal, electrical, and physical properties [117 - 124] (Fig. **4**).

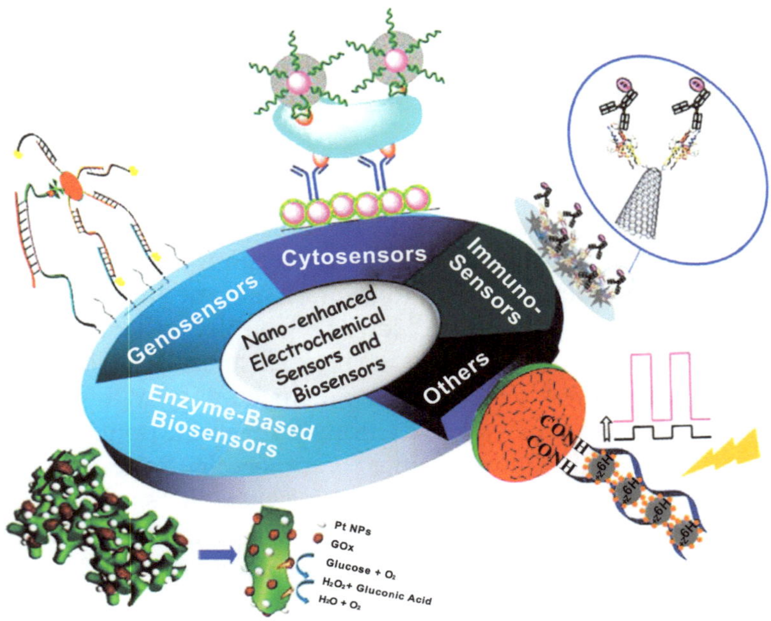

Fig. (3). Schematic illustration of electrochemical sensors and biosensors based on nanomaterials and nanostructures. Enzyme-based biosensors, genosensors, immunosensors, and cytosensors are shown. (reprinted with permission Ref 114).

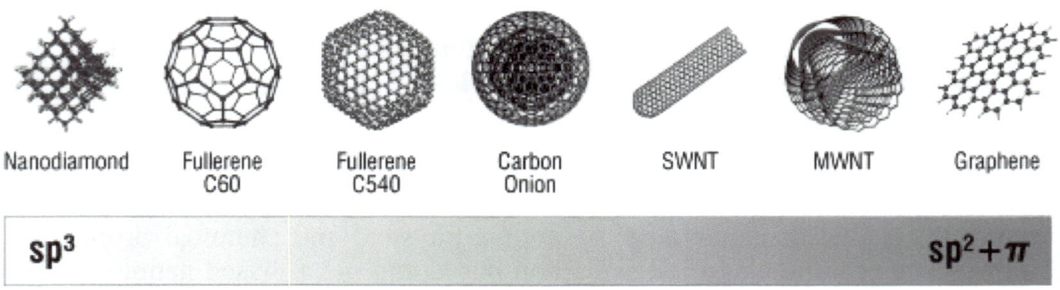

Fig. (4). Hybridization states of carbon-based nanomaterials (reprinted with permission Ref 125).

On the other hand, inorganic nanomaterials such as titanium dioxide nanomaterials [126 - 128], iron oxides [129 - 131], platinum nanoparticles [132 - 134] and palladium-based nanomaterials [116, 135 - 137] and especially gold and silver nanoparticles [138 - 152] are a promising area of today and future researches. They have good chemical and mechanical stability and are biocompatible with the biological materials. Gold nanoparticles are widely used in nanobiosensing due to their good electron transfer property and desirable biocompatibility [138 - 152].

Applications of Electrochemical Nanobiosensors in Pharmaceutical Analysis

A drug is any chemical you take affecting the way humans or animal bodies work, changing their physical or mental state. Drugs are available at different dosage forms for delivering the active compounds in the system namely oral solutions: suspensions, syrups, elixirs, emulsions, powders and other miscellaneous solutions; solid dosage forms such as capsules, tablets; semi-solid dosage forms such as ointments, suppositories [153 - 158]. In the following sections, the use of nanobiosensors in detecting various kinds of drugs is discussed.

a. Antibiotics, Antibacterials and Antimicrobials

Antibiotics are generally used for treatment or prevention of bacterial infection in the healing range. There are different kinds of antibiotics, based on the bacterial spectrum, broad or narrow; based on the route of administration, they can be oral, injectable or topical; related with the type of activity, they can be bactericidal or bacteriostatic [159 - 164]. The first biosensor for the detection of penicillin, β-lactam antibiotics, was prepared using penicillinase enzyme with a potentiometric sensor [165]. In their work, Chen *et al.* in 2010, developed a biosensor for the determination of penicillin, β-lactam antibiotics which are the most often used antibiotics for treating various bacterial infections. The authors developed amperometric penicillin biosensor based on multi-walled carbon nanotubes, hematein, and β-lactamase using glassy carbon electrode as the transducer. Penicillin was detected with a 50 nM detection limit, at pH 6.8, 1 mM PBS containing 0.1 M NaCl [166]. Some of the selected studies related to the detection of antibiotics using biosensors are summarized in Table **1**.

Table 1. Some of the selected studies related the detection of antibiotics using biosensors and nanobiosensors.

Analyte	Transducer	Method	Detection Limit	Ref
Ciprofloxacin	dsDNA/GCE	DPV	9 μM	[167]
Ciprofloxacin	MWCNT/DNA	CV	0.33 μg mL^{-1}	[168]
Gatifloxacin	MWCNT-PE	DPV	4.5×10^{-9} M	[169]
Gatifloxacin Moxifloxacin Sparfloxacin	dsDNA/GCE	DPV	5.0×10^{-8} M 8.0×10^{-8} M 8.0×10^{-8} M	[170]
Gemifloxacin	dsDNA/PGE	DPV	0.23 μg mL^{-1}	[171]

(Table 1) contd.....

Analyte	Transducer	Method	Detection Limit	Ref
Gemifloxacin	dsDNA/PGE	DPV	0.23 μg/mL	[172]
Lomefloxacin	nano-SnO_2/DHP /GCE	CV	NS	[173]
Metronidazole	dsDNA/GCE	DPV	0.02 μM	[174]
Moxifloxacin	CPE (in solution)	CV, DPV	NS	[175]
Nitrofurantoin	HMDE (in solution)	CV, DPV	NS	[176]
Nitrofurazone	dsDNA/GCE	DPV	8.0×10^{-7} M	[177]
Penicillin	MWCNTs)/ hematein/ β-lactamase/GC	CA	50 nM	[166]
Pipemidic Acid	GE/Tyr	FIA	18 nM	[178]
Rifampicin	DNA/CPE	DPV	NS	[179]
Sulfadiazine	MWCNT/DNA/GCE	CV	NS	[180]
3-Nitrophenyl ferrocene	GCE (in solution)	CV	NS	[181]

b. Cardiac Stimulant

Cardiovascular disease generally refers to conditions that involve narrowed or blocked blood vessels that can lead to a heart attack, chest pain (angina) or stroke. Heart diseases can be primarily grouped into three major disorders: cardiac failure, ischemia and cardiac arrhythmia. Cardiac stimulant is a pharmacological agent that increases the action of the heart [182 - 184].

There are many drugs prescribed for heart disease. Mostly used drugs are; ACE inhibitors, Aldosterone inhibitors, Angiotensin II receptor blockers, Beta-blockers, Calcium channel blockers, Cholesterol-lowering drugs, Digoxin, Diuretics, Inotropic therapy, Potassium or magnesium Vasodilators, Warfarin, *etc*.

Mirmomtaz *et al*. in 2009, prepared a biosensor based on a ds-DNA-modified pencil graphite electrode for the detection of amirolide [185]. Using electrochemical biosensor, dopamine was analyzed with different biocomponent based biosensors such as DNA, horseradish peroxidase, tyrosinase *etc*. In their work, Ribeiro *et al*. in 2013, established an enzymatic biosensor for dopamine determination using multiwall carbon nanotubes paste electrode. The authors used square wave voltammetric method and with 2μM limit of detection value, dopamine which belongs to Cardiovascular and Renal Drugs /Drugs for heart failure/Positive inotropics, was analyzed with an enzymatic biosensor [186]. Some of the selected studies related with the detection of cardiovascular drugs using biosensors are summarized in Table **2**.

Table 2. Some of the selected studies related with the detection of cardiovascular drugs using biosensors and nanobiosensors.

Analyte	Transducer	Method	Detection Limit	Ref
Amiloride	dsDNA/PGE	DPV	0.5 μM	[185]
Dopamine	MWCNT/HRP/CP	SWV	2 μM	[186]
Dopamine	Au/CDs-CS/GCE	DPV	0.001 μM	[187]
Dopamine	Au/CA/tyr–MWCNT–Nafion	DPV CA	50 nM 3 nM	[188]
Dopamine	PA-MNPs	AMP EIS	7.25 μM 14.1 μM	[189]
Dopamine	DNA/PAMAM/MWNT-Chit/Au	DPV	0.03 μM	[190]
Dopamine	AuNPs@SiO_2	CV	$2.0x10^{-8}$M	[191]

c. Antineoplastic Agents

Antineoplastics or Anticancer drugs are the drugs that prevent or inhibit the maturation and proliferation of neoplasms. Antineoplastic agents travel the body and destroy cancer cells. Antineoplastic agents have several side effects. These drugs also destroy the body's normal cells in addition to cancerous cells. Antineoplastic drugs are not only used prominently in different types of cancers but also in conjunction with surgery, radiotherapy and immunotherapy in the combined modality approach [192 - 195]. Antineoplastic agents can be classified as:

Alkylating Agents: These group members are Nitrogen mustards (Melphalan, Cyclophosphamide, Ifosfamide); Nitrosoureas; Alkylsulfonates; Ethyleneimines; Triazene; Methyl Hydrazines; Platinum Coordination complexes (Cisplatin, Carboplatin, Oxaliplatin).

Antimetabolites: The members of these group can be ordered as Folate Antagonists (Methotrexate); Purine antagonists; Pyrimidine antagonists (5-Fluorouracil, Cytarabibe).

Natural Products: This group can be subdivided as: Plant Products, Vinca Alkaloids (Vincristine, Vinblastine); Taxanes (Paclitaxel, Docetaxel); Epipodophyllotoxins (Etoposide); Camptothecins (Irinotecan), Microorganism Products, Antibiotics (Doxorubicin, Bleomycin); Enzymes (L-Asparaginase).

Miscellaneous: There are different drugs such as Hydroxyurea, Imatinib Mesylate, Rituximab, Epirubicin, Bortezomib, Zoledronic Acid, Geftinib, Leucovorin, Pamidronate, Gemcitabine.

Hormones and Antagonists: This group contains different type of hormones or antogonists such as: Corticosteroids (Prednisone, Dexamethasone); Estrogens (Ethinyloestradiol); Antiestrogens (Tamoxifen); Progesteron derivative (Megestrol Acetate); Androgen (Testosterone propionate); Antiandrogen (Flutamide, Bicalutamide); Aromatase inhibitor (Letrozole, Anastrazole); 5-alpha reductase inhibitor (Finasteride); GnRH Analogue (Leuprolide, Buserelin); Growth Hormone, glucagon and insulin inhibitor (Octreotide) [240 - 243]. Some of the selected studies related with the detection of anticancer drugs using biosensors are summarized in Table **3**.

Table 3. Some of the selected studies related with the detection of anticancer drugs using biosensors and nanobiosensors.

Analyte	Transducer	Method	Detection Limit	Ref
Adriamycin	(ds-DNA)/Ag $_{nano}$/PPAA/MWCNTs–COOH	DPV	$3.2x10^{-12}$ M	[196]
Berberine	DNA/GCE	DPV	$1.45x10^{-5}$ M	[197]
Bleomycin	DNA/PGE	EIS	NS	[198]
Cisplatin	EDTA MT /HMDE	CV	NS	[199]
Cladribine	dsDNA/GCE	DPV	NS	[200]
Clofarabine	dsDNA-GCE	DPV	0.08 μM	[201]
Cyclophosphamide	DNA/PGE DNA/CPE DNA/HMDE	DPV, ACV	NS	[202]
Dacarbazine	DNA/SPCE	CV, DPV	2.0×10^{-8} M	[203]
Daunorubicin	DNA-MWCNTs/GCE	CV	NS	[204]
Daunomycin Doxorubicin Idarubicine Mitoksantron	SPE/AuNPs/pTTBA/dsDNA/CL/AuNPs	AMP	5.5 ± 0.3 fM 3.6 ± 0.2 fM 2.2 ± 0.1 fM 1.2 ± 0.05 fM	[205]
Doxorubicin	HMDE (in solution)	SWV	NS	[206]
Echinomycin	Guanine modified GCE	Adsorptive transfer stripping CV	NS	[207]

(Table 3) contd.....

Analyte	Transducer	Method	Detection Limit	Ref
Flutamide	HMDE (in solution)	CV, DPV, SWV	NS	[208]
Fulvestrant	dsDNA/PGE	DPV	0.41 μg mL^{-1}	[209]
Furazolidone	MWCNT/DNA/GCE	LSV	NS	[210]
Hexakis (imidazole) manganese(II) terephthalate	DNA/GCE	CV	NS	[211]
Imatinib	DPV	dsDNA/GCE	NS	[212]
Irinotecan	dsDNA/ PGE	DPV	-	[213]
Lapatinib	dsDNA/GCE	DPV	NS	[214]
Leuprolide	dsDNA/PGE	DPV	0.06 ppm	[215]
Methotrexate	DNA/GCE	Potentiometric stripping analysis	5.63×10^{-9} M	[216]
Methotrexate	dsDNA/GC	SWV	5.0×10^{-9} mol L^{-1}	[217]
Methotrexate	dsDNA/GCE	DPV	NS	[218]
Mitomycin C	dsDNA/Chit-IL-PGE	DPV, CV, EIS	NS	[219]
Mitomycin C	DNA/CSPE DNA/CPE	SWV, DPV	33 ng mL^{-1} 19 ng mL^{-1}	[220]
Mitomycin C	dsDNA immobilized GO/PGE	DPV	NS	[221]
Mitomycin C	DNA/Chit/CNT/PGE	DPV	11.01 μg mL^{-1}	[222]
Mitomycin C	dsDNA/PGE	DPV	0.23 μg mL^{-1}	[223]
Mitomycin C	PVF/Au	DPV	90 ng/mL	[224]
Methotrexate	GCE (in solution)	CV	NS	[225]
Methotrexate	DNA/GCE	GCE	$5.0x10^{-9}$ M	[226]
Methotrexate	dsDNA/GCE(ox)	SWV	8.0×10^{-9} M	[227]
Mitoxantrone	GCE (in solution)	LSV	NS	[228]
Morin	PTFE-DNA/GCE	LSV	NS	[229]
paclitaxel	Cysteamine/DNA/SWCNT-film-modified AuE	SWV	8.86×10^{-6} M	[230]

(Table 3) contd.....

Analyte	Transducer	Method	Detection Limit	Ref
Paclitaxel	DNA/SAM/AuE	CV	1.2×10^{-8} M	[231]
Peroxynitrite	dsDNA/CPE	DPV	NS	[232]
Rutin	DNA/CPIE	DPV	1.3×10^{-9} M	[233]
Rutin	DNA/MWCNTs–COOH/GCE	DPV	NS	[234]
Rutin	DNA/MWNTs–COOH/Fe_3O_4 /GCE	DPV	NS	[235]
Sildenafil	DNA/GCE	Constant current potentiometry, DPV	NS	[236]
Sophoridine	GCE (in solution)	DPV	NS	[237]
Spermidine	DNA/PPy nanofiber /PtE	DPV	0.02 μM	[238]
Spermidine	dsDNA/AuE	DPV	0.72 μM	[239]
Sulfadiazine	MWCNT-GCE/DNA	CV	0.03 μg mL^{-1}	[240]
Tamoxifen	dsDNA/CPE	Zero current potentiometry	1.1×10^{-7} M	[241]
Tamoxifen	HRP/PANI/PtE	CA	0.07 ng mL^{-1}	[242]
Tamoxifen	dsDNA/CP	ZCP	1.1×10^{-7} M	[243]
Topotecan	dsDNA/SWCNT/PGE	DPV	NS	[244]
Taxol	Cysteamine/DNA/SWNTs-film/ Au	SWV	$8.86x10^{-6}$ M	[245]
Taxol	DNA/SAM/Au	EIS	$1.2x\ 10^{-8}$ M	[246]
Taxol	dsDNA/PGE	DPV	8.0×10^{-8} M	[247]
quercetin	p-AMT/PG	DPV	2.2 μM	[248]
Quinacrine	NA/polyallylamine(PAA) polyion complex membrane	CA	10 μM	[249]
2-nitroacridone	dsDNA/GCE	CV	NS	[250]
4-nitrophenylferrocene	GCE (in solution)	CV	NS	[251]
5-fluorourasil ortho and meta compound	DNA/AuE	CV	NS	[252]

(Table 3) contd.....

Analyte	Transducer	Method	Detection Limit	Ref
6-thioguanine	dsDNA/PGE	DPV	4.60 µg mL^{-1}	[253]
6-Thioguanin	HMDE	SWV	2.1 nM	[254]
6-thioguanine	dsDNA/AET/ AuE (SAM)	DPSV, SWV	2.0×10^{-7} M 1.2×10^{-7} M	[255]
6-mercaptopurine	dsDNA/GNs/ Chit/GCE	DPV	NS	[256]
Quercetin	dsDNA/GCE	DPV	NS	[257]

d. Antiviral Drugs

Antiviral drugs are one class of antimicrobials, a larger group which also includes antibiotics, anti-fungal and anti-parasitic drugs. They are a class of medication used specifically for treating viral infections. There are several different kinds of antiviral agents that are prescribed to fight viral infections. Like antibiotics, specific antivirals are used for specific viruses [258 - 262].

These medications can be divided into systemic antivirals and topical antivirals. Systemic antivirals are usually given orally in pill form. However, they sometimes may be administered intravenously (into a vein). Topical antivirals are usually available in different forms, including creams, ointments, gels, patches and sprays.

In their work, Zhou *et al.* in 2010, developed an amperometric method for the biosensing of Magnolol which has antiviral, anti-stress, anti-anxiety, strong antioxidants and antitumor activities. Using glassy carbon electrode, they reached 1.5×10^{-9} M detection limit at pH 7.0 PBS [263]. Lin *et al.* in 2008, studied interaction of Aloe-emodin which has Chemotherapeutic Drugs/ Antimicrobials/Antiviral activity. Using (ds-DNA)/GCE with differential pulse voltammetric method, detection limit of $6.7\text{x}10^{-8}$ M was reached [264]. Dogan Topal *et al.* in 2009, studied antiviral drug efavirenz using differential pulse voltammetric method at dsDNA/PGE. 0.599 ppm detection limit was obtained using this method at pH 4.8 acetate buffer [265].

e. Antidepressant and Antipsychotic Drugs

Antidepressants are drugs that treat depression and improve the symptoms. The main types of antidepressants are tricyclic antidepressants, selective serotonin reuptake inhibitors, monoamine oxidase inhibitors and other atypical antidepressants (*e.g.* miscellaneous antidepressants, phenylpiperazine antidepressants, tetracyclic antidepressants) [266 - 270].

Antipsychotic drugs act by blocking dopamine receptors in the brain and interfering with dopaminergic transmission. These drugs are used to treat psychosis, including schizophrenia and mania [266 - 270].

Antipsychotic drugs are classified as typical or atypical. Atypical antipsychotics are the recently developed drugs that are useful in patients that do not respond to treatment with other typical antipsychotics. Other type of antipsychotics drugs are also available in the pharmaceutical market which named as miscellaneous antipsychotic agents, phenothiazine antipsychotics, psychotherapeutic combinations and thioxanthenes [266 - 270]. Some of the selected studies related with the detection of antipsychotic and antidepressant drugs using biosensors are summarized in Table **4**.

Table 4. Some of the selected studies related with the detection of antipsychotic and antidepressant drugs using biosensors and nanobiosensors.

Analyte	Transducer	Method	Detection Limit	Reference
Aripiprazole	GCE	CV, DPV	NS	[271]
Amitriptyline	GRT/SiAlNb/ DNA/CPE	DPV	0.12 μM	[272]
Buzepide methiodide	DNA/GCE	DPV	NS	[273]
Chlorpromazine hydrochloride	DNA/MWCNTs-modified AuE	PQCI	NS	[274]
Promethazine	DNA/AuE	CV	3.8×10^{-10} M	[275]
Promethazine	dsDNA/GCE_{ox}	DPV	3.0×10^{-10} M	[276]

f. Antiemetic Drugs

The nausea and vomiting medicines are also known as antiemetics, antinauseants. These drugs are effective against vomiting and nausea. Antiemetics are typically used to treat motion sickness and the side effects of opioid analgesics, general anaesthetics, and chemotherapy directed against cancer [277 - 279]. They may be used for severe cases of gastroenteritis, especially if the patient is dehydrated. These drugs can be classified as: Anticholinergic drugs: Hyoscine, Dicyclomine; H1 Antihistaminics: Promethadine, Diphenhydramine, Cinnarizine, Cyclizine, Doxylamine; Neuroleptics: Haloperidol, Prochlorperidol, Chlorpromazine; Prokinetic drug: Domperidone, Cisapride, Mosapride, Metochlopramide; 5-HT3 Antagonist: Granisteron, Ondansteron; Adjuvant Antiemetics: Dexamethasone, Benzodiazepine, Cannabinoids. An electrochemical biosensor for thalidomide which is a sedative and anti-emetic drug is reported by Oliveira, *et al.* in 2009, They detected the interaction of DNA with Thalidomide in solution phase using glassy carbon electrode in pH 4.5 Acetate buffer solution [279].

g. Adrenergic Agonist

Adrenergic drugs are medications that stimulate certain nerves in the human body. They do this either by mimicking the action of the chemical messengers epinephrine and norepinephrine or by stimulating their release. They represent a family of agents which have adrenaline-like effects, acting directly or indirectly on any of the five G-protein-coupled adrenergic receptors: $alpha_1$, $alpha_2$, $beta_1$, $beta_2$ and $beta_3$ [280 - 287].

These drugs act directly on one or more adrenergic receptors. According to receptor selectivity they are two types: Non-selective and Selective.

Non-selective: drugs act on one or more receptors; these are: Adrenaline (almost all adrenergic receptors); Noradrenaline (acts on α_1, α_2 and β_1); Isoprenaline (acts on β_1, β_2, β_3) and Dopamine (acts on α_1, α_2, β_1, D_1, D_2) [280 - 287].

Selective: drugs which act on a single receptor only; these are further classified into α selective and β selective. These sub group can be divided as: α_1 selective: Phenylephrine, Methoxamine, Midodrine, Oxymetazoline; α_2 selective: α-Methyl dopa, clonidine, brimonidine; β_1 selective: Dobutamine; β_2 selective: Salbutamol, Albuterol, Terbutaline, Salmeterol, Formoterol, Pirbuterol [280 - 287].

Some of the selected studies related with the detection of Adrenergic drugs using biosensors are summarized in Table **5**.

Table 5. Some of the selected studies related with the detection of Adrenergic drugs using biosensors and nanobiosensors.

Analyte	Transducer	Method	Detection Limit	Ref
Epinephrine	CPE (IL/ZnO/NP/CP)	SWV	0.06 M	[282]
Glutamate	GluOx/CMWCNT/AuNP/CHIT Au	CV	1.6 μM	[283]
Isoprenaline	GCE (in solution)	DPSV	NS	[284]
Norepinephrine	AuNPs/DNA/GCE	DPV	5 nM	[285]
Ractopamine	DNA/GCE	DPV	NS	[286]
Serotonin Dopamine	DNA–PPy_{ox}/CFE	DPV	7.0×10^{-9} M 5.0×10^{-8} M	[287]

h. Antiparasitic

Antiparasitics are a class of medications which are indicated for the treatment of parasitic diseases. These type of parasites can be orders as helminths, amoeba, ectoparasites, parasitic fungi, and protozoa, among others. Antiparasitics target the parasitic agents of the infections by destroying them or inhibiting their growth; they are usually effective against a limited number of parasites within a particular class [288 - 292]. Antiparasitics are one of the antimicrobial drugs which include antibiotics that target bacteria, and antifungals that target fungi [288 - 292]. Antiparasitic drugs can be divided in mainly two parts: Anthelmintics, Scabicides. Anthelmintics are agents that destroy or expel parasitic worms (helminths). Interestingly, most anthelmintics were developed for use in veterinary medicine. Scabicides are agents that kill scabies mites and mite eggs. Pediculicides are agents used to kill head lice [288 - 292].

In their study, Radi *et al.* in 2014, investigated interaction of DNA with Nitazoxanide which is an antiparasitic. Nitazoxanide was detected using dsDNA/SPCE, at pH 4.0 and 7.4 PBS, with an LOD value of 0.025 μM [291]. Catalán *et al.* in 2010, developed a biosensor based on GCE/Chit-CNT/DNA transducer, for 2-(2-nitrophenyl)-1H-benzimidazole and N-benzoyl-2-(2-nitrophenyl)-benzimidazole. Using differential pulse voltammetry 2-(--nitrophenyl)-1H-benzimidazole was detected with 10.4 μM LOD value and N-benzoyl-2-(2-nitrophenyl)-benzimidazole was detected with 18.7 μM LOD value at pH 4.7 phosphate buffer [292].

i. Antimalarial

Antimalarial drugs are used for the treatment and prevention of malaria infection. Most antimalarial drugs target the erythrocytic stage of malaria infection, which is the phase of infection that causes symptomatic illness. Treatment of the acute blood stage infection is necessary for malaria caused by all malaria species. In their study, Gu and Hasebe in 2012, developed a DNA based biosensor named as DNA–Cu(II)/PAA/GCE, using amperometric technique, at pH 7.0 PBS Quinacrine detection was achieved with LOD value of 10 μM [293]. In another study, Artesunate, which is an Antimicrobials/Antimalarial drug, detected using graphene, polyaniline and horse radish peroxidase enzyme using CV and EIS methods. Artesunate was determined from Parenteral artesunate, human urine, human serum and human plasma with a sensitivity value of 0.15 $\mu A.ng.mL^{-1}$ [294].

j. Vitamins, Minerals, Antioxidants, Nutritional Drugs

There are 13 vitamins that are essential to human body working. Knowledge of the different types and understanding the purpose of these vitamins are important for good health. There are two types of vitamins: fat-soluble and water-soluble. The fat soluble vitamins: They are stored in human fat cells, consequently requiring fat in order to be absorbed. These vitamins are Vitamin A, D, E and K. The water soluble vitamins: They are not stored in human body; therefore, they need to be replenished daily. These vitamins are Vitamin B1 (Thiamin), B2 (Riboflavin), B3 (Niacin), B5 (Pantothenic Acid), B6 (Pyridoxine), B7 (Biotin), B9 (Folic Acid (Folate)), B12 and vitamin C [295 - 298]. Some of the selected studies related with the detection of vitamins, minerals, antioxidants, nutritional drugs using biosensors are summarized in Table **6**.

Table 6. Some of the selected studies related with the detection of vitamins, minerals, antioxidants, nutritional drugs using biosensors and nanobiosensors.

Analyte	Transducer	Method	Detection Limit	Reference
Aspartame	AOX/CaE/CoPC/SPCE	CA	0.2 μM	[298]
Biotin (B8)	streptavidin-modified magnetic microbeads	SWV	20 mgL^{-1}	[299]
Catechin	Lacc/Au NPs-encapsulated dendrimers /GCE	CA	NS	[300]
Folic acid	DNA/PGE	DPV	1.06×10^{-8} M	[301]
Hydrazine	AuNPs/ssDNA/AuE	CV	0.56 μM	[302]
Lysine	LO/DAO/HRP/Os Polymer/GA	CA	NS	[303]
Lysine	LO/PPyox/Pt E	CA	4 μM	[304]
Riboflavin (B2)	DNA/PGE	DPV DPASV	0.34 μg/ mL 0.076 ng/mL	[305]
Tyrosine	PAO/Ferrocene/SPCE	CA	2.0 ± 0.18 μM	[306]
Vitamin B6	DNA/GCE	CV	NS	[307]
Vitamin B1	dsDNA/ PMWCNTPE	DPV	0.44 μg mL^{-1}	[308]
Vitamin B2 Vitamin B6 Vitamin C	PEDOT/GC	DPV	0.02 μM 0.3 μM 0.5 μM	[309]

k. Analgesic, Anti-inflammatory Drugs

Analgesics work at the level of the nerves, either by blocking the signal from the peripheral nervous system, or by distorting the interpretation by the central nervous system. Analgesics should be dosed adequately to ensure that the pain is at least tolerable, and frequently enough to avoid the anxiety that accompanies the anticipated return of pain. The term Analgesics encompasses a class of drugs that are designed to relieve pain without causing the loss of consciousness [310 - 312]. The different classes of analgesic drugs are narcotic and non-narcotic analgesics.

The narcotic analgesics, also termed as opioids, are all derived from opium such as morphine, codeine, and a number of semi-synthetics (meperidine, propoxyphene) and others. Non-narcotic analgesics are classified as: Nonsteroidal anti-inflammatory drugs, selective COX-2 inhibitors, Acetaminophen, salicylates, topical analgesics that are applied on the skin and the others [310 - 312]. Some of the selected studies related with the detection of analgesic and anti-inflammatory drugs using biosensors are summarized in Table 7.

Table 7. Some of the selected studies related with the detection of analgesic and anti-inflammatory drugs using biosensors and nanobiosensors.

Analyte	Transducer	Method	Detection Limit	Reference
Acetaminophen	Banana tissue/ polypyrrole CE	DPV	0.01 μM	[313]
Acetaminophen	Hematoxylin/GCE	DPV	0.14 μM	[314]
Aspirin	DNA/PPy/Pt	DPV	$5.24x10^{-6}$ μM	[315]
Amitriptyline	Graphite (SiO_2/Al_2O_3 $/Nb_2O_5$/DNA)	DPV	0.12 μM	[316]
Cocaine	DNA/Au	EIS	0.21 nM	[317]
Cocaine	Aptamer AuNPs/GO/SPCE	DPV	1 nM	[318]
Codeine Morphine	dsDNA/MWCNTs–PDDA/PGE	DPV	0.041 μg mL^{-1} 0.043 μg mL^{-1}	[319]
Diclofenac	DNA/GO/GCE	DPV	NS	[320]
Eserine Neostigmine	TTF–TCNQ–ionic/ionic liquid gel	CA	7.6 μM	[321]
Levetiracetam	HRP/ SPCE	CA	1.75 10^{-5} M	[322]
Promethazine	MWCN/SiAlNb/DNA	SWV	$5.9x10^{-6}$ M	[323]
Salicylic acid	ssDNA/PPy_{ox}/PtE	DPV	9.19×10^{-2} μM	[324]
Salicylic acid Aspirin	DNA/PPy_{ox}/PtE	DPV	8.59×10^{-2} μM 5.24×10^{-6} M	[325]
1-methyl-2-piperidinoethylester of 2-hexoxyphenylcarbamic acid	DNA/SPE	DPV	0.06 μM 0.016 μM	[326]

l. Pesticides

Plants, insects, bacteria, fungi and other organisms are a natural part of the environment. Some can benefit people, while others can be pests that you may need or want to control. Human can choose from many different methods to control a pest. One method is to use a pesticide. There are many different types of pesticides, each is meant to be effective against specific pests. The term "-cide" comes from the Latin word "to kill." Many pesticides can be grouped into chemical families such as: organochlorines, organophosphates, carbamates, thiocarbamate, dithiocarbamates, phenoxy and benzoic acid herbicides, triazines, ureas, and chloroacetanilides, phenoxy compounds *etc* [327, 328]. In electrochemical biosensing of pesticides, since these can inhibit enzymes, mainly electrochemical enzymatic biosensors are used [329, 330]. In their work, Ribeiro *et al.* in 2010, developed a biosensor for the determination of paraquat using gold nanoparticles and DNA modified gold electrode as the transducer. Using differential pulse voltammetry, paraquat was determined with the limit of detection of 1.4×10^{-6} M [331]. In another work, Yang *et al.* in 2013, detected methyl parathion and carbofuran with a limit of detection values of 5×10^{-14} M and 5×10^{-13} M, respectively, using platinum nanoparticles based biosensor. Acetylcholinesterase was immobilized in platinum nanoparticles, carboxylic graphene, nafion modified glassy carbon electrode surface, and the detection of parathion and carbofuran were achieved using inhibition of acetylcholinesterase enzyme [332]. Some of the selected studies related with the detection of pesticides using biosensors are summarized in Table **8**.

Table 8. Some of the selected studies related with the detection of pesticides using biosensors and nanobiosensors.

Analyte	Transducer	Method	Detection Limit	Ref
Carbaryl	AChE/GR/PANI	CA	20 ng.mL^{-1}	[333]
Carbaryl Formetanate Propoxur Ziram	Lacc–TYR–AuNPs–CS/GPE	SWV	0.02 μM 0.22 μM 0.19 μM 1.68 nM	[334]
Carbaryl Methomyl	AChE/GC/MWCNT/PANI	CA	1.4 μM 0.95 μM	[335]
Carbaryl Monocrotophos	AChE/GA/ILGR/Gel/GCE	DPV	5.3 fM 0.46 fM	[336]
Carbaryl Malathion	AChE/CS/PB-CS/ERGO-AuNPs-β-CD/GCE	CA	4.14 pg.mL^{-1} 1.15 pg.mL^{-1}	[337]
Carbaryl	AChE/TiO_2-G/GCE	CA	0.3 ng.mL^{-1}	[338]
Carbofuran	AChE/PAMAM-Au/CNTs/GCE	DPV	4.0 nM	[339]
Carbofuran	AChE/Fe_3O_4–CH/GCE	CV	3.6 nM	[340]
Carbofuran	AChE-CS/NiO NPs-CGR-NF	CA	0.5 pM	[341]

(Table 8) contd.....

Analyte	Transducer	Method	Detection Limit	Ref
Carbofuran Methyl Parathion	AChE–CS/SnO_2NPs–CGR–NF/GCE	DPV	0.5 pM 0.05 pM	[342]
Carbofuran Ziram	Lacc/PB/GPE	SWV	0.1 μM 5.2 nM	[343]
Carbosulfan	AChE/Pt/ZnO/Chitosan	CV	0.24 nM	[344]
Chlorfenvinphos	AChE/CoPC-SPCE	CA	10 μM	[345]
Chlorpyrifos Malathion	AChE/ZnS/Pin5COOH/AuE	CA	1.5 nM 0.1 nM	[346]
Chlorpyrifos Malathion Carbofuran Isoprocarb	AChE/CPBA/AuNPs/RGO-CS/GCE	CA	0.1 ppb 0.5 ppb 0.05 ppb 0.5 ppb	[347]
Chlorpyrifos Profenofos Malathion	Tyr/GR/PtNPs/GCE	CA	0.2 ppb 0.8 ppb 3 ppb	[348]
Chlorpyrifos	AChE/[BMIM][BF_4] MWCNT gel/CP	CA	4 nM	[349]
Chlorpyrifos Methamidophos	AChE/OMC–CS/Fe_3O_4–CS/SPCE	DPV	0.05 $\mu g.L^{-1}$ 1 $\mu g.L^{-1}$	[350]
Chlorpyrifos	Tyr/SPCE/GA/IrOx–BSA	CA	0.003 μM	[351]
Chlorpyrifos Methyl Parathion	AChE/AuNPs-CSs/BDD	DPV	0.13 pM 0.49 pM	[352]
Chlorpyrifos	AChE-CLDH/GN-AuNPs/GCE	DPV	0.05 $g.L^{-1}$	[353]
Dichlorvos	AChE/[BSmim]HSO_4-AuNPs-porous carbon/BDD	DPV	0.30 pM	[354]
Dichlorvos	AChE–ERGO–Nafion/GCE	CA	2.0 $ng.mL^{-1}$	[355]
Dimethoate	AChE/MWCNT-ß-CD/GCE	CV	2 nM	[356]
Formetanate	Lacc/AuNPs/AuE	SWV	0.095 μM	[357]
Furadan	AChE/ B-f-Fe@AuMNPs/ GR-SA/GCE	CA	0.01 ppb	[358]
Malathion	AChE/PEDOT-MWCNTs/FTO	CA	1 fM	[359]
Malathion	AChE/MWCNT/CA/NPG	CA	0.5 $ng.mL^{-1}$	[360]
Malathion	AChE/CuONWs/SWCNTs/GCE	DPV	0.1 ppb	[361]
Malathion Chlorpyrifos Monocrotophos Carbofuran	AChE/Chit-PB-MWNTs-HGNs/Au	CA	0.05 nM 0.05 nM 0.1 nM 2.5 nM	[362]
Malathion Chlorpyrifos Monocrotophos Endosulfan	AChE/Fe_3O_4NPs/cMWCNT/ITO	CV	0.1 nM	[363]

(Table 8) contd.....

Analyte	Transducer	Method	Detection Limit	Ref
Malaoxon Chlorpyriphos-Oxon Ethyl Paraoxon	AChE/CoPC/SPE	CA	0.5 nM 5 pM 5 nM	[364]
Monocrotophos	AChE/Au–PDDA–PB/GCE	CA	0.8 pg.mL^{-1}	[365]
Paraoxon	AChE/OPH/MWCNT/$(PEI/DNA)_2$	CV	0.5 µM	[366]
Paraoxon	BChE/poly(TTBO)/AgNWs	CA	0.212 µM	[367]
Paraoxon Malaoxon	AChE/SWCNTs/ Co phtalocyanine/SPCE	CA	3 ppb 2 ppb	[368]
Paraoxon Dichlorvos	AChE/GR/CdSe@ZnS/ITO	PEC	0.61 fM 2.5 pM	[369]
Paraoxon Malaoxon Aldicarb Carbofuran	AChE/CB/TC-0–gNPs/GCE	CA	0.05 nM 0.1 nM 0.01 µM 0.1 nM	[370]
Phosmet	AChE/PVA-AWP/Fe–Ni NP	CA	0.1 nM	[371]
Phoxim	AChE/CChit/AgNC/RGO/ GCE	DPV	81 pM	[372]
Tyramine	Tyr/SWCNT-COOH/SPCE	CA	0.62 µM	[373]
Triazophos	AChE/AuDMBG/RGO/GCE	CA	0.35 ppb	[374]

CONCLUSIONS

This chapter gives a general description of biosensors with special emphasis to electrochemical sensors, describing their main components and operating principle. The role of each part of biosensor, receptor and transducer, including their sub-classification is given. Enzymatic sensors, immunosensors and DNA sensors are also described. A brief introduction of nanomaterials employed in electrochemical biosensors also is given.

The main focus of the chapter is the different class of drugs, their role and brief description of application of nanobiosensors for their analyses. Some analytical performance parameters of representative nanobiosensors are given (in tables) for some class of drugs. The results obtained so far for drug detection using electrochemical nanobiosensors clearly show the capability of these devices to be used as simple analytical tools and alternative to conventional techniques which request professional users in addition to their high cost. Although the reported results look promising, the real application of these devices would request further technological development in terms of integration with sampling and easy to use readers that would send these out of the laboratory.

ABBREVIATIONS

AChE	Acetylcholinesterase
AgNC	Silver nanocluster
AgNWs	Silver nanowires
AuE	Gold electrode
AuNPs	Gold nanoparticles
BChE	Butyrylcholinesterase
BDD	Boron Doped Diamond
B-f-Fe@AuMNPs	boronic acid-functionalized Fe@Au magnetic nanoparticles
BSA	Bovine Serum Albumine
CA	Chronoamperometry
CB	Carbon black
CChit	Carboxylic chitosan
CGR	Carboxylic graphene
CLDH	Alcined layered doublehydroxide
CNC	carbon nanostructure-chitosan composite
CoPC	Cobalt (II) phthalocyanine
CPBA	3-carboxyphenylboronic
CS	Chitosan
CuO NWs	Copper oxide nanowires
Cys	Cysteamine
DHP	dihexadecylphosphate
DMBG	Dimethylbiguanide
dsDNA	double stranded DNA
ERGO	Electrochemical reduced graphene oxide
FTO	fluorine doped tin oxide
GA	Glutaraldehyde
GCE	Glassy carbon electrode
Gel	Gelatin
GPE	Graphene doped carbon paste
GR	Graphene
HGNs	Hollow gold nanospheres
HRP	Horseradish peroxidase
ILGR	Ionic liquid functionalized graphene
IrOx	Iridium oxide Nanoparticles

ITO	Indium tin oxide
Lacc	Laccase
MWCNT	Multiwalled carbon nanotube
NC	Nanocomposite
NF	Nafion
NPG	nanoporous gold
OMC	ordered mesoporous carbon
OPH	Organophosphate hydrolase
PAMAM	polyamidoamine
PANI	Polyaniline
PB	Prussian blue
PDDA	Poly (dimethyl diallyl ammonium chloride)
PEI	Polyethyleneimine
PGE	pencil graphite electrode
Pin5COOH	poly(indole-5-carboxylic acid)
poly(TTBO)	Poly(5,6-bis(octyloxy)-4,7-di(thieno [3][3,2-b]thiophen-2- yl) benzo[c] [1, 2, 5]oxoadiazole) /
PPy	Polypyrolle
PVA-AWP	Azide-unit waterpendant polyvinyl alcohol
SA	Sodium alginate
SnO_2NPs	SnO_2nanoparticles
SPCE	Screen printed carbon electrode
TC-0	11,17,23-Tetra-tert-butyl-25,26,27,28-tetrakis-1-(2′-hydroxyethyl)-N-(3″,4″-dihydro xyphenyl) amidocarbonyl)-methoxy) -2,8,14,20-tetrathiacalix arene in 1,3-alternate conformation
Tyr	Tyrosinase
β-CD	β-cyclodextrin

CONSENT FOR PUBLICATION

Not applicable.

CONFLICT OF INTEREST

The authors declares no conflict of interest, financial or otherwise.

ACKNOWLEDGEMENTS

Nanobiosensors and Bioelectronics Group acknowledges the support from

MINECO (project MAT2014-52485-P and Severo Ochoa Program, Grant SEV-2013-0295) and Secretaria d'Universitats i Recerca del Departament d'Economia i Coneixement de la Generalitat de Catalunya (2014 SGR 260).

REFERENCES

[1] Eggins, R. *Chemical sensors and biosensors Brain John Wiley Sons Sussex*; Brain John Wiley Sons: Sussex, **2002**.

[2] Coulet, P.R. *What is a Biosensor? Chapter 1Biosensor principles and applications*; Blum, L.J.; Coulet, P.R., Eds.; Marcel Dekker Inc.: New York, **1991**, pp. 1-6.

[3] Turner, A.P. *Biosensors: Fundamentals and Applications*; Turner, A.P.; Karube, I.; Wilson, G.S., Eds.; Oxford University Press: Oxford, **1987**, pp. 5-7.

[4] Scheller, F.; Schubet, F. *Biosensors*; Elsevier: Amsterdam, **1992**.

[5] Rogers, K.R.; Mulchandani, A.; Zhou, W. *Biosensor and Chemical Sensor Technology Process Monitoring and Control*; American Chemical Society: Washington, DC, **1995**, p. 2.

[6] Brett, A.M. *Biosensors and modern biospecific analytical techniques*; Gorton, L., Ed.; Elsevier: Amsterdam, Netherlands, **2005**, p. 179.
[http://dx.doi.org/10.1016/S0166-526X(05)44004-0]

[7] Buerk, D.G. *Biosensors: Theory and applications*; CRC Press, **1995**.

[8] Turner, A; Karube, I; Wilson, GS *Biosensors: fundamentals and applications*; , **1987**.
[http://dx.doi.org/10.1016/S0003-2670(00)85361-1]

[9] Turner, A.P. Biosensors: sense and sensibility. *Chem. Soc. Rev.,* **2013**, *42*(8), 3184-3196.
[http://dx.doi.org/10.1039/c3cs35528d] [PMID: 23420144]

[10] Turner, A.P. Biosensors. *Curr. Opin. Biotechnol.,* **1994**, *5*(1), 49-53.
[http://dx.doi.org/10.1016/S0958-1669(05)80069-2] [PMID: 7764643]

[11] Mascini, M; Palcihetti, I; Marrazza, G DNA electrochemical biosensors. *Fresenius J Anal Chemistry,* **2001**, *369*, 15-22.

[12] Sharma, S.K.; Sehgal, N.; Kumar, A. Biomolecules for development of biosensor and their applications. *Curr. Appl. Phys.,* **2003**, *3*, 307-316.
[http://dx.doi.org/10.1016/S1567-1739(02)00219-5]

[13] Zhai, J.; Cui, H.; Yang, R. DNA based biosensors. *Biotechnol. Adv.,* **1997**, *15*(1), 43-58.
[http://dx.doi.org/10.1016/S0734-9750(97)00003-7] [PMID: 14539378]

[14] Macholan, L. Recent progress in developing enzyme and tissue-based membrane electrodes. *Acta Biotechnol.,* **1987**, *69*, 547-553.
[http://dx.doi.org/10.1002/abio.370070615]

[15] D'Souza, S.F. Microbial biosensors. *Biosens. Bioelectron.,* **2001**, *16*(6), 337-353.
[http://dx.doi.org/10.1016/S0956-5663(01)00125-7] [PMID: 11672648]

[16] Clark, L.C., Jr; Lyons, C. Electrode systems for continuous monitoring in cardiovascular surgery. *Ann. N. Y. Acad. Sci.,* **1962**, *102*, 29-45.
[http://dx.doi.org/10.1111/j.1749-6632.1962.tb13623.x] [PMID: 14021529]

[17] Wang, J. Electrochemical glucose biosensors. *Chem. Rev.,* **2008**, *108*(2), 814-825.
[http://dx.doi.org/10.1021/cr068123a] [PMID: 18154363]

[18] Wang, J. Glucose biosensors: 40 years of advances and challenges. *Electroanalysis,* **2001**, *13*(12), 983.
[http://dx.doi.org/10.1002/1521-4109(200108)13:12<983::AID-ELAN983>3.0.CO;2-#]

[19] Newman, J.D.; Turner, A.P. Home blood glucose biosensors: a commercial perspective. *Biosens. Bioelectron.,* **2005**, *20*(12), 2435-2453.

[http://dx.doi.org/10.1016/j.bios.2004.11.012] [PMID: 15854818]

[20] D'Orazio, P. Biosensors in clinical chemistry. *Clin. Chim. Acta,* **2003**, *334*(1-2), 41-69. [http://dx.doi.org/10.1016/S0009-8981(03)00241-9] [PMID: 12867275]

[21] Malhotra, B.D.; Chaubey, A. Biosensors for clinical diagnostics industry. *Sensor Actuat B,* **2003**, *91*, 117-127. [http://dx.doi.org/10.1016/S0925-4005(03)00075-3]

[22] Bahadır, E.B.; Sezgintürk, M.K. Electrochemical biosensors for hormone analyses. *Biosens. Bioelectron.,* **2015**, *68*, 62-71. [http://dx.doi.org/10.1016/j.bios.2014.12.054] [PMID: 25558874]

[23] Yoo, E.H.; Lee, S.Y. Glucose biosensors: an overview of use in clinical practice. *Sensors (Basel),* **2010**, *10*(5), 4558-4576. [http://dx.doi.org/10.3390/s100504558] [PMID: 22399892]

[24] Luong, J.H.; Bouvrette, P.; Male, K.B. Developments and applications of biosensors in food analysis. *Trends Biotechnol.,* **1997**, *15*(9), 369-377. [http://dx.doi.org/10.1016/S0167-7799(97)01071-8] [PMID: 9293036]

[25] Mulchandani, A.; Bassi, A.S. Principles and applications of biosensors for bioprocess monitoring and control. *Crit. Rev. Biotechnol.,* **1995**, *15*(2), 105-124. [http://dx.doi.org/10.3109/07388559509147402] [PMID: 7641291]

[26] Song, S.; Xu, H.; Fan, C. Potential diagnostic applications of biosensors: current and future directions. *Int. J. Nanomedicine,* **2006**, *1*(4), 433-440. [http://dx.doi.org/10.2147/nano.2006.1.4.433] [PMID: 17722277]

[27] Luong, J.H.; Mulchandani, A.; Guilbault, G.G. Developments and applications of biosensors. *Trends Biotechnol.,* **1988**, *6*(12), 310-316. [http://dx.doi.org/10.1016/0167-7799(88)90026-1] [PMID: 9293036]

[28] Rogers, K.R.; Williams, L.R. Biosensors for environmental monitoring: a regulatory perspective. *TrAC Trends in Analytical Chemistry,* **1995**, *14*(7), 289-294. [http://dx.doi.org/10.1016/0165-9936(95)97054-5]

[29] Higgins, I.J.; Lowe, C.R. Introduction to the principles and applications of biosensors. *Philos. Trans. R. Soc. Lond. B Biol. Sci.,* **1987**, *316*(1176), 3-11. [http://dx.doi.org/10.1098/rstb.1987.0013] [PMID: 2889231]

[30] Ravindra, N.M.; Prodan, C.; Fnu, S.; Padronl, I.; Sikha, S.K. Advances in the manufacturing, types, and applications of biosensors. *JOM,* **2007**, *59*(12), 37-43. [http://dx.doi.org/10.1007/s11837-007-0150-0]

[31] Pérez-López, B.; Merkoçi, A. Nanomaterials based biosensors for food analysis applications. *Trends Food Sci. Technol.,* **2011**, *22*(11), 625-639. [http://dx.doi.org/10.1016/j.tifs.2011.04.001]

[32] Albareda-Sirvent, M.; Merkoçi, A.; Alegret, S. Pesticide determination in tap water and juice samples using disposable amperometric biosensors made using thick-film technology. *Anal. Chim. Acta,* **2001**, *442*(1), 35-44. [http://dx.doi.org/10.1016/S0003-2670(01)01017-0]

[33] Pérez, B.; Pumera, M.; del Valle, M.; Merkoçi, A.; Alegret, S. Glucose biosensor based on carbon nanotube epoxy composites. *J. Nanosci. Nanotechnol.,* **2005**, *5*(10), 1694-1698. [http://dx.doi.org/10.1166/jnn.2005.400] [PMID: 16245530]

[34] Rawson, D.M.; Willmer, A.J.; Turner, A.P. Whole-cell biosensors for environmental monitoring. *Biosensors,* **1989**, *4*(5), 299-311. [http://dx.doi.org/10.1016/0265-928X(89)80011-2] [PMID: 2775317]

[35] Grieshaber, D.; MacKenzie, R.; Vörös, J.; Reimhult, E. Electrochemical Biosensors - Sensor

Principles and Architectures. *Sensors (Basel),* **2008**, *8*(3), 1400-1458. [http://dx.doi.org/10.3390/s8031400] [PMID: 27879772]

[36] Pancrazio, J.J.; Whelan, J.P.; Borkholder, D.A.; Ma, W.; Stenger, D.A. Development and application of cell-based biosensors. *Ann. Biomed. Eng.,* **1999**, *27*(6), 697-711. [http://dx.doi.org/10.1114/1.225] [PMID: 10625143]

[37] Švitel, J.; Tkác, J.; Voštiar, I.; Navrátil, M.; Štefuca, V.; Bučko, M.; Gemeiner, P. Gluconobacter in biosensors: applications of whole cells and enzymes isolated from Gluconobacter and Acetobacter to biosensor construction. *Biotechnol. Lett.,* **2006**, *28*(24), 2003-2010. [http://dx.doi.org/10.1007/s10529-006-9195-3] [PMID: 17072528]

[38] Erickson, D.; Mandal, S.; Yang, A.H.; Cordovez, B. Nanobiosensors: optofluidic, electrical and mechanical approaches to biomolecular detection at the nanoscale. *Microfluid. Nanofluidics,* **2008**, *4*(1-2), 33-52. [http://dx.doi.org/10.1007/s10404-007-0198-8] [PMID: 18806888]

[39] Turner, A.P. Tech.Sight. Biochemistry. Biosensors--sense and sensitivity. *Science,* **2000**, *290*(5495), 1315-1317. [http://dx.doi.org/10.1126/science.290.5495.1315] [PMID: 11185408]

[40] Leech, D. Affinity biosensors. *Chem. Soc. Rev.,* **1994**, *23*(3), 205-213. [http://dx.doi.org/10.1039/cs9942300205]

[41] Cosnier, S. Affinity biosensors based on electropolymerized films. *Electroanalysis,* **2005**, *17*(19), 1701-1715. [http://dx.doi.org/10.1002/elan.200503308]

[42] Newman, J.D.; White, S.F.; Tothill, I.E.; Turner, A.P. Catalytic materials, membranes, and fabrication technologies suitable for the construction of amperometric biosensors. *Anal. Chem.,* **1995**, *67*(24), 4594-4599. [http://dx.doi.org/10.1021/ac00120a027]

[43] Kwong, WK *Catalytic biosensors: novel analytical tools and their applications in sports, food and environmental monitoring.,* **2000**.

[44] Xu, F.; Zhen, G.; Yu, F.; Kuennemann, E.; Textor, M.; Knoll, W. Combined affinity and catalytic biosensor: *in situ* enzymatic activity monitoring of surface-bound enzymes. *J. Am. Chem. Soc.,* **2005**, *127*(38), 13084-13085. [http://dx.doi.org/10.1021/ja050818q] [PMID: 16173702]

[45] Singh, M.; Verma, N.; Garg, A.K.; Redhu, N. Urea biosensors. *Sens. Actuators B Chem.,* **2008**, *134*(1), 345-351. [http://dx.doi.org/10.1016/j.snb.2008.04.025]

[46] Dhawan, G.; Sumana, G.; Malhotra, B.D. Recent developments in urea biosensors. *Biochem. Eng. J.,* **2009**, *44*(1), 42-52. [http://dx.doi.org/10.1016/j.bej.2008.07.004]

[47] Amine, A.; Mohammadi, H.; Bourais, I.; Palleschi, G. Enzyme inhibition-based biosensors for food safety and environmental monitoring. *Biosens. Bioelectron.,* **2006**, *21*(8), 1405-1423. [http://dx.doi.org/10.1016/j.bios.2005.07.012] [PMID: 16125923]

[48] Wilson, G.S.; Hu, Y. Enzyme-based biosensors for *in vivo* measurements. *Chem. Rev.,* **2000**, *100*(7), 2693-2704. [http://dx.doi.org/10.1021/cr990003y] [PMID: 11749301]

[49] Katz, E.; Bückmann, A.F.; Willner, I. Self-powered enzyme-based biosensors. *J. Am. Chem. Soc.,* **2001**, *123*(43), 10752-10753. [http://dx.doi.org/10.1021/ja0167102] [PMID: 11674014]

[50] Acha, V.; Andrews, T.; Huang, Q.; Sardar, D.K.; Hornsby, P.J. Tissue-based biosensors.*Recognition Receptors in Biosensors*; Springer New York, **2010**, pp. 365-381.

[http://dx.doi.org/10.1007/978-1-4419-0919-0_9]

[51] Wijesuriya, D.C.; Rechnitz, G.A. Biosensors based on plant and animal tissues. *Biosens. Bioelectron.,* **1993**, *8*(3-4), 155-160.
[http://dx.doi.org/10.1016/0956-5663(93)85027-L] [PMID: 8395191]

[52] Luppa, P.B.; Sokoll, L.J.; Chan, D.W. Immunosensors--principles and applications to clinical chemistry. *Clin. Chim. Acta,* **2001**, *314*(1-2), 1-26.
[http://dx.doi.org/10.1016/S0009-8981(01)00629-5] [PMID: 11718675]

[53] Morgan, C.L.; Newman, D.J.; Price, C.P. Immunosensors: technology and opportunities in laboratory medicine. *Clin. Chem.,* **1996**, *42*(2), 193-209.
[PMID: 8595711]

[54] Skládal, P. Advances in electrochemical immunosensors. *Electroanalysis,* **1997**, *9*(10), 737-745.
[http://dx.doi.org/10.1002/elan.1140091002]

[55] Song, S.; Wang, L.; Li, J.; Fan, C.; Zhao, J. Aptamer-based biosensors. *TrAC Trends in Analytical Chemistry,* **2008**, *27*(2), 108-117.
[http://dx.doi.org/10.1016/j.trac.2007.12.004]

[56] Taylor, R.F.; Marenchic, I.G.; Spencer, R.H. Antibody-and receptor-based biosensors for detection and process control. *Anal. Chim. Acta,* **1991**, *249*(1), 67-70.
[http://dx.doi.org/10.1016/0003-2670(91)87009-V]

[57] Pancrazio, J.J.; Whelan, J.P.; Borkholder, D.A.; Ma, W.; Stenger, D.A. Development and application of cell-based biosensors. *Ann. Biomed. Eng.,* **1999**, *27*(6), 697-711.
[http://dx.doi.org/10.1114/1.225] [PMID: 10625143]

[58] Eldefrawi, M.E.; Sherby, S.M.; Andreou, A.G.; Mansour, N.A.; Annau, Z.; Blum, N.A.; Valdes, J.J. Acetylcholine receptor-based biosensor. *Anal. Lett.,* **1988**, *21*(9), 1665-1680.
[http://dx.doi.org/10.1080/00032718808066519]

[59] North, J.R. Immunosensors: antibody-based biosensors. *Trends Biotechnol.,* **1985**, *3*(7), 180-186.
[http://dx.doi.org/10.1016/0167-7799(85)90119-2]

[60] Hahn, S.; Mergenthaler, S.; Zimmermann, B.; Holzgreve, W. Nucleic acid based biosensors: the desires of the user. *Bioelectrochemistry,* **2005**, *67*(2), 151-154.
[http://dx.doi.org/10.1016/j.bioelechem.2004.07.006] [PMID: 16019267]

[61] Zhai, J.; Cui, H.; Yang, R. DNA based biosensors. *Biotechnol. Adv.,* **1997**, *15*(1), 43-58.
[http://dx.doi.org/10.1016/S0734-9750(97)00003-7] [PMID: 14539378]

[62] Belli, S.L.; Rechnitz, G.A. Prototype potentiometric biosensor using intact chemoreceptor structures. *Anal. Lett.,* **1986**, *19*(3-4), 403-416.
[http://dx.doi.org/10.1080/00032718608064506]

[63] Schöning, M.J.; Schroth, P.; Schütz, S. The Use of Insect Chemoreceptors for the Assembly of Biosensors Based on Semiconductor Field□Effect Transistors. *Electroanalysis,* **2000**, *12*(9), 645-652.
[http://dx.doi.org/10.1002/1521-4109(200005)12:9<645::AID-ELAN645>3.0.CO;2-O]

[64] Belli, S.L.; Rechnitz, G.A. Biosensors based on native chemoreceptors. *Fresenius Z. Anal. Chem.,* **1988**, *331*(3-4), 439-447.
[http://dx.doi.org/10.1007/BF00481924]

[65] Briegel, A.; Wong, M.L.; Hodges, H.L.; Oikonomou, C.M.; Piasta, K.N.; Harris, M.J.; Fowler, D.J.; Thompson, L.K.; Falke, J.J.; Kiessling, L.L.; Jensen, G.J. New insights into bacterial chemoreceptor array structure and assembly from electron cryotomography. *Biochemistry,* **2014**, *53*(10), 1575-1585.
[http://dx.doi.org/10.1021/bi5000614] [PMID: 24580139]

[66] D'Souza, S.F. Immobilization and stabilization of biomaterials for biosensor applications. *Appl. Biochem. Biotechnol.,* **2001**, *96*(1-3), 225-238.
[http://dx.doi.org/10.1385/ABAB:96:1-3:225] [PMID: 11783889]

[67] Arroyo, M. Inmovilización de enzimas. Fundamentos, métodos y aplicaciones. *Ars. Pharm.,* **1998,** *39*(2), 23-39.

[68] Asgher, M.; Shahid, M.; Kamal, S.; Iqbal, H.M. Recent trends and valorization of immobilization strategies and ligninolytic enzymes by industrial biotechnology. *J. Mol. Catal., B Enzym.,* **2014**, *101*, 56-66.
[http://dx.doi.org/10.1016/j.molcatb.2013.12.016]

[69] Bowers, L.D.; Carr, P.W. *Immobilized enzymes in analytical chemistry*; Springer Berlin Heidelberg, **1980**, pp. 89-129.

[70] Brady, D.; Jordaan, J. Advances in enzyme immobilisation. *Biotechnol. Lett.,* **2009**, *31*(11), 1639-1650.
[http://dx.doi.org/10.1007/s10529-009-0076-4] [PMID: 19590826]

[71] Carr, P.W.; Bowers, L.D. *Immobilized enzymes in analytical and clinical chemistry*; Wiley, **1980**.

[72] Jesionowski, T.; Zdarta, J.; Krajewska, B. Enzyme immobilization by adsorption:a review. *Adsorption,* **2014**, *20*, 801-821.
[http://dx.doi.org/10.1007/s10450-014-9623-y]

[73] Putzbach, W; Ronkainen, NJ Immobilization techniques in the fabrication of nanomaterial-based electrochemical biosensors: A review. *Sensors,* **2013**, *13*(4), 4811-4840.
[http://dx.doi.org/10.3390/s130404811]

[74] Jesionowski, T.; Zdarta, J.; Krajewska, B. Enzyme immobilization by adsorption: a review. *Adsorption,* **2014**, *20*(5-6), 801-821.
[http://dx.doi.org/10.1007/s10450-014-9623-y]

[75] Sassolas, A.; Blum, L.J.; Leca-Bouvier, B.D. Immobilization strategies to develop enzymatic biosensors. *Biotechnol. Adv.,* **2012**, *30*(3), 489-511.
[http://dx.doi.org/10.1016/j.biotechadv.2011.09.003] [PMID: 21951558]

[76] Georgopoulos, K.N.; Siontorou, C.G. A planar bilayer lipid membrane biosensor based on enzymes in the service of environmental monitoring. *WIT Transactions on Ecology and the Environment,* **2016**, *209*, 91-102.
[http://dx.doi.org/10.2495/WP160091]

[77] Yuan, J.; Wen, D.; Gaponik, N.; Eychmüller, A. Enzyme-encapsulating quantum dot hydrogels and xerogels as biosensors: multifunctional platforms for both biocatalysis and fluorescent probing. *Angew. Chem. Int. Ed. Engl.,* **2013**, *52*(3), 976-979.
[http://dx.doi.org/10.1002/anie.201205791] [PMID: 23172829]

[78] Guan, H.; Liu, X.; Wang, W. Encapsulation of tyrosinase within liposome bioreactors for developing an amperometric phenolic compounds biosensor. *J. Solid State Electrochem.,* **2013**, *17*(11), 2887-2893.
[http://dx.doi.org/10.1007/s10008-013-2181-5]

[79] Hasanzadeh, M.; Shadjou, N.; Eskandani, M.; Soleymani, J.; Jafari, F.; de la Guardia, M. Dendrimer-encapsulated and cored metal nanoparticles for electrochemical nanobiosensing. *TrAC Trends in Analytical Chemistry,* **2014**, *53*, 137-149.
[http://dx.doi.org/10.1016/j.trac.2013.09.015]

[80] Kang, A.; Park, J.; Ju, J.; Jeong, G.S.; Lee, S.H. Cell encapsulation *via* microtechnologies. *Biomaterials,* **2014**, *35*(9), 2651-2663.
[http://dx.doi.org/10.1016/j.biomaterials.2013.12.073] [PMID: 24439405]

[81] Chang, T.M. Microencapsulation of enzymes, cells, and genetically engineered microorganisms. *Methods Mol. Med.,* **1999**, *18*, 315-330.
[PMID: 21370187]

[82] Cosnier, S. Biomolecule immobilization on electrode surfaces by entrapment or attachment to

electrochemically polymerized films. A review. *Biosens. Bioelectron.,* **1999**, *14*(5), 443-456. [http://dx.doi.org/10.1016/S0956-5663(99)00024-X] [PMID: 10451912]

[83] Kennedy, J.F.; White, C.A. Principles of immobilization of enzymes. In: *Handbook of enzyme biotechnology*; , **1985**; Vol. 2, pp. 147-207.

[84] Trevan, M.D. Enzyme immobilization by entrapment. *Methods Mol. Biol.,* **1988**, *3*, 491-494. [PMID: 21400185]

[85] Gerard, M; Chaubey, A; Malhotra, BD Application of conducting polymers to biosensors Biosens. *Bioelectron,* **2002**, *17*(5), 345-359.

[86] Lei, W.; Si, W.; Xu, Y.; Gu, Z.; Hao, Q. Conducting polymer composites with graphene for use in chemical sensors and biosensors. *Mikrochim. Acta,* **2014**, *181*(7-8), 707-722. [http://dx.doi.org/10.1007/s00604-014-1160-6]

[87] Ates, M. A review study of (bio)sensor systems based on conducting polymers. *Mater. Sci. Eng. C,* **2013**, *33*(4), 1853-1859. [http://dx.doi.org/10.1016/j.msec.2013.01.035] [PMID: 23498205]

[88] Putzbach, W.; Ronkainen, N.J. Immobilization techniques in the fabrication of nanomaterial-based electrochemical biosensors: a review. *Sensors (Basel),* **2013**, *13*(4), 4811-4840. [http://dx.doi.org/10.3390/s130404811] [PMID: 23580051]

[89] Zhou, Z.; Piepenbreier, F.; Marthala, V.R.; Karbacher, K.; Hartmann, M. Immobilization of lipase in cage-type mesoporous organosilicas *via* covalent bonding and crosslinking. *Catal. Today,* **2015**, *243*, 173-183. [http://dx.doi.org/10.1016/j.cattod.2014.07.047]

[90] Brena, B.; González-Pombo, P.; Batista-Viera, F. Immobilization of enzymes: a literature survey. In: *Immobilization of Enzymes and Cells: Third Edition*; , **2013**; pp. 15-31.

[91] Kim, M.H.; Park, S.; Kim, Y.H.; Won, K.; Lee, S.H. Immobilization of formate dehydrogenase from Candida boidinii through cross-linked enzyme aggregates. *J. Mol. Catal., B Enzym.,* **2013**, *97*, 209-214. [http://dx.doi.org/10.1016/j.molcatb.2013.08.020]

[92] Bezbradica, D.I.; Mateo, C.; Guisan, J.M. Novel support for enzyme immobilization prepared by chemical activation with cysteine and glutaraldehyde. *J. Mol. Catal., B Enzym.,* **2014**, *102*, 218-224. [http://dx.doi.org/10.1016/j.molcatb.2014.02.021]

[93] Barbosa, O.; Ortiz, C.; Berenguer-Murcia, Á.; Torres, R.; Rodrigues, R.C.; Fernandez-Lafuente, R. Glutaraldehyde in bio-catalysts design: a useful crosslinker and a versatile tool in enzyme immobilization. *RSC Advances,* **2014**, *4*(4), 1583-1600. [http://dx.doi.org/10.1039/C3RA45991H]

[94] López-Gallego, F.; Guisán, J.M.; Betancor, L. Glutaraldehyde-mediated protein immobilization. In: *Immobilization of Enzymes and Cells: Third Edition*; , **2013**; pp. 33-41. [http://dx.doi.org/10.1007/978-1-62703-550-7_3]

[95] Stepankova, V.; Bidmanova, S.; Koudelakova, T.; Prokop, Z.; Chaloupkova, R.; Damborsky, J. Strategies for stabilization of enzymes in organic solvents. *ACS Catal.,* **2013**, *3*(12), 2823-2836. [http://dx.doi.org/10.1021/cs400684x]

[96] Xueji, Z.; Ju, H.; Wang, J. *Electrochemical sensors, biosensors and their biomedical applications*; Elsevier: San Diego, **2008**.

[97] Pohanka, M.; Skládal, P. Electrochemical biosensors–principles and applications. *J. Appl. Biomed.,* **2008**, *6*(2), 57-64.

[98] Thévenot, D.R.; Toth, K.; Durst, R.A.; Wilson, G.S. Electrochemical biosensors: recommended definitions and classification. *Biosens. Bioelectron.,* **2001**, *16*(1-2), 121-131. [PMID: 11261847]

[99] Ronkainen, N.J.; Halsall, H.B.; Heineman, W.R. Electrochemical biosensors. *Chem. Soc. Rev.,* **2010**, *39*(5), 1747-1763.
[http://dx.doi.org/10.1039/b714449k] [PMID: 20419217]

[100] Luo, X.; Morrin, A.; Killard, A.J.; Smyth, M.R. Application of nanoparticles in electrochemical sensors and biosensors. *Electroanalysis,* **2006**, *18*(4), 319-326.
[http://dx.doi.org/10.1002/elan.200503415]

[101] Aizawa, M.A. Principles and applications of electrochemical and optical biosensors. *Anal. Chim. Acta,* **1991**, *250*, 249-256.
[http://dx.doi.org/10.1016/0003-2670(91)85073-2]

[102] Windmiller, J.R.; Wang, J. Wearable electrochemical sensors and biosensors: a review. *Electroanalysis,* **2013**, *25*(1), 29-46.
[http://dx.doi.org/10.1002/elan.201200349]

[103] Ramanathan, K.; Danielsson, B. Principles and applications of thermal biosensors. *Biosens. Bioelectron.,* **2001**, *16*(6), 417-423.
[http://dx.doi.org/10.1016/S0956-5663(01)00124-5] [PMID: 11672656]

[104] Minunni, M.; Mascini, M.; Guilbault, G.G.; Hock, B. The quartz Crystal Microbalance as Biosensor: A Status Report on Its Future. *Anal. Lett.,* **1995**, *28*, 749-764.
[http://dx.doi.org/10.1080/00032719508001422]

[105] O'Sullivan, C.K.; Guilbault, G.G. Commercial Quartz Crystal Microbalances- theory and applications. *Biosens. Bioelectron.,* **1999**, *14*, 663-670.
[http://dx.doi.org/10.1016/S0956-5663(99)00040-8]

[106] Estevez, M.C.; Otte, M.A.; Sepulveda, B.; Lechuga, L.M. Trends and challenges of refractometric nanoplasmonic biosensors: a review. *Anal. Chim. Acta,* **2014**, *806*, 55-73.
[http://dx.doi.org/10.1016/j.aca.2013.10.048] [PMID: 24331040]

[107] Caruso, F.; Rodda, E.; Furlong, D.N.; Niikura, K.; Okahata, Y. Quartz crystal microbalance study of DNA immobilization and hybridization for nucleic Acid sensor development. *Anal. Chem.,* **1997**, *69*(11), 2043-2049.
[http://dx.doi.org/10.1021/ac961220r] [PMID: 21639244]

[108] Zhou, X.C.; Huang, L.Q.; Li, S.F. Microgravimetric DNA sensor based on quartz crystal microbalance: comparison of oligonucleotide immobilization methods and the application in genetic diagnosis. *Biosens. Bioelectron.,* **2001**, *16*(1-2), 85-95.
[http://dx.doi.org/10.1016/S0956-5663(00)00136-6] [PMID: 11261857]

[109] Qian, F.; Asanov, A.N. Oldham PBA total internal reflection fluorescence biosensor for aluminum (III). *Microchem. J.,* **2001**, *70*, 63-68.
[http://dx.doi.org/10.1016/S0026-265X(00)00162-4]

[110] Liu, X.; Song, D.; Zhang, Q.; Tian, Y.; Zhang, H. An optical surface plasmon resonance biosensor for determination of tetanus toxin. *Talanta,* **2004**, *62*(4), 773-779.
[http://dx.doi.org/10.1016/j.talanta.2003.09.025] [PMID: 18969362]

[111] Fan, X.; White, I.M.; Shopova, S.I.; Zhu, H.; Suter, J.D.; Sun, Y. Sensitive optical biosensors for unlabeled targets: a review. *Anal. Chim. Acta,* **2008**, *620*(1-2), 8-26.
[http://dx.doi.org/10.1016/j.aca.2008.05.022] [PMID: 18558119]

[112] Özkan, S.A. Electroanalytical Methods.*Pharmaceutical Analysis And Their Validation*; Hnb Pub.: USA, **2012**.

[113] Özkan, S.A.; Kauffmann, J.M.; Zuman, P. Electroanalysis.*Biomedical And Pharmaceutical Sciences (Main Title), (Voltammetry, Amperometry, Biosensors, Applications)*; Springer-Verlag Berlin Heidelberg, **2015**.

[114] Zhu, C.; Yang, G.; Li, H.; Du, D.; Lin, Y. Electrochemical sensors and biosensors based on

nanomaterials and nanostructures. *Anal. Chem.,* **2015**, *87*(1), 230-249. [http://dx.doi.org/10.1021/ac5039863] [PMID: 25354297]

[115] Hong, G.; Diao, S.; Antaris, A.L.; Dai, H. Carbon nanomaterials for biological imaging and nanomedicinal therapy. *Chem. Rev.,* **2015**, *115*(19), 10816-10906. [http://dx.doi.org/10.1021/acs.chemrev.5b00008] [PMID: 25997028]

[116] Chen, A.; Ostrom, C. Palladium-based nanomaterials: Synthesis and electrochemical applications. *Chem. Rev.,* **2015**, *115*(21), 11999-12044. [http://dx.doi.org/10.1021/acs.chemrev.5b00324] [PMID: 26402587]

[117] Tiwari, J.N.; Vij, V.; Kemp, K.C.; Kim, K.S. Engineered carbon-nanomaterial-based electrochemical sensors for biomolecules. *ACS Nano,* **2016**, *10*(1), 46-80. [http://dx.doi.org/10.1021/acsnano.5b05690] [PMID: 26579616]

[118] Ambrosi, A.; Chua, C.K.; Bonanni, A.; Pumera, M. Electrochemistry of graphene and related materials. *Chem. Rev.,* **2014**, *114*(14), 7150-7188. [http://dx.doi.org/10.1021/cr500023c] [PMID: 24895834]

[119] Balasubramanian, K.; Burghard, M. Biosensors based on carbon nanotubes. *Anal. Bioanal. Chem.,* **2006**, *385*(3), 452-468. [http://dx.doi.org/10.1007/s00216-006-0314-8] [PMID: 16568294]

[120] Fang, Y.; Wang, E. Electrochemical biosensors on platforms of graphene. *Chem. Commun. (Camb.),* **2013**, *49*(83), 9526-9539. [http://dx.doi.org/10.1039/c3cc44735a] [PMID: 24025792]

[121] Chung, C.; Kim, Y.K.; Shin, D.; Ryoo, S.R.; Hong, B.H.; Min, D.H. Biomedical applications of graphene and graphene oxide. *Acc. Chem. Res.,* **2013**, *46*(10), 2211-2224. [http://dx.doi.org/10.1021/ar300159f] [PMID: 23480658]

[122] Wang, Y.; Li, Z.; Wang, J.; Li, J.; Lin, Y. Graphene and graphene oxide: biofunctionalization and applications in biotechnology. *Trends Biotechnol.,* **2011**, *29*(5), 205-212. [http://dx.doi.org/10.1016/j.tibtech.2011.01.008] [PMID: 21397350]

[123] Merkoçi, A. Carbon nanotubes in analytical sciences. *Mikrochim. Acta,* **2006**, *152*(3-4), 157-174. [http://dx.doi.org/10.1007/s00604-005-0439-z]

[124] Pérez-López, B.; Merkoçi, A. Carbon nanotubes and graphene in analytical sciences. *Mikrochim. Acta,* **2012**, *179*(1-2), 1-16. [http://dx.doi.org/10.1007/s00604-012-0871-9]

[125] Mauter, M.S.; Elimelech, M. Environmental applications of carbon-based nanomaterials. *Environ. Sci. Technol.,* **2008**, *42*(16), 5843-5859. [http://dx.doi.org/10.1021/es8006904] [PMID: 18767635]

[126] Bai, J.; Zhou, B. Titanium dioxide nanomaterials for sensor applications. *Chem. Rev.,* **2014**, *114*(19), 10131-10176. [http://dx.doi.org/10.1021/cr400625j] [PMID: 24922605]

[127] Wang, W.; Xie, Y.; Xia, C.; Du, H.; Tian, F. Titanium dioxide nanotube arrays modified with a nanocomposite of silver nanoparticles and reduced graphene oxide for electrochemical sensing. *Mikrochim. Acta,* **2014**, *181*(11-12), 1325-1331. [http://dx.doi.org/10.1007/s00604-014-1258-x]

[128] Li, X.; Yao, J.; Liu, F.; He, H.; Zhou, M.; Mao, N.; Zhang, Y. Nickel/Copper nanoparticles modified TiO 2 nanotubes for non-enzymatic glucose biosensors. *Sens. Actuators B Chem.,* **2013**, *181*, 501-508. [http://dx.doi.org/10.1016/j.snb.2013.02.035]

[129] Laurent, S.; Forge, D.; Port, M.; Roch, A.; Robic, C.; Vander Elst, L.; Muller, R.N. Magnetic iron oxide nanoparticles: synthesis, stabilization, vectorization, physicochemical characterizations, and biological applications. *Chem. Rev.,* **2008**, *108*(6), 2064-2110. [http://dx.doi.org/10.1021/cr068445e] [PMID: 18543879]

[130] Hsing, I.; Xu, Y.; Zhao, W. Micro-and Nano-Magnetic Particles for Applications in Biosensing. *Electroanalysis,* **2007**, *19*(7-8), 755-768.
[http://dx.doi.org/10.1002/elan.200603785]

[131] Hasanzadeh, M.; Shadjou, N.; de la Guardia, M. Iron and iron-oxide magnetic nanoparticles as signal-amplification elements in electrochemical biosensing. *TrAC Trends in Analytical Chemistry,* **2015**, *72*, 1-9.
[http://dx.doi.org/10.1016/j.trac.2015.03.016]

[132] Hrapovic, S.; Liu, Y.; Male, K.B.; Luong, J.H. Electrochemical biosensing platforms using platinum nanoparticles and carbon nanotubes. *Anal. Chem.,* **2004**, *76*(4), 1083-1088.
[http://dx.doi.org/10.1021/ac035143t] [PMID: 14961742]

[133] Claussen, J.C.; Kumar, A.; Jaroch, D.B.; Khawaja, M.H.; Hibbard, A.B.; Porterfield, D.M.; Fisher, T.S. Nanostructuring platinum nanoparticles on multilayered graphene petal nanosheets for electrochemical biosensing. *Adv. Funct. Mater.,* **2012**, *22*(16), 3399-3405.
[http://dx.doi.org/10.1002/adfm.201200551]

[134] Yang, M.; Yang, Y.; Liu, Y.; Shen, G.; Yu, R. Platinum nanoparticles-doped sol-gel/carbon nanotubes composite electrochemical sensors and biosensors. *Biosens. Bioelectron.,* **2006**, *21*(7), 1125-1131.
[http://dx.doi.org/10.1016/j.bios.2005.04.009] [PMID: 15885999]

[135] Chen, A.; Ostrom, C. Palladium-based nanomaterials: Synthesis and electrochemical applications. *Chem. Rev.,* **2015**, *115*(21), 11999-12044.
[http://dx.doi.org/10.1021/acs.chemrev.5b00324] [PMID: 26402587]

[136] Thiagarajan, S.; Yang, R.F.; Chen, S.M. Palladium nanoparticles modified electrode for the selective detection of catecholamine neurotransmitters in presence of ascorbic acid. *Bioelectrochemistry,* **2009**, *75*(2), 163-169.
[http://dx.doi.org/10.1016/j.bioelechem.2009.03.014] [PMID: 19409863]

[137] Li, Z.; Wang, X.; Wen, G.; Shuang, S.; Dong, C.; Paau, M.C.; Choi, M.M. Application of hydrophobic palladium nanoparticles for the development of electrochemical glucose biosensor. *Biosens. Bioelectron.,* **2011**, *26*(11), 4619-4623.
[http://dx.doi.org/10.1016/j.bios.2011.04.057] [PMID: 21612909]

[138] Zeng, S.; Yong, K.T.; Roy, I.; Dinh, X.Q.; Yu, X.; Luan, F. A review on functionalized gold nanoparticles for biosensing applications. *Plasmonics,* **2011**, *6*(3), 491.
[http://dx.doi.org/10.1007/s11468-011-9228-1]

[139] Shan, C.; Yang, H.; Han, D.; Zhang, Q.; Ivaska, A.; Niu, L. Graphene/AuNPs/chitosan nanocomposites film for glucose biosensing. *Biosens. Bioelectron.,* **2010**, *25*(5), 1070-1074.
[http://dx.doi.org/10.1016/j.bios.2009.09.024] [PMID: 19883999]

[140] Saha, K.; Agasti, S.S.; Kim, C.; Li, X.; Rotello, V.M. Gold nanoparticles in chemical and biological sensing. *Chem. Rev.,* **2012**, *112*(5), 2739-2779.
[http://dx.doi.org/10.1021/cr2001178] [PMID: 22295941]

[141] Peng, P.; Hu, A.; Gerlich, A.P.; Zou, G.; Liu, L.; Zhou, Y.N. Joining of silver nanomaterials at low temperatures: processes, properties, and applications. *ACS Appl. Mater. Interfaces,* **2015**, *7*(23), 12597-12618.
[http://dx.doi.org/10.1021/acsami.5b02134] [PMID: 26005792]

[142] Liu, R.; Zhang, Y.; Zhang, S.; Qiu, W.; Gao, Y. Silver enhancement of gold nanoparticles for biosensing: from qualitative to quantitative. *Appl. Spectrosc. Rev.,* **2014**, *49*(2), 121-138.
[http://dx.doi.org/10.1080/05704928.2013.807817]

[143] Aragay, G.; Pino, F.; Merkoçi, A. Nanomaterials for sensing and destroying pesticides. *Chem. Rev.,* **2012**, *112*(10), 5317-5338.
[http://dx.doi.org/10.1021/cr300020c] [PMID: 22897703]

[144] Xia, N.; Zhang, L. Nanomaterials-based sensing strategies for electrochemical detection of

microRNAs. *Materials (Basel),* **2014**, *7*(7), 5366-5384.
[http://dx.doi.org/10.3390/ma7075366] [PMID: 28788133]

[145] Guo, S.; Wen, D.; Zhai, Y.; Dong, S.; Wang, E. Platinum nanoparticle ensemble-on-graphene hybrid nanosheet: one-pot, rapid synthesis, and used as new electrode material for electrochemical sensing. *ACS Nano,* **2010**, *4*(7), 3959-3968.
[http://dx.doi.org/10.1021/nn100852h] [PMID: 20568706]

[146] Bonanni, A.; del Valle, M. Use of nanomaterials for impedimetric DNA sensors: a review. *Anal. Chim. Acta,* **2010**, *678*(1), 7-17.
[http://dx.doi.org/10.1016/j.aca.2010.08.022] [PMID: 20869498]

[147] Wang, F.; Hu, S. Electrochemical sensors based on metal and semiconductor nanoparticles. *Mikrochim. Acta,* **2009**, *165*(1-2), 1-22.
[http://dx.doi.org/10.1007/s00604-009-0136-4]

[148] Campbell, F.W.; Compton, R.G. The use of nanoparticles in electroanalysis: an updated review. *Anal. Bioanal. Chem.,* **2010**, *396*(1), 241-259.
[http://dx.doi.org/10.1007/s00216-009-3063-7] [PMID: 19730834]

[149] Wang, J. Nanomaterial-based electrochemical biosensors. *Analyst (Lond.),* **2005**, *130*(4), 421-426.
[http://dx.doi.org/10.1039/b414248a] [PMID: 15846872]

[150] Murphy, C.J.; Gole, A.M.; Hunyadi, S.E.; Stone, J.W.; Sisco, P.N.; Alkilany, A.; Kinard, B.E.; Hankins, P. Chemical sensing and imaging with metallic nanorods. *Chem. Commun. (Camb.),* **2008**, (5), 544-557.
[http://dx.doi.org/10.1039/B711069C] [PMID: 18209787]

[151] Chen, A.; Chatterjee, S. Nanomaterials based electrochemical sensors for biomedical applications. *Chem. Soc. Rev.,* **2013**, *42*(12), 5425-5438.
[http://dx.doi.org/10.1039/c3cs35518g] [PMID: 23508125]

[152] Pumera, M.; Ambrosi, A.; Bonanni, A.; Chng, E.L.; Poh, H.L. Graphene for electrochemical sensing and biosensing. *TrAC Trends in Analytical Chemistry,* **2010**, *29*(9), 954-965.
[http://dx.doi.org/10.1016/j.trac.2010.05.011]

[153] Reynolds, J.E. *Martindale: the extra pharmacopoeia*; The Pharmaceutical Press: London, UK, **1982**.

[154] Lachman, L.; Lieberman, H.A.; Kanig, J.L.; Publications, L.F., Eds. *The Theory and Practice of Industrial Pharmacy,* 3rd ed; , **1986**.

[155] Wermuth, C.G.; Publications, E., Eds. *The Practice of Medicinal Chemistry,* 2nd ed; , **2003**.

[156] Pharmaceutical Principles of Solid Dosage Forms. J T. Carstensen, Technomic Publishing Company, **1993**.

[157] Tenjarla, S. Microemulsions: an overview and pharmaceutical applications. *Crit. Rev. Ther. Drug Carrier Syst.,* **1999**, *16*(5), 461-521.
[http://dx.doi.org/10.1615/CritRevTherDrugCarrierSyst.v16.i5.20] [PMID: 10635455]

[158] Cole, E.T. Liquid filled hard gelatin capsules. *Pharm Tech,* **1989**, *13*(9), 124-140.

[159] Lancini, G.; Parenti, F.; Gallo, G.G. The antibiotics In: *Antibiotics*; Springer: USA, **1995**.

[160] Helms, S.E.; Bredle, D.L.; Zajic, J.; Jarjoura, D.; Brodell, R.T.; Krishnarao, I. Oral contraceptive failure rates and oral antibiotics. *J. Am. Acad. Dermatol.,* **1997**, *36*(5 Pt 1), 705-710.
[http://dx.doi.org/10.1016/S0190-9622(97)80322-2] [PMID: 9146531]

[161] Appelbaum, P.C.; Hunter, P.A. The fluoroquinolone antibacterials: past, present and future perspectives. *Int. J. Antimicrob. Agents,* **2000**, *16*(1), 5-15.
[http://dx.doi.org/10.1016/S0924-8579(00)00192-8] [PMID: 11185413]

[162] Fischetti, V.A. Bacteriophage lysins as effective antibacterials. *Curr. Opin. Microbiol.,* **2008**, *11*(5), 393-400.

[http://dx.doi.org/10.1016/j.mib.2008.09.012] [PMID: 18824123]

[163] Dorman, H.J.; Deans, S.G. Antimicrobial agents from plants: antibacterial activity of plant volatile oils. *J. Appl. Microbiol.,* **2000**, *88*(2), 308-316. [http://dx.doi.org/10.1046/j.1365-2672.2000.00969.x] [PMID: 10736000]

[164] Coates, A.; Hu, Y.; Bax, R.; Page, C. The future challenges facing the development of new antimicrobial drugs. *Nat. Rev. Drug Discov.,* **2002**, *1*(11), 895-910. [http://dx.doi.org/10.1038/nrd940] [PMID: 12415249]

[165] Fernandes, J.C.; Kubota, L.T.; Oliveira-Neto, G. Potentiometric biosensor for I-ascorbic acid based on ascorbate oxidase of natural source immobilized on ethylene-vinylacetate membrane. *Anal. Chim. Acta,* **1999**, *385*, 3-12. [http://dx.doi.org/10.1016/S0003-2670(98)00327-4]

[166] Chen, B.; Ma, M.; Su, X. An amperometric penicillin biosensor with enhanced sensitivity based on co-immobilization of carbon nanotubes, hematein, and β-lactamase on glassy carbon electrode. *Anal. Chim. Acta,* **2010**, *674*(1), 89-95. [http://dx.doi.org/10.1016/j.aca.2010.06.014] [PMID: 20638504]

[167] Nawaz, H.; Rauf, S.; Akhtar, K.; Khalid, A.M. Electrochemical DNA biosensor for the study of ciprofloxacin-DNA interaction. *Anal. Biochem.,* **2006**, *354*(1), 28-34. [http://dx.doi.org/10.1016/j.ab.2006.04.004] [PMID: 16707087]

[168] Fotouhi, L.; Atoofi, Z.; Heravi, M.M. Interaction of ciprofloxacin with DNA studied by spectroscopy and voltammetry at MWCNT/DNA modified glassy carbon electrode. *Talanta,* **2013**, *103*, 194-200. [http://dx.doi.org/10.1016/j.talanta.2012.10.032] [PMID: 23200377]

[169] Brahman, P.K.; Dar, R.A.; Tıwarı, S.; Pıtre, K.S. *Rev. Anal. Chem.,* **2012**, *31*, 83-92. [http://dx.doi.org/10.1515/revac-2011-0032]

[170] Radi, A.E.; Wahdan, T.; Anwar, Z.; Mostafa, H. Electrochemical and spectroscopic studies on the interaction of gatifloxacin, moxifloxacin and sparfloxacin with DNA and their analytical applications. *Electroanal.,* **2010**, *22*, 2665-2671. [http://dx.doi.org/10.1002/elan.201000285]

[171] Bozal, B.; Topal, B.D.; Uslu, B.; Ozkan, S.A.; Aboul-Enein, H.Y. Electrochemical study of the interaction between the antibacterial drug gemifloxacin and dsDNA using pencil graphite electrode. *Curr. Anal. Chem.,* **2012**, *8*(4), 528-533. [http://dx.doi.org/10.2174/157341112803216735]

[172] Bozal, B.; Topal, B.D.; Uslu, B.; Ozkan, S.A.; Aboul-Enein, H.Y. Electrochemical study of the interaction between the antibacterial drug gemifloxacin and dsDNA using pencil graphite electrode. *Curr. Anal. Chem.,* **2012**, *8*(4), 528-533. [http://dx.doi.org/10.2174/157341112803216735]

[173] Zhang, Y.; Cai, Y.; Su, S.; Ni, Y. Electrochemical Studies of Oxidation of Lomefloxacin and Interaction with Calf Thymus DNA at Nano-SnO2/DHP Modified Electrode. *Electroanalysis,* **2006**, *18*(15), 1479-1484. [http://dx.doi.org/10.1002/elan.200603543]

[174] Jiang, X.; Lin, X. Voltammetry of the interaction of metronidazole with DNA and its analytical applications. *Bioelectrochemistry,* **2006**, *68*(2), 206-212. [http://dx.doi.org/10.1016/j.bioelechem.2005.08.003] [PMID: 16253570]

[175] Mallappa, M.; Gowda, B.G.; Mahesh, R.T. Mechanism of interaction of antibacterial drug moxifloxacin with herring sperm DNA: electrochemical and spectroscopic studies. *Pharma Chem.,* **2014**, *6*, 398-405.

[176] Yola, M.L.; Ozaltin, N. Electrochemical studies on the interaction of an antibacterial drug nitrofurantoin with DNA. *J. Electroanal. Chem.,* **2011**, *653*, 56-60. [http://dx.doi.org/10.1016/j.jelechem.2011.01.003]

[177] Ni, Y.; Wang, P.; Kokot, S. Voltammetric investigation of DNA damage induced by nitrofurazone and short-lived nitro-radicals with the use of an electrochemical DNA biosensor. *Biosens. Bioelectron.,* **2012**, *38*(1), 245-251.
[http://dx.doi.org/10.1016/j.bios.2012.05.034] [PMID: 22717476]

[178] Bertolino, F.A.; De Vito, I.E.; Messina, G.A.; Fernández, H.; Raba, J. Microfluidic-enzymatic biosensor with immobilized tyrosinase for electrochemical detection of pipemidic acid in pharmaceutical samples. *J. Electroanal. Chem.,* **2011**, *651*(2), 204-210.
[http://dx.doi.org/10.1016/j.jelechem.2010.12.007]

[179] Girousi, S.T.; Gherghi, I.Ch.; Karava, M.K. DNA-modified carbon paste electrode applied to the study of interaction between rifampicin (RIF) and DNA in solution and at the electrode surface. *J. Pharm. Biomed. Anal.,* **2004**, *36*(4), 851-858.
[http://dx.doi.org/10.1016/j.jpba.2004.08.034] [PMID: 15533679]

[180] Fotouhi, L.; Hashkavayi, A.B.; Heravi, M.M. Interaction of sulfadiazine with DNA on a MWCNT modified glassy carbon electrode: determination of DNA. *Int. J. Biol. Macromol.,* **2013**, *53*, 101-106.
[http://dx.doi.org/10.1016/j.ijbiomac.2012.11.021] [PMID: 23178367]

[181] Ali, S.; Altaf, A.A.; Badshah, A.; Lal, B.; Kamal, S.; Ullah, S. DNA interaction, Antibacterial and Antifungal studies of 3-nitrophenylferrocene. *J. Chem. Soc. Pak.,* **2013**, *34*, 3.

[182] Montague, D.; Rosas, R.; Bohr, D.F. Bradykinin: Vascular relaxant, cardiac stimulant. *Science,* **1963**, *141*(3584), 907-908.
[http://dx.doi.org/10.1126/science.141.3584.907] [PMID: 14043332]

[183] Reyes, A.J. Cardiovascular drugs and serum uric acid. *Cardiovasc. Drugs Ther.,* **2003**, *17*(5-6), 397-414.
[http://dx.doi.org/10.1023/B:CARD.0000015855.02485.e3] [PMID: 15107595]

[184] Sutter, M.C.; Wang, Y.X. Recent cardiovascular drugs from Chinese medicinal plants. *Cardiovasc. Res.,* **1993**, *27*(11), 1891-1901.
[http://dx.doi.org/10.1093/cvr/27.11.1891] [PMID: 8287391]

[185] Mirmomtaz, E.; Ensafi, A.A.; Soleimanian-Zad, S. Determination of amiloride using a ds-DN--modified pencil graphite electrode based on guanine and adenine signals. *Electrochim. Acta,* **2009**, *54*, 1141-1146.
[http://dx.doi.org/10.1016/j.electacta.2008.09.003]

[186] Ribeiro, F.A.; Tarley, C.R.; Borges, K.B.; Pereira, A.C. *Sens. Actuators B Chem.,* **2013**, *185*, 743-754.
[http://dx.doi.org/10.1016/j.snb.2013.05.072]

[187] Huang, Q.; Zhang, H.; Hu, S.; Li, F.; Weng, W.; Chen, J.; Wang, Q.; He, Y.; Zhang, W.; Bao, X. A sensitive and reliable dopamine biosensor was developed based on the Au@carbon dots-chitosan composite film. *Biosens. Bioelectron.,* **2014**, *52*, 277-280.
[http://dx.doi.org/10.1016/j.bios.2013.09.003] [PMID: 24064477]

[188] Canbay, E.; Akyilmaz, E. Design of a multiwalled carbon nanotube-Nafion-cysteamine modified tyrosinase biosensor and its adaptation of dopamine determination. *Anal. Biochem.,* **2014**, *444*, 8-15.
[http://dx.doi.org/10.1016/j.ab.2013.09.019] [PMID: 24090870]

[189] Chandra, S.; Arora, K.; Bahadur, D. Impedimetric biosensor based on magnetic nanoparticles for electrochemical detection of dopamine. *Mater. Sci. Eng. B,* **2012**, *177*(17), 1531-1537.
[http://dx.doi.org/10.1016/j.mseb.2012.04.012]

[190] Liu, X.; Peng, Y.; Qu, X.; Ai, S.; Han, R.; Zhu, X. Multi-walled carbon nanotube-chitosan/poly (amidoamine)/DNA nanocomposite modified gold electrode for determination of dopamine and uric acid under coexistence of ascorbic acid. *J. Electroanal. Chem.,* **2011**, *654*(1), 72-78.
[http://dx.doi.org/10.1016/j.jelechem.2011.01.024]

[191] Yu, D.; Zeng, Y.; Qi, Y.; Zhou, T.; Shi, G. A novel electrochemical sensor for determination of dopamine based on AuNPs@SiO2 core-shell imprinted composite. *Biosens. Bioelectron.,* **2012**, *38*(1),

270-277.
[http://dx.doi.org/10.1016/j.bios.2012.05.045] [PMID: 22742811]

[192] Young, R.C.; Ozols, R.F.; Myers, C.E. The anthracycline antineoplastic drugs. *N. Engl. J. Med.,* **1981**, *305*(3), 139-153.
[http://dx.doi.org/10.1056/NEJM198107163050305] [PMID: 7017406]

[193] Robinson, M.J.; Osheroff, N. Effects of antineoplastic drugs on the post-strand-passage DNA cleavage/religation equilibrium of topoisomerase II. *Biochemistry,* **1991**, *30*(7), 1807-1813.
[http://dx.doi.org/10.1021/bi00221a012] [PMID: 1847075]

[194] Teicher, B.A.; Lazo, J.S.; Sartorelli, A.C. Classification of antineoplastic agents by their selective toxicities toward oxygenated and hypoxic tumor cells. *Cancer Res.,* **1981**, *41*(1), 73-81.
[PMID: 7448778]

[195] Matney, T.S.; Nguyen, T.V.; Connor, T.H.; Dana, W.J.; Theiss, J.C. Genotoxic classification of anticancer drugs. *Teratog. Carcinog. Mutagen.,* **1985**, *5*(5), 319-328.
[http://dx.doi.org/10.1002/tcm.1770050502] [PMID: 2867614]

[196] Zhang, Y.; Zhang, K.; Ma, H. Electrochemical DNA biosensor based on silver nanoparticles/poly(3-(3-pyridyl) acrylic acid)/carbon nanotubes modified electrode. *Anal. Biochem.,* **2009**, *387*(1), 13-19.
[http://dx.doi.org/10.1016/j.ab.2008.10.043] [PMID: 19032931]

[197] Tian, X.; Song, Y.; Dong, H.; Ye, B. Interaction of anticancer herbal drug berberine with DNA immobilized on the glassy carbon electrode. *Bioelectrochemistry,* **2008**, *73*(1), 18-22.
[http://dx.doi.org/10.1016/j.bioelechem.2008.02.005] [PMID: 18455966]

[198] Erdem, A.; Congur, G. Impedimetric detection of *in situ* interaction between anti-cancer drug bleomycin and DNA. *Int. J. Biol. Macromol.,* **2013**, *61*, 295-301.
[http://dx.doi.org/10.1016/j.ijbiomac.2013.07.012] [PMID: 23892034]

[199] Krizkova, S.; Adam, V.; Petrlova, J.; Zitka, O.; Stejskal, K.; Zehnalek, J.; Kizek, R. A Suggestion of Electrochemical Biosensor for Study of Platinum (II)□DNA Interactions. *Electroanalysis,* **2007**, *19*(2□3), 331-338.
[http://dx.doi.org/10.1002/elan.200603737]

[200] Pontinha, A.D.; Satana, H.; Diculescu, V.C.; Oliveira□Brett, A.M. Anodic Oxidation of Cladribine and *In situ* Evaluation of DNA□Cladribine Interaction. *Electroanalysis,* **2011**, *23*(11), 2651-2657.
[http://dx.doi.org/10.1002/elan.201100320]

[201] Satana, H.E.; Pontinha, A.D.; Diculescu, V.C.; Oliveira-Brett, A.M. Nucleoside analogue electrochemical behaviour and *in situ* evaluation of DNA-clofarabine interaction. *Bioelectrochemistry,* **2012**, *87*, 3-8.
[http://dx.doi.org/10.1016/j.bioelechem.2011.07.004] [PMID: 21856252]

[202] Palaska, P.; Aritzoglou, E.; Girousi, S. Sensitive detection of cyclophosphamide using DNA-modified carbon paste, pencil graphite and hanging mercury drop electrodes. *Talanta,* **2007**, *72*(3), 1199-1206.
[http://dx.doi.org/10.1016/j.talanta.2007.01.013] [PMID: 19071745]

[203] Radi, A.E.; Eissa, A.; Nassef, H.M. Voltammetric and spectroscopic studies on the binding of the antitumor drug dacarbazine with DNA. *J. Electroanal. Chem.,* **2014**, *717–718*, 24-28.
[http://dx.doi.org/10.1016/j.jelechem.2014.01.007]

[204] Yan, W.; Shen, X.C.; Zhang, Z.L.; Chen, C.; Pang, D.W. Electrochemical behavior of daunorubicin at DNA□MWCNT bioconjugates modified glassy carbon electrodes. *Anal. Lett.,* **2005**, *38*, 2579-2595.
[http://dx.doi.org/10.1080/00032710500369778]

[205] Chandra, P.; Zaidi, S.A.; Noh, H-B.; Shim, Y.B. Separation and simultaneous detection of anticancer drugs in a microfluidic device with an amperometric biosensor. *Biosens. Bioelectron.,* **2011**, *28*(1), 326-332.
[http://dx.doi.org/10.1016/j.bios.2011.07.038] [PMID: 21820886]

[206] Huska, D.; Adam, V.; Babula, P.; Hrabeta, J.; Stiborova, M.; Eckschlager, T.; Trnkova, L.; Kizek, R.

Square-wave voltammetry as a tool for investigation of doxorubicin interactions with DNA isolated from neuroblastoma cells. *Electroanal.,* **2009**, *21*, 487-494. [http://dx.doi.org/10.1002/elan.200804429]

[207] Li, N.; Guo, L.; Jiang, J.; Yang, X. Interaction of echinomycin with guanine: electrochemistry and spectroscopy studies. *Biophys. Chem.,* **2004**, *111*(3), 259-265. [http://dx.doi.org/10.1016/j.bpc.2004.06.007] [PMID: 15501569]

[208] Temerk, Y.; Ibrahim, H. Electrochemical studies and spectroscopic investigations on the interaction of an anticancer drug flutamide with DNA and its analytical applications. *J. Electroanal. Chem.,* **2015**, *736*, 1-7. [http://dx.doi.org/10.1016/j.jelechem.2014.10.019]

[209] Dogan-Topal, B.; Ozkan, S.A. Electrochemical determination of anticancer drug fulvestrant at dsDNA modified pencil graphite electrode. *Electrochim. Acta,* **2011**, *56*, 4433-4438. [http://dx.doi.org/10.1016/j.electacta.2011.01.066]

[210] Fotouhi, L.; Bahmani, F. MWCNT modified glassy carbon electrode: Probing furazolidone-DNA interactions and DNA determination. *Electroanal.,* **2013**, *25*, 757-764. [http://dx.doi.org/10.1002/elan.201200495]

[211] Zhang, S.S.; Niu, S.Y.; Qu, B.; Jie, G.F.; Xu, H.; Ding, C.F. Studies on the interaction mechanism between hexakis(imidazole) manganese(II) terephthalate and DNA and preparation of DNA electrochemical sensor. *J. Inorg. Biochem.,* **2005**, *99*(12), 2340-2347. [http://dx.doi.org/10.1016/j.jinorgbio.2005.08.016] [PMID: 16219357]

[212] Diculescu, V.C.; Vivan, M.; Oliveira-Brett, A.M. Voltammetric behavior of antileukemia drug glivec. Part III: *In situ* DNA oxidative damage by the glivec electrochemical metabolite. *Electroanal.,* **2006**, *18*, 1963-1970. [http://dx.doi.org/10.1002/elan.200603602]

[213] Topkaya, S.N.; Aydinlik, S.; Aladag, N.; Ozsoz, M.; Ozkan-Ariksoysal, D. Different DNA immobilization strategies for the interaction of anticancer drug irinotecan with DNA based on electrochemical DNA biosensors. *Comb. Chem. High Throughput Screen.,* **2010**, *13*(7), 582-589. [http://dx.doi.org/10.2174/138620731104070582] [PMID: 20426745]

[214] Dogan-Topal, B.; Bozal-Palabiyik, B.; Ozkan, S.A.; Uslu, B. Investigation of anticancer drug lapatinib and its interaction with dsDNA by electrochemical and spectroscopic techniques. *Sens. Actuators B Chem.,* **2014**, *194*, 185-194. [http://dx.doi.org/10.1016/j.snb.2013.12.088]

[215] Dogan-Topal, B.; Ozkan, S.A. A novel sensitive electrochemical DNA biosensor for assaying of anticancer drug leuprolide and its adsorptive stripping voltammetric determination. *Talanta,* **2011**, *83*(3), 780-788. [http://dx.doi.org/10.1016/j.talanta.2010.10.049] [PMID: 21147320]

[216] Rafique, B.; Khalid, A.M.; Akhtar, K.; Jabbar, A. Interaction of anticancer drug methotrexate with DNA analyzed by electrochemical and spectroscopic methods. *Biosens. Bioelectron.,* **2013**, *44*, 21-26. [http://dx.doi.org/10.1016/j.bios.2012.12.028] [PMID: 23384765]

[217] Wang, F.; Wu, Y.; Liu, J.; Ye, B. DNA Langmuir–Blodgett modified glassy carbon electrode as voltammetric sensor for determinate of methotrexate. *Electrochim. Acta,* **2009**, *54*(5), 1408-1413. [http://dx.doi.org/10.1016/j.electacta.2008.09.027]

[218] Pontinha, A.D.; Jorge, S.M.; Chiorcea Paquim, A.M.; Diculescu, V.C.; Oliveira-Brett, A.M. *In situ* evaluation of anticancer drug methotrexate-DNA interaction using a DNA-electrochemical biosensor and AFM characterization. *Phys. Chem. Chem. Phys.,* **2011**, *13*(12), 5227-5234. [http://dx.doi.org/10.1039/c0cp02377a] [PMID: 21359288]

[219] Eksin, E.; Muti, M.; Erdem, A. Chitosan/Ionic liquid composite electrode for electrochemical monitoring of the surface-confined interaction between mitomycin C and DNA. *Electroanal.,* **2013**, *25*, 2321-2329.

[220] Ozkan, D.; Karadeniz, H.; Erdem, A.; Mascini, M.; Ozsoz, M. Electrochemical genosensor for mitomycin C-DNA interaction based on guanine signal. *J. Pharm. Biomed. Anal.*, **2004**, *35*(4), 905-912.
[http://dx.doi.org/10.1016/j.jpba.2004.03.001] [PMID: 15193735]

[221] Erdem, A.; Muti, M.; Papakonstantinou, P.; Canavar, E.; Karadeniz, H.; Congur, G.; Sharma, S. Graphene oxide integrated sensor for electrochemical monitoring of mitomycin C-DNA interaction. *Analyst (Lond.)*, **2012**, *137*(9), 2129-2135.
[http://dx.doi.org/10.1039/c2an16011k] [PMID: 22439135]

[222] Canavar, P.E.; Eksin, E.; Erdem, A. Electrochemical monitoring of the interaction between mitomycin C and DNA at chitosan carbon nanotube composite modified electrodes. *Turk. J. Chem.*, **2015**, *39*, 1-12.
[http://dx.doi.org/10.3906/kim-1402-11]

[223] Karadeniz, H.; Alparslan, L.; Erdem, A.; Karasulu, E. Electrochemical investigation of interaction between mitomycin C and DNA in a novel drug-delivery system. *J. Pharm. Biomed. Anal.*, **2007**, *45*(2), 322-326.
[http://dx.doi.org/10.1016/j.jpba.2007.05.005] [PMID: 17590560]

[224] Kuralay, F.; Erdem, A.; Abacı, S.; Ozyuruk, H.; Yıldız, A. Characterization of poly (vinylferrocenium) coated surfaces and their applications in DNA sensor technology. *J. Appl. Electrochem.*, **2010**, *40*, 2039-2050.
[http://dx.doi.org/10.1007/s10800-010-0185-8]

[225] Hajian, R.; Tavakol, M. Interaction of anticancer drug methotrexate with ds-DNA analyzed by spectroscopic and electrochemical methods. *E-J. Chem.*, **2012**, *9*, 471-480.
[http://dx.doi.org/10.1155/2012/378674]

[226] Rafique, B.; Khalid, A.M.; Akhtar, K.; Jabbar, A. Interaction of anticancer drug methotrexate with DNA analyzed by electrochemical and spectroscopic methods. *Biosens. Bioelectron.*, **2013**, *44*, 21-26.
[http://dx.doi.org/10.1016/j.bios.2012.12.028] [PMID: 23384765]

[227] Wang, F.; Wu, Y.; Liu, J.; Ye, B. DNA Langmuir–Blodgett modified glassy carbon electrode as voltammetric sensor for determinate of methotrexate. *Electrochim. Acta*, **2009**, *54*, 1408-1413.
[http://dx.doi.org/10.1016/j.electacta.2008.09.027]

[228] Li, N.; Ma, Y.; Yang, C.; Guo, L.; Yang, X. Interaction of anticancer drug mitoxantrone with DNA analyzed by electrochemical and spectroscopic methods. *Biophys. Chem.*, **2005**, *116*(3), 199-205.
[http://dx.doi.org/10.1016/j.bpc.2005.04.009] [PMID: 15893412]

[229] Wang, F.; Xu, Y.; Zhao, J.; Hu, S. Electrochemical oxidation of morin and interaction with DNA. *Bioelectrochemistry*, **2007**, *70*(2), 356-362.
[http://dx.doi.org/10.1016/j.bioelechem.2006.05.003] [PMID: 16787762]

[230] Cheng, W.X.; Peng, D.Y.; Lu, C.H.; Fang, C.W. Direct electrochemical behavior of the Cysteamine/DNA/SWNTs-film-modified Au electrode and its interaction with taxol. *Russ. J. Electrochem.*, **2008**, *44*, 1133-1139.
[http://dx.doi.org/10.1134/S1023193508090103]

[231] Mehdinia, A.; Habib Kazemi, S.; Zahra Bathaie, S.; Alizadeh, A.; Shamsipur, M.; Mousavi, M.F. Electrochemical studies of DNA immobilization onto the azide-terminated monolayers and its interaction with taxol. *Anal. Biochem.*, **2008**, *375*(2), 331-338.
[http://dx.doi.org/10.1016/j.ab.2008.01.006] [PMID: 18243122]

[232] de la Fuente, E.; Villagra, G.; Bollo, S. Electrochemical nucleic acid biosensors for the detection of interaction between peroxynitrite and DNA. *Electroanal.*, **2007**, *19*, 1518-1523.
[http://dx.doi.org/10.1002/elan.200703885]

[233] Wang, Y.; Xiong, H.; Zhang, X.; Wang, S. Received Detection of rutin at DNA modified carbon paste electrode based on a mixture of ionic liquid and paraffin oil as a binder. *Mikrochim. Acta*, **2010**, *170*,

27-32.
[http://dx.doi.org/10.1007/s00604-010-0380-7]

[234] Niu, S.; Zhao, M.; Hu, L.; Zhang, S. Carbon nanotube-enhanced DNA biosensor for DNA hybridization detection using rutin-Mn as electrochemical indicator. *Sens. Actuators B Chem.,* **2008**, *135*(1), 200-205.
[http://dx.doi.org/10.1016/j.snb.2008.08.022]

[235] Bian, C.L.; Zeng, Q.X.; Yang, L.J.; Xıong, H.Y.; Zhang, X.H.; Wang, S.F. Voltammetric studies of the interaction of rutin with DNA and its analytical applications on the MWNTs–COOH/Fe3O4 modified electrode. *Sens. Actuators B Chem.,* **2011**, *156*, 615-620.
[http://dx.doi.org/10.1016/j.snb.2011.02.007]

[236] Rauf, S.; Nawaz, H.; Akhtar, K.; Ghauri, M.A.; Khalid, A.M. Studies on sildenafil citrate (Viagra) interaction with DNA using electrochemical DNA biosensor. *Biosens. Bioelectron.,* **2007**, *22*(11), 2471-2477.
[http://dx.doi.org/10.1016/j.bios.2006.09.023] [PMID: 17085032]

[237] Liu, H.; Zhao, D.; Zhao, A.; Zou, K.; Li, T.; Wu, Y. Electrochemical studies of anticancer herbal drug sophoridine and its interaction with DNA. *J. Chin. Chem. Soc. (Taipei),* **2014**, *61*, 897-902.
[http://dx.doi.org/10.1002/jccs.201300584]

[238] Ghanbari, Kh.; Bathaie, S.Z.; Mousavi, M.F. Electrochemically fabricated polypyrrole nanofiber-modified electrode as a new electrochemical DNA biosensor. *Biosens. Bioelectron.,* **2008**, *23*(12), 1825-1831.
[http://dx.doi.org/10.1016/j.bios.2008.02.029] [PMID: 18406598]

[239] Mehdinia, A.; Kazemi, S.H.; Bathaie, S.Z.; Alizadeh, A.; Shamsipur, M.; Mousavi, M.F. Electrochemical DNA nano-biosensor for the study of spermidine-DNA interaction. *J. Pharm. Biomed. Anal.,* **2009**, *49*(3), 587-593.
[http://dx.doi.org/10.1016/j.jpba.2008.12.035] [PMID: 19186020]

[240] Fotouhi, L.; Hashkavayi, A.B.; Heravi, M.M. Interaction of sulfadiazine with DNA on a MWCNT modified glassy carbon electrode: determination of DNA. *Int. J. Biol. Macromol.,* **2013**, *53*, 101-106.
[http://dx.doi.org/10.1016/j.ijbiomac.2012.11.021] [PMID: 23178367]

[241] Guo, X.X.; Song, Z.J.; Sun, J.J.; Song, J.F. Interaction of calf thymus dsDNA with anti-tumor drug tamoxifen studied by zero current potentiometry. *Biosens. Bioelectron.,* **2011**, *26*(10), 4001-4005.
[http://dx.doi.org/10.1016/j.bios.2011.03.021] [PMID: 21524901]

[242] Radhapyari, K.; Kotoky, P.; Khan, R. Detection of anticancer drug tamoxifen using biosensor based on polyaniline probe modified with horseradish peroxidase. *Mater. Sci. Eng. C,* **2013**, *33*(2), 583-587.
[http://dx.doi.org/10.1016/j.msec.2012.09.021] [PMID: 25427459]

[243] Guo, X.X.; Song, Z.J.; Sun, J.J.; Song, J.F. Interaction of calf thymus dsDNA with anti-tumor drug tamoxifen studied by zero current potentiometry. *Biosens. Bioelectron.,* **2011**, *26*(10), 4001-4005.
[http://dx.doi.org/10.1016/j.bios.2011.03.021] [PMID: 21524901]

[244] Congur, G.; Erdem, A.; Mese, F. Electrochemical investigation of the interaction between topotecan and DNA at disposable graphite electrodes. *Bioelectrochemistry,* **2015**, *102*, 21-28.
[http://dx.doi.org/10.1016/j.bioelechem.2014.11.003] [PMID: 25461757]

[245] Cheng, W.X.; Peng, D.Y.; Lu, C.H.; Fang, C.W. Direct electrochemical behavior of the Cysteamine/DNA/SWNTs-film-modified Au electrode and its interaction with taxol. *Russ. J. Electrochem.,* **2008**, *44*(9), 1052-1057.
[http://dx.doi.org/10.1134/S1023193508090103]

[246] Mehdinia, A.; Habib Kazemi, S.; Zahra Bathaie, S.; Alizadeh, A.; Shamsipur, M.; Mousavi, M.F. Electrochemical studies of DNA immobilization onto the azide-terminated monolayers and its interaction with taxol. *Anal. Biochem.,* **2008**, *375*(2), 331-338.
[http://dx.doi.org/10.1016/j.ab.2008.01.006] [PMID: 18243122]

[247] Tajik, S.; Taher, M.A.; Beitollahi, H.; Torkzadeh-Mahani, M. Electrochemical determination of the anticancer drug taxol at a ds-DNA modified pencil-graphite electrode and its application as a label-free electrochemical biosensor. *Talanta,* **2015**, *134*, 60-64. [http://dx.doi.org/10.1016/j.talanta.2014.10.063] [PMID: 25618641]

[248] Muti, M.; Gençdağ, K.; Nacak, F.M.; Aslan, A. Electrochemical polymerized 5-amino-2-mercato-1,3,4-thiadiazole modified single use sensors for detection of quercetin. *Colloids Surf. B Biointerfaces,* **2013**, *106*, 181-186. [http://dx.doi.org/10.1016/j.colsurfb.2013.01.018] [PMID: 23434710]

[249] Gu, T.; Hasebe, Y. Novel amperometric assay for drug-DNA interaction based on an inhibitory effect on an electrocatalytic activity of DNA-Cu(II) complex. *Biosens. Bioelectron.,* **2012**, *33*(1), 222-227. [http://dx.doi.org/10.1016/j.bios.2012.01.005] [PMID: 22305389]

[250] Chen, J.; Zhang, J.; Zhuang, Q.; Chen, J.; Lin, X. Electrochemical studies of the interaction of 2-nitroacridone with DNA and determination of DNA. *Electroanal.,* **2007**, *19*, 1765-1772. [http://dx.doi.org/10.1002/elan.200703929]

[251] Shah, A.; Zaheer, M.; Qureshi, R.; Akhter, Z.; Nazar, M.F. Voltammetric and spectroscopic investigations of 4-nitrophenylferrocene interacting with DNA. *Spectrochim. Acta A Mol. Biomol. Spectrosc.,* **2010**, *75*(3), 1082-1087. [http://dx.doi.org/10.1016/j.saa.2009.12.061] [PMID: 20080435]

[252] Qin, D.A.; Cai, X.Q.; Miao, Q.; Wang, Z.H.; Hu, M.L. Cyclic voltammetry and molecular docking study of the interactions of two derivatives of 5-fluorouracil with DNA. *Int. J. Electrochem. Sci.,* **2014**, *9*, 1608-1620.

[253] Eksin, E.; Congur, G.; Mese, F.; Erdem, A. Electrochemical monitoring of surface confined interaction between 6-Thioguanine and DNA by using single-use graphite electrode. *J. Electroanal. Chem.,* **2014**, *733*, 33-38. [http://dx.doi.org/10.1016/j.jelechem.2014.08.012]

[254] Mirmomtaz, E.; Ensafi, A.A. Voltammetric determination of trace quantities of 6-thioguanine based on the interaction with DNA at a mercury electrode. *Electrochim. Acta,* **2009**, *54*(18), 4353-4358. [http://dx.doi.org/10.1016/j.electacta.2009.03.004]

[255] Wang, W.; Wang, S.F.; Xie, F. An electrochemical sensor of non-electroactive drug 6-thioguanine based on the dsDNA/AET/Au. *Sens. Actuators B Chem.,* **2006**, *120*, 238-244. [http://dx.doi.org/10.1016/j.snb.2006.02.012]

[256] Tang, W.; Zhang, M.; Zeng, X. Establishment of dsDNA/GNs/chit/GCE biosensor and electrochemical study on interaction between 6-mercaptopurine and DNA. *Biomed. Mater. Eng.,* **2014**, *24*(1), 1071-1077. [PMID: 24211998]

[257] Oliveira-Brett, A.M.; Diculescu, V.C. Electrochemical study of quercetin-DNA interactions: part II. *In situ* sensing with DNA biosensors. *Bioelectrochemistry,* **2004**, *64*(2), 143-150. [http://dx.doi.org/10.1016/j.bioelechem.2004.05.002] [PMID: 15296787]

[258] De Clercq, E. Antiviral drugs in current clinical use. *J. Clin. Virol.,* **2004**, *30*(2), 115-133. [http://dx.doi.org/10.1016/j.jcv.2004.02.009] [PMID: 15125867]

[259] De Clercq, E. Strategies in the design of antiviral drugs. *Nat. Rev. Drug Discov.,* **2002**, *1*(1), 13-25. [http://dx.doi.org/10.1038/nrd703] [PMID: 12119605]

[260] De Clercq, E.; Holý, A. Acyclic nucleoside phosphonates: a key class of antiviral drugs. *Nat. Rev. Drug Discov.,* **2005**, *4*(11), 928-940. [http://dx.doi.org/10.1038/nrd1877] [PMID: 16264436]

[261] Biron, K.K. Antiviral drugs for cytomegalovirus diseases. *Antiviral Res.,* **2006**, *71*(2-3), 154-163. [http://dx.doi.org/10.1016/j.antiviral.2006.05.002] [PMID: 16765457]

[262] Balfour, H.H., Jr Antiviral drugs. *N. Engl. J. Med.,* **1999**, *340*(16), 1255-1268. [http://dx.doi.org/10.1056/NEJM199904223401608] [PMID: 10210711]

[263] Zhou, C.; Dong, Y.; Li, Z.; Xu, X.; Liu, Z. Electrochemistry of magnolol and interaction with DNA. *J. Electroanal. Chem.,* **2010**, *642*, 115-119. [http://dx.doi.org/10.1016/j.jelechem.2010.02.020]

[264] Lin, X.H.; Wan, H.Y.; Zhang, Y.F.; Chen, J.H. Studies of the interaction between Aloe-emodin and DNA and preparation of DNA biosensor for detection of PML-RARalpha fusion gene in acute promyelocytic leukemia. *Talanta,* **2008**, *74*(4), 944-950. [http://dx.doi.org/10.1016/j.talanta.2007.07.038] [PMID: 18371732]

[265] Dogan-Topal, B.; Uslu, B.; Ozkan, S.A. Voltammetric studies on the HIV-1 inhibitory drug Efavirenz: the interaction between dsDNA and drug using electrochemical DNA biosensor and adsorptive stripping voltammetric determination on disposable pencil graphite electrode. *Biosens. Bioelectron.,* **2009**, *24*(8), 2358-2364. [http://dx.doi.org/10.1016/j.bios.2008.12.005] [PMID: 19135352]

[266] Bye, C.; Clubley, M.; Peck, A.W. Drowsiness, impaired performance and tricyclic antidepressants drugs. *Br. J. Clin. Pharmacol.,* **1978**, *6*(2), 155-162. [http://dx.doi.org/10.1111/j.1365-2125.1978.tb00841.x] [PMID: 678393]

[267] Sulser, F.; Vetulani, J.; Mobley, P.L. Mode of action of antidepressant drugs. *Biochem. Pharmacol.,* **1978**, *27*(3), 257-261. [http://dx.doi.org/10.1016/0006-2952(78)90226-5] [PMID: 202286]

[268] Andrade, S.E.; Raebel, M.A.; Brown, J.; Lane, K.; Livingston, J.; Boudreau, D.; Rolnick, S.J.; Roblin, D.; Smith, D.H.; Willy, M.E.; Staffa, J.A.; Platt, R. Use of antidepressant medications during pregnancy: a multisite study. *Am. J. Obstet. Gynecol.,* **2008**, *198*(2), 194.e1-194.e5. [http://dx.doi.org/10.1016/j.ajog.2007.07.036] [PMID: 17905176]

[269] Lieberman, J.A.; Stroup, T.S.; McEvoy, J.P.; Swartz, M.S.; Rosenheck, R.A.; Perkins, D.O.; Keefe, R.S.; Davis, S.M.; Davis, C.E.; Lebowitz, B.D.; Severe, J.; Hsiao, J.K. Clinical Antipsychotic Trials of Intervention Effectiveness (CATIE) Investigators. Effectiveness of antipsychotic drugs in patients with chronic schizophrenia. *N. Engl. J. Med.,* **2005**, *353*(12), 1209-1223. [http://dx.doi.org/10.1056/NEJMoa051688] [PMID: 16172203]

[270] Seeman, P.; Lee, T. Antipsychotic drugs: direct correlation between clinical potency and presynaptic action on dopamine neurons. *Science,* **1975**, *188*(4194), 1217-1219. [http://dx.doi.org/10.1126/science.1145194] [PMID: 1145194]

[271] Kurbanoglu, S.; Dogan-Topal, B.; Hlavata, L.; Labuda, J.; Ozkan, S.A.; Uslu, B. Electrochemical investigation of an interaction of the antidepressant drug aripiprazole with original and damaged calf thymus dsDNA. *Electrochim. Acta,* **2015**, *169*, 233-240. [http://dx.doi.org/10.1016/j.electacta.2015.04.087]

[272] Marco, J.P.; Borges, K.B.; Tarley, C.R.; Ribeiro, E.S.; Pereira, A.C. Development and application of an electrochemical biosensor based on carbon paste and silica modified with niobium oxide, alumina and DNA (SiO2/Al2O3/Nb2O5/DNA) for amitriptyline determination. *J. Electroanal. Chem.,* **2013**, *704*, 159-168. [http://dx.doi.org/10.1016/j.jelechem.2013.06.021]

[273] Kalanur, S.S.; Seetharamappa, J.; Prashanth, S.N. Interaction of an antidepressant buzepide methiodide with DNA immobilized on the glassy carbon electrode. *Colloids Surf. B Biointerfaces,* **2011**, *82*(2), 438-442. [http://dx.doi.org/10.1016/j.colsurfb.2010.09.017] [PMID: 21030222]

[274] Guo, M.; Chen, J.; Nie, L.; Yao, S. Electrostatic assembly of calf thymus DNA on multi-walled carbon nanotube modified gold electrode and its interaction with chlorpromazine hydrochloride. *Electrochim. Acta,* **2004**, *49*, 2637-2643. [http://dx.doi.org/10.1016/j.electacta.2004.02.014]

[275] Wei, X.; Hao, Q.; Zhou, Q.; Wu, J.; Lu, L.; Wang, X.; Yang, X. Interaction between promethazine hydrochloride and DNA and its application in electrochemical detection of DNA hybridization. *Electrochim. Acta,* **2008**, *53*, 7338-7343. [http://dx.doi.org/10.1016/j.electacta.2008.04.007]

[276] Yang, Z.S.; Zhao, J.; Zhang, D.P.; Liu, Y.C. Electrochemical determination of trace promethazine hydrochloride by a pretreated glassy carbon electrode modified with DNA. *Anal. Sci.,* **2007**, *23*(5), 569-572. [http://dx.doi.org/10.2116/analsci.23.569] [PMID: 17495403]

[277] Moertel, C.G.; Reitemeier, R.J.; Gage, R.P. A controlled clinical evaluation of antiemetic drugs. *JAMA,* **1963**, *186*(2), 116-118. [http://dx.doi.org/10.1001/jama.1963.03710020036011] [PMID: 14056524]

[278] Wood, C.D. Antimotion sickness and antiemetic drugs. *Drugs,* **1979**, *17*(6), 471-479. [http://dx.doi.org/10.2165/00003495-197917060-00003] [PMID: 467277]

[279] Oliveira, S.C.; Chiorcea-Paquim, A.M.; Ribeiro, S.M.; Melo, A.T.; Vivan, M.; Oliveira-Brett, A.M. *In situ* electrochemical and AFM study of thalidomide-DNA interaction. *Bioelectrochemistry,* **2009**, *76*(1-2), 201-207. [http://dx.doi.org/10.1016/j.bioelechem.2009.03.003] [PMID: 19386555]

[280] Ariëns, E.J. The structure-activity relationships of beta adrenergic drugs and beta adrenergic blocking drugs. *Ann. N. Y. Acad. Sci.,* **1967**, *139*(3), 606-631. [http://dx.doi.org/10.1111/j.1749-6632.1967.tb41232.x] [PMID: 5342360]

[281] Potter, D.E.; Rowland, J.M. Adrenergic drugs and intraocular pressure: effects of selective β-adrenergic agonists. *Exp. Eye Res.,* **1978**, *27*(6), 615-625. [http://dx.doi.org/10.1016/0014-4835(78)90029-5] [PMID: 33056]

[282] Sadeghi, R.; Karimi-Maleh, H.; Bahari, A.; Taghavi, M. A novel biosensor based on ZnO nanoparticle/1, 3-dipropylimidazolium bromide ionic liquid-modified carbon paste electrode for square-wave voltammetric determination of epinephrine. *Phys. Chem. Liquids,* **2013**, *51*(6), 704-714. [http://dx.doi.org/10.1080/00319104.2013.782547]

[283] Batra, B.; Pundir, C.S. An amperometric glutamate biosensor based on immobilization of glutamate oxidase onto carboxylated multiwalled carbon nanotubes/gold nanoparticles/chitosan composite film modified Au electrode. *Biosens. Bioelectron.,* **2013**, *47*, 496-501. [http://dx.doi.org/10.1016/j.bios.2013.03.063] [PMID: 23628843]

[284] Ni, Y.; Wei, M.; Kokot, S. Electrochemical and spectroscopic study on the interaction between isoprenaline and DNA using multivariate curve resolution-alternating least squares. *Int. J. Biol. Macromol.,* **2011**, *49*(4), 622-628. [http://dx.doi.org/10.1016/j.ijbiomac.2011.06.022] [PMID: 21729719]

[285] Lu, L.P.; Wang, S.Q.; Lin, X.Q. Fabrication of layer-by-layer deposited multilayer films containing DNA and gold nanoparticle for norepinephrine biosensor. *Anal. Chim. Acta,* **2004**, *519*, 161-166. [http://dx.doi.org/10.1016/j.aca.2004.05.062]

[286] Ni, Y.; Wang, Y.; Kokot, S. Voltammetric, UV–Vis spectrometric and fluorescence study of the interaction of ractopamine and DNA with the aid of multivariate curve resolution-alternating least square. *Electroanal.,* **2010**, *22*, 2216-2224. [http://dx.doi.org/10.1002/elan.200900596]

[287] Jiang, X.; Lin, X. Overoxidized polypyrrole film directed DNA immobilization for construction of electrochemical micro-biosensors and simultaneous determination of serotonin and dopamine. *Anal. Chim. Acta,* **2005**, *537*, 145-151. [http://dx.doi.org/10.1016/j.aca.2005.01.049]

[288] Kayser, O.; Kiderlen, A.F.; Croft, S.L. Natural products as antiparasitic drugs. *Parasitol. Res.,* **2003**, *90*(2) Suppl. 2, S55-S62.

[http://dx.doi.org/10.1007/s00436-002-0768-3] [PMID: 12937967]

[289] Liu, L.X.; Weller, P.F. Antiparasitic drugs. *N. Engl. J. Med.*, **1996**, *334*(18), 1178-1184. [http://dx.doi.org/10.1056/NEJM199605023341808] [PMID: 8602186]

[290] Hayeshi, R; Masimirembwa, C; Mukanganyama, S; Ungell, ALB The potential inhibitory effect of antiparasitic drugs and natural products on P-glycoprotein mediated efflux. *European Journal of Pharmaceutical Sciences,* **2006**, *29*(1), 70-81.

[291] Radi, A.E.; El-Naggar, A.E.; Nassef, H.M. Electrochemical and spectral studies on the interaction of the antiparasitic drug nitazoxanide with DNA. *Electrochim. Acta,* **2014**, *129*, 259-265. [http://dx.doi.org/10.1016/j.electacta.2014.02.092]

[292] Catalán, M.; Álvarez-Lueje, A.; Bollo, S. Electrochemistry of interaction of 2-(2-nitropheny-)-benzimidazole derivatives with DNA. *Bioelectrochemistry,* **2010**, *79*(2), 162-167. [http://dx.doi.org/10.1016/j.bioelechem.2010.02.002] [PMID: 20304709]

[293] Gu, T.; Hasebe, Y. Novel amperometric assay for drug-DNA interaction based on an inhibitory effect on an electrocatalytic activity of DNA-Cu(II) complex. *Biosens. Bioelectron.,* **2012**, *33*(1), 222-227. [http://dx.doi.org/10.1016/j.bios.2012.01.005] [PMID: 22305389]

[294] Radhapyari, K.; Kotoky, P.; Das, M.R.; Khan, R. Graphene-polyaniline nanocomposite based biosensor for detection of antimalarial drug artesunate in pharmaceutical formulation and biological fluids. *Talanta,* **2013**, *111*, 47-53. [http://dx.doi.org/10.1016/j.talanta.2013.03.020] [PMID: 23622524]

[295] Seddon, J.M.; Ajani, U.A.; Sperduto, R.D. Eye Disease Case-Control Study Group. Dietary carotenoids, vitamins A, C, and E, and advanced age-related macular degeneration. *JAMA,* **1994**, *272*(18), 1413-1420. [http://dx.doi.org/10.1001/jama.1994.03520180037032] [PMID: 7933422]

[296] Sies, H.; Stahl, W. Vitamins E and C, beta-carotene, and other carotenoids as antioxidants. *Am. J. Clin. Nutr.,* **1995**, *62*(6) Suppl., 1315S-1321S. [PMID: 7495226]

[297] Sies, H.; Stahl, W.; Sundquist, A.R. Antioxidant functions of vitamins. Vitamins E and C, beta-carotene, and other carotenoids. *Ann. N. Y. Acad. Sci.,* **1992**, *669*(1), 7-20. [http://dx.doi.org/10.1111/j.1749-6632.1992.tb17085.x] [PMID: 1444060]

[298] Radulescu, M.C.; Bucur, B.; Bucur, M.P.; Radu, G.L. Bienzymatic biosensor for rapid detection of aspartame by flow injection analysis. *Sensors (Basel),* **2014**, *14*(1), 1028-1038. [http://dx.doi.org/10.3390/s140101028] [PMID: 24412899]

[299] Kergaravat, S.V.; Gómez, G.A.; Fabiano, S.N.; Laube Chávez, T.I.; Pividori, M.I.; Hernández, S.R. Biotin determination in food supplements by an electrochemical magneto biosensor. *Talanta,* **2012**, *97*, 484-490. [http://dx.doi.org/10.1016/j.talanta.2012.05.003] [PMID: 22841112]

[300] Rahman, M.A.; Noh, H.B.; Shim, Y.B. Direct electrochemistry of laccase immobilized on au nanoparticles encapsulated-dendrimer bonded conducting polymer: application for a catechin sensor. *Anal. Chem.,* **2008**, *80*(21), 8020-8027. [http://dx.doi.org/10.1021/ac801033s] [PMID: 18841943]

[301] Mirmoghtadaie, L.; Ensafi, A.A.; Kadivar, M.; Norouzi, P. Highly selective electrochemical biosensor for the determination of folic acid based on DNA modified-pencil graphite electrode using response surface methodology. *Mater. Sci. Eng. C,* **2013**, *33*(3), 1753-1758. [http://dx.doi.org/10.1016/j.msec.2012.12.090] [PMID: 23827633]

[302] Chang, G.; Luo, Y.; Lu, W.; Hu, J.; Liao, F.; Sun, X. Immobilization of Au nanoparticles on Au electrode for hydrazine detection: Using thiolated single-stranded DNA as a linker. *Thin Solid Films,* **2011**, *519*, 6130-6134. [http://dx.doi.org/10.1016/j.tsf.2011.03.049]

[303] Boka, B.; Korozs, M.; Nánási, M.; Adányi, N. Novel Amperometric Tri□Enzyme Biosensor for Lysine Determination in Pharmaceutical Products and Food Samples. *Electroanal.,* **2015**, *27*(3), 817-824.
[http://dx.doi.org/10.1002/elan.201400600]

[304] Guerrieri, A.; Ciriello, R.; Cataldi, T.R. A novel amperometric biosensor based on a co-crosslinked L-lysine-α-oxidase/overoxidized polypyrrole bilayer for the highly selective determination of L-lysine. *Anal. Chim. Acta,* **2013**, *795*, 52-59.
[http://dx.doi.org/10.1016/j.aca.2013.07.036] [PMID: 23998537]

[305] Ensafi, A.A.; Heydari-Bafrooei, E.; Amini, M. DNA-functionalized biosensor for riboflavin based electrochemical interaction on pretreated pencil graphite electrode. *Biosens. Bioelectron.,* **2012**, *31*(1), 376-381.
[http://dx.doi.org/10.1016/j.bios.2011.10.050] [PMID: 22099958]

[306] Calvo-Pérez, A.; Domínguez-Renedo, O.; Alonso-Lomillo, M.A.; Arcos-Martínez, M.J. Disposable amperometric biosensor for the determination of tyramine using plasma amino oxidase. *Mikrochim. Acta,* **2013**, *180*(3-4), 253-259.
[http://dx.doi.org/10.1007/s00604-012-0926-y]

[307] Liu, S.Q.; Cao, M.L.; Dong, S.L. Electrochemical and ultraviolet-visible spectroscopic studies on the interaction of deoxyribonucleic acid with vitamin B6. *Bioelectrochemistry,* **2008**, *74*(1), 164-169.
[http://dx.doi.org/10.1016/j.bioelechem.2008.07.004] [PMID: 18722826]

[308] Brahman, P.K.; Dar, R.A.; Pitre, K.S. DNA-functionalized electrochemical biosensor for detection of vitamin B1 using electrochemically treated multiwalled carbon nanotube paste electrode by voltammetric methods. *Sens. Actuators B Chem.,* **2013**, *177*, 807-812.
[http://dx.doi.org/10.1016/j.snb.2012.11.073]

[309] Nie, T.; Xu, J.K.; Lu, L.M.; Zhang, K.X.; Bai, L.; Wen, Y.P. Electroactive species-doped poly(3,4-ethylenedioxythiophene) films: enhanced sensitivity for electrochemical simultaneous determination of vitamins B2, B6 and C. *Biosens. Bioelectron.,* **2013**, *50*, 244-250.
[http://dx.doi.org/10.1016/j.bios.2013.06.057] [PMID: 23871872]

[310] Jaanus, SD; Lesher, GA Anti-inflammatory drugs. *Clinical ocular pharmacology,* **1995**, 303-335.

[311] Lehmann, J.M.; Lenhard, J.M.; Oliver, B.B.; Ringold, G.M.; Kliewer, S.A. Peroxisome proliferator-activated receptors α and γ are activated by indomethacin and other non-steroidal anti-inflammatory drugs. *J. Biol. Chem.,* **1997**, *272*(6), 3406-3410.
[http://dx.doi.org/10.1074/jbc.272.6.3406] [PMID: 9013583]

[312] Kowalski, M.L.; Makowska, J.S.; Blanca, M.; Bavbek, S.; Bochenek, G.; Bousquet, J.; Bousquet, P.; Celik, G.; Demoly, P.; Gomes, E.R.; Niżankowska-Mogilnicka, E.; Romano, A.; Sanchez-Borges, M.; Sanz, M.; Torres, M.J.; De Weck, A.; Szczeklik, A.; Brockow, K. Hypersensitivity to nonsteroidal anti-inflammatory drugs (NSAIDs) - classification, diagnosis and management: review of the EAACI/ENDA(#) and GA2LEN/HANNA. *Allergy,* **2011**, *66*(7), 818-829.
[http://dx.doi.org/10.1111/j.1398-9995.2011.02557.x] [PMID: 21631520]

[313] Rounaghi, G.; Kakhki, R.M. Preparation and electrochemical application of a new biosensor based on plant tissue/polypyrrole for determination of acetaminophen. *Bull. Mater. Sci.,* **2012**, *35*(5), 811-816.
[http://dx.doi.org/10.1007/s12034-012-0355-x]

[314] Nasirizadeh, N.; Zare, H.R. Differential pulse voltammetric simultaneous determination of noradrenalin and acetaminophen using a hematoxylin biosensor. *Talanta,* **2009**, *80*(2), 656-663.
[http://dx.doi.org/10.1016/j.talanta.2009.07.044] [PMID: 19836534]

[315] Yousef Elahi, M.; Bathaie, S.Z.; Kazemi, S.H.; Mousavi, M.F. DNA immobilization on a polypyrrole nanofiber modified electrode and its interaction with salicylic acid/aspirin. *Anal. Biochem.,* **2011**, *411*(2), 176-184.
[http://dx.doi.org/10.1016/j.ab.2011.01.006] [PMID: 21236237]

[316] Marco, J.P.; Borges, K.B.; Tarley, C.R.; Ribeiro, E.S.; Pereira, A.C. Development and application of an electrochemical biosensor based on carbon paste and silica modified with niobium oxide, alumina and DNA (SiO2/Al2O3/Nb2O5/DNA) for amitriptyline determination. *J. Electroanal. Chem.*, **2013**, *704*, 159-168. [http://dx.doi.org/10.1016/j.jelechem.2013.06.021]

[317] Sheng, Q.; Liu, R.; Zhang, S.; Zheng, J. Ultrasensitive electrochemical cocaine biosensor based on reversible DNA nanostructure. *Biosens. Bioelectron.*, **2014**, *51*, 191-194. [http://dx.doi.org/10.1016/j.bios.2013.07.053] [PMID: 23962705]

[318] Jiang, B.; Wang, M.; Chen, Y.; Xie, J.; Xiang, Y. Highly sensitive electrochemical detection of cocaine on graphene/AuNP modified electrode *via* catalytic redox-recycling amplification. *Biosens. Bioelectron.*, **2012**, *32*(1), 305-308. [http://dx.doi.org/10.1016/j.bios.2011.12.010] [PMID: 22204778]

[319] Ensafi, A.A.; Heydari-Bafrooei, E.; Rezaei, B. Different interaction of codeine and morphine with DNA: a concept for simultaneous determination. *Biosens. Bioelectron.*, **2013**, *41*, 627-633. [http://dx.doi.org/10.1016/j.bios.2012.09.039] [PMID: 23079341]

[320] Wei, L.; Borowiec, J.; Zhu, L.; Zhang, J. Electrochemical investigation on the interaction of diclofenac with DNA and its application to the construction of a graphene-based biosensor. *J. Solid State Electrochem.*, **2012**, *16*, 3817-3823. [http://dx.doi.org/10.1007/s10008-012-1815-3]

[321] Zamfir, L.G.; Rotariu, L.; Bala, C. Acetylcholinesterase biosensor for carbamate drugs based on tetrathiafulvalene-tetracyanoquinodimethane/ionic liquid conductive gels. *Biosens. Bioelectron.*, **2013**, *46*, 61-67. [http://dx.doi.org/10.1016/j.bios.2013.02.018] [PMID: 23500478]

[322] Alonso-Lomillo, M.A.; Domínguez-Renedo, O.; Hernández-Martín, A.; Arcos-Martínez, M.J. Horseradish peroxidase covalent grafting onto screen-printed carbon electrodes for levetiracetam chronoamperometric determination. *Anal. Biochem.*, **2009**, *395*(1), 86-90. [http://dx.doi.org/10.1016/j.ab.2009.08.004] [PMID: 19665443]

[323] Marco, J.P.; Borges, K.B.; Tarley, C.R.; Ribeiro, E.S.; Pereira, A.C. Development of a simple, rapid and validated square wave voltametric method for determination of promethazine in raw material and pharmaceutical formulation using DNA modified multiwall carbon nanotube paste electrode. *Sens. Actuators B Chem.*, **2013**, *177*, 251-259. [http://dx.doi.org/10.1016/j.snb.2012.11.005]

[324] Abedi, M.; Bathaie, S.Z.; Mousavi, B.M. Interaction between DNA and some salicylic acid derivatives and characterization of their DNA targets. *Electroanal.*, **2013**, *25*, 2547-2556. [http://dx.doi.org/10.1002/elan.201300342]

[325] Yousef Elahi, M.; Bathaie, S.Z.; Kazemi, S.H.; Mousavi, M.F. DNA immobilization on a polypyrrole nanofiber modified electrode and its interaction with salicylic acid/aspirin. *Anal. Biochem.*, **2011**, *411*(2), 176-184. [http://dx.doi.org/10.1016/j.ab.2011.01.006] [PMID: 21236237]

[326] Vaníčková, M.; Lehotay, J.; Cizmárik, J.; Labuda, J. Kinetic study of the degradation of a potential local anesthetic drug in serum using the DNA-based electrochemical biosensor. *Bioelectrochemistry*, **2005**, *66*(1-2), 125-127. [http://dx.doi.org/10.1016/j.bioelechem.2004.03.011] [PMID: 15833712]

[327] Ware, G.W. *Pesticides. Theory and application*; WH Freeman Co., **1983**.

[328] McEwen, F.L.; Stephenson, G.R. *The use and significance of pesticides in the environment*; John Wiley Sons, **1979**.

[329] Ivanov, A.N.; Evtugyn, G.A.; Lukachova, L.V.; Karyakina, E.E.; Budnikov, H.C.; Kiseleva, S.G.; Karyakin, A.A. New polyaniline-based potentiometric biosensor for pesticides detection. *IEEE Sens.*

J., **2003**, *3*(3), 333-340.
[http://dx.doi.org/10.1109/JSEN.2003.814647]

[330] Ivanov, A.N.; Younusov, R.R.; Evtugyn, G.A.; Arduini, F.; Moscone, D.; Palleschi, G. Acetylcholinesterase biosensor based on single-walled carbon nanotubes--Co phtalocyanine for organophosphorus pesticides detection. *Talanta,* **2011**, *85*(1), 216-221.
[http://dx.doi.org/10.1016/j.talanta.2011.03.045] [PMID: 21645691]

[331] Ribeiro, J.A.; Carreira, C.A.; Lee, H.J.; Silva, F.; Martins, A.; Pereira, C.M. Voltammetric determination of paraquat at DNA–gold nanoparticle composite electrodes. *Electrochim. Acta,* **2010**, *55*(27), 7892-7896.
[http://dx.doi.org/10.1016/j.electacta.2010.03.058]

[332] Yang, L.; Wang, G.; Liu, Y. An acetylcholinesterase biosensor based on platinum nanoparticles-carboxylic graphene-nafion-modified electrode for detection of pesticides. *Anal. Biochem.,* **2013**, *437*(2), 144-149.
[http://dx.doi.org/10.1016/j.ab.2013.03.004] [PMID: 23499967]

[333] Li, Y.; Zhang, Y.; Han, G.; Xiao, Y.; Li, M.; Zhou, W. An Acetylcholinesterase Biosensor Based on Graphene/Polyaniline Composite Film for Detection of Pesticides. *Chin. J. Chem.,* **2016**, *34*(1), 82-88.
[http://dx.doi.org/10.1002/cjoc.201500747]

[334] Oliveira, T.M.; Barroso, M.F.; Morais, S.; Araújo, M.; Freire, C.; de Lima-Neto, P.; Correia, A.N.; Oliveira, M.B.; Delerue-Matos, C. Sensitive bi-enzymatic biosensor based on polyphenoloxidases-gold nanoparticles-chitosan hybrid film-graphene doped carbon paste electrode for carbamates detection. *Bioelectrochemistry,* **2014**, *98*, 20-29.
[http://dx.doi.org/10.1016/j.bioelechem.2014.02.003] [PMID: 24642204]

[335] Cesarino, I.; Moraes, F.C.; Lanza, M.R.; Machado, S.A. Electrochemical detection of carbamate pesticides in fruit and vegetables with a biosensor based on acetylcholinesterase immobilised on a composite of polyaniline-carbon nanotubes. *Food Chem.,* **2012**, *135*(3), 873-879.
[http://dx.doi.org/10.1016/j.foodchem.2012.04.147] [PMID: 22953799]

[336] Zheng, Y.; Liu, Z.; Jing, Y.; Li, J.; Zhan, H. An acetylcholinesterase biosensor based on ionic liquid functionalized graphene–gelatin-modified electrode for sensitive detection of pesticides. *Sens. Actuators B Chem.,* **2015**, *210*, 389-397.
[http://dx.doi.org/10.1016/j.snb.2015.01.003]

[337] Zhao, H.; Ji, X.; Wang, B.; Wang, N.; Li, X.; Ni, R.; Ren, J. An ultra-sensitive acetylcholinesterase biosensor based on reduced graphene oxide-Au nanoparticles-β-cyclodextrin/Prussian blue-chitosan nanocomposites for organophosphorus pesticides detection. *Biosens. Bioelectron.,* **2015**, *65*, 23-30.
[http://dx.doi.org/10.1016/j.bios.2014.10.007] [PMID: 25461134]

[338] Wang, K.; Li, H.N.; Wu, J.; Ju, C.; Yan, J.J.; Liu, Q.; Qiu, B. TiO2-decorated graphene nanohybrids for fabricating an amperometric acetylcholinesterase biosensor. *Analyst (Lond.),* **2011**, *136*(16), 3349-3354.
[http://dx.doi.org/10.1039/c1an15227k] [PMID: 21738917]

[339] Qu, Y.; Sun, Q.; Xiao, F.; Shi, G.; Jin, L. Layer-by-Layer self-assembled acetylcholinesterase/PAMAM-Au on CNTs modified electrode for sensing pesticides. *Bioelectrochemistry,* **2010**, *77*(2), 139-144.
[http://dx.doi.org/10.1016/j.bioelechem.2009.08.001] [PMID: 19733130]

[340] Jeyapragasam, T.; Saraswathi, R. Electrochemical biosensing of carbofuran based on acetylcholinesterase immobilized onto iron oxide–chitosan nanocomposite. *Sens. Actuators B Chem.,* **2014**, *191*, 681-687.
[http://dx.doi.org/10.1016/j.snb.2013.10.054]

[341] Yang, L.; Wang, G.; Liu, Y.; Wang, M. Development of a biosensor based on immobilization of acetylcholinesterase on NiO nanoparticles-carboxylic graphene-nafion modified electrode for detection of pesticides. *Talanta,* **2013**, *113*, 135-141.

[http://dx.doi.org/10.1016/j.talanta.2013.03.025] [PMID: 23708635]

[342] Zhou, Q.; Yang, L.; Wang, G.; Yang, Y. Acetylcholinesterase biosensor based on SnO2 nanoparticles-carboxylic graphene-nafion modified electrode for detection of pesticides. *Biosens. Bioelectron.*, **2013**, *49*, 25-31.
[http://dx.doi.org/10.1016/j.bios.2013.04.037] [PMID: 23708814]

[343] Oliveira, T.M.; Fátima Barroso, M.; Morais, S.; Araújo, M.; Freire, C.; de Lima-Neto, P.; Correia, A.N.; Oliveira, M.B.; Delerue-Matos, C. Laccase-Prussian blue film-graphene doped carbon paste modified electrode for carbamate pesticides quantification. *Biosens. Bioelectron.*, **2013**, *47*, 292-299.
[http://dx.doi.org/10.1016/j.bios.2013.03.026] [PMID: 23587791]

[344] Nesakumar, N.; Sethuraman, S.; Krishnan, U.M.; Rayappan, J.B. Electrochemical acetylcholinesterase biosensor based on ZnO nanocuboids modified platinum electrode for the detection of carbosulfan in rice. *Biosens. Bioelectron.*, **2016**, *77*, 1070-1077.
[http://dx.doi.org/10.1016/j.bios.2015.11.010] [PMID: 26562329]

[345] Crew, A.; Lonsdale, D.; Byrd, N.; Pittson, R.; Hart, J.P. A screen-printed, amperometric biosensor array incorporated into a novel automated system for the simultaneous determination of organophosphate pesticides. *Biosens. Bioelectron.*, **2011**, *26*(6), 2847-2851.
[http://dx.doi.org/10.1016/j.bios.2010.11.018] [PMID: 21163641]

[346] Chauhan, N.; Narang, J.; Pundir, C.S. Immobilization of rat brain acetylcholinesterase on ZnS and poly(indole-5-carboxylic acid) modified Au electrode for detection of organophosphorus insecticides. *Biosens. Bioelectron.*, **2011**, *29*(1), 82-88.
[http://dx.doi.org/10.1016/j.bios.2011.07.070] [PMID: 21873044]

[347] Liu, Q.; Shi, J.; Zeng, L.; Wang, T.; Cai, Y.; Jiang, G. Evaluation of graphene as an advantageous adsorbent for solid-phase extraction with chlorophenols as model analytes. *J. Chromatogr. A*, **2011**, *1218*(2), 197-204.
[http://dx.doi.org/10.1016/j.chroma.2010.11.022] [PMID: 21163486]

[348] Liu, T.; Xu, M.; Yin, H.; Ai, S.; Qu, X.; Zong, S. A glassy carbon electrode modified with graphene and tyrosinase immobilized on platinum nanoparticles for sensing organophosphorus pesticides. *Mikrochim. Acta*, **2011**, *175*(1-2), 129.
[http://dx.doi.org/10.1007/s00604-011-0665-5]

[349] Zamfir, L.G.; Rotariu, L.; Bala, C. A novel, sensitive, reusable and low potential acetylcholinesterase biosensor for chlorpyrifos based on 1-butyl-3-methylimidazolium tetrafluoroborate/multiwalled carbon nanotubes gel. *Biosens. Bioelectron.*, **2011**, *26*(8), 3692-3695.
[http://dx.doi.org/10.1016/j.bios.2011.02.001] [PMID: 21377346]

[350] Zhang, Q.; Xu, Q.; Guo, Y.; Sun, X.; Wang, X. Acetylcholinesterase biosensor based on the mesoporous carbon/ferroferric oxide modified electrode for detecting organophosphorus pesticides. *RSC Advances*, **2016**, *6*(29), 24698-24703.
[http://dx.doi.org/10.1039/C5RA21799G]

[351] Mayorga-Martinez, C.C.; Pino, F.; Kurbanoglu, S.; Rivas, L.; Ozkan, S.A.; Merkoçi, A. Iridium oxide nanoparticle induced dual catalytic/inhibition based detection of phenol and pesticide compounds. *J. Mater. Chem. B Mater. Biol. Med.*, **2014**, *2*(16), 2233-2239.
[http://dx.doi.org/10.1039/C3TB21765E]

[352] Wei, M.; Zeng, G.; Lu, Q. Determination of organophosphate pesticides using an acetylcholinesterase-based biosensor based on a boron-doped diamond electrode modified with gold nanoparticles and carbon spheres. *Mikrochim. Acta*, **2014**, *181*(1-2), 121-127.
[http://dx.doi.org/10.1007/s00604-013-1078-4]

[353] Zhai, C.; Guo, Y.; Sun, X.; Zheng, Y.; Wang, X. An acetylcholinesterase biosensor based on graphene-gold nanocomposite and calcined layered double hydroxide. *Enzyme Microb. Technol.*, **2014**, *58-59*, 8-13.
[http://dx.doi.org/10.1016/j.enzmictec.2014.02.004] [PMID: 24731819]

[354] Wei, M.; Wang, J. A novel acetylcholinesterase biosensor based on ionic liquids-AuNPs-porous carbon composite matrix for detection of organophosphate pesticides. *Sens. Actuators B Chem.,* **2015**, *211*, 290-296. [http://dx.doi.org/10.1016/j.snb.2015.01.112]

[355] Wu, S.; Huang, F.; Lan, X.; Wang, X.; Wang, J.; Meng, C. Electrochemically reduced graphene oxide and Nafion nanocomposite for ultralow potential detection of organophosphate pesticide. *Sens. Actuators B Chem.,* **2013**, *177*, 724-729. [http://dx.doi.org/10.1016/j.snb.2012.11.069]

[356] Du, D.; Wang, M.; Cai, J.; Zhang, A. Sensitive acetylcholinesterase biosensor based on assembly of β-cyclodextrins onto multiwall carbon nanotubes for detection of organophosphates pesticide. *Sens. Actuators B Chem.,* **2010**, *146*(1), 337-341. [http://dx.doi.org/10.1016/j.snb.2010.02.053]

[357] Ribeiro, F.W.; Barroso, M.F.; Morais, S.; Viswanathan, S.; de Lima-Neto, P.; Correia, A.N.; Oliveira, M.B.; Delerue-Matos, C. Simple laccase-based biosensor for formetanate hydrochloride quantification in fruits. *Bioelectrochemistry,* **2014**, *95*, 7-14. [http://dx.doi.org/10.1016/j.bioelechem.2013.09.005] [PMID: 24161938]

[358] Dong, J.; Liu, T.; Meng, X.; Zhu, J.; Shang, K.; Ai, S.; Cui, S. Amperometric biosensor based on immobilization of acetylcholinesterase *via* specific binding on biocompatible boronic acid-functionalized Fe@ Au magnetic nanoparticles. *J. Solid State Electrochem.,* **2012**, *16*(12), 3783-3790. [http://dx.doi.org/10.1007/s10008-012-1812-6]

[359] Kaur, N.; Thakur, H.; Prabhakar, N. Conducting polymer and multi-walled carbon nanotubes nanocomposites based amperometric biosensor for detection of organophosphate. *J. Electroanal. Chem.,* **2016**, *775*, 121-128. [http://dx.doi.org/10.1016/j.jelechem.2016.05.037]

[360] Ding, J.; Zhang, H.; Jia, F.; Qin, W.; Du, D. Assembly of carbon nanotubes on a nanoporous gold electrode for acetylcholinesterase biosensor design. *Sens. Actuators B Chem.,* **2014**, *199*, 284-290. [http://dx.doi.org/10.1016/j.snb.2014.04.012]

[361] Huo, D.; Li, Q.; Zhang, Y.; Hou, C.; Lei, Y. A highly efficient organophosphorus pesticides sensor based on CuO nanowires–SWCNTs hybrid nanocomposite. *Sens. Actuators B Chem.,* **2014**, *199*, 410-417. [http://dx.doi.org/10.1016/j.snb.2014.04.016]

[362] Zhai, C.; Sun, X.; Zhao, W.; Gong, Z.; Wang, X. Acetylcholinesterase biosensor based on chitosan/prussian blue/multiwall carbon nanotubes/hollow gold nanospheres nanocomposite film by one-step electrodeposition. *Biosens. Bioelectron.,* **2013**, *42*, 124-130. [http://dx.doi.org/10.1016/j.bios.2012.10.058] [PMID: 23202341]

[363] Chauhan, N.; Pundir, C.S. An amperometric acetylcholinesterase sensor based on Fe3O4 nanoparticle/multi-walled carbon nanotube-modified ITO-coated glass plate for the detection of pesticides. *Electrochim. Acta,* **2012**, *67*, 79-86. [http://dx.doi.org/10.1016/j.electacta.2012.02.012]

[364] Mishra, R.K.; Dominguez, R.B.; Bhand, S.; Muñoz, R.; Marty, J.L. A novel automated flow-based biosensor for the determination of organophosphate pesticides in milk. *Biosens. Bioelectron.,* **2012**, *32*(1), 56-61. [http://dx.doi.org/10.1016/j.bios.2011.11.028] [PMID: 22221795]

[365] Wu, S.; Lan, X.; Zhao, W.; Li, Y.; Zhang, L.; Wang, H.; Han, M.; Tao, S. Controlled immobilization of acetylcholinesterase on improved hydrophobic gold nanoparticle/Prussian blue modified surface for ultra-trace organophosphate pesticide detection. *Biosens. Bioelectron.,* **2011**, *27*(1), 82-87. [http://dx.doi.org/10.1016/j.bios.2011.06.020] [PMID: 21752626]

[366] Zhang, Y.; Arugula, M.A.; Wales, M.; Wild, J.; Simonian, A.L. A novel layer-by-layer assembled multi-enzyme/CNT biosensor for discriminative detection between organophosphorus and non-

organophosphrus pesticides. *Biosens. Bioelectron.*, **2015**, *67*, 287-295. [http://dx.doi.org/10.1016/j.bios.2014.08.036] [PMID: 25192873]

[367] Turan, J.; Kesik, M.; Soylemez, S.; Goker, S.; Coskun, S.; Unalan, H.E.; Toppare, L. An effective surface design based on a conjugated polymer and silver nanowires for the detection of paraoxon in tap water and milk. *Sens. Actuators B Chem.*, **2016**, *228*, 278-286. [http://dx.doi.org/10.1016/j.snb.2016.01.034]

[368] Ivanov, A.N.; Younusov, R.R.; Evtugyn, G.A.; Arduini, F.; Moscone, D.; Palleschi, G. Acetylcholinesterase biosensor based on single-walled carbon nanotubes--Co phtalocyanine for organophosphorus pesticides detection. *Talanta,* **2011**, *85*(1), 216-221. [http://dx.doi.org/10.1016/j.talanta.2011.03.045] [PMID: 21645691]

[369] Li, X.; Zheng, Z.; Liu, X.; Zhao, S.; Liu, S. Nanostructured photoelectrochemical biosensor for highly sensitive detection of organophosphorous pesticides. *Biosens. Bioelectron.*, **2015**, *64*, 1-5. [http://dx.doi.org/10.1016/j.bios.2014.08.006] [PMID: 25173731]

[370] Evtugyn, G.A.; Shamagsumova, R.V.; Padnya, P.V.; Stoikov, I.I.; Antipin, I.S. Cholinesterase sensor based on glassy carbon electrode modified with Ag nanoparticles decorated with macrocyclic ligands. *Talanta,* **2014**, *127*, 9-17. [http://dx.doi.org/10.1016/j.talanta.2014.03.048] [PMID: 24913851]

[371] El-Moghazy, A.Y.; Soliman, E.A.; Ibrahim, H.Z.; Noguer, T.; Marty, J.L.; Istamboulie, G. Ultra-sensitive biosensor based on genetically engineered acetylcholinesterase immobilized in poly (vinyl alcohol)/Fe-Ni alloy nanocomposite for phosmet detection in olive oil. *Food Chem.*, **2016**, *203*, 73-78. [http://dx.doi.org/10.1016/j.foodchem.2016.02.014] [PMID: 26948591]

[372] Li, X.; Zheng, Z.; Liu, X.; Zhao, S.; Liu, S. Nanostructured photoelectrochemical biosensor for highly sensitive detection of organophosphorous pesticides. *Biosens. Bioelectron.*, **2015**, *64*, 1-5. [http://dx.doi.org/10.1016/j.bios.2014.08.006] [PMID: 25173731]

[373] Apetrei, I.M.; Apetrei, C. The biocomposite screen-printed biosensor based on immobilization of tyrosinase onto the carboxyl functionalised carbon nanotube for assaying tyramine in fish products. *J. Food Eng.*, **2015**, *149*, 1-8. [http://dx.doi.org/10.1016/j.jfoodeng.2014.09.036]

[374] Ju, K.J.; Feng, J.X.; Feng, J.J.; Zhang, Q.L.; Xu, T.Q.; Wei, J.; Wang, A.J. Biosensor for pesticide triazophos based on its inhibition of acetylcholinesterase and using a glassy carbon electrode modified with coral-like gold nanostructures supported on reduced graphene oxide. *Mikrochim. Acta,* **2015**, *182*(15-16), 2427-2434. [http://dx.doi.org/10.1007/s00604-015-1584-7]

CHAPTER 9

Optical Sensor Arrays for Pharmaceutical and Biomedical Analyses

Pavel Anzenbacher, Jr[*, 1] and **Mehmet Gokhan Caglayan**[2]

[1] *Center for Photochemical Sciences and Department of Chemistry, Bowling Green State University, Bowling Green, OH, USA*

[2] *Faculty of Pharmacy, Department of Analytical Chemistry, Ankara University, Ankara, Turkey*

Abstract: Similarly to the sense of taste or smell, which is derived from triggering multiple different signals in receptors of human tongue or in the nasal cavity, also arrays of cross-reactive sensors give a rise to a characteristic response pattern specific for each different analytes. While the artificial receptors and sensors are rarely selective, the fingerprint-like response pattern is analyte specific and substitutes the sensor selectivity. Thus, the sensing of a particular analyte or a group of analytes using cross-reactive sensor arrays eliminates the need for analyte-specific sensors, which are difficult to design and synthesize. Response patterns from optical sensor arrays can be utilized in the development of low-cost analytical methods suitable for on-site analyses with high sensitivity, which is especially useful in pharmaceutical and biomedical areas that require the detection of very low concentration of analytes in complex biological samples. This chapter focuses on the use of optical sensor arrays in pharmaceutical and biomedical analyses and pattern recognition, and quantitative analysis of the results using chemometric methods.

Keywords: Array sensing, Artificial neural networks, Biomedical analysis, Chemical sensors, Chiral analysis, Chemometric, Colorimetric, Cross-reactive sensor arrays, Fluorescence, Hierarchical cluster analysis, Linear discriminant analysis, Multivariate analysis, Optical sensors, Pharmaceutical analysis, Principal component analysis, Support vector machines.

INTRODUCTION

A chemical sensor is a device that provides analytical information about a sample [1, 2]. Depending on an application, this information can provide qualitative identification or more frequently – and preferably - the chemical sensors provide a quantitative analysis of one or more chemical species in sample. The species of

* **Corresponding author Pavel Anzenbacher Jr:** Center for Photochemical Sciences and Department of Chemistry, Bowling Green State University, Bowling Green, OH, USA; Tel: (419) 372 20 80; Fax: (419) 372 98 09; E-mail: pavel@bgsu.edu

Atta-ur-Rahman, Sibel A. Ozkan & Rida Ahmed (Eds.)

interest (analyte) can be molecules, ions, or biologically important species such as DNA, amino acids and proteins. In general, a chemical sensor consists of two main parts: a *receptor* and a *transducer*. The function of the receptor is to recognize the analyte while the transducer provides observable change in properties that can be transformed into analytical signals. Depending on the molecular interactions between the receptor and the analyte, the receptors show different degrees of affinity and selectivity toward an analyte. The interaction between an analyte and a receptor induces changes in the chemical properties of the transducer. Owing to the changes in the properties of the transducer, it is generally possible to obtain information about an analyte by transforming the chemical changes into measurable signals from the sensor. Such a signal can be a change in the optical properties (*e.g.*, fluorescence and color), electrochemical (*e.g.*, voltammetry and redox properties), electrical (*e.g.*, conductivity and current), magnetic or a signal based on mass or heat.

There are several types of interactions frequently utilized in molecular recognition [3]. The strongest is electrostatic interaction that occurs between charged species. However, the strength of electrostatic forces is highly affected by the medium. For instance, this interaction is stronger in a medium with a smaller dielectric constant, and vice versa. This type of interaction is also affected by the pH of the medium because depending on the pH the molecules may be in their neutral or ionized forms. Despite being weaker than electrostatic interaction, hydrogen bonding is also significant in many chemical sensing applications due to the orientation of functional groups. This dimensionality feature provides certain degrees of selectivity. However, being a strong hydrogen bonding agent, water reduces the efficacy of the hydrogen bonding between a chemical sensor and analyte, thus interfering with the molecular recognition process. Another interaction that has a moderate strength is observed between metal ions and electron donors, which results in the formation of coordination bonds. The π-π interaction can also be effective when aromatic rings are parallel to each other. Despite being considered the weakest molecular interaction, the overall effect of van der Waals forces can be strong depending on the molecular shape. In particular, complementary shapes produce a considerable number of van der Waals interactions making a remarkable contribution to molecular recognition. Finally, hydrophobic interaction should also be considered especially in aqueous media [4]. Molecular recognition occurs as a sum of different types of interactions that are affected by internal factors (*e.g.*, molecular shapes and pKa) and external factors (*e.g.*, solvent type). Therefore, when designing a chemical sensor, these interactions should be thoroughly considered.

The molecular design of a receptor and the corresponding sensor is a key step to obtain an effective sensor. Traditionally, a sensor is designed to be completely selective to a single analyte, which is known as the 'lock and key' approach [5, 6]. The lock and key design is inspired by biological systems that employ high substrate selectivity in enzyme reactions. In addition to selectivity that is the main concern for this type of design, the resulting sensor must also display other properties such as sensitivity and chemical stability.

Although analyte-specific sensors are highly successful in chemical sensing, they also have certain disadvantages. Firstly, the lock and key design requires time and effort associated with the design and synthesis to incorporate parameters that should be taken into account such as molecular interactions, complementary shapes and solvent effect. Secondly, structurally similar compounds can also bind to the receptor and unfortunately real-world samples frequently contain those having subtle differences in their molecular structure. This further complicates the design process particularly when the challenge of preventing interferences is considered. Finally, it is impossible to design a sensor for an analyte with an unknown exact structure, which is the case in many biomolecules [7].

An alternative in molecular sensing strategy is to turn the drawbacks of the lock and key design to an advantage through cross-reactive sensing, which is also inspired by biological systems. Cross-reactive sensing mimics mammalian olfactory and gustatory systems. These systems employ several semi-selective sensing elements that respond differentially to odors and foods. Specific response patterns generated by the sensing elements in the nose or tongue are then transmitted to the brain to be stored in memory. The next time that the same response pattern is produced in the nose or tongue, the brain matches the "fingerprint" that belongs to that food or odor. Similarly, in cross reactive sensor arrays, instead of using one sensing element with strong affinity to a particular analyte, total response generated from the entire array is used as the fingerprint for the analyte. This approach is very useful in chemical sensing for several reasons. First, high selectivity of the sensors is not required; therefore, the design and synthesis of sensor materials are much easier compared to analyte-specific sensors. Second, not only one analyte but also a group of analytes can be determined using the same array. Third, a multiplexed analysis can be conducted simultaneously with the array methodology. Finally, biomolecules with an unknown exact structure can be sensed contrary to the case in lock and key sensors.

Although the design of cross-reactive sensor elements does not require as much effort as lock and key designs, this does not mean that the sensor can respond to all types of chemicals in a sample. A certain degree of selectivity is still required and molecular interactions between sensors and analytes should be considered. Highly cross-reactive sensing elements provide a lower level of information, and an accurate analysis can only be achieved using the total response produced by a high number of sensing elements, which might be more laborious and expensive compared to the lock and key approach. On the other hand, partially selective sensing elements can produce a high level of information in a shorter time and is sufficient for the accurate analysis of chemical species. In addition, the discriminatory power of sensor arrays is higher in partially selective sensing elements. Although there are still some researchers that employ highly cross reactive sensing elements such as common colorimetric and fluorimetric dyes in chemical sensing, most of the effort on the subject is made towards the development of novel partially selective sensing elements.

Different types of materials are used in cross-reactive sensor arrays including supramolecular sensors, nanomaterials, polymeric materials and nucleic acids. Supramolecular chemistry is based on principles such as host-guest chemistry, molecular self-assembly, molecular recognition and preorganization that have all been used in various chemical sensing applications. The supramolecular chemistry related to sensing is also a part of new discipline called "supramolecular analytical chemistry" [8]. Supramolecules are one of the most remarkable materials in cross-reactive sensor arrays [9 - 18]. In supramolecular sensing, different strategies are applied; *e.g.*, direct sensing, indicator displacement assays and indicator uptake. In direct sensing, an analyte is directly captured by the receptor, and the binding of the analyte produces an observable signal. This can be signal enhancement or quenching or band shift in fluorescence arrays and a change in the color intensity or color change in colorimetric arrays. Different mechanisms have an influence on signal production; for example, photo-induced electron transfer from an analyte to a sensing element might occur causing fluorescence quenching, or the binding of an analyte may prevent photo-induced electron transfer from an electron donator group to fluorophore of the same sensor resulting in fluorescence enhancement. In direct sensing, fluorophore/chromophore groups are an integral part of the sensor molecule.

On the other hand, an indicator can be used as a fluorescence or absorbance source as such in indicator displacement assay and indicator uptake assay. Here, the sensor is composed of the receptor and an indicator, and the competition between the analyte and indicator to bind the receptor produces an observable optical signal, whether is fluorescence or a change in color.

Owing to their outstanding features, nanomaterials such as, for example, gold nanoparticles or quantum dots have become increasingly important for their use as sensors. This popularity comes from the different and tunable chemical and physical properties of nanomaterials compared to bulk materials. This is due to the quantum confinement effect, which is related to the materials having a comparable size to the wavelengths of their electrons. The chemical and physical properties of nanomaterials are size- and shape–dependent and therefore easy to modify. They also have a high surface-to-volume ratio and their surface can easily be modified since they have different affinities to certain chemical groups such as thiol, phospine and amine. These tunable properties make nanomaterials a good candidate for sensing elements in cross reactive sensor arrays and recently some optical sensor arrays utilizing nanoparticles have been reported, especially in bio-related fields [19 - 30].

Simple chemical dyes have also been used in optical sensor arrays [31 - 35]. Such dyes may also respond to many analyte species as their colors/fluorescences are affected by a wide range of molecular interactions such as proton transfer, electron transfer, hydrogen bonding, π-π interaction and dipolar interactions. However, due to their poor selectivity, generally, a higher number of chemical dyes is required to elicit required information (response pattern) on analytes [32]. On the other hand, chemical dyes have certain advantages, such as low price and commercial availability, which eliminates the need for designing and preparative steps that would otherwise be required.

In addition to the above-mentioned supramolecular sensors, nanomaterials and chemical dyes, there is a wide variety of sensing elements used in optical sensor arrays that do not fall into the above categories. The new examples of sensor materials in this rapidly developing field include conjugated polymers [36 - 39], ionic liquids [40 - 42] and molecularly imprinted photonic polymers [43 - 46]. Furthermore, the green chemistry also impacted the development of sensors by reducing or eliminating health and environmental hazards in the manufacture or use of chemicals [47]. In green chemistry, there are 12 principles to follow when designing a chemical method. It is easier to follow these principles in designing and synthesizing cross-reactive sensors because they allow more flexibility for the modification of synthesis steps and the final product [48 - 52]. Since partial selectivity is required and a variety of sensing materials are available, it is possible to design 'the greenest sensor' through the use of safer chemicals and reaction conditions, less hazardous chemical synthesis, renewable starting materials and degradable products.

Chemometric Analysis of Optical Array Sensors

Optical sensor arrays generate a large volume of multidimensional data. The response pattern of an analyte in an optical sensor array is a collection of optical responses of individual sensing elements. These responses can be measured by simple instruments such as cameras, office scanners for colorimetric output and ultraviolet (UV) scanners. Multivariate responses obtained from arrays require chemometric techniques such as pattern recognition to extract the information about the analyte. Chemometric methods provide chemical information through the analysis of chemical data based on mathematics and statistics [66, 67]. Chemometry software that employs multivariate analyses allow clustering samples according to similarities in their optical responses, classifying samples based on the known class of chemicals and calibration of a single component through multivariate regression techniques. To date, several pattern recognition techniques have been reported in the literature. In this paper, we present those that are most frequently used in the optical array sensing. Table **1** summarizes the studies that used optical sensor arrays in pharmaceutical and biomedical analyses. According to the results of this review, linear discriminant analysis (LDA) and principal component analysis (PCA) are the most commonly used data analysis techniques in array sensing, followed by support vector machines (SVM), hierarchical cluster analysis (HCA) and artificial neural networks (ANN). However, before delving into these techniques, it is necessary to address the preprocessing of data as the first stage of data analysis.

Table 1. Examples of optical sensor array applications in pharmaceutical and biomedical analyses.

Sensors	Analytes	Det.	Chemo.	Ref.
Nanographene oxide – DNA ensemble	Proteins: SubA, Fib, Hb, CC, Lyso, HRP, BSA, Lip, casein	Fluo	LDA	[53]
Combinatorial peptide – metallocenter	Tripeptides	Abs	PCA	[54]
Conjugated polymers	Proteins: α-Am, BSA, Cht, CC, Fer, β-Gal, Hb, His, HSA, Myo, Pap, PhosA, PhosB, RibA,	Fluo	LDA	[36]
Supramolecular complex: conjugated polymer – green fluorescent protein	Sixteen different mammalian cell types	Fluo	HCA, LDA	[38]
Graphene oxide - aptamer	Proteins: different types of Hb	Fluo	LDA	[55]
Metal – ligand complex	Pharmaceutical: adrafinil	Fluo	LDA,SVM	[56]
Cucurbit[n]uril derivatives	Pharmaceutical: OTC Drugs	Fluo	LDA,SVM	[57]
Gold nanoparticle – fluorescent polymer complex	Proteins: SubA, Lip, PhosA, PhosB, β-Gal	Fluo	LDA	[22]
Cu(II)-complexes	ee amino acids: L-Val, D-Val, L-Leu, D-Leu,, L-Tle, D-Tle,	Abs.	PCA	[58]
Combinatorial peptide	ATP, GTP	Abs.	PCA	[59]
Enzyme amplified-gold nanoparticle complex	Proteins: α-Am, BSA, CC, Fer, HSA, Lip, Lyso, Myo, PhosB,	Fluo	LDA	[60]

(Table 1) contd.....

Sensors	Analytes	Det.	Chemo.	Ref.
Polysulfonated fluorescent dyes	Pharmaceuticals: Purine derivative drugs	Fluo	HCA, PCA	[34]
Palladium - dye complexes	Short oligodeoxynucleotides	Abs	LDA	[31]
Cucurbiturils	Pharmaceutical: Date-Rape Drug γ- Hydroxybutyrate	Abs Fluo	-	[61]
BINOL based macrocycle	ee: Ket, Ibu	Fluo	LDA,SVM	[62]
Calix [4]pyrrole derivatives	Pharmaceuticals: Ket, Flu, Dic, Ibu, Art, Nap, Sal, L-Ala, L-Tyr, Mef	Fluo	PCA, LDA	[63]
Gold nanoparticle-fluorescent polymer complex	Cell surface- Drug mechanism	Fluo	HCA,LDA	[21]
Cationic conjugated oligoelectrolyte- single-stranded DNA complex	Bacteria: *E.coli*, *L.acidophilus*, *Sporomusa*, *R.palustris*, *S.mutans*	Fluo	RDA	[64]
Gold nanoparticle-polymer complex	Proteins: HSA, CC, PhosA, PhosB, Hb, Lyso, RibA,Myo, Fib, α-At, IGG	Fluo	LDA	[20]
Cp*Rh – GCR, Azophloxine, NBB	ADP, ATP, pyrophosphate	Abs	ANN, LDA	[65]
Photo-initiated polymerization of molecularly imprinted films	Pharmaceuticals: CIP, OFL, DOP, AZI, PEN	SPR	-	[44]
Fluorescent dyes – transition metal complexes	Peptides: Car, HCar, KD, BK	Fluo	PCA, LDA	[35]

Abbreviations: Det–Detection mode; Chemo–Chemometric method; ee–enantiomeric excess; SubA–Subtilisin A; Fib–Fibrinogen; Hb–Hemoglobin; CC– Cytochrome *c*; Lyso–Lysozyme; HRP–Horseradish peroxidase; BSA–Bovine serum albumi; Lip–Lipase; α-Am–α-Amylase; Cht–α-chymotrypsin; Fer–Ferritin; β-Gal–β-galactosidase; His–Histone; HSA–Human serum albumin; Myo–Myoglobin; Pap–Papain; PhosA–Acid phosphatase; PhosB–Alkaline phosphatase; RibA–ribonuclease A; OTC–Over the counter; CC551–Cytochrome *c*551; Ferre–Ferredoxin; Val–Valine; Leu–Leucine; Tle–tert-Leucine; ATP–Adenosine triphosphate; GTP–Guanosine-5'-triphosphate; Ket–Ketoprofen; Ibu–Ibuprofen; Flu–Flurbiprofen; Dic–Diclofenac; Art–Artesunat;Nap–Naproxen; Sal–Salycyclic acid; Ala–Alanine; L-Tyrosine; Mef–Mefenamic acid; RibA–Ribonuclease A; α-At–α-Antitrypsin; IGG–Immunoglobulin G; GCR–Glycine cresol red; NBB–Naphthol Blue Black; LDA–Linear discriminant analysis; PCA–Principal component analysis; HCA–Hierarchical cluster analysis; RDA–Regularized discriminant analysis; ANN–Artificial neural networks; CIP–Ciprofloxacin; OFL–Ofloxacin; DOP–Dopamine; AZI–Azithromycin; PEN–Penicillin; SPR–Surface Plasmon Resonance; Car–Carnosine; HCar–Homocarnosine; KD–kallidin; BK– bradykinin

An analyst should first examine the dataset to see whether there is any missing data(points). Where possible, this data should then be filled up with the mean value of the row or column to prevent errors during the multivariate analysis. In cases where there are several missing data in a row or column, removing that row or column is considered the best approach [68].

One of the frequently used preprocessing step is mean centering in which the mean of the column is subtracted from each data. The resulting column is then centered on the origin with a zero mean. Mean centering is an effective way of preprocessing particularly in cases of baseline problems or small datasets [69].

Another prominent preprocessing method used in optical array sensing is scaling. In variance scaling, following the mean centering step, each data is divided by the standard deviation of the column. This time, the resulting column has a unit variance and a zero mean. Scaling is useful in the analysis of low-response samples but noisy data may result in significant errors [70].

Baseline correction is another common pre-processing method and it is especially valuable in optical measurements where matrices displaying high backgrounds. The background response is subtracted from each data point and this must be carried out separately for each color channel (red, green, blue, *etc.*). In addition, there are a number of preprocessing techniques such as linearization, signal averaging and maximum scaling.

Principal Component Analysis

PCA is a multivariate analysis technique that chemists are most familiar with since it is one of the oldest chemometric methods and almost all modern statistical software includes PCA packages.

It is difficult to find patterns in multivariate data. However, PCA identifies patterns by reducing the dimensions of data without loss of information. This is achieved by creating new orthogonal (perpendicular) axes and transforming the data using these axes.

Calculation begins with the mean-centering of data from all dimensions. Then the covariance matrix between dimensions is calculated. The covariance matrix shows the variance of dimensions from the mean with respect to each other. In the next stage, the eigenvectors are calculated for covariance matrix. When multiplied by the covariance matrix, an eigenvector provides the integer multiple of that matrix, which is called an eigenvalue. When eigenvectors of covariance matrices are plotted, new perpendicular axes are generated. The calculated eigenvector and eigenvalues of covariance matrices have generally different weights. The highest eigenvalue is the most significant. However, the lowest eigenvalues can be disregarded due to their insignificance. Generally, eigenvectors with the highest two or three eigenvalues, *i.e.*, the first two or three principal components, are sufficient to demonstrate the patterns in a dataset. The results of PCA can be graphically presented as a score plot for analytes and loading plot for optical responses.

Linear Discriminant Analysis

LDA begins with the development of a model for identifying the relationship between observations and known groups. Therefore, LDA is also known as a supervised pattern recognition method. The aim of LDA is to maximize class discrimination by increasing the ratio of between-class to within-class variance. In mathematical modeling, LDA draws the best linear boundaries between classes to separate them. This is achieved by projecting a feature space into a subspace that is smaller than the dimensions of the sample. Furthermore, LDA reduces the dimensionality of data using a discriminant function. After developing the discriminant function, the model is tested for validation. This can be performed by predicting one sample at a time using other samples as a training set, a process known as cross-validation. Cross-validation is complete once every sample in the dataset is predicted. At the end of cross-validation, the classification ability of the discriminant function can be evaluated and expressed in percentage. Alternatively, LDA can be validated using a test set. In this technique, a known set that is not used in LDA is assumed as unknown. The prediction of the test set validates LDA in terms of its classification ability. The last validation technique for supervised pattern recognition is bootstrap, which is based on testing several internal test sets produced from a training set used in LDA. The main difference between cross-validation and bootstrap is that in the latter, testing is performed on randomly mixed data from a training set rather than testing the data from one sample at a time. At the end of the bootstrap validation, classification ability is presented as an average prediction ability for internal test sets.

The results of LDA are generally presented as a graphical output by plotting discriminant scores against canonical factors. This type of graphical representation visualizes the clustering ability of LDA and the resolution of samples.

Support Vector Machines

SVM is a supervised method based on machine learning. It is used for both clustering and regression. SVM is especially useful for separating classes by non-linear boundaries, for which other methods such as LDA fails to produce a solution. Recently, SVM has received considerable interest in cross-reactive sensor arrays since it performs very well on problems where classical methods fail; such as poorly distributed, high-dimensional data or low ratio of samples to dimensionality of the input data. For overlapping classes, discriminant analysis that creates linear hyperplanes to separate classes causes poor classification. Here, SVM can make a successful classification by transformation the data into a feature space in which classes can be linearly separated. Basis expansion in a feature space provides a better separation of classes and then translates it to the original input space (Fig. **1**). The step after transformation is finding hyperplanes that have the maximum distance to the nearest data point, called the maximum margin. However, accurate classification is more important than the maximization of margins since in some cases, the maximum margin does not necessarily present the most accurate classification. Therefore, SVM chooses the hyperplane that not only has the maximum margin but also correctly classifies the samples.

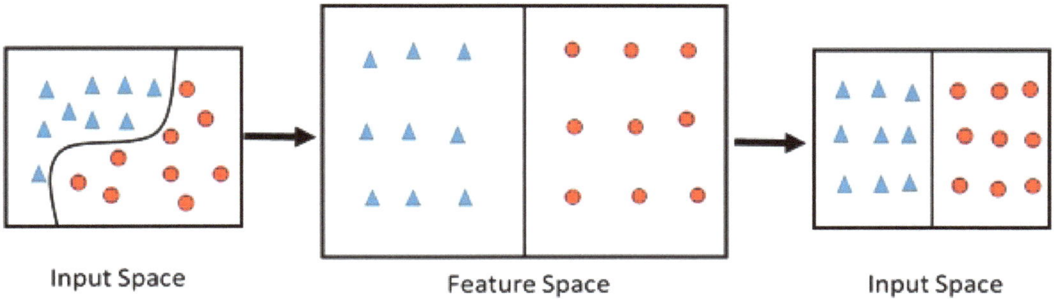

Fig. (1). Process of data classification in SVM.

Hierarchical Cluster Analysis

Although LDA and PCA successfully discriminate the responses obtained from sensor arrays, particularly in complex cases, other chemometric techniques are also employed. HCA is one of these techniques that clusters samples according to their similarities or differences.

HCA begins with a search for the closest sample pair based on their distance. There are different types of distance, with the most known being the Euclidean metric, which refers to the distance between two data points with coordinates (a, b) and (c, d) and can be calculated using the following equation:

$$d((a,b), (c,d)) = \sqrt{(a-c)^2 + (b-d)^2} \quad \textbf{(1)}$$

The second common method for distance calculation is Manhattan distance that can be determined for the above-mentioned two data points as follows:

$$d((a,b), (c,d)) = |a-c| + |b-d| \quad \textbf{(2)}$$

Another method for hierarchical clustering is through the calculation of similarity index, which is also called Jackard's similarity index. After calculating the similarity index or distances, the pair with the highest similarity index, *i.e.*, the closest pair, is clustered. Then, new distances or similarity indexes between this pair and other data points need to be calculated. There are different ways of performing this calculation with the most preferred being maximum linkage clustering. In this method, while calculating the distance between a merged pair and other data points, the farthest point of the merged pair is used and for this reason the method is also known as farthest neighbor clustering. For example, if one of the data points in a pair has distance z to a data point outside the pair, and the other data point in the pair has distance t to that point and $z>t$, then z is the distance between the pair and that point. In each step, the closest samples are paired and new distances are calculated until all samples are clustered. Finally, clustering is being visualized using a tree diagram called dendrogram.

Neural Networks

Similar to the design of cross-reactive chemical sensor molecules that mimics the receptor elements of the sensory system, another chemometric method, artificial neural network (ANN), was inspired by the neural system of humans. ANN receives inputs from transducers and processes them to produce outputs. This process can be very simple or complicated depending on the data obtained and analyte composition. The resulting output is a non-linear integration of the inputs multiplied by their weights [71]. ANN is especially useful in cases where classical techniques fail to evaluate non-linear relationships.

Optical Array Sensing in Pharmaceutical and Biomedical Analyses

In pharmaceutical industry, the need for analysis is determined by regulatory requirements in order to assure the quality of a drug. The desired quality of a drug can be provided by carefully organizing every procedure during manufacturing. The crystal form and stereo-chemical structure of a final product has to be specified [72]. The Food and Drug Administration of the United States of America listed the policies in the development of new stereo-isomeric drugs as follows [73]:

> " - *Appropriate manufacturing and control procedures should be used to assure stereoisomeric composition of a product, with respect to identity, strength, quality and purity. Manufacturers should notify compendia of these specifications and tests.*

> *- Pharmacokinetic evaluations that do not use a chiral assay will be misleading if the disposition of the enantiomers is different. Therefore, techniques to quantify individual stereoisomers in pharmacokinetic samples should be available early. If the pharmacokinetics of the enantiomers are demonstrated to be the same or to exist as a fixed-ratio in the target population, an achiral assay or an assay that monitors one of the enantiomers may be used, subsequently.*"

Approximately half the drugs sold in the global market are chiral and about a quarter are single enantiomers [74]. The use of single enantiomers has certain advantages such as having high efficacy due to their more selective pharmacologic profiles, less side effects, lower drug interaction and simpler pharmacokinetics because of the difference in their metabolism rates [75]. Moreover, one enantiomer can have serious adverse effects on patients while the other can have therapeutic effects. Therefore, the synthesis of pure enantiomers is one of the important goals in pharmaceutical industry. Regardless of whether the final product is a racemate or contains a single enantiomer, pre-clinical and clinical studies for single enantiomers are necessary in the drug development stage [76]. Consequently, development of chiral assays for pharmaceuticals are required in pharmaceutical industry. Chiral analysis of pharmaceuticals are mostly performed by liquid chromatography and capillary electrophoresis [77 - 87] but nuclear magnetic resonance (NMR) spectroscopy or circular dichroism (CD) are also utilized [88 - 98]. Recently, optical array-based sensing has been used for the enantiomeric excess analysis [62, 99 - 102]. Considering the time and number of samples required for assaying chiral compounds, simple optical array methods offer rapid determination of enantiomeric excess using low concentrations of samples.

As an example for the use of sensors in pharmaceutical analysis, Anzenbacher and Ema *et al.* (2016) developed fluorescent macrocyclic sensors for the determination of enantiomeric excess of ibuprofen, ketoprofen and some other carboxylates [62] (Fig. **2**). Intrinsic chirality of naphthalene auxiliary, which serves here also as a fluorophore, and arrangement of the hydrogen-bond donors within the macrocyclic cavity provide recognition ability and enantioselectivity [103]. The X-ray structure revealed that the fluorophore of the chemical sensor is close to the binding cavity which is important for the production of fluorescence response upon analyte binding. The fluorescence of two sensors (**S1** and **S2** in Fig. **2**) were quenched by the analyte binding through photoinduced electron transfer while the remaining two sensors (**S3** and **S4** in Fig. **2**) exhibited enhanced fluorescence due to their electron-rich fluorophores. Fluorescence titrations revealed that chemosensors produce different response patterns for the enantiomers of the same analytes allowing 100% discrimination of responses by LDA in array experiments. In the same study, another chemometric technique, SVM, was successfully employed for the quantitative determination of enantiomeric excess.

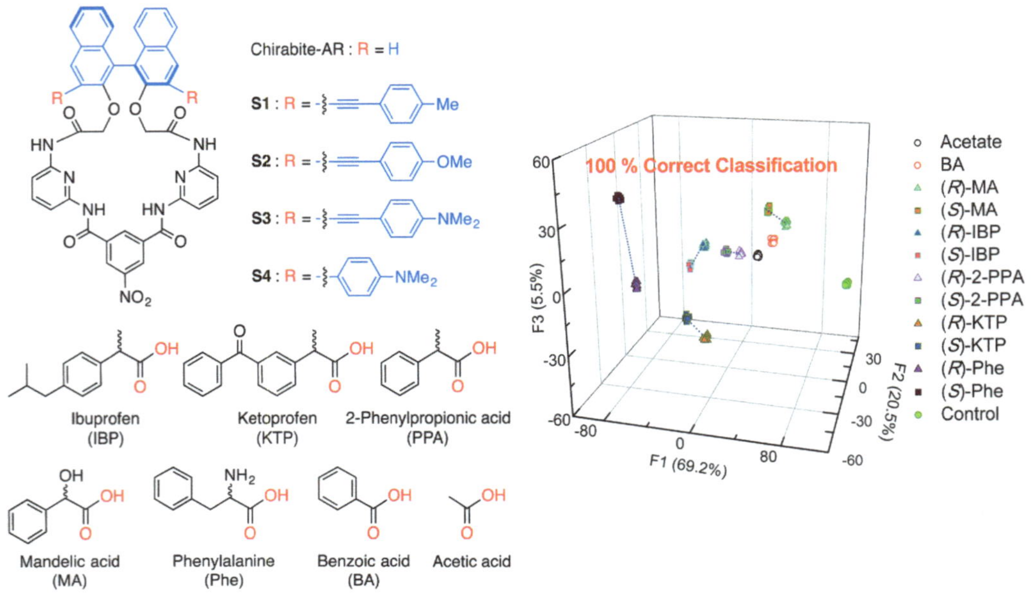

Fig. (2). Left: Structures of macrocyclic sensors (top) and analytes (bottom); Right: Linear discriminant analysis of analytes showing 100% accurate classification by employing macrocyclic sensors (Reproduced from Ref [62]. with permission from the Royal Society of Chemistry)

Optical sensor arrays are not only useful for chiral pharmaceutical analysis but they also enable high-throughput sensing of biologically important chiral molecules. As an example, analysis of chiral amines including certain amino acids such as valine and phenylalanine were proposed by Shcherbakova *et al.* (2016) [102]. In this method, self-assembly of 2-formylphenylboronic acid with a chiral diol and a chiral amine produced two diastereomeric iminoboronates. The application of dynamic covalent self-assembly in fluorescence array resulted in the formation of different patterns for R and S enantiomers of amino acids as well as some other amines and amino alcohols.

Asymmetric synthesis of chiral pharmaceuticals or biologically important molecules is an effective way of obtaining predominantly one enantiomer [104]. In this type of synthesis, either chiral catalysts or chiral auxiliaries are frequently used [105]. Toward this end, the favored are chromatographic methods [106, 107]. The advantages of optical array-based sensing over chromatographic techniques such as lower cost, shorter analysis time and the fact that they are amenable to HTS make them ideal for the analysis of enantiomeric excess to determine the optimal chiral catalyst or auxiliary. Recently, the Anslyn *et al.* (2014) demonstrated the advantages of HTS amenable optical array-based sensing in asymmetric synthesis through the development of an enantioselective indicator displacement assay for asymmetrically synthesized α-amino acids [108].

Chiral analysis is not the only application of optical sensor arrays in pharmaceutical analyses. Abused and counterfeit drugs are another area of interest. Over-the-counter (OTC) available drugs are one of these products and supramolecular sensors composed of two cucurbit[n]uril type receptors have been utilized in fluorimetric sensor arrays (Fig. **3**) [57]. Sensor molecules have affinities toward amines, ammonium and pyridinium salts. Fluorescence quenching and amplification of these probes depend on the nature of the analytes. Analytes containing pyridine moieties or nitro groups quench the fluorescence of probes. Aliphatic amines may also induce quenching by photoinduced electron transfer. On the other hand, analytes without quencher moieties amplify fluorescence since complex formation between cucurbit[n]uril type receptors and analytes increases the rigidity of probes and decreases vibrational and rotational modes that promote non-radiative decays. In the study, Minami *et al.* (2012) [13] used LDA to classify and identify the differential responses produced against several OTC drugs. The quantitative ability of optical sensor arrays was also tested by analyzing doxylamine, pseudoephedrine and phenylephrine in urine. The SVM results of the quantitative analysis were reported to be satisfactory with low errors. Finally, urine samples collected at different time intervals from a volunteer that had ingested doxylamine were analyzed. The pharmacokinetic profile of doxylamine was clearly identified regardless of electrolyte fluctuations or existence of doxylamine metabolites in urine. Fluorimetric sensor arrays offer several advantages over classical methods for OTC drugs, such as urine analysis not requiring a sample preparation step, a lower limit of detection values, easier pharmacokinetic analysis and prevention of false positive urine drug screening due to cross-reactivity of certain OTC drugs; *e.g.*, doxylamine and methadone. This study demonstrated the great potential of optical sensor arrays in pharmaceutical analysis. By using only two cross-reactive probes, it was possible to discriminate and quantify several pharmaceuticals in a complex biological matrix and explore their pharmacokinetic profile. The results also indicate that optical array sensing offers a solution for another problem in pharmaceutical industry; detection of the impurity and degradation of products by providing an accurate analysis of active pharmaceutical ingredients. Pharmaceutical industry requires the analysis of enormous number of samples in limited time during research & development, analytical development and quality control stages and optical array sensing fulfills these requirements through HTS.

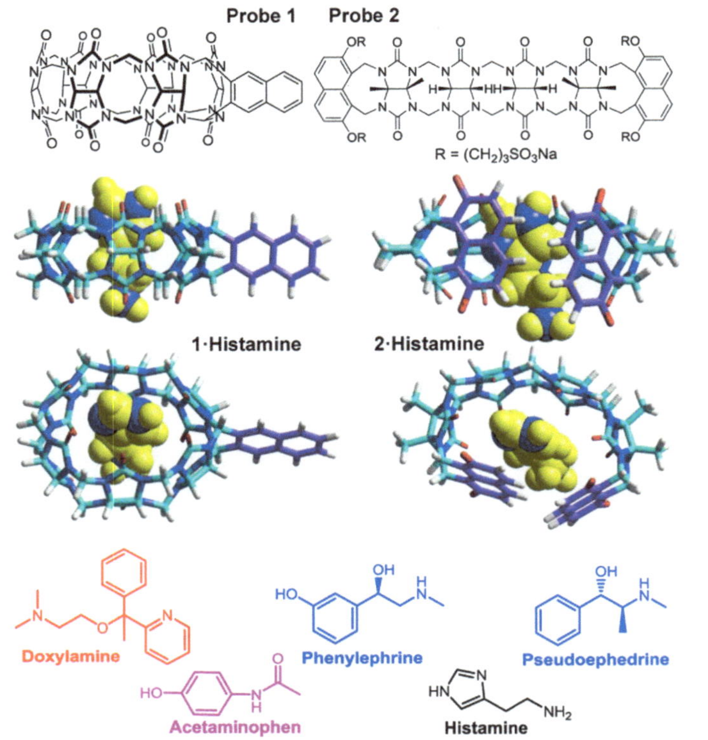

Fig. (3). Top: Structures of cucurbit[n]uril type probes. Center: Complex between probes and histamine Bottom: Analytes associated with over-the-counter cold remedies such as NyQuil. (This figure was reprinted (adapted) with permission from [57] Copyright (2013) American Chemical Society)

Anzenbacher *et al.* (2016) developed another fluorimetric sensor array for an active pharmaceutical ingredient, adrafinil [56]. Adrafinil is used for the treatment of sleep disorders; however, it is also used off-label by athletes, soldiers and students as a nootropic agent and a performance enhancer in sports. Since it is nootropic (also called smart drug), counterfeits are frequently sold through online and black market sales. It is reported that approximately 50% of online sold drugs are counterfeit [109]. In addition, since adrafinil is prohibited in sports and large number of biological samples such as urine and blood are tested in doping control centers, the development of simple HTS methods for testing adrafinil in biological matrices has a great significance. In the array proposed by Anzenbacher *et al.* (2016) [56], two probes based on metal-ligand complexes were used. For the Cu complex, indicator displacement assay using coumarin 343 fluorescent indicators was employed for a turn-on response. In an indicator displacement assay, the sensor first binds to an indicator causing a change in the signal. Then, the analyte displaces the indicator resulting in signal recovery. Displacement can only occur if the affinity constant of the analyte is greater than the indicator. This provides increased selectivity to cross-reactive sensors. On the other hand, fluorescence quenching is observed in the Zn complex in the introduction of analytes. Binding constants derived from fluorescence titrations indicate different affinities and several possible interferants including metabolites of adrafinil. One important aspect of the study by Anzenbacher *et al.* (2016) [56] was the materials used in the construction of the sensor array. The sensor array was constructed on a chromatography paper using a commercial wax printer. Simple, inexpensive and disposable paper-based optical sensor arrays were developed for doping analysis. For this type of analysis, the only instrumentation required was a wax printer and a UV scanner. Hydrophobic barriers (wax) were printed on papers to construct the arrays. The paper-based sensor arrays were tested for the qualitative and quantitative analysis of adrafinil in urine (Fig. **4**). Adrafinil and possible interferants were successfully classified using LDA. For the quantitative analysis, SVM was used with a high correlation coefficient and low errors. The limit of detection for the array fulfilled the World Anti-Doping Agency's minimum performance requirement for analytical assays used in doping analyses. As a result, a simple but effective paper-based fluorimetric sensor array was proposed for an active pharmaceutical ingredient.

An effective way of diagnosing a disease is to measure specific proteins (biomarkers) in blood since their reduced or increased amounts are an indicator of certain diseases [110]. It is known that blood serum has more than 20,000 proteins ranging in concentration from g/L to pg/L with the total protein concentration being approximately 1 mM [19]. In quantification of a specific protein, analyte-specific agents, *i.e.*, antibodies, are often applied using various techniques. Enzyme-linked immunosorbent assay (ELISA) is one of the most frequently used methods for the determination of proteins. In ELISA, enzyme-conjugated antibodies interact with analytes immobilized on a surface. Then, a substrate is added to the conjugated enzyme to produce an observable product following enzyme-substrate interaction. The measured enzyme activity is related to the amount of the analyte. However, despite being a very convenient assay for protein analysis, ELISA requires antibodies which are very expensive and difficult to prepare. In addition, even though non-specific interactions of antibodies can be reduced, such interactions are unavoidable and result in lower accuracies in analysis. Obtaining a specific antibody for each protein is also an exhaustive process, which is not a concern in the design of cross-reactive sensor arrays. Furthermore, other techniques for protein quantification such as sodium dodecyl sulfate polyacrylamide gel electrophoresis and matrix-assisted laser desorption ionization mass spectrometry have the disadvantages of longer analysis time, low accuracy and expensive instrumentation [19]. Therefore, optical sensor arrays offer a feasible alternative to these techniques through differential sensing using a variety of sensing elements such as fluorescent protein-nanoparticle system [19, 21], DNA-nanoparticles [111], polymers [36], fluorescent polymer-nanoparticle system [22] and porphyrins [112].

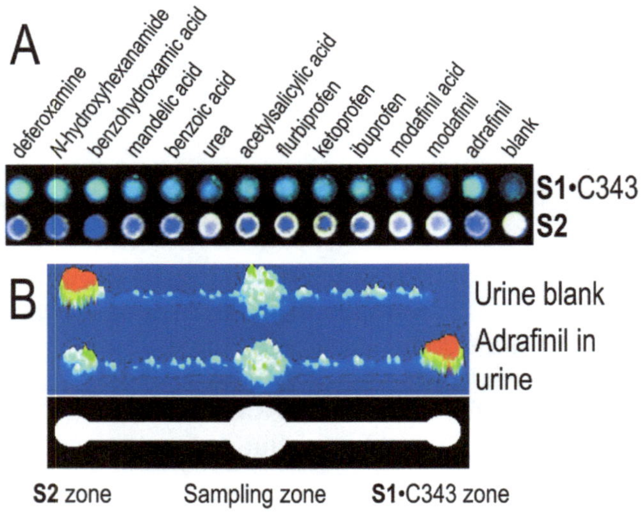

Fig. (4). Paper microzone array (**A**). Fluorescence image from a paper microzone array comprising two sensors (S1_C343 and S2) and fourteen analytes, obtained as a sum of the RGB channels collected in measurements. Microfluidic paper analytical device (μPAD) (**B**). Top: 3D intensity fluorescence image of the two μPADs for the sensing of adrafinil in urine. Bottom: Schematics of the μPAD layout. (This figure was reproduced from Ref [56]. with permission from the Royal Society of Chemistry)

Anslyn *et al.* (2005) achieved peptide sensing using metal-centered and peptide-armed sensing elements [54]. The authors constructed sensor arrays using different peptide arms that produced response patterns for tripeptide analytes. Colorimetric signals were produced by indicator-uptake assay, which is a good choice when high sensitivity is required as in biological analyses. In indicator uptake assays, first, analytes at low concentrations were introduced to bind to the receptors. Then, an indicator was used to occupy the empty sites on the receptor. The number of empty sites on a receptor depends on the interaction between the analyte and the receptor. If this interaction is weak, then more sites are available for indicator occupation, and vice versa. Moreover, weak affinity of a receptor to an analyte can result in the replacement of the analyte with an indicator resulting in a larger indicator uptake. Therefore, receptor analyte affinity determines the level of indicator uptake. Anslyn *et al.* (2015) [54] conducted their study using an Orange G indicator and the indicator-uptake levels of tripeptides were measured by a charge-coupled device (CCD). Multivariate data produced by CCD were evaluated by PCA.

The sensing strategy in a conjugated nanoparticle system is mainly based on the displacement of fluorescent or colored products with analyte biomolecules. Then, optical responses from recovered fluorescent or colored molecules are determined by LDA [19, 21, 22, 111] or HCA [21]. Differential responses from an optical sensor array allow the qualitative discrimination of a mixture of proteins or quantitative discrimination of a protein at different concentrations. The Rotello Research Group conducted extensive research on protein sensing with conjugated nanoparticles and conjugated polymers [19 - 22, 36, 60]. They used fluorescent proteins as an indicator for screening drug mechanism on human breast cancer cells [21]. The sensor developed by the authors comprised non-covalently bonded benzyl terminated gold nanoparticles-fluorescent protein complexes and responded to changes in cell surfaces and cell deaths, allowing the determination of the cancer drug mechanism and effective drug dose (Fig. **5**). This sensing strategy is similar to that employed in indicator displacement assays. Firstly, fluorescent proteins non-covalently bind to gold nanoparticles and fluorescence quenching was observed. Then, cell surfaces with higher affinity towards gold nanoparticles are replaced with fluorescent proteins and their fluorescence is recovered. The affinities of cell surfaces depend on drug treatment; therefore, drug-induced physicochemical changes can be detected using these optical sensor arrays.

In recent years, miniaturization has become an important goal in pharmaceutical and biomedical analyses [113 - 116]. The main objectives of miniaturization are reducing chemical (sample, reagent, and solvent) consumption and increasing the speed of analysis (high-throughput analysis) without losing sensitivity. Furthermore, chemicals used in bioanalysis are generally too expensive and reducing their amount in analysis means reducing the cost, which is another factor favoring the miniaturization. In this area, most of the effort has been devoted to the development of miniaturized systems in separation sciences since optical methods are generally subject to interferences in biological analyses. However, among optical methods, cross-reactive sensor arrays are a strong option since they are not affected by interferences and they are amenable to high-throughput analysis and miniaturized systems. Studies employing paper microplates [56] and glass nanowell plates [117] in optical array sensing can be given as the examples of miniaturization work in the literature.

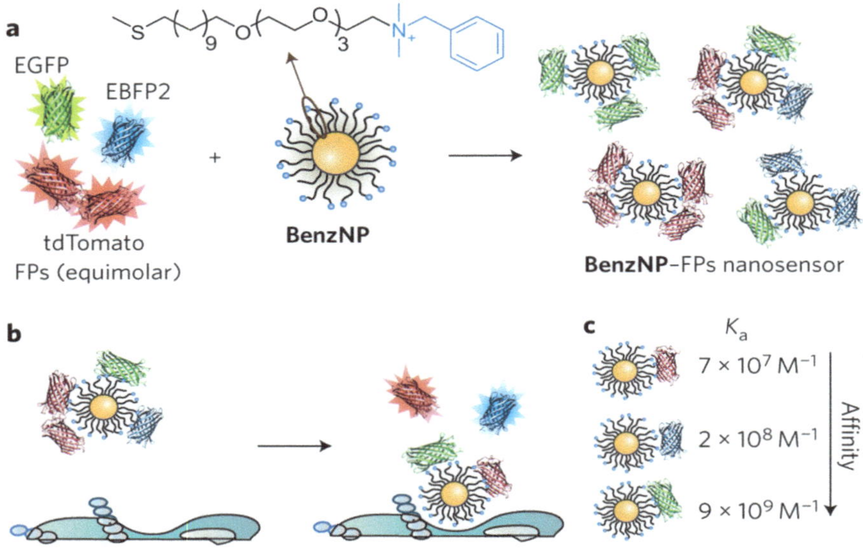

Fig. (5). Assembly and working principle of nanosensors. **a**, Fabrication of nanosensors comprising benzyl terminated gold nanoparticle and fluorescent complexes. **b**, Sensing strategy **c**, Differential affinity of BenzNP to tdTomato (red), EBFP2 (blue) and EGFP (green) protein. (This figure was reproduced with permission from the Royal Society of Chemistry, see ref [21].)

CONCLUSION

Sensor arrays are powerful tools for the analysis of a wide variety of analytes including environmental and biological samples. In this chapter, we only focused on optical sensor arrays in pharmaceutical and biomedical analyses; however, array-based sensing has been rapidly growing in almost all areas of analytical chemistry particularly considering the recent developments in supramolecular analytical chemistry, nanoscience, biochemistry and chemometrics. Optical sensor arrays offer many advantages over classical sensing strategies. Firstly, the instrumentation required for optical sensor arrays (UV scanners and CCD cameras) is generally easier to obtain and operate and with the technological progress that has been made, it is now possible to use even mobile phones for the diagnosis or detection of analytes in complex biological materials. Optical methods are among the most popular detection techniques for miniaturized and on-site analyses. Secondly, sensing elements used in arrays are partly selective (cross-reactive) and therefore, their synthesis is much easier than that of classical sensors with a lock and key design. Furthermore, flexibility in sensor design allows employing the green chemistry principles in sensing. The introduction of nanomaterials as a cross-reactive sensor element has improved array methodologies due to their easily tunable and controllable properties. Finally, user-friendly software in chemometrics allows easy classification and visualization of analytes with a large amount of multivariate data.

Although the use of array methodologies is not so common in pharmaceutical analyses, it has a great potential of fulfilling the expectations of pharmaceutical industry and researchers, and recently, considerable improvement has been made in this area. Studies have demonstrated that optical arrays can be effectively used in different areas of pharmaceutical and biomedical analyses such as enantiomeric excess, detection of drug-abuse, pharmacokinetic profiling and diagnosis.

CONSENT FOR PUBLICATION

Not applicable.

CONFLICT OF INTEREST

The authors declares no conflict of interest, financial or otherwise.

ACKNOWLEDGEMENTS

Declared none.

REFERENCES

[1] Bănică, F-G. *What are Chemical Sensors? Chemical Sensors and Biosensors*; John Wiley & Sons, Ltd, **2012**, pp. 1-20.
[http://dx.doi.org/10.1002/9781118354162]

[2] Askim, J.R.; Mahmoudi, M.; Suslick, K.S. Optical sensor arrays for chemical sensing: the optoelectronic nose. *Chem. Soc. Rev.,* **2013**, *42*(22), 8649-8682.
[http://dx.doi.org/10.1039/c3cs60179j] [PMID: 24091381]

[3] Persch, E.; Dumele, O.; Diederich, F. Molecular recognition in chemical and biological systems. *Angew. Chem. Int. Ed. Engl.,* **2015**, *54*(11), 3290-3327.
[http://dx.doi.org/10.1002/anie.201408487] [PMID: 25630692]

[4] Ariga, K.; Kunitake, T. *Supramolecular chemistry - Fundamentals and applications: Advanced textbook*; , **2006**, pp. 1-208.

[5] Collins, BE; Wright, AT; Anslyn, EV Combining molecular recognition, optical detection, and chemometric analysis. In: *Creative Chemical Sensor Systems: Topics in Current Chemistry*; Springer, **2007**; Vol. 277, pp. 181-218.
[http://dx.doi.org/10.1007/128_2007_114]

[6] Le, N.D.; Rana, S.; Rotello, V.M. Chemical nose sensors: an alternative strategy for cancer diagnosis. *Expert Rev. Mol. Diagn.,* **2013**, *13*(2), 111-113.
[http://dx.doi.org/10.1586/erm.12.143] [PMID: 23477550]

[7] Anzenbacher, P.; Palacios, M.A. *Array-Based Sensors. Chemosensors*; John Wiley & Sons, Inc., **2011**, pp. 345-368.

[8] Anslyn, E.V. Supramolecular analytical chemistry. *J. Org. Chem.,* **2007**, *72*(3), 687-699.
[http://dx.doi.org/10.1021/jo0617971] [PMID: 17253783]

[9] Chang, K-C.; Minami, T.; Koutnik, P.; Savechenkov, P.Y.; Liu, Y.; Anzenbacher, P., Jr Anion binding

modes in meso-substituted hexapyrrolic calix[4]pyrrole isomers. *J. Am. Chem. Soc.,* **2014**, *136*(4), 1520-1525.
[http://dx.doi.org/10.1021/ja411391c] [PMID: 24392650]

[10] Ghanem, E.; Hopfer, H.; Navarro, A.; Ritzer, M.S.; Mahmood, L.; Fredell, M.; Cubley, A.; Bolen, J.; Fattah, R.; Teasdale, K.; Lieu, L.; Chua, T.; Marini, F.; Heymann, H.; Anslyn, E.V. Predicting the composition of red wine blends using an array of multicomponent Peptide-based sensors. *Molecules,* **2015**, *20*(5), 9170-9182.
[http://dx.doi.org/10.3390/molecules20059170] [PMID: 26007178]

[11] Liu, Y.; Bonizzoni, M. A supramolecular sensing array for qualitative and quantitative analysis of organophosphates in water. *J. Am. Chem. Soc.,* **2014**, *136*(40), 14223-14229.
[http://dx.doi.org/10.1021/ja507905r] [PMID: 25246316]

[12] Minaker, S.A.; Daze, K.D.; Ma, M.C.; Hof, F. Antibody-free reading of the histone code using a simple chemical sensor array. *J. Am. Chem. Soc.,* **2012**, *134*(28), 11674-11680.
[http://dx.doi.org/10.1021/ja303465x] [PMID: 22703116]

[13] Minami, T.; Esipenko, N.A.; Zhang, B.; Kozelkova, M.E.; Isaacs, L.; Nishiyabu, R.; Kubo, Y.; Anzenbacher, P., Jr Supramolecular sensor for cancer-associated nitrosamines. *J. Am. Chem. Soc.,* **2012**, *134*(49), 20021-20024.
[http://dx.doi.org/10.1021/ja3102192] [PMID: 23194337]

[14] Montes-Navajas, P.; Baumes, L.A.; Corma, A.; Garcia, H. Dual-response colorimetric sensor array for the identification of amines in water based on supramolecular host–guest complexation. *Tetrahedron Lett.,* **2009**, *50*(20), 2301-2304.
[http://dx.doi.org/10.1016/j.tetlet.2009.02.189]

[15] Koshi, Y.; Nakata, E.; Yamane, H.; Hamachi, I. A fluorescent lectin array using supramolecular hydrogel for simple detection and pattern profiling for various glycoconjugates. *J. Am. Chem. Soc.,* **2006**, *128*(32), 10413-10422.
[http://dx.doi.org/10.1021/ja0613963] [PMID: 16895406]

[16] Palacios, M.A.; Nishiyabu, R.; Marquez, M.; Anzenbacher, P., Jr Supramolecular chemistry approach to the design of a high-resolution sensor array for multianion detection in water. *J. Am. Chem. Soc.,* **2007**, *129*(24), 7538-7544.
[http://dx.doi.org/10.1021/ja0704784] [PMID: 17530846]

[17] Chen, X-X.; Jiang, Y-B.; Anslyn, E.V. A racemate-rules effect supramolecular polymer for ee determination of malic acid in the high ee region. *Chem. Commun. (Camb.),* **2016**, *52*(85), 12669-12671.
[http://dx.doi.org/10.1039/C6CC06716F] [PMID: 27722581]

[18] Diehl, K.L.; Bachman, J.L.; Chapin, B.M.; Edupuganti, R.; Rogelio Escamilla, P.; Gade, A.M. *Design and Synthesis of Synthetic Receptors for Biomolecule Recognition. Synthetic Receptors for Biomolecules: Design Principles and Applications*; Chapter 2. The Royal Society of Chemistry, **2015**, pp. 39-85.
[http://dx.doi.org/10.1039/9781782622062-00039]

[19] De, M; Rana, S; Akpinar, H; Miranda, OR; Arvizo, RR; Bunz, UHF; Rotello, VM Sensing of proteins in human serum using conjugates of nanoparticles and green fluorescent protein. *Nat Chem.,* **2009**, *1*, 461-465.
[http://dx.doi.org/10.1038/nchem.334]

[20] Moyano, D.F.; Rana, S.; Bunz, U.H.; Rotello, V.M. Gold nanoparticle-polymer/biopolymer complexes for protein sensing. *Faraday Discuss.,* **2011**, *152*(0), 33-42.
[http://dx.doi.org/10.1039/c1fd00024a] [PMID: 22455037]

[21] Rana, S.; Le, N.D.; Mout, R.; Saha, K.; Tonga, G.Y.; Bain, R.E.; Miranda, O.R.; Rotello, C.M.; Rotello, V.M. A multichannel nanosensor for instantaneous readout of cancer drug mechanisms. *Nat. Nanotechnol.,* **2015**, *10*(1), 65-69.

[http://dx.doi.org/10.1038/nnano.2014.285] [PMID: 25502312]

[22] You, C.C.; Miranda, O.R.; Gider, B.; Ghosh, P.S.; Kim, I-B.; Erdogan, B.; Krovi, S.A.; Bunz, U.H.; Rotello, V.M. Detection and identification of proteins using nanoparticle-fluorescent polymer 'chemical nose' sensors. *Nat. Nanotechnol.,* **2007**, *2*(5), 318-323. [http://dx.doi.org/10.1038/nnano.2007.99] [PMID: 18654291]

[23] Ghasemi, F.; Hormozi-Nezhad, M.R.; Mahmoudi, M. A colorimetric sensor array for detection and discrimination of biothiols based on aggregation of gold nanoparticles. *Anal. Chim. Acta,* **2015**, *882*, 58-67. [http://dx.doi.org/10.1016/j.aca.2015.04.011] [PMID: 26043092]

[24] Kong, H.; Liu, D.; Zhang, S.; Zhang, X. Protein sensing and cell discrimination using a sensor array based on nanomaterial-assisted chemiluminescence. *Anal. Chem.,* **2011**, *83*(6), 1867-1870. [http://dx.doi.org/10.1021/ac200076c] [PMID: 21323355]

[25] Li, D.; Dong, Y.; Li, B.; Wu, Y.; Wang, K.; Zhang, S. Colorimetric sensor array with unmodified noble metal nanoparticles for naked-eye detection of proteins and bacteria. *Analyst (Lond.),* **2015**, *140*(22), 7672-7677. [http://dx.doi.org/10.1039/C5AN01267H] [PMID: 26446513]

[26] Kong, H.; Lu, Y.; Wang, H.; Wen, F.; Zhang, S.; Zhang, X. Protein discrimination using fluorescent gold nanoparticles on plasmonic substrates. *Anal. Chem.,* **2012**, *84*(10), 4258-4261. [http://dx.doi.org/10.1021/ac300718p] [PMID: 22545857]

[27] Liu, J.; Li, G.; Yang, X.; Wang, K.; Li, L.; Liu, W.; Shi, X.; Guo, Y. Exciton energy transfer-based quantum dot fluorescence sensing array: "chemical noses" for discrimination of different nucleobases. *Anal. Chem.,* **2015**, *87*(2), 876-883. [http://dx.doi.org/10.1021/ac503819e] [PMID: 25495103]

[28] Wang, M.; Ye, H.; You, L.; Chen, X. A Supramolecular Sensor Array Using Lanthanide-Doped Nanoparticles for Sensitive Detection of Glyphosate and Proteins. *ACS Appl. Mater. Interfaces,* **2016**, *8*(1), 574-581. [http://dx.doi.org/10.1021/acsami.5b09607] [PMID: 26651854]

[29] Kahn, N.; Lavie, O.; Paz, M.; Segev, Y.; Haick, H. Dynamic Nanoparticle-Based Flexible Sensors: Diagnosis of Ovarian Carcinoma from Exhaled Breath. *Nano Lett.,* **2015**, *15*(10), 7023-7028. [http://dx.doi.org/10.1021/acs.nanolett.5b03052] [PMID: 26352191]

[30] Song, E.; Choi, J-W. Multi-analyte detection of chemical species using a conducting polymer nanowire-based sensor array platform. *Sens. Actuators B Chem.,* **2015**, *215*, 99-106. [http://dx.doi.org/10.1016/j.snb.2015.03.039]

[31] Gao, J.; Granzhan, A.; Qian, X.; Severin, K. Pattern-based sensing of short oligodeoxynucleotides with palladium-dye complexes. *Chem. Commun. (Camb.),* **2010**, *46*(30), 5515-5517. [http://dx.doi.org/10.1039/c0cc00389a] [PMID: 20442970]

[32] LaGasse, M.K.; Rankin, J.M.; Askim, J.R.; Suslick, K.S. Colorimetric sensor arrays: Interplay of geometry, substrate and immobilization. *Sens. Actuators B Chem.,* **2014**, *197*, 116-122. [http://dx.doi.org/10.1016/j.snb.2014.01.102]

[33] Sheini, A.; Khajehsharifi, H.; Shahbazy, M.; Kompany-Zareh, M. A chemosensor array for the colorimetric identification of some carboxylic acids in human urine samples. *Sens. Actuators B Chem.,* **2017**, *242*, 288-298. [http://dx.doi.org/10.1016/j.snb.2016.11.008]

[34] Köstereli, Z.; Severin, K. Array-based sensing of purine derivatives with fluorescent dyes. *Org. Biomol. Chem.,* **2015**, *13*(35), 9231-9235. [http://dx.doi.org/10.1039/C5OB01201E] [PMID: 26223489]

[35] Rochat, S.; Gao, J.; Qian, X.; Zaubitzer, F.; Severin, K. Cross-reactive sensor arrays for the detection of peptides in aqueous solution by fluorescence spectroscopy. *Chemistry,* **2010**, *16*(1), 104-113.

[http://dx.doi.org/10.1002/chem.200902202] [PMID: 19938007]

[36] Miranda, O.R.; You, C-C.; Phillips, R.; Kim, I-B.; Ghosh, P.S.; Bunz, U.H.; Rotello, V.M. Array-based sensing of proteins using conjugated polymers. *J. Am. Chem. Soc.,* **2007**, *129*(32), 9856-9857. [http://dx.doi.org/10.1021/ja0737927] [PMID: 17658813]

[37] Han, J.; Bender, M.; Hahn, S.; Seehafer, K.; Bunz, U.H. Polyelectrolyte Complexes Formed from Conjugated Polymers: Array-Based Sensing of Organic Acids. *Chemistry,* **2016**, *22*(10), 3230-3233. [http://dx.doi.org/10.1002/chem.201504447] [PMID: 26748536]

[38] Rana, S.; Elci, S.G.; Mout, R.; Singla, A.K.; Yazdani, M.; Bender, M.; Bajaj, A.; Saha, K.; Bunz, U.H.; Jirik, F.R.; Rotello, V.M. Ratiometric Array of Conjugated Polymers-Fluorescent Protein Provides a Robust Mammalian Cell Sensor. *J. Am. Chem. Soc.,* **2016**, *138*(13), 4522-4529. [http://dx.doi.org/10.1021/jacs.6b00067] [PMID: 26967961]

[39] Wu, J.; Tan, C.; Chen, Z.; Chen, Y.Z.; Tan, Y.; Jiang, Y. Fluorescence array-based sensing of nitroaromatics using conjugated polyelectrolytes. *Analyst (Lond.),* **2016**, *141*(11), 3242-3245. [http://dx.doi.org/10.1039/C6AN00678G] [PMID: 27169808]

[40] Zhu, W.; Li, W.; Yang, H.; Jiang, Y.; Wang, C.; Chen, Y.; Li, G. A rapid and efficient way to dynamic creation of cross-reactive sensor arrays based on ionic liquids. *Chemistry,* **2013**, *19*(35), 11603-11612. [http://dx.doi.org/10.1002/chem.201300789] [PMID: 23873515]

[41] Chen, S.; Wei, L.; Chen, X-W.; Wang, J-H. Suspension Array of Ionic Liquid or Ionic Liquid-Quantum Dots Conjugates for the Discrimination of Proteins and Bacteria. *Anal. Chem.,* **2015**, *87*(21), 10902-10909.
[http://dx.doi.org/10.1021/acs.analchem.5b02453] [PMID: 26424154]

[42] Galpothdeniya, W.I.; Regmi, B.P.; McCarter, K.S.; de Rooy, S.L.; Siraj, N.; Warner, I.M. Virtual colorimetric sensor array: single ionic liquid for solvent discrimination. *Anal. Chem.,* **2015**, *87*(8), 4464-4471.
[http://dx.doi.org/10.1021/acs.analchem.5b00714] [PMID: 25822878]

[43] Xu, D.; Zhu, W.; Wang, C.; Tian, T.; Cui, J.; Li, J.; Wang, H.; Li, G. Molecularly imprinted photonic polymers as sensing elements for the creation of cross-reactive sensor arrays. *Chemistry,* **2014**, *20*(50), 16620-16625.
[http://dx.doi.org/10.1002/chem.201404101] [PMID: 25331488]

[44] Luo, Q.; Yu, N.; Shi, C.; Wang, X.; Wu, J. Surface plasmon resonance sensor for antibiotics detection based on photo-initiated polymerization molecularly imprinted array. *Talanta,* **2016**, *161*, 797-803. [http://dx.doi.org/10.1016/j.talanta.2016.09.049] [PMID: 27769483]

[45] Qiu, H.; Luo, C.; Sun, M.; Lu, F.; Fan, L.; Li, X. A chemiluminescence array sensor based on graphene-magnetite-molecularly imprinted polymers for determination of benzenediol isomers. *Anal. Chim. Acta,* **2012**, *744*, 75-81.
[http://dx.doi.org/10.1016/j.aca.2012.07.022] [PMID: 22935377]

[46] Silverio, O.V.; So, R.C.; Elnar, K.J.S.; Malapit, C.A.; Nepomuceno, M.C.M. Development of dieldrin, endosulfan, and hexachlorobenzene-imprinted polymers for dye-displacement array sensing. *Journal of Applied Polymer Science,* **2017**, *134*(2), 44401.
[http://dx.doi.org/10.1002/app.44401]

[47] de la Guardia, M.; Garrigues, S. *Handbook of Green Analytical Chemistry*; Wiley-VCH Verlag GmbH & Co. KGaA, **2012**.

[48] Brett Christopher, M.A. Novel sensor devices and monitoring strategies for green and sustainable chemistry processes. *Pure Appl. Chem.,* **2007**, *79*(11), 1969-1980.

[49] Di Carlo, G.; Curulli, A.; Toro, R.G.; Bianchini, C.; De Caro, T.; Padeletti, G.; Zane, D.; Ingo, G.M. Green synthesis of gold-chitosan nanocomposites for caffeic acid sensing. *Langmuir,* **2012**, *28*(12), 5471-5479.
[http://dx.doi.org/10.1021/la204924d] [PMID: 22385276]

[50] Ganguly, M.; Jana, J.; Mondal, C.; Pal, A.; Pal, T. Green synthesis of highly fluorescent Au(I)@Ag2/Ag3-thiolate core-shell particles for selective detection of cysteine and Pb(II). *Phys. Chem. Chem. Phys.,* **2014**, *16*(34), 18185-18197. [http://dx.doi.org/10.1039/C4CP01782J] [PMID: 25052962]

[51] Rehman, A.; Zeng, X. Ionic liquids as green solvents and electrolytes for robust chemical sensor development. *Acc. Chem. Res.,* **2012**, *45*(10), 1667-1677. [http://dx.doi.org/10.1021/ar200330v] [PMID: 22891895]

[52] Shen, L.Q.; Melnikov, F.; Roethle, J.; Gudibanda, A.; Judson, R.S.; Zimmerman, J.B. Coupled molecular design diagrams to guide safer chemical design with reduced likelihood of perturbing the NRF2-ARE antioxidant pathway and inducing cytotoxicity. *Green Chem.,* **2016**, *18*(23), 6387-6394. [http://dx.doi.org/10.1039/C6GC02073A]

[53] Pei, H.; Li, J.; Lv, M.; Wang, J.; Gao, J.; Lu, J.; Li, Y.; Huang, Q.; Hu, J.; Fan, C. A graphene-based sensor array for high-precision and adaptive target identification with ensemble aptamers. *J. Am. Chem. Soc.,* **2012**, *134*(33), 13843-13849. [http://dx.doi.org/10.1021/ja305814u] [PMID: 22849568]

[54] Wright, A.T.; Anslyn, E.V.; McDevitt, J.T. A differential array of metalated synthetic receptors for the analysis of tripeptide mixtures. *J. Am. Chem. Soc.,* **2005**, *127*(49), 17405-17411. [http://dx.doi.org/10.1021/ja055696g] [PMID: 16332090]

[55] Lin, M.; Li, W.; Wang, Y.; Yang, X.; Wang, K.; Wang, Q.; Wang, P.; Chang, Y.; Tan, Y. Discrimination of hemoglobins with subtle differences using an aptamer based sensing array. *Chem. Commun. (Camb.),* **2015**, *51*(39), 8304-8306. [http://dx.doi.org/10.1039/C5CC00929D] [PMID: 25876962]

[56] Caglayan, M.G.; Sheykhi, S.; Mosca, L.; Anzenbacher, P., Jr Fluorescent zinc and copper complexes for detection of adrafinil in paper-based microfluidic devices. *Chem. Commun. (Camb.),* **2016**, *52*(53), 8279-8282. [http://dx.doi.org/10.1039/C6CC03640F] [PMID: 27293080]

[57] Minami, T.; Esipenko, N.A.; Akdeniz, A.; Zhang, B.; Isaacs, L.; Anzenbacher, P., Jr Multianalyte sensing of addictive over-the-counter (OTC) drugs. *J. Am. Chem. Soc.,* **2013**, *135*(40), 15238-15243. [http://dx.doi.org/10.1021/ja407722a] [PMID: 24000805]

[58] Folmer-Andersen, J.F.; Kitamura, M.; Anslyn, E.V. Pattern-based discrimination of enantiomeric and structurally similar amino acids: an optical mimic of the mammalian taste response. *J. Am. Chem. Soc.,* **2006**, *128*(17), 5652-5653. [http://dx.doi.org/10.1021/ja061313i] [PMID: 16637629]

[59] McCleskey, S.C.; Griffin, M.J.; Schneider, S.E.; McDevitt, J.T.; Anslyn, E.V. Differential receptors create patterns diagnostic for ATP and GTP. *J. Am. Chem. Soc.,* **2003**, *125*(5), 1114-1115. [http://dx.doi.org/10.1021/ja021230b] [PMID: 12553782]

[60] Miranda, O.R.; Chen, H-T.; You, C-C.; Mortenson, D.E.; Yang, X-C.; Bunz, U.H.; Rotello, V.M. Enzyme-amplified array sensing of proteins in solution and in biofluids. *J. Am. Chem. Soc.,* **2010**, *132*(14), 5285-5289. [http://dx.doi.org/10.1021/ja1006756] [PMID: 20329726]

[61] Baumes, L.A.; Buaki Sogo, M.; Montes-Navajas, P.; Corma, A.; Garcia, H. A colorimetric sensor array for the detection of the date-rape drug γ-hydroxybutyric acid (GHB): a supramolecular approach. *Chemistry,* **2010**, *16*(15), 4489-4495. [http://dx.doi.org/10.1002/chem.200903127] [PMID: 20309968]

[62] Akdeniz, A.; Minami, T.; Watanabe, S.; Yokoyama, M.; Ema, T.; Anzenbacher, P. Determination of enantiomeric excess of carboxylates by fluorescent macrocyclic sensors. *Chem. Sci. (Camb.),* **2016**, *7*(3), 2016-2022. [http://dx.doi.org/10.1039/C5SC04235F]

[63] Liu, Y.; Minami, T.; Nishiyabu, R.; Wang, Z.; Anzenbacher, P., Jr Sensing of carboxylate drugs in urine by a supramolecular sensor array. *J. Am. Chem. Soc.,* **2013**, *135*(20), 7705-7712. [http://dx.doi.org/10.1021/ja4015748] [PMID: 23656505]

[64] Duarte, A.; Chworos, A.; Flagan, S.F.; Hanrahan, G.; Bazan, G.C. Identification of bacteria by conjugated oligoelectrolyte/single-stranded DNA electrostatic complexes. *J. Am. Chem. Soc.,* **2010**, *132*(36), 12562-12564. [http://dx.doi.org/10.1021/ja105747b] [PMID: 20731391]

[65] Buryak, A.; Zaubitzer, F.; Pozdnoukhov, A.; Severin, K. Indicator displacement assays as molecular timers. *J. Am. Chem. Soc.,* **2008**, *130*(34), 11260-11261. [http://dx.doi.org/10.1021/ja8037118] [PMID: 18680249]

[66] Ciosek, P.; Wróblewski, W. Sensor arrays for liquid sensing--electronic tongue systems. *Analyst (Lond.),* **2007**, *132*(10), 963-978. [http://dx.doi.org/10.1039/b705107g] [PMID: 17893798]

[67] Hibbert, D.B. 4.12 - Chemometric Analysis of Sensory Data A2 - Brown, Steven D.*Comprehensive Chemometrics*; Tauler, R.; Walczak, B., Eds.; Elsevier: Oxford, **2009**, pp. 377-424. [http://dx.doi.org/10.1016/B978-044452701-1.00010-7]

[68] Otto, M. *Chemometrics: Statistics and Computer Application in Analytical Chemistry*; Wiley-VCH, **1998**, p. 376.

[69] Brereton, R.G. *Applied Chemometrics for Scientists*; , **2007**, pp. 1-379. [http://dx.doi.org/10.1002/9780470057780]

[70] Zeaiter, M.; Rutledge, D. 3.04 - Preprocessing Methods A2 - Brown, Steven D.*Comprehensive Chemometrics*; Tauler, R.; Walczak, B., Eds.; Elsevier: Oxford, **2009**, pp. 121-231. [http://dx.doi.org/10.1016/B978-044452701-1.00074-0]

[71] Marini, F. 3.14 - Neural Networks A2 - Brown, Steven D.*Comprehensive Chemometrics*; Tauler, R.; Walczak, B., Eds.; Elsevier: Oxford, **2009**, pp. 477-505. [http://dx.doi.org/10.1016/B978-044452701-1.00128-9]

[72] Ahuja, S. Modern pharmaceutical analysis: an overview. *Sep Sci Technol (San Diego, CA, U S),* **2001**, *3*, 1-22.

[73] US-FDA. *Development of New Stereoisomeric Drugs,* http://www.fda.gov/Drugs/GuidanceComplianceRegulatoryInformation/Guidances/ucm122883.htm **2014**.

[74] Hutt, A.J. The development of single-isomer molecules: why and how. *CNS Spectr.,* **2002**, *7*(4) Suppl. 1, 14-22. [http://dx.doi.org/10.1017/S1092852900028558] [PMID: 15131488]

[75] McConathy, J.; Owens, M.J. Stereochemistry in Drug Action. *Prim. Care Companion J. Clin. Psychiatry,* **2003**, *5*(2), 70-73. [http://dx.doi.org/10.4088/PCC.v05n0202] [PMID: 15156233]

[76] Schmitt, U.; Branch, S.K.; Holzgrabe, U. Chiral separations by cyclodextrin-modified capillary electrophoresis - determination of the enantiomeric excess. *J. Sep. Sci.,* **2002**, *25*, 959-974. [http://dx.doi.org/10.1002/1615-9314(20021101)25:15/17<959::AID-JSSC959>3.0.CO;2-Q]

[77] Ward, T.J.; Ward, K.D. Chiral separations: a review of current topics and trends. *Anal. Chem.,* **2012**, *84*(2), 626-635. [http://dx.doi.org/10.1021/ac202892w] [PMID: 22066781]

[78] Ali, I.; Suhail, M.; Al-Othman, Z.A.; Alwarthan, A.; Aboul-Enein, H.Y. Enantiomeric resolution of multiple chiral centres racemates by capillary electrophoresis. *Biomed. Chromatogr.,* **2016**, *30*(5), 683-694. [http://dx.doi.org/10.1002/bmc.3691] [PMID: 26840015]

[79] Arai, T. Chiral separation of pharmaceuticals possessing a carboxy moiety. *J. Chromatogr. B Biomed.*

Sci. Appl., **1998**, *717*(1-2), 295-311.
[http://dx.doi.org/10.1016/S0378-4347(98)00246-1] [PMID: 9832251]

[80] Aturki, Z.; Rocco, A.; Rocchi, S.; Fanali, S. Current applications of miniaturized chromatographic and electrophoretic techniques in drug analysis. *J. Pharm. Biomed. Anal.,* **2014**, *101*, 194-220.
[http://dx.doi.org/10.1016/j.jpba.2014.03.041] [PMID: 24784002]

[81] Aturki, Z.; Schmid, M.G.; Chankvetadze, B.; Fanali, S. Enantiomeric separation of new cathinone derivatives designer drugs by capillary electrochromatography using a chiral stationary phase, based on amylose tris(5-chloro-2-methylphenylcarbamate). *Electrophoresis,* **2014**, *35*(21-22), 3242-3249.
[http://dx.doi.org/10.1002/elps.201400085] [PMID: 24854346]

[82] Cavazzini, A.; Marchetti, N.; Guzzinati, R.; Pierini, M.; Ciogli, A.; Kotoni, D. Enantioseparation by ultra-high-performance liquid chromatography. *TrAC Trends in Analytical Chemistry.,* **2014**, *63*, 95-103.
[http://dx.doi.org/10.1016/j.trac.2014.06.026]

[83] Jiang, J.; Tian, L.; Huang, Y.; Yan, Y.; Li, Y. Enantioselective and sensitive determination of carvedilol in human plasma using chiral stationary-phase column and reverse-phase liquid chromatography with tandem mass spectrometry. *J. Chromatogr. B Analyt. Technol. Biomed. Life Sci.,* **2014**, *960*, 92-97.
[http://dx.doi.org/10.1016/j.jchromb.2014.04.016] [PMID: 24792533]

[84] Prokhorova, A.F.; Shapovalova, E.N.; Shpigun, O.A. Chiral analysis of pharmaceuticals by capillary electrophoresis using antibiotics as chiral selectors. *J. Pharm. Biomed. Anal.,* **2010**, *53*(5), 1170-1179.
[http://dx.doi.org/10.1016/j.jpba.2010.07.005] [PMID: 20675089]

[85] Van Eeckhaut, A.; Michotte, Y. Chiral separations by capillary electrophoresis: Recent developments and applications. *Electrophoresis,* **2006**, *27*(14), 2880-2895.
[http://dx.doi.org/10.1002/elps.200500375] [PMID: 16688697]

[86] Wolrab, D.; Frühauf, P.; Moulisová, A.; Kuchař, M.; Gerner, C.; Lindner, W.; Kohout, M. Chiral separation of new designer drugs (Cathinones) on chiral ion-exchange type stationary phases. *J. Pharm. Biomed. Anal.,* **2016**, *120*, 306-315.
[http://dx.doi.org/10.1016/j.jpba.2015.12.023] [PMID: 26765268]

[87] Zhang, C.; Liu, H.; Liu, D.; Wang, L.; Gao, J.; Zhou, Z.; Wang, P. Enantiomeric Separations of Pyriproxyfen and its Six Chiral Metabolites by High-Performance Liquid Chromatography. *Chirality,* **2016**, *28*(3), 245-252.
[http://dx.doi.org/10.1002/chir.22568] [PMID: 26773961]

[88] Wenzel, T.J. *Discrimination of Chiral Compounds Using NMR Spectroscopy*; Wiley-VCH Verlag GmbH & Co. KGaA, **2007**, p. 576.

[89] Wolf, C.; Bentley, K.W. Chirality sensing using stereodynamic probes with distinct electronic circular dichroism output. *Chem. Soc. Rev.,* **2013**, *42*(12), 5408-5424.
[http://dx.doi.org/10.1039/c3cs35498a] [PMID: 23482984]

[90] de Andrés, F.; Castañeda, G.; Ríos, A. Achiral liquid chromatography with circular dichroism detection for the determination of carnitine enantiomers in dietary supplements and pharmaceutical formulations. *J. Pharm. Biomed. Anal.,* **2010**, *51*(2), 478-483.
[http://dx.doi.org/10.1016/j.jpba.2009.02.018] [PMID: 19303234]

[91] Dorottya, B.; Sandor, G. Recent Advances in the Impurity Profiling of Drugs. *Curr. Pharm. Anal.,* **2008**, *4*(4), 215-230.
[http://dx.doi.org/10.2174/157341208786306199]

[92] Lecoeur-Lorin, M.; Delépée, R.; Ribet, J-P.; Morin, P. Chiral analysis of milnacipran by a nonchiral HPLC--circular dichroism: improvement of the linearity of dichroic response by temperature control. *J. Sep. Sci.,* **2008**, *31*(16-17), 3009-3014.
[http://dx.doi.org/10.1002/jssc.200800291] [PMID: 18785147]

[93] Lorin, M.; Delepee, R.; Maurizot, J-C.; Ribet, J-P.; Morin, P. Sensitivity improvement of circular dichroism detection in HPLC by using a low-pass electronic noise filter: Application to the enantiomeric determination purity of a basic drug. *Chirality,* **2007**, *19*(2), 106-113. [http://dx.doi.org/10.1002/chir.20352] [PMID: 17096379]

[94] Lough, W.J. *Alternative analytical techniques for determination or isolation of drug enantiomers. Drug Stereochemistry. Drugs and the Pharmaceutical Sciences*; CRC Press, **2012**, pp. 167-170.

[95] Nageswara Rao, R.; Santhakumar, K. Cyclodextrin assisted enantiomeric recognition of emtricitabine by 19F NMR spectroscopy. *New J. Chem.,* **2016**, *40*(10), 8408-8417. [http://dx.doi.org/10.1039/C6NJ00356G]

[96] Redondo, J.; Capdevila, A.; Ciudad, S. Determination of the enantiomeric purity of the antiasthmatic drug montelukast by means of 1H NMR spectroscopy. *Chirality,* **2013**, *25*(11), 780-786. [http://dx.doi.org/10.1002/chir.22213] [PMID: 23893787]

[97] Salsbury, J.S.; Isbester, P.K. Quantitative 1H NMR method for the routine spectroscopic determination of enantiomeric purity of active pharmaceutical ingredients fenfluramine, sertraline, and paroxetine. *Magn. Reson. Chem.,* **2005**, *43*(11), 910-917. [http://dx.doi.org/10.1002/mrc.1642] [PMID: 16052604]

[98] Samakashvili, S.; Salgado, A.; Scriba, G.K.; Chankvetadze, B. Comparative enantioseparation of ketoprofen with trimethylated α-, β-, and γ-cyclodextrins in capillary electrophoresis and study of related selector-selectand interactions using nuclear magnetic resonance spectroscopy. *Chirality,* **2013**, *25*(2), 79-88. [http://dx.doi.org/10.1002/chir.22111] [PMID: 23255364]

[99] Leung, D.; Anslyn, E.V. Transitioning enantioselective indicator displacement assays for α-amino acids to protocols amenable to high-throughput screening. *J. Am. Chem. Soc.,* **2008**, *130*(37), 12328-12333. [http://dx.doi.org/10.1021/ja8038079] [PMID: 18714993]

[100] Mei, X.; Wolf, C. Determination of enantiomeric excess and concentration of unprotected amino acids, amines, amino alcohols, and carboxylic acids by competitive binding assays with a chiral scandium complex. *J. Am. Chem. Soc.,* **2006**, *128*(41), 13326-13327. [http://dx.doi.org/10.1021/ja0636486] [PMID: 17031923]

[101] Shabbir, S.H.; Regan, C.J.; Anslyn, E.V. Molecular recognition and self-assembly special feature: A general protocol for creating high-throughput screening assays for reaction yield and enantiomeric excess applied to hydrobenzoin. *Proc. Natl. Acad. Sci. USA,* **2009**, *106*(26), 10487-10492. [http://dx.doi.org/10.1073/pnas.0809530106] [PMID: 19332790]

[102] Shcherbakova, E.G.; Brega, V.; Minami, T.; Sheykhi, S.; James, T.D.; Anzenbacher, P., Jr Toward Fluorescence-Based High-Throughput Screening for Enantiomeric Excess in Amines and Amino Acid Derivatives. *Chemistry,* **2016**, *22*(29), 10074-10080. [http://dx.doi.org/10.1002/chem.201601614] [PMID: 27271215]

[103] Pu, L. Fluorescence of organic molecules in chiral recognition. *Chem. Rev.,* **2004**, *104*(3), 1687-1716. [http://dx.doi.org/10.1021/cr030052h] [PMID: 15008630]

[104] Jaroch, S.; Weinmann, H.; Zeitler, K. Asymmetric organocatalysis. *ChemMedChem,* **2007**, *2*(9), 1261-1264. [http://dx.doi.org/10.1002/cmdc.200700109] [PMID: 17607798]

[105] Christmann, M.; Brase, S., Eds. *Asymmetric Synthesis - The Essentials*; 355.Wiley-VCH Verlag GmbH & Co. KGaA, **2008**, p.

[106] Welch, C.J.; Grau, B.; Moore, J.; Mathre, D.J. Use of chiral HPLC-MS for rapid evaluation of the yeast-mediated enantioselective bioreduction of a diaryl ketone. *J. Org. Chem.,* **2001**, *66*(20), 6836-6837. [http://dx.doi.org/10.1021/jo015966n] [PMID: 11578248]

[107] Welch, C.J.; Szczerba, T.; Perrin, S.R. Some recent high-performance liquid chromatography separations of the enantiomers of pharmaceuticals and other compounds using the Whelk-O 1 chiral stationary phase. *J. Chromatogr. A,* **1997**, *758*, 93-98.
[http://dx.doi.org/10.1016/S0021-9673(96)00569-9]

[108] Jo, H.H.; Lin, C-Y.; Anslyn, E.V. Rapid optical methods for enantiomeric excess analysis: from enantioselective indicator displacement assays to exciton-coupled circular dichroism. *Acc. Chem. Res.,* **2014**, *47*(7), 2212-2221.
[http://dx.doi.org/10.1021/ar500147x] [PMID: 24892802]

[109] Kwok, K.; Taylor, L.S., Eds. *Raman spectroscopy for the analysis of counterfeit tablets*; John Wiley & Sons Ltd., **2012**.
[http://dx.doi.org/10.1002/9781119962328.ch9a]

[110] Hardy, J.; Selkoe, D.J. The amyloid hypothesis of Alzheimer's disease: progress and problems on the road to therapeutics. *Science,* **2002**, *297*(5580), 353-356.
[http://dx.doi.org/10.1126/science.1072994] [PMID: 12130773]

[111] Yang, X.; Li, J.; Pei, H.; Li, D.; Zhao, Y.; Gao, J.; Lu, J.; Shi, J.; Fan, C.; Huang, Q. Pattern recognition analysis of proteins using DNA-decorated catalytic gold nanoparticles. *Small,* **2013**, *9*(17), 2844-2849.
[http://dx.doi.org/10.1002/smll.201202772] [PMID: 23447503]

[112] Zhou, H.; Baldini, L.; Hong, J.; Wilson, A.J.; Hamilton, A.D. Pattern recognition of proteins based on an array of functionalized porphyrins. *J. Am. Chem. Soc.,* **2006**, *128*(7), 2421-2425.
[http://dx.doi.org/10.1021/ja056833c] [PMID: 16478197]

[113] Berrade, L.; Garcia, A.E.; Camarero, J.A. Protein microarrays: novel developments and applications. *Pharm. Res.,* **2011**, *28*(7), 1480-1499.
[http://dx.doi.org/10.1007/s11095-010-0325-1] [PMID: 21116694]

[114] Gómez-Hens, A.; Aguilar-Caballos, M.P. Modern analytical approaches to high-throughput drug discovery. *TrAC Trends in Analytical Chemistry.,* **2007**, *26*(3), 171-182.
[http://dx.doi.org/10.1016/j.trac.2006.12.001]

[115] Morbioli, G.G.; Mazzu-Nascimento, T.; Aquino, A.; Cervantes, C.; Carrilho, E. Recombinant drugs-on-a-chip: The usage of capillary electrophoresis and trends in miniaturized systems - A review. *Anal. Chim. Acta,* **2016**, *935*, 44-57.
[http://dx.doi.org/10.1016/j.aca.2016.06.019] [PMID: 27543014]

[116] Sierra, I.; Marina, M.L.; Pérez-Quintanilla, D.; Morante-Zarcero, S.; Silva, M. Approaches for enantioselective resolution of pharmaceuticals by miniaturised separation techniques with new chiral phases based on nanoparticles and monolithis. *Electrophoresis,* **2016**, *37*(19), 2538-2553.
[http://dx.doi.org/10.1002/elps.201600131] [PMID: 27434636]

[117] Anzenbacher, P., Jr; Liu, Y.L.; Kozelkova, M.E. Hydrophilic polymer matrices in optical array sensing. *Curr. Opin. Chem. Biol.,* **2010**, *14*(6), 693-704.
[http://dx.doi.org/10.1016/j.cbpa.2010.08.011] [PMID: 20817542]

CHAPTER 10

Applications of Ionic Liquids in Chemical Science

Imran Ali[1,*], **Zeid A. Alothman**[2], **Abdulrahman Alwarthan**[2] and **Hassan Y. Aboul-Enein**[3,*]

[1] *Department of Chemistry, Jamia Millia Islamia (Central University), New Delhi – 110025, India*

[2] *Department of Chemistry, College of Science, King Saud University, Riyadh 11451, Kingdom of Saudi Arabia*

[3] *Pharmaceutical and Medicinal Chemistry Department, Pharmaceutical and Drug Industries Research Division, National Research Centre, Dokki, Giza12622, Egypt*

Abstract: The ionic liquids are very important solvents in chemical science. These are termed as melts, designer solvents, electrolytes, ionic glasses, fused salts, ionic fluid, and green solvents. The ionic liquids are being used in the different research and industrial areas of chemical science. The most important applications of the ionic liquids in the chemical science include extraction, separation, electro-analysis and spectroscopic studies. This chapter describes the utility of the ionic liquids in the chemical science. Besides, the efforts were made to highlight the toxicity and future of the ionic liquids.

Keywords: Electrochemistry, Extraction, Future perspectives, Ionic liquids, Separation, Spectroscopy, Toxicity.

INTRODUCTION

There are many organic and inorganic solvents being used for a variety of purposes. During the development of science and technology, especially the chemical science, the researchers realise the toxicity and limitations of these solvents. The most serious problems are handling and the toxicity of these classical solvents. Besides, these cannot be moduled as per the requirements. Additionally, the recycling of these solvent is another issue, especially in the commercial industries. These limitations compelled researchers to explore new alternatives. In this need, ionic liquids were seen to fulfil the demands of the researchers.

* **Corresponding authors Hassan Y. Aboul-Enein:** Pharmaceutical and Medicinal Chemistry Department, Pharmaceutical and Drug Industries Research Division, National Research Centre, Dokki, Giza 12622, Egypt; Tel: +201003678948; Fax: +20233370931; E-mail: haboulenein@yahoo.com;
Imran Ali: Department of Chemistry, Jamia Millia Islamia (Central University), New Delhi – 110025, India; Tel: +9211458226; E-mails: drimran_ali@yahoo.com, drimran.chiral@gmail.com

The ionic liquids (ILs) are the organic compound with ions weakly coordinated, which are present as liquids at a small temperature (<100 °C). These have infinitely small haze pressure. These are prepared by ions and small life ions. These are termed as melts, designer solvents, electrolytes, ionic glasses, fused salts, ionic fluid and green solvents. Also termed as solvents for hope, trendy solvents and eco-friendly solvents, are a concern for researchers. The ionic liquids demonstrate compound and thermal stability, non-flammability, elevated ionic conductivity and a broad electro-chemical latent pane. These sorts of solvents are evaluated as solvents or co-catalysts in different reactions together with inorganic production, polymerization [1, 2], organic catalysis, biocatalysis [3 - 16]. Ionic liquids are used as electrolytes, sealants, electric battery fluids. The methylimidazolium and pyridinium ions are a starter for the growth of ionic liquids. Most significant ionic liquids are molecules involving 1-butyl-3-5-dimethylpyridinium bromide, 1-ethyl-3-methylimidazolium cation and 1-ethy--3-methylimidazolium dicyanamide.

In each chemical procedure, scientists alert their attention on the growth of eco-friendly or environment benevolent procedures and analyses. It is owing to the increasing numeral of environment toxins due to increase in the world population and industrial growth. The ionic liquids participate in such sort of procedures and contribute to a research boom. These days, the explosive organic molecules are considered to be conventional engineering solvents. The scientists are in the interest of replacement of the usual solvents with non-volatile ionic liquids to stop the release of explosive organic molecules, a main basis of environment contamination. The ionic liquids were first practical as the "red oil" created through Friedel–Crafts reactions catalyzed by $AlCl_3$ [17]. First viable procedure was developed by BASF AG (Ludwigshafen, Germany) to search acids using three liquids [18]. Numerous additional procedures produced by the similar industry are in the pilot stage [19]. The ionic liquids were examined in engineering procedures. The ionic liquids may be hydrophobic and hydrophilic as per the structure of cations and anions, respectively [3, 20]. The anions seem more significant in decisive water miscibility of these liquids [20].

Significance of the Ionic Liquids

The ionic liquids have huge preponderance of appliance. A number of significant factors are summed up in this article. Approximately, fifty percent of viable pharmaceuticals are ionic liquids since the amalgamation of a pharmaceutically active cation and pharmaceutically dynamic anion directs the creation of energetic ionic liquid [21,22]. A number of ionic liquids like 1-alkylpyridinium chlorides are utilized in the dissolution of cellulose [23, 24]. The valorization of cellulose - conversion to extra precious chemicals, is attained using ionic liquids. The envoy products are sorbitol, glucose esters and alkylgycosides [25]. The ionic liquids are significant for the extraction of chemicals like ethanol, biodiesel and bio-fuels from algae [26]. The ionic liquids also play a significant role in the partition of gases that is 1-butyl-3-methylimidazolium chloride is applied in the partition of hydrogen from ammonia borane [27]. As well, the ionic liquids and amines are evaluated for incarcerate carbon dioxide CO_2 and cleanse usual gas [28]. These liquids are also utilized for the revival of nuclear energy that is 1-butyl-3-methylimidazolium chloride. The non-aqueous electrolyte medium is utilized for the revival of uranium and other metal ions from spent nuclear fuel and other sources [29 - 31]. Similarly, protonated betaine bis-(trifluoromethanesulfonyl) imide is evaluated like a solvent for uranium oxides [32]. In addition, *N*-methyl-*N*-propylpiperidinium bis-(trifluoromethylsulfonyl)imide, and *N*-butyl-*N*-methylpyrrolidinium bis-(trifluoromethylsulfonyl)imide are evaluated for the electro-deposition of Eu and U metals, respectively [33, 34]. In addition, ionic liquids are projected as sorbents to capture carbon dioxide. There are a variety of recompenses over conventional sorbents like the presently leading amine based techniques. Super base derived protic ionic liquids are utilized for the sorption of carbon dioxide as well [35].

Applications of the Ionic Liquids

As cited in the above paragraphs, the ionic liquids are extremely significant class of the molecule. These days, the ionic liquids are accepted owing to their special features. Different applications of ionic liquids are discussed in the following sub-sections.

Extraction

Ionic liquids are important molecules being used in the extraction processes of the different chemical analyses. The non-volatilities make these molecules ideal solvents, and, hence are ideal in liquid-liquid extraction. Besides, their different partition coefficients make them preferable solvents in various chemical analyses. Moreover, these liquids can be prepared from inexpensive precursors. Also, their properties can be moduled by adjusting the concentration of cation and anions. That is why, these liquids can be prepared as per the requirements of liquid-liquid extraction. The viscosity, capacity, thermal stability and selectivity of these liquid ions can be changed as per the requirement, which make them ideal solvents. Briefly, these solvents are of variable properties and can be used under varied conditions.

It was observed that the significant variables to ascertain the efficiency of liquid-liquid separation procedure using ionic liquids when distribution coefficients. [BMIm][PF_6] (mainly admired) was selected as a solvent by Carda-Broch *et al.*

[36]. The authors used ionic liquid:water (Pil:water) and ionic liquid:heptane (Pil:heptane) systems to determine partition coefficients for forty compounds with acid and base groups. Besides, other neutral molecules and antioxidants were also studied. The technique used was liquid chromatography. As per the authors, the Pil:water values measured at pH 2, 5.1 and 10 were difficult. The partition coefficients were compared to those obtained with the corresponding Poct:water coefficients. The noticeable differences in the partitioning performance of basic, acidic and neutral molecules were seen. The association between Pil/water and Poct/water systems was dissimilar to that reported beforehand. [BMIm][PF_6] solvent variables were designed for the ionic liquid and water and ionic liquid/heptane systems by linear free energy solvation methods. Low basicity of the [BMIm][PF_6] solvent in comparison to octanol was seen. Furthermore, high consistency of ions in the ionic liquids phase was also shown by the regression analysis achieved. It has also been reported that phenoxide ions are extra powerfully associated with [BMIm][PF_6] in comparison to mainly other ionized compounds.

Some studies were carried out for determining the partition coefficients of ILs with water [37]. The results showed closed association with partition coefficients of the solutes in water-1-octanol system. These observations dictated the applicability of ionic liquids in liquid-liquid extraction. These liquids have been found to break azeotropic systems and, hence, show an extraordinary presentation enabling extractive distillation phenomenon. It requires less energy in comparison to the straight procedure using 1,2-ethanediol [38].

Jork *et al.* [39] described in his experiments that ionic liquids showed competitiveness with conventional solvents for extraction purpose. The authors used ethanol, water, THF and water mixture containing different ionic liquids for establishing their findings. Visser *et al.* [40] carried out liquid-liquid extraction from water containing polycyclic N-alkylisoquinolinium cations ([Cnisoq]$^+$) and *bis*-(perfluoroethylsulfonyl)imide anion ([BETI]$^-$. The distribution ratios of chloro-benzenes were acceptable. Interesting results were presented by Nobuoka *et al.* [41] on chiral separation of DL-tyrosine. The author used isoquinoliniums and *bis*-(trifluoromethanesulfonyl)imide anion for the purpose. Cull *et al.* [42] reported ionic liquid extraction of erythromycin-A and 1,3-dicyanobenzene. The authors used BMIm][PF_6] ionic liquid. The authors compared this extraction using butyl acetate and the results were comparable. Extractions were carried out in the pH range of 5-9. The results slightly vary at higher pH. Matsumoto *et al.* [43] described similar extraction efficiencies of lactic acid using imidazolium based ionic liquids having tri-n-butyl phosphate extractant and conventional solvent. The authors claimed that their results may be used *in situ* extractive fermentation.

Normally, amino acids are insoluble in ionic liquid. For extraction of amino acids, crown ethers are added in to the ionic liquids. Carda-Broch *et al.* [44] used crown ether and ionic liquids for the extraction of amino acids. Amino acid cations form complexes with crown ether and get extracted. Smirnova *et al.* [45] used dicyclohexano-18-crown-6 ($DC_{18}C_6$) crown ether and [BMIm][PF_6] for the extraction of alanine, glycine and tryptophan amino acids. The authors studied the effect of crown ether and amino acid amount, pH, amount of water and organic solvents and in the company of some cations on amino acids. Amino acid to crown ether ratio was 1:1 alanine, glycine, leucine and tryptophan while this ratio was 1:2 for arginine and lysine amino acids. According to the investigators, the described method was used to recuperate amino acids from the samples of the pharmaceuticals and fermentation soup. Chun *et al.* [46] used 1-alkyl-3-metyl-imidazolium hexa-fluoro-phosphates in combination of crown ether. The authors changed alkyl group by butyl to nonyl for better removal. This was useful for the extraction of chlorides of alkali metals with dicyclohexano-18-crown-6 ($DC_{18}C_6$) solutions. It was observed that the extraction effectiveness, usually, vanished on chain length of 1-alkyl group enlarged. In the same circumstances, removal of chlorides of alkali metal from $DC_{18}C_6$ solutions with nitrobenzene, chloroform and 1-octanol was untraceable. The extraction efficiency order of $DC_{18}C_6$ was K(I) > Rb(I) > Cs(I) > Na(I) > Li(I). K(I)/Rb(I) and K(I)/Cs(I) selectivity showed common increase with the major improvement; as the alkyl group was stretched out. Selectivity and efficiency of Alkali metal cation removal were unaltered by disparity of the water phase anion from chloride to sulphate or nitrate for $DC_{18}C_6$ in [C_8MIm][PF_6]. Luo *et al.* [47] used a series of N-alkyl aza-18-crown-6 ethers in ionic liquids for recyclable extractants for strontium and caesium. It was observed that the pH dependent complexation ability of these ethers permitted for a simplistic strip procedure was developed so that together macrocyclic-ligands and ionic liquids could be used.

The fixing of macrocyclic-ligands and ionic liquids help in removal, was extremely selective for strontium. Dietz and Dzielawa *et al.* [48] described two methods of strontium transport in acidic nitrate media to a chain of 1-alkyl-3-methylimidazolium based ionic liquids, having dicyclohexano-18-crown-6 ($DCH_{18}C_6$), that could be moved from cation exchange to the nitrate of strontium to complex of crown ether leading to partition [49]. Besides, crown ethers calix [4]arene have been used as additives for the extraction of various molecules using ionic liquids. These ligands have pyridine moiety with molecular structure similar to the unit of 3-methylimidazolium hexa-fluoro-phosphate. It is not soluble in ionic liquids and indicated 230 fold extractability for silver from water solution in comparison to $CHCl_3$ [50, 51]. These authors described the extraction of silver, copper, cobalt, nickel and zinc metal ions. It was observed that silver was transferred by calix [4]arene in aqueous solution in the ionic liquids *via* a cation exchange mechanism. Furthermore, the authors used slope-analysis and Job's methods for determining formation ration, which was recorded as 1:1 ratio for silver. Removal of silver from calix [4]arene in ionic liquids was controlled by pH. Similarly, Luo *et al.* [52] reported the extraction of cesium cation from water solution to ionic liquids (hydrophobic) without the preface of an anion (organophilic) in water phase. It was confirmed by using calix [4]arenebis(tert-octylbenzo-crown-6) (BOBCalixC_6) as an extractant.

Liu *et al.* [53] used 1-octyl-3-methylimidazolium hexafluorophosphate [C_8MIm][PF_6] for the removal by liquid-phase micro-extraction (LPME) technique. The compounds extracted were polycyclic aromatic hydrocarbons (PAHs). The authors selected these compounds because of non-volatility and adequate viscosity of the used ionic liquids. The reported compounds get extracted just by using 3.0 μL drop of [C_8MIm][PF_6]. Much high enhancement was obtained for polycyclic aromatic hydrocarbons in comparison to 1-octanol as a solvent. As per the authors, because of low volatility of polycyclic aromatic hydrocarbons direct immersion provided higher enhancement factors as the technique was used for liquid chromatography. It was observed that for 30 minutes immersion, the enhancement factor, correlation coefficient, and reproducibility were at 42 to 166, 0.9169 to 0.9976 and 2.8 to 12.0% respectively. The authors claimed this method as of high potentiality in sample preparation. Liu *et al.* [54] used 1-hexyl-3-methyl-imidazolium hexa-fluoro-phosphate [C_6MIm][PF_6] as removal solvent for 4-*tert*-octylphenol and 4-nonylphenol. As per the authors, the method was molecule reliant owing to the dissimilar partition coefficients and the comparatively great viscosity of ionic liquid; in spite of much big size of drops on the needle of micro-syringe in comparison to the usual solvents 1-octanol and carbon tetrachloride. Detection limit and enhancement factor were 0.3 μg/L and 163 for 4-nonylphenol respectively. These values were 0.7 μg/L and 130 for 4-tertoctylphenol, respectively. This process was useful for the water of river, tap water and sewage treatment plant effluent to monitor 4-ter--octylphenol and 4-nonylphenol, respectively. The recoveries reported by the authors were in the range of 90% to 113%; after spiking 10.0 μg/L sample.

Visser *et al.* [55] used 1-alkyl-3-methylimidazolium hexa-fluoro-phosphate [CnMIm][PF_6] in liquid-liquid removal of metal ions from in water. The authors described the partition of metal ions in the systems as constant with conventional liquid-liquid extraction. Metal ion likeness for hydrophobic medium required the attendance of an extractant. The authors developed 1-(2-pyridylazo)-2-naphthol and 1-(2-thiazolylazo)-2-naphthol and inorganic modifiers such as CN^-, OCN^-, SCN^- and halides for partition different metal ions among [C_4Mim][PF_6] and [C_6Mim][PF_6] and an water. 1-(2-Pyridylazo)-2-naphthol and 1-(2-thiazolylaz-)-2-naphthol indicated pH dependent relative extraction of Fe(III), Cd(II), Co(II) and Ni(II). The partitioning was two times high in pH range of 1 to 13. It was also observed that the affect of halides on partitioning of Hg^{2+} complexes increased in the order of $F^- < Cl^- < Br^- < I^-$. However, SCN^- had maximum effect on increasing the partitioning of Hg(II). pH and temperature of the experimental water phase were significant to control the effectiveness of the extraction of metal ions *i.e.* Cu(II), Zn(II), Cr(VI). Wei *et al.* [56] used dithizone for metal chelation to obtain impartial metal dithizone complexes with metal ions for extracting the different metal cations in water using [BMIm][PF_6] ionic liquid. It was observed that the distribution of metal dithiozonates between [BMIm][PF_6] and water was pH dependent. Extraction competence of metal complexes could be exploited by adjusting the pH value in the system. So, removal, separation and pre-concentration of metal cations with [BMIm][PF_6] and water could be obtained by adjusting the pH of the extraction system. Initially, the results shows that the use of [BMIm][PF_6] as an exchange solvent to restore usual organic solvents in liquid-liquid removal of metal ions was talented.

Visser and Rogers [57] produced hydrophobic ionic liquid by altering alkyl chain lengths in the 1-alkyl-3-methylimidazolium cation, in blend with PF6 or $N(SO_2CF3)_2$ anions. It was appropriate for use in liquid-liquid extractions of significant materials such as actinides. Actinides were extracted by using this method by adding additional modifiers. The used modifiers were octyl(phenyl)-N,N-di-isobutyl-carbamoyl-methyl phosphine oxide. The authors claimed this method useful for actinide chemistry and considered a fresh group of materials with adaptable solvent character. Whitehead *et al.* [58] reported the extraction of gold and silver metal ions from ore. The authors used 1-butyl-3-methylimidazolium sulfate of hydrogen ionic liquid [BMIm][HSO_4] with $FeSO_4$ as an oxidant. The percentage recoveries of gold and silver were 85 and 60, at room temperature in 50 hours, respectively. Gold extraction efficiency was almost equal to those obtained with H_2SO_4/thiourea/$Fe_2(SO_4)_3$ system. However, the percentage recoveries of copper, lead, iron and zinc heavy metal cations were low. The authors recycled ionic liquids for discriminating strip of gold and silver on activated charcoal. These recycled ionic liquids were used for four conjunctive cycles. Nakashima *et al.* [59] used octyl(phenyl)-N,N-di-isobutyl-carba-oyl-methyl phosphine oxide dissolved in an ionic liquid [BMIm][PF_6], which increased the extraction percentage and selectivity for lanthanide ions in comparison to conservative organic solvents. Furthermore, the resurgence of lanthanides removed into ionic liquids could be determined by using some strip solutions having complexing reagents [60].

Chun *et al.* [61] reported the advantages of some additives for enhanced extraction efficiencies. The authors reported that luckily large values of the distribution coefficients were obtained by means of ionic liquids as removal solvents for the extraction of metal ions using crown ethers. Visser *et al.* [62] developed some hydrophobic ionic liquids intended to extract mercury and cadmium in water. These were synthesized using thiourea, urea and thioether substituted alkyl groups to imidazoles and combining the consequential cationic species with [PF_6]. These ionic liquids were analyzed to evaluate their metal ions extraction capacities. The extraction efficiency increased by several magnitudes by using liquid-liquid extraction mercury and cadmium in water. This efficiency was independent when the ionic liquids were utilized as the solitary removing segment or doped into various [1-alkyl-3-methyl-imidazolium][PF_6] (alkyl = n-C4-C8) ionic liquids, which formed 1:1 complex. It was seen that the metal ions sharing ratios extended on increasing the length of the alkyl chain in 1:1 ratio. However, an increase in the TSIL/[BMIm][PF_6] ratio resulted in high distribution of mercury and cadmium. Finally, thiourea and urea derivative cations gave high distribution ratios. And for mercury was higher in comparison to cadmium. A transformation in water phase *i.e.* pH did not encourage the strip of metal cations from the extract phase. Holbrey *et al.* [63] synthesized ionic liquids having bis-imidazolium-cation involving a small ethylene glycol and long decylglycol spacers. 1,1'-[1,2-Ethanediylbis(oxy-1,2-ethanediyl)] bis[3-methyl-1H-imidazolium-1-yl] imide [$N(Tf)_2$] was prepared from the equivalent chloride salt. It was interesting to note that the ethylene glycol function radically increase the allocation ratio of Hg^{+2}. Contrarily, there was no effect on cesium metal ion in water solution to the hydrophobic ionic liquids. It may be considered as an instance of pH dependency partitioning and strip of mercury in ionic liquid aqueous two phase systems. The crystal arrangement of the related Hg^{+2} carbene complex revealed the likelihood of a carbene removal mechanism.

Eber *et al.* [64] reported an easy method for the cleaning of fuels by extracting sulphur and nitrogen compounds in gasoline and diesel oil. The authors reported that ionic liquids might be an option to common hydro-desulfurization for profound desulfurization decreasing the values of sulphur to 10 ppm or lower. According to the authors, the selective removal features of ionic liquids, particularly sulphur compounds that are difficult to remove by dibenzothiophene derivatives always present in distillates of diesel oil. [BMIm][C_8SO_3] and [EMIm][C_2SO_3] ionic liquids are used for the purpose. This is due to the fact that these are freely available and inexpensive. The extraction *via* ionic liquids has the main advantages that there is no effect on the octane number. Similarly, Zhang *et al.* [65] also reported that [$EMImBF^4$], [$BMImBF_4$], [$BMImPF_6$] and heavy trimethyl amine hydrochloride ($AlCl_3$-TMAC) indicated elevated selectivity, mainly near aromatic sulphur and nitrogen molecules for extractive desulfurization and denitrogenation. The reported ionic liquids were recyclable by simple distillation and water disarticulation. The sorbed aromatic sulphur molecules were quantitatively recovered. Aromatic molecules with high π-

electron density were found to be positively sorbed. It was found significant that alkyl replacement on the aromatic rings decreased the sorption ability due to steric result. The cation and anion structures and dimension in the ionic liquids are significant variables distressing the sorption ability for aromatic molecules. The sulphur and nitrogen containing compounds at small amount were removed from fuels without joint obstruction. Finally, $AlCl_3$-TMAC ionic liquid was established to have amazingly high sorption capacity for aromatic compounds. As per Huang *et al.* [66], CuCl based ionic liquids may be good extracts in gasoline industries. These are useful and effective for desulfurization of gasoline working as extraction sorbents. One pot process was projected by Lo *et al.* [67] for sulphur having molecules in the light oils. Both removal and oxidation of sulphur substances from light oil concurrently showed an advantage that the process augmented the desulfurization amount by approximately an order of the scale in relation to that of only extraction by ionic liquids. The ionic liquids are recyclable at ambient temp., which may be regenerated after the job is over and reused without any loss of action. Briefly, ionic liquids have more advantages in comparison to conventional solvents being applied in liquid-liquid extraction phenomenon. The exceptional features may be moduled to permit accomplishment of high effectiveness and selectivity of separation.

Separation

The ionic liquids have immense significance in analytical science owing to their solvent system additives and stationary phase modifier nature in both chromatography and capillary electrophoresis. The following sub-sections describe the application of ionic liquids in chromatographic and capillary electrophoretic modalities.

Liquid Chromatography

The capacity to manufacture ionic liquids with diverse solubilities in aqueous and organic solvents makes them ideal for use in chromatography. The ionic liquids are thick and, therefore, applied in chromatography as additives [68]. It is owing to the reality that the ionic liquids are frequently miscible with ordinary solvents applied in high performance liquid chromatography like ethanenitrile or methanol. The viscosities reduce after addition with high performance liquid chromatography solvents [69]. Thus, the ionic liquids may be applied as additives in moderately elevated ratios with water or organic phases. Poole and coworkers [70] evaluated the application of the different ionic liquids, together with propylammonium and nitrate ethylammonium nitrate. Danielson and Waichigo [71] determined the application of ethylammonium methanoate like mobile phases for HPLC for the separation of both amines and aromatic carboxylic acids.Ionic liquids have been used as solvent systems additives [72 - 79]. Zheng *et al.* [72] utilized 1-ethyl-3-methyl-imidazolium tetrafluoroborate, 1-butyl-3-met-yl-imidazolium tetrafluoroborate, 1-octyl-3-methyl-imidazolium methylsulfate ([OMIm][MS]) and 1-ethyl-3-methylimidazolium methylsulfate ([EMIm][MS]) sorts of ionic liquids for the partition of 2,4,6-trichlorophenol and 2-chlorophenol. The imidazolium ionic liquids make liquid clathrates with aromatic hydrocarbons, efficiently generating comprehensive structure which may be modified by separation if utilized as solvent systems additives [80]. Marszall [76] utilized alkyl imidazolium ionic liquids and experienced that the ionic liquids condensed the retention of basic molecule to a larger size in comparison to alkyl ammonium solvent system additives. Moreover, the protection of silanol groups by the ionic liquids progressed to the peak form and to the resolution of separation [74, 76, 79]. Almost all ionic liquid additives indicated enhanced peak shape in comparison to those achieved by using the ordinary ammonium ethanoate solvent system additive. When $[C_4\ mim]^+$ and $[C_2\ mim]^+$ were matched with methyl sulfate $[C_1SO_4]^-$ anion, fine separation was obtained for the separating 2,4,6-trichlorophenol and 2-chlorophenol in high performance liquid chromatography using 80 percent methanol-water solvent system [72]. Nevertheless, it was observed when balancing with the $[BF_4]^-$ anion, there was little to no enhancement in separation.

Zhang *et al.* [75] observed that the accumulation of ionic liquids [1-alkyl-3-methyl-imidazolium] to the solvent system created the peaks with a smaller amount of tailing and low retention volume for the resolution of ephedrines by HPLC. This examination was accredited to the protection of remaining silanol groups by imidazolium ionic liquids. These effects were supplementary when the ionic liquids with larger alkyl chains were applied, most probably owing to long alkyl chains protecting the silanols in comparison to small alkyl chains. Zheng and Row [81] reported the separation of unresolved hypoxanthine and guanine using 1-octyl-3-methyl-imidazolium methyl sulfate ($[C_8\ mim][C_1SO_4]$) as solvent system additive. The communications of the cations with silanol groups of the stationary phase surface participate with the polar groups of the analytes. This protected the remaining silanol groups improving the peak shapes and dropping the retention times for the analytes. Additionally, the silanol groups on the stationary phase, the cations of the ionic liquids modified by the connections amongst the analytes and stationary phases in high performance liquid chromatography by the sorption of ionic liquid cation to octadecyl group of stationary phase [73, 81]. Xiao *et al.* [73] reported the effect of 1,3-dialky--imidazolium ionic liquids on the analysed amines. These authors evaluated the feat of ionic liquid with that of more normally used tetra-butyl-ammonium bromide ($[N_{4\,4\,4\,4}]Br$). Ruiz-Angel *et al.* [82] studied the effects of triethylamine and $[C_2mim][BF_4]$ on the retention factor, shape of the peaks and resolution of some β-blockers.

It was established that chaotropic anions have encouraging effect on the solvation of cationic analytes [83]. The chaotropic anions only reduce retention but do not advance peak shape [83, 84]. The ionic liquids are not limited to utilize as solvent system modifiers, but may also be applied as chemical hop stationary phases for high performance liquid chromatography [85 - 89]. Immobilized 1-butyl-3-methyl-imidazolium bromide ([C_4 mim]Br) is utilized as a stationary phase to divide twenty eight small aromatic test molecules in HPLC [85]. Subsequently, Stalcup and coworkers [86] analyzed the stationary phase and evaluated it with industrial accessible quaternary amine silica column. It was observed that mobilized [C_4mim]Br showed high hydrophobic interactions in comparison to industrial stationary phase at all solvent systems symphony. The partition coefficients for the molecules in [C_4mim]Br were established using shake flask method. These were found to be associated with the retention of the molecule described, representing the utility of HPLC for resolving partition coefficients of the molecules with the ionic liquids. Also, the same authors [86] established that stationary phase (ionic liquid) also capable of going through ion-exchange interactions since the ionic potency of the solvent system was customized. The other authors [87 - 99] were also of the same opinion that the ionic liquid stationary phase may analyze molecules under ion exchange modes of HPLC, representing the adaptability and tune capability of the ionic liquids in analytical science.

Peng *et al.* [100] mobilized 1-octyl-3-methyl-imidazolium hexafluoro-phosphate ([C_8MIM][PF_6]) ionic liquid on poly-propylene empty fiber and used for empty fiber liquid phase micro-separation. The molecules extracted were 2,4-dichlorophenol, 3-chorophenol, 2,4,6-trichlorophenol and 4-chlorophenol. These were separated into the ionic liquid membrane. Further, these were reverse extracted in 10 μL NaOH acceptor solution in the lumen of the hollow fiber. The solution was loaded on to high performance liquid chromatography. The chemically modified silica particles, with ionic liquid, had been used as fixed phase HPLC for the analysis of alkaloids [101].

Capillary Electrophoresis

Capillary electrophoresis (CE) is a particular analytical technique depending on difference in the electrophoric mobilities of the molecules *via* a running buffer. The capillary electrophoretic analysis methods involve ionic liquids as background electrolytes. The additives have significantly improved electrophoretic separations [102]. Capillary wall has also been tailored with the ionic liquids [103]. López-Pastor *et al.* [104] reviewed the applications of ionic liquids in capillary electrophoresis separations. The authors showed that the ionic liquids interacted and played a crucial role in the alteration of electrophoretic analysis.

The general requirements of the capillary electrophoresis are running buffers and additives, low viscosity, small molar volume, low vapor pressure, relatively high permittivity, wide liquid range, good solvating power for cations and anions and analytes, high non-toxicity and chemical stability.

Yanes *et al.* [103] utilized alkyl-imidazolium cations corresponding with a range of anions as background electrolyte for the analysis of polyphenols in grape seed extracts. These authors analysed that pHs of the background electrolyte were inadequately acidic to make sure that the molecules were in their neutral forms. The authors also utilized the ionic liquids loaded on to capillary for the analyses purpose. Further, the same authors [105] utilized $[N_{2222}][BF_4]$ as the background electrolyte. This was alike to the earlier work [103], apart from that large electroosmotic flow for the similar concentration. The authors experienced that the delocalized charge of $[C_2mim][BF_4]$ with $[N_{2\,2\,2\,2}][BF_4]$ formed small relationship constants, accountable of elevated immigration of the ions. As per this manuscript, the relationship of polyphenols with free cations of the ionic liquids was more overriding than the cations sorbed on to the walls of the instrumental capillary. The effectual immigration of the molecules increased with the augment in $[C_2mim][BF_4]$ concentration. The analysis of the polyphenols was enhanced by non-consistant alkyl chain from ethyl to butyl.

Yue and Shi [106] explained the analysis of flavonoids; extracted from Chinese herbal medicines by $[C_n\ mim]^+$ ionic liquid additives. The authors reported that the hydrogen bondings among C-2 proton of imidazolium cation and phenols were accountable for the analysis. The authors also analyzed that the analysis was enhanced by escalating the extent of the alkyl chain (from C_2 to C_4) of the imidazolium cation. François *et al.* [107] utilized $[C_4mim][NTf_2\ ]$ for separating suprofen, carprofen, naproxen and ketoprofen. The authors reported that the electroosmotic mobility of the background electrolyte and the electrophoresis mobility of suprofen, carprofen, naproxen and ketoprofen noticeably reduced. According to the investigators, it was owing to the sorption of the ionic liquid cations on capillary wall; It affected the active covering of ionic liquid. Further, the investigators described that ionic liquid sorbed on walls of the capillary (at low concentrations) worked as stationary phase, accountable for augment immigration times. Further, capillary wall became completely covered at elevated amount of ionic liquid. As a consequence, the ion pair exchanged between 2-ary--propanoic acids and complimentary ionic liquid cations were conquered. Chieu *et al.* [108] reported S-$[CHTA]^+[Tf_2N]^-$, chiral ionic liquid as electrolyte and chiral selector for capillary electrophoresis separations. The investigators determined a diversity of pharmaceutical products *viz.* flurbiprofen, warfarin, ibuprofen atenolol, ketoprofen, propranol and indoprofen using ionic liquid. For ibuprofen, another chiral molecule viz. sodium cholate was used for the enantiomeric separation. In case of furbiprofen, in addition to S-$[CHTA]^+$ $[Tf_2N]^-$ and sodium cholate, 3rd impartial chiral selector, 1-S-octyl-β-D-thiogluco-pyranoside was also required. Likewise, nucleoside monophosphates and phenols were separated using $[C_4mim][BF_4]$ ionic liquid [109].

Gas Chromatography

The ionic liquids are used as the fixed phases in gas chromatography. These have good capacities to interact with a variety of molecules and that is why tested in fixed phases in gas chromatography. The solvation process may be used to determine the input from characteristic inter-molecular bondings to the gas liquid partition coefficient. Good wetting capacity and viscosity of ionic liquids made them ideal candidates for coating on chromatographic media in silica capillaries. Therefore, Armstrong, *et al.* [110] used 1-butyl-3-methyl-imidazolium hexa-fluoro-phosphate and the equivalent chloride salts as fixed phases in gas chromatography. The authors determined Rohrschneider-McReynolds constants for both ionic liquid and well-known and commercially available polysiloxane stationary phase. Ionic liquid stationary phases appear to be twofold in the nature, acting as a low polarity stationary phase to non-polar molecules. However, the nature of the anions may have an important consequence together with the solubilising aptitude and the selectivity of the fixed phases based on ionic liquids. The unusual properties of the ionic liquids made them valuable supports in numerous research areas of analytical science.

Anderson and Armstrong *et al.* [111] used low thermal strength and showed adverse retention performance for a few classes of the compound. The authors synthesized 1-benzyl-3-methyl-imidazolium trifluoro-methane-sulfonate and 1-(--methoxyphenyl)-3-methyl-imidazolium trifluoro-methane-sulfonate, based on large imidazolium cations with trifluoro-methane-sulfonate anions. The authors used Abraham solvation parameter model to evaluate solvation features. The solvation parameter model and correlation correlated amongst the structures of cation and degree of ionic liquid retention. The developed ionic liquids showed fine thermal stability up to 260 °C. These generated regular peak shapes, exhibiting double nature selectivity behaviour. The ionic liquid based fixed phases showed various retention behaviours for the different molecules in comparison to common methyl phenyl polysiloxane stationary phase of gas chromatography. The dissimilarity in selectivity was because of the single solvation feature of the ionic liquid. These permitted to the author to predict hopeful ending concerning these ionic liquid as practical double nature stationary phases in gas chromatography. Letcher *et al.* [112] determined the activity coefficients on $[C_8MIm][BF_4]$ at 298.15 K and 323.15 K. The result showed that $[C_8MIM][BF_4]$ ionic liquid indicated better solvent for parting of benzene and alkanes. Similarly, Warren *et al.* [113] reported the activity constants for hydrocarbon at endless dilution in 1-methyl-3-octyl-imidazolium-chloride. The activity constants at unlimited dilution of the linear and branched C_1 to C_6 alcohols, acetonitrile, acetone, alkylethers, ethylacetate, and chloromethanes in ionic liquid 4-methyl-N-butyl-pyridinium tetrafluoro-borate were ascertained [114]. The ionic liquids 1-methyl-3-ethyl-imidazolium bis-(trifluoromethyl-sulfonyl) amide and 1,2-dimethyl-3-ethyl-imidazolium bis-(trifluoro-methyl-sulfonyl) amide were used a stationary phase in gas chromatography [115].

Gas chromatography involving ionic liquids has been used for the chiral separations. The task can be achieved on dissolved chiral selector an achiral ionic liquid and the ionic liquids themselves as chiral reagent. The configuration of the stereogenic center can be moduled in ionic liquid chiral selectors due to synthetic nature. It is useful to determine the chiral recognition mechanism. Berthod *et al.* [116] reported the chiral separations. The author prepared chiral stationary phase in gas chromatography by dissolving permethylated-β-cyclodextrin in $[BMIm][PF_6]$ ionic liquid. The chiral columns with this chiral elector with and without ionic liquid were compared in terms of their separation efficiencies. Total sixty four racemates were tested and only twenty one compounds were resolved. It was observed that perhaps the imidazolium ion pair made an inclusion complex within cyclodextrin cavity, responsible for chiral separation mechanism. Similarly out of eighty compounds tested, only sixteen could be resolved. Generally, retention factors on ionic liquids based chiral columns were lower in comparison to the conventional chiral stationary phases. Besides, the peak efficiencies available with the ionic liquids were considerably higher in comparison to those obtained with the common chiral columns. Ding *et al.* [117] reported the chiral separations of diols, sulfoxides, epoxides, alcohols, and acetylated amines on ionic liquid chiral fixed phase in gas chromatography.

Planeta and Roth *et al.* [118] used $[BMIm][PF_6]$ ionic liquid for fixed liquid in capillary column chromatography along with supercritical CO_2 as the mobile phase. The stationary phase was enclosed in open tubular capillary columns. The studied compounds were α-ionone, anisole, benzil, azulene, napthalene, pyrene and vertrole. The analytical data calculated by usual solvation theory showed near values of the solubility parameter of carbon dioxide extended $[BMIm][PF_6]$. According to the authors, it established the feasibility of open tubular supercritical fluid chromatography to offer partition data on small volatility solutes in this system.

The ionic liquids are used as stationary and mobile phases in counter current chromatography. High viscosity of ionic liquids does not allow them as mobile phases in counter current chromatography. Therefore, these are mixed with other organic molecules such as ethanol, methanol, acetonitrile, 1-propanol, and 2-propanol. Before using as mobile phase. Besides, water is also used as the thinner for these ionic liquids. The most commonly used mobile phase in counter current chromatography is 40:20:40 mixture of water, acetonitrile, and [BMIm][PF_6]. Berthod and Carda-Broch *et al.* [119] used octanol-water system for partitioning thirty eight aromatic compounds having acidic, basic and neutral properties. The results were comparable with those obtained with ionic liquids. The authors describe some hurdles in using ionic liquids in current chromatography. These include high back pressure, limitation of UV absorbance and non-volatality. The non volatility make them unfit with mass detector.

Electrochemistry

Ionic liquids have been used in electrochemistry and some reviews have appeared in the literature [120]. ILs have been divided in three categories in electrochemistry *i.e.* i) systems based on aluminium trichloride and organic salts [1-butylpyridinium chloride (BPCl), 1-alkyl-3-methyl-imidazolium chloride (CnMIm)Cl, and their derivatives, ii) systems based on organic cations as in (i) and anions BF_4, PF_6 and SbF_6 and iii) systems based on the above cited organic cations with anions CF_3SO_3, $(CF_3SO_2)_2N$ type. The ionic liquids prepared as per the latter ions are steady in ambient situation with only low water uptake.

A good electrolyte has having chemical stability, large electrochemical windows, high conductivity, and excellent thermal and negligible evaporation. These features are prominent in the ionic liquids and, hence, the ionic liquids are capable materials in a variety of electrochemical strategies, such as, fuel cells, sensors, batteries and electrochromic windows. It was observed that Li^+ cell with the ionic liquids electrolyte performed well in conditions of cell routine and cycle stability [121 - 123].

Good conductivities of the ionic liquids ranging from 10^{-3} to 10^{-2} Ω^{-1} cm^{-1} make them ideal solvents for small temperature electrodeposition studies. These are useful to the elements, which may not be obtained from aqueous solutions *i.e.* Al, Po, Si and Ti. Silicon has been electrodeposited at nano scale in the RTIL 1-buty--1-methyl-pyrrolidinium bis-(trifluoromethylsulfonyl)-imide soaked with silicon tetra chloride [124]. A silicon layer of 100 nm thickness exhibited a band gap of 1.0 ± 0.2 eV, which is revealed by *in situ* current/voltage tunnelling spectroscopy, demonstrating semiconducting electrodeposited silicon. These systems are limited to laboratory only but industrial applications are to be evolved. Besides, this is not applicable to noble elements such as germanium. 1-Methyl-3-(2,6-(S)-dimethylocten-2-yl)-imidazolium tetrafluoro-borate [MDIM][BF4] micro droplet and thin film deposits are studied in enjoin to measure the capability of ions in water to partition into the ionic liquids [125].

The ionic liquids are beneficial for the electrochemical strength in comparison to water, contributing the opportunity of advanced actuation voltages for electrically forced materials. [EMIm][N(Tf)$_2$] is studied as an electrolyte for electromechanical actuators with polypyrrole [126]. The actuator presentation in IL electrolytes was considerably improved in comparison to conventional organic or aqueous electrolytes. Bennett and Leo *et al.* [127] used 1-ethyl-3-met-yl-imidazolium trifluoro-methane-sulfonate IL, which was found as a feasible solvent for Nafion polymer sensors and actuators. The results indicated Nafion transducers solvated with this IL and enhanced stability in air than in water. However, ILs have limited speed response than in water. That is why amperometric oxygen gas sensors were made using 1-n-butyl-3-me-hyl-imidazolium hexafluoro-phosphate [BMIm][PF$_6$] [128] and 1-ethyl-3-met-yl-imidazolium tetrafluoro-borate [EMIm][BF$_4$] [129] porous polyethylene membrane as a firm state electrolyte. The developed sensor was capable of working at room temperature. It showed a wide detection range, elevated stability, elevated sensitivity, and admirable reproducibility. Brinz and Simon *et al.* [130] used an electro-chemical sensor with ILs as an electrolyte for the examination of fire detector. Imidazole, pyridine derivatives and 1-ethyl-3-methyl-imidazolium tetrafluoroborate [EMIm][BF4],1-ethyl-3-methyl-imidazolium chloride [EMIm][Cl] were applied as electrolytes. These were used on a non-woven stuff comprising silicates and polymers. The detection was achieved amperometrically and potentiometrically by two or three electrode systems. Lee and Chou [131] used ethanol sensor with detection limit of 0.13% and response time of 336 seconds.

The capability of hydrophilic ions to separated from aqueous phase into ILs phase was unforeseen and exposed to depend on, at exact interface of ion with ILs, type of ion, and concentration of biased electrolyte. The voltammetric response achieved for the reduction of $Fe(CN)_3^{-6}$ partitioned into micro droplet and slim film deposit of $MDIM^+BF^{-4}$ showed discerning sorption of ferricyanide into the ILs phase. ILs have been found effective to enumerate concentration of $Fe(CN)_3^{-6}$. The electrochemical property of some probes in ILs are almost similar to the other organic solvents [132]. The mechanism of the electro-reductive creation of the organohalogenated compounds is almost similar in [BMIm][PF$_6$] to the usual organic solvents. The uses of ILs in electro-chemical blend of conductive polymers, electro-chemical and electro-chromic analyses of combined electro-chemical and chemically prepared conducting polymers and manufacture of conducting polymer electro-chromic procedure have been studied [133].

Spectroscopy

The ionic liquids are used in spectroscopic applications. It is due to their unique feature such as good solvation power. These are quite effective solvent for metal ion complexes, which are unstable in other solvents. These have been used in UV-Vis. and IR spectroscopy. 1-Methyl-3-ethylimidazolium chloride based ionic liquids are used for recording spectra of Ir, Os, Ru *etc.* elements. Air and water steady ionic liquids have good solvating features jointly with big spectroscopic transparency. The UV cut-off wavelengths of alkyl-imidazolium based ionic liquids are in the range of 230 to 250 nm. This made them appropriate solvents for spectroscopic dimensions, particularly in visible range. Appleby *et al.* [134] used ionic liquids for salvation of inorganic, organic, and organometallic compounds.

Seddon *et al.* [135] described 1-methyl-3-ethyl-imidazolium halide-aluminum(III) halide and N-butylpyridinium chloride-aluminum(III) chloride as the solvents for recording the spectra of transition metal halide complexes. Billard *et al.* [136] analysed the significance of the ionic liquids in spectroscopy. The authors described the stability of some species in the ionic liquids such as Eu(II); stable in [BMIm][PF_6]. Lu *et al.* [137] studied [BMIm][PF_6] and carbon dioxide as a solvent in spectroscopy. This study was carried out from 35 to 50 °C. Besides, the pressure of the carbon dioxide ranged from 0 to 230 bar. As per the authors, the results were comparable with the usual solvents. The authors also claimed that these can be used in combination with the usual solvents.

Spectroscopic intermolecular interactions have been studied using ionic liquids as spectroscopic solvents. Alvaro *et al.* [138] used [BMIm][PF_6] as a spectroscopic solvent for some photo-chemical reactions. The authors studied hydrogen abstraction, oxygen quenching, electron transfer and energy transfer. The results were interesting with respect to molecular diffusion reproduction dispersal controlled phenomenon, weak charge transfer interactions, amazing low oxygen solubility and extended life span of triplet excited state. The authors claimed that ionic liquids are useful for hydrogen transfer, electron transfer and energy transfer

In spite of low volatility, ionic liquids have been used in mass spectroscopy. But these are useful in high vacuum mass spectroscopy. The ionic liquids have outstanding solubility features and vacuum constancy in comparison to other usual solvents and solutions. Some ionic matrices create uniform solutions of high ion peak strength and equal or low recognition limit in comparison to the presently used matrixes. However, these differ broadly in their capability to create gas phase ions. The ionic liquids are useful for making homogeneous paste of the compounds to be studied, with no vapour pressure. These features of the ionic liquids have been used in MALDI-MS analysis. This has directed assisted qualitative and quantitative analyses of the molecules as compared to the traditional solid matrices. Some workers are doing experiments in this direction using TSILs for MALDI-MS techniques. Armstrong *et al.* [139] used these approaches for analyzing peptides, poly(ethylene glycol) and proteins. The ionic liquids were made up of equal ratios of traditional MALDI matrixes *i.e.* 2,5-dihydroxybenzoic acid, sinapinic acid, α-cyano-4-hydroxycinnamic acid *etc.* Zabet-Moghaddam *et al.* [140] used ionic liquids for the qualitative and quantitative study of small molecular weight substances. The authors studied sugars, amino acids and vitamins. The authors reported that ionic liquids were used effectively to substitute solid matrixes, being used for the examination of biologically attractive small molecular weight molecules.

Carda-Broch *et al.* [141] used ionic liquids for the determination of molecular weight of DNA using TOF-MS. The three studied oligo-nucleotides were d(pT)(10), d(pC)(11) and d(pC)(12) along with some ionic matrices prepared using various bases connected with two acids *i.e.* 2,5-dihydroxy-benzoic acid and 3-hydroxy-picolinic acid. As per the authors, the results showed that the most excellent ionic matrix increased the ion peak strength of the oligo-nucleotides in comparison to the usual molecular matrix in the experimental setting reported. As per the authors, the ionic liquids gave signal to noise ratio 10 times higher than the equivalent molecular matrix. Mank *et al.* [142] reported the evaluation of MALDI-MS performance using matrix 2,5-dihydroxybenzoic acid butylamine. The authors compared the results when MALDI-MS was used with matrixes 3,5-dimethoxy-4-hydroxy-cinnamic acid triethyl-amine and α-cyano-4-hydr-xy-cinnamic acid butyl amine. The results were comparable. Furthermore, the authors studied ionic liquid MALDI-MS matrix 2,5-dihydroxy benzoic acid butyl amine and compared the results with ionic liquid MALDI-MS matrix 3,5-dimethoxy-4-hydroxycinnamic acid triethylamine, α-cyano-4-hydroxycinnamic acid butylamine and the frequently used solid MALDI matrix 2,5-dihydroxybenzoic acid and α-cyano-4-hydroxycinnamic acid. The authors calculated reproducibility in terms of standard deviations.

The ionic liquids are used in electrospray ionization mass spectrometry, owing to the fact that these ionic liquids are very conductive liquids, low volatile and capable of generating instrumental jet, which is difficult with the normal solvents. The ionic energies of the ionic liquids suggest the use of ionic liquids as ion beam basis with a greatly wider choice of mass/charge and chemical molecules [143, 144]. Dyson *et al.* [145] reported that pure ionic liquids do not necessitate incessant sample injection and the occurrence of a molecular solvent assists the analysis. Jackson and Duckworth *et al.* [146] reported that ionic liquids studied in undiluted form showed comparable signal to noise ratios results with those practical in usual, diluted ES-MS. Ionic liquids were examined in undiluted appearance, where the outcomes indicated that signal to noise ratio for minor compounds were similar to those seen in usual, diluted ES-MS. As per the authors this advance could be willingly functional for mass spectrometric analyses of the ionic liquids and ionic impurity/additive dissolved within particularly those which are active solvent. Karkkainen *et al.* [147] reported purity of microwave prepared ionic liquids using mass spectroscopy. Zhao [148] reported the analysis of a ruthenium catalyst immersed in an ionic.

TOXICITY

Currently, the toxicity of the ionic liquids for micro organisms and superior flora and fauna is described. The concerned person should consult the review of Quijano and co-workers [149]. In this review the authors discussed toxicity matter of the ionic liquids. Pham *et al.* [150] also reviewed the toxic property of the ionic liquids on *Pseudokirchneriella subcapitata, Daphnia magna, Vibrio fischeri, Lemma minor* and zebrafish *Danio rerio.* The authors accomplished that the ionic liquids are incompletely toxic for most of the flora and fauna. Further, the authors explain the toxic behavior of ionic liquids owing to their hydrophobicity. The hydrophobicity is directly related to toxicity [151 - 153]. This also described that the toxicity owing to anion is extremely small [154]. In addition, a number of authors also described the toxicity of ionic liquids. Lee *et al.* [155] reported the [BMIM][PF6] ionic liquid on *E. coli* expansion at amount as low as 0.5% (v/v). Sendovski *et al.* [156] described [BMIM][PF6] ionic liquid poisonous for *S. cerevisiae*. Not much reports are available in the literature on the toxicity of ionic liquids. It needs more investigation.

FUTURE PROSPECTIVE

On the basis of the above discussion, the ionic liquids are extensively required in materials and chemical science, technologies, engineering and industries. The precise applications of the ionic liquids are of immense apprehension in supercapacitors, batteries, hydrogen fuel cells and ultra-capacitors areas. The ionic liquids can be applied as solvent system, stationary phase and modifiers in chromatography. Likewise, the ionic liquids may be used as background electrolytes, additives and chemicals for covering capillary walls. The ionic liquids may be mixed with a diversity of solvents given those outstanding expectations in analytical science. The ionic liquids may be the solvents in enantiomeric separations due to their exclusive properties. In addition, non-volatile behavior made these ionic liquids environmental friendly solvents in the analytical science. This can be predicted that the ionic liquids can restore regularly organic solvents in hope. Certainly, the ionic liquids will be utilized by analytical scientists in near future.

Though the ionic liquids are considered as the environmental friendly solvents but their toxicities for flora and fauna liquids limit their applications the different areas of science, technology, industry and engineering. Thus, there is an immense requirement to harmonize the ionic liquids to make these harmless. This may be obtained by altering the alkyl chain or the cation/anion characteristics but preserving the main features. Therefore, the modification of the ionic liquids is a huge and potential face for the researchers and the engineers.

CONCLUSION

The ionic liquids are very important class of compounds. These are gaining importance in a wide range of science and technology. This article is restricted to the use of these liquids in chemical science. These have utilities in extraction, separation science, electrochemistry and spectroscopy. This is due to the fact that the ionic liquids have unique solvent properties. During the write up of this article it was apprehended that the applications of the ionic liquids are not completely developed. Still there is plenty of uses for ionic liquids in chemical science. There is a need to synthesize ionic liquids. This feature of ionic liquids makes them ideal solvents in chemical science. There is no serious limitation of the ionic liquids and, hence, they have bright future. Briefly, the future of the ionic liquids in the chemical science is vivid.

CONSENT FOR PUBLICATION

Not applicable.

CONFLICT OF INTEREST

The authors declares no conflict of interest, financial or otherwise.

ACKNOWLEDGEMENTS

The authors extend their appreciation to the International Scientific Partnership Program ISPP at King Saud University for funding this research work through ISPP# 0037.

REFERENCES

[1] Seddon, K. Ionic liquids: designer solvents for green synthesis. *Tce.,* **2002**, *730*, 33-35.

[2] Gordon, C.M. New developments in catalysis using ionic liquids. *Appl. Catal. A Gen.,* **2001**, *222*, 101-117.
[http://dx.doi.org/10.1016/S0926-860X(01)00834-1]

[3] Houlton, S. Ionic liquids: the route to cleaner and more efficient fine chemical synthesis. *Chemical*

Week, **2004**, s10-s11.

[4] Zhao, H.; Malhotra, S.V. Applications of ionic liquids in organic synthesis. *Aldrichim Acta,* **2002**, *35*, 75-83. [http://dx.doi.org/10.1016/S0003-2670(01)01543-4]

[5] Earle, M.; Forestier, A.; Olivier-Bourbigou, H.; Wasserscheid, P. Organic synthesis.*Ionic Liquids in Synthesis*; Wasserscheid, P.; Welton, T., Eds.; Wiley-VCH Verlag: Weinheim, **2003**, pp. 174-288.

[6] Jain, N.; Kumar, A.; Chauhan, S.; Chauhan, S.M. Chemical and biochemical transformations in ionic liquids. *Tetrahedron,* **2005**, *61*, 1015-1060. [http://dx.doi.org/10.1016/j.tet.2004.10.070]

[7] Endres, F.; Welton, T. Inorganic synthesis.*Ionic Liquids in Synthesis*; Wasserscheid, P.; Welton, T., Eds.; Wiley-VCH Verlag: Weinheim, **2003**, pp. 289-318.

[8] Husum, T.L.; Jorgensen, C.T.; Christensen, M.W.; Kirk, O. Enzyme catalysed synthesis in ambient temperature ionic liquids. *Biocatal. Biotransform.,* **2001**, *19*, 331-338. [http://dx.doi.org/10.3109/10242420109003648]

[9] Kragl, U.; Eckstein, M.; Kaftzik, N. Enzyme catalysis in ionic liquids. *Curr. Opin. Biotechnol.,* **2002**, *13*(6), 565-571. [http://dx.doi.org/10.1016/S0958-1669(02)00353-1] [PMID: 12482515]

[10] Park, S.; Kazlauskas, R.J. Biocatalysis in ionic liquids - advantages beyond green technology. *Curr. Opin. Biotechnol.,* **2003**, *14*(4), 432-437. [http://dx.doi.org/10.1016/S0958-1669(03)00100-9] [PMID: 12943854]

[11] Sheldon, R.A.; Lau, R.M.; Sorgedrager, M.J.; Rantwijk, F.V.; Seddon, K.R. Biocatalysis in ionic liquids. *Green Chem.,* **2002**, *4*, 147-151. [http://dx.doi.org/10.1039/b110008b]

[12] van Rantwijk, F.; Madeira Lau, R.; Sheldon, R.A. Biocatalytic transformations in ionic liquids. *Trends Biotechnol.,* **2003**, *21*(3), 131-138. [http://dx.doi.org/10.1016/S0167-7799(03)00008-8] [PMID: 12628370]

[13] Kubisa, P. Application of ionic liquids as solvents for polymerization processes. *Prog. Polym. Sci.,* **2004**, *29*, 3-12. [http://dx.doi.org/10.1016/j.progpolymsci.2003.10.002]

[14] Carmichael, A.J.; Haddleton, D.M. Polymer synthesis in ionic liquids.*Ionic Liquids in Synthesis*; Wasserscheid, P.; Welton, T., Eds.; Wiley-VCH Verlag: Weinheim, **2003**, pp. 319-335.

[15] Seddon, K.R. Ionic liquids for clean technology. *J. Chem. Technol. Biotechnol.,* **1997**, *68*, 351-356. [http://dx.doi.org/10.1002/(SICI)1097-4660(199704)68:4<351::AID-JCTB613>3.0.CO;2-4]

[16] Welton, T. Room-temperature ionic liquids—solvents for synthesis and catalysis. *Chem. Rev.,* **1999**, *99*(8), 2071-2084. [http://dx.doi.org/10.1021/cr980032t] [PMID: 11849019]

[17] Wilkes, J.S. A short history of ionic liquids—from molten salts to neoteric solvents. *Green Chem.,* **2002**, *4*, 73-80. [http://dx.doi.org/10.1039/b110838g]

[18] Freemantle, M. BASF's smart ionic liquid. *Chem. Eng. News,* **2003**, *2001*(13), 9.

[19] Parkinson, G. Ionic liquids make an environmental splash. *Chem. Eng. Prog.,* **2004**, *100*, 7-9.

[20] Huddleston, J.G.; Visser, A.E.; Reichert, W.M.; Willauer, H.D.; Broker, G.A.; Rogers, R.D. Characterization and comparison of hydrophilic and hydrophobic room temperature ionic liquids incorporating the imidazolium cation. *Green Chem.,* **2001**, *3*, 156-164. [http://dx.doi.org/10.1039/b103275p]

[21] Postleb, F.; Stefanik, D.; Seifert, H.; Giernoth, R. Bionic Liquids: Imidazolium-based Ionic Liquids

with Antimicrobial Activity. *Z. Naturforsch. B,* **2013**, *68*, 1123-1128. [http://dx.doi.org/10.5560/znb.2013-3150]

[23] Swatloski, R.P.; Spear, S.K.; Holbrey, J.D.; Rogers, R.D. Dissolution of Cellose with Ionic Liquids. *J. Am. Chem. Soc.,* **2002**, *124*, 4974-4975. [http://dx.doi.org/10.1021/ja025790m] [PMID: 11982358]

[24] Graenacher, C. Manufacture and Application of New Cellulose Solutions and Cellulose Derivatives. Produced therefrom 1934. US 1934/1943176

[25] Ignatyev, I.; Doorslaer, C.V.; Mertens, P.G.; Binnemans, K.; de Vos, D.E. Synthesis of glucose esters from cellulose in ionic liquids. *Holzforschung,* **2011**, *66*, 417-425.

[26] Teixeira, R.E. Energy-efficient extraction of fuel and chemical feedstocks from algae. *Green Chem.,* **2012**, *14*, 419-427. [http://dx.doi.org/10.1039/c2gc16225c]

[27] Karkamkar, A.; Aardahl, C.; Autrey, T. Recent developments on hydrogen release from ammonia borane. *Material Matters.,* **2007**, *2*, 6-9.

[28] Barghi, S.H.; Adibi, M.; Rashtchian, D. An experimental study on permeability, diffusivity, and selectivity of CO2 and CH4 through [bmim][PF6] ionic liquid supported on an alumina membrane: Investigation of temperature fluctuations effects. *J. Membr. Sci.,* **2010**, *362*, 346-352. [http://dx.doi.org/10.1016/j.memsci.2010.06.047]

[29] Giridhar, P.; Venkatesan, K.A.; Srinivasan, T.G.; Rao, P.R. Electrochemical behavior of uranium(VI) in 1-butyl-3-methylimidazolium chloride and thermal characterization of uranium oxide deposit. *Electrochim. Acta,* **2007**, *52*, 3006-3012. [http://dx.doi.org/10.1016/j.electacta.2006.09.038]

[30] Jayakumar, M.; Venkatesan, K.A.; Srinivasan, T.G. Electrochemical behavior of fission palladium in 1-butyl-3-methylimidazolium chloride. *Electrochim. Acta,* **2007**, *52*, 7121-7127. [http://dx.doi.org/10.1016/j.electacta.2007.05.049]

[31] Jayakumar, M.; Venkatesan, K.A.; Srinivasan, T.G.; Rao, P.R. Extraction-Electrodeposition (EX-EL) process for the recovery of fission palladium from high-level liquid waste. *J. Appl. Electrochem.,* **2009**, *39*, 1955-1962. [http://dx.doi.org/10.1007/s10800-009-9905-3]

[32] Rao, C.J.; Venkatesan, K.A.; Nagarajan, K.; Srinivasan, T.G. Dissolution of uranium oxides and electrochemical behavior of U(VI) in task specific ionic liquid. *Radiochim. Acta,* **2008**, *96*, 403409. [http://dx.doi.org/10.1524/ract.2008.1508]

[33] Rao, C.J.; Venkatesan, K.A.; Nagarajan, K.; Srinivasan, T.G.; Rao, P.R. Electrodeposition of metallic uranium at near ambient conditions from room temperature ionic liquid. *J. Nucl. Mater.,* **2011**, *408*, 25-29. [http://dx.doi.org/10.1016/j.jnucmat.2010.10.022]

[34] Rao, J.C.; Venkatesan, K.A.; Nagarajan, K.; Srinivasan, T.G.; Rao, P.R. Electrochemical behavior of europium (III) in N-butyl-N-methylpyrrolidinium bis(trifluoromethylsulfonyl)imide. *Electrochim. Acta,* **2009**, *54*, 4718-4725. [http://dx.doi.org/10.1016/j.electacta.2009.03.074]

[35] Wang, C.; Luo, H.; Jiang, D.E.; Li, H.; Dai, S. Carbon dioxide capture by superbase-derived protic ionic liquids. *Angew. Chem. Int. Ed. Engl.,* **2010**, *49*(34), 5978-5981. [http://dx.doi.org/10.1002/anie.201002641] [PMID: 20632428]

[36] Carda-Broch, S.; Berthod, A.; Armstrong, D.W. Solvent properties of the 1-butyl-3-methylimidazolium hexafluorophosphate ionic liquid. *Anal. Bioanal. Chem.,* **2003**, *375*(2), 191-199. [http://dx.doi.org/10.1007/s00216-002-1684-1] [PMID: 12560962]

[37] Huddleston, J.G.; Willauer, H.D.; Swatloski, R.P.; Visser, A.E.; Rogers, R.D. Room temperature ionic liquids as novel media for "clean" liquid-liquid extraction. *Chem. Commun. (Camb.),* **1998**, *16*, 1765-

1766.
[http://dx.doi.org/10.1039/A803999B]

[38] Seiler, M.; Jork, C.; Kavarnou, A.; Arlt, W.; Hirsch, R. Separation of azeotropic mixtures using hyperbranched polymers or ionic liquids. *AIChE J.,* **2004**, *50*, 2439-2454.
[http://dx.doi.org/10.1002/aic.10249]

[39] Jork, C.; Seiler, M.; Beste, Y-A.; Arlt, W. Influence of ionic liquids on the phase behavior of aqueous azeotropic systems. *J. Chem. Eng. Data,* **2004**, *49*, 852-857.
[http://dx.doi.org/10.1021/je034183r]

[40] Visser, A.E.; Holbrey, J.D.; Rogers, R.D. Hydrophobic ionic liquids incorporating N-alkylisoquinolinium cations and their utilization in liquid-liquid separations. *Chem. Commun. (Camb.),* **2001**, *23*(23), 2484-2485.
[http://dx.doi.org/10.1039/b109340c] [PMID: 12240026]

[41] Nobuoka, K.; Korenaga, W.; Ishikawa, Y. Environmentally safe ionic liquids for chiral resolution. 2004. Japan Patent JP 2004277351

[42] Cull, S.G.; Holbrey, J.D.; Vargas-Mora, V.; Seddon, K.R.; Lye, G.J. Room-temperature ionic liquids as replacements for organic solvents in multiphase bioprocess operations. *Biotechnol. Bioeng.,* **2000**, *69*(2), 227-233.
[http://dx.doi.org/10.1002/(SICI)1097-0290(20000720)69:2<227::AID-BIT12>3.0.CO;2-0] [PMID: 10861402]

[43] Matsumoto, M.; Mochiduki, K.; Fukunishi, K.; Kondo, K. Extraction of organic acids using imidazolium-based ionic liquids and their toxicity to *Lactobacillus rhamnosus. Separ. Purif. Tech.,* **2004**, *40*, 97-101.
[http://dx.doi.org/10.1016/j.seppur.2004.01.009]

[44] Carda-Broch, S.; Berthod, A.; Armstrong, D.W. Solvent properties of the 1-butyl-3-methylimidazolium hexafluorophosphate ionic liquid. *Anal. Bioanal. Chem.,* **2003**, *375*(2), 191-199.
[http://dx.doi.org/10.1007/s00216-002-1684-1] [PMID: 12560962]

[45] Smirnova, S.V.; Torocheshnikova, I.I.; Formanovsky, A.A.; Pletnev, I.V. Solvent extraction of amino acids into a room temperature ionic liquid with dicyclohexano-18-crown-6. *Anal. Bioanal. Chem.,* **2004**, *378*(5), 1369-1375.
[http://dx.doi.org/10.1007/s00216-003-2398-8] [PMID: 14747900]

[46] Chun, S.; Dzyuba, S.V.; Bartsch, R.A. Influence of structural variation in room-temperature ionic liquids on the selectivity and efficiency of competitive alkali metal salt extraction by a crown ether. *Anal. Chem.,* **2001**, *73*(15), 3737-3741.
[http://dx.doi.org/10.1021/ac010061v] [PMID: 11510842]

[47] Luo, H.; Dai, S.; Bonnesen, P.V. Solvent extraction of sr^{2+} and cs^{+} based on room-temperature ionic liquids containing monoaza-substituted crown ethers. *Anal. Chem.,* **2004**, *76*(10), 2773-2779.
[http://dx.doi.org/10.1021/ac035473d] [PMID: 15144187]

[48] Dietz, M.L.; Dzielawa, J.A. Ion-exchange as a mode of cation transfer into room-temperature ionic liquids containing crown ethers: implications for the 'greenness' of ionic liquids as diluents in liquid-liquid extraction. *Chem. Commun. (Camb.),* **2001**, *20*(20), 2124-2125.
[http://dx.doi.org/10.1039/b104349h] [PMID: 12240194]

[49] Dietz, M.l.; Dzielawa, J.A.; Laszak, I.; Young, B.A.; Jensen, M.P. Influence of solvent structural variations on the mechanism of facilitated ion transfer into room-temperature ionic liquids. *Green Chem.,* **2003**, *5*, 682-685.
[http://dx.doi.org/10.1039/B310507P]

[50] Shimojo, K.; Goto, M. First application of calixarenes as extractants in room-temperature ionic liquids. *Chem. Lett.,* **2004**, *33*, 320-321.
[http://dx.doi.org/10.1246/cl.2004.320]

[51] Shimojo, K.; Goto, M. Solvent extraction and stripping of silver ions in room-temperature ionic liquids containing calixarenes. *Anal. Chem.,* **2004**, *76*(17), 5039-5044. [http://dx.doi.org/10.1021/ac049549x] [PMID: 15373439]

[52] Luo, H.; Dai, S.; Bonnesen, P.V.; Buchanan, A.C., III; Holbrey, J.D.; Bridges, N.J.; Rogers, R.D. Extraction of cesium ions from aqueous solutions using calix[4]arene-bis(tert-octylbenzo-crown-6) in ionic liquids. *Anal. Chem.,* **2004**, *76*(11), 3078-3083. [http://dx.doi.org/10.1021/ac049949k] [PMID: 15167785]

[53] Liu, J-F.; Jiang, G-B.; Chi, Y-G.; Cai, Y-Q.; Zhou, Q-X.; Hu, J-T. Use of ionic liquids for liquid-phase microextraction of polycyclic aromatic hydrocarbons. *Anal. Chem.,* **2003**, *75*(21), 5870-5876. [http://dx.doi.org/10.1021/ac034506m] [PMID: 14588028]

[54] Liu, J-F.; Chi, Y-G.; Jiang, G-B.; Tai, C.; Peng, J-F.; Hu, J-T. Ionic liquid-based liquid-phase microextraction, a new sample enrichment procedure for liquid chromatography. *J. Chromatogr. A,* **2004**, *1026*(1-2), 143-147. [http://dx.doi.org/10.1016/j.chroma.2003.11.005] [PMID: 14763740]

[55] Visser, A. E.; Swatloski, R. P.; Griffin, S. T.; Hartman, D. H.; Rogers, R. D. Liquid/liquid extraction of metal ions in room temperature ionic liquids. *Sep. Sci. and Tech.,* **2001**, *36*(5-6), 785-804.

[56] Wei, G-T.; Yang, Z.; Chen, C-J. Room temperature ionic liquid as a novel medium for liquid/liquid extraction of metal ions. *Anal. Chim. Acta,* **2003**, *488*, 183-192. [http://dx.doi.org/10.1016/S0003-2670(03)00660-3]

[57] Visser, A.E.; Rogers, R.D. Room-temperature ionic liquids: New solvents for f-element separations and associated solution chemistry. *J. Solid State Chem.,* **2003**, *171*, 109-113. [http://dx.doi.org/10.1016/S0022-4596(02)00193-7]

[58] Whitehead, J.A.; Lawrance, G.A.; McCluskey, A. "Green" leaching: Recyclable and selective leaching of gold-bearing ore in an ionic liquid. *Green Chem.,* **2004**, *6*, 313-315. [http://dx.doi.org/10.1039/B406148A]

[59] Nakashima, K.; Kubota, F.; Maruyama, T.; Goto, M. Ionic liquids as a novel solvent for lanthanide extraction. *Anal. Sci.,* **2003**, *19*(8), 1097-1098. [http://dx.doi.org/10.2116/analsci.19.1097] [PMID: 12945658]

[60] Dai, S.; Ju, Y.H.; Barnes, C.E. Solvent extraction of strontium nitrate by a crown ether using room-temperature ionic liquids. In: *J. Chem. Soc. Dalton Trans.: Inorg. Chem*; 1201-1202., **1999**; 8, pp.

[61] Chun, S.; Dzyuba, S.V.; Bartsch, R.A. Influence of structural variation in room-temperature ionic liquids on the selectivity and efficiency of competitive alkali metal salt extraction by a crown ether. *Anal. Chem.,* **2001**, *73*(15), 3737-3741. [http://dx.doi.org/10.1021/ac010061v] [PMID: 11510842]

[62] Visser, A.E.; Swatloski, R.P.; Reichert, W.M.; Mayton, R.; Sheff, S.; Wierzbicki, A.; Davis, J.H., Jr; Rogers, R.D. Task-specific ionic liquids incorporating novel cations for the coordination and extraction of Hg^{2+} and Cd^{2+}: synthesis, characterization, and extraction studies. *Environ. Sci. Technol.,* **2002**, *36*(11), 2523-2529. [http://dx.doi.org/10.1021/es0158004] [PMID: 12075816]

[63] Holbrey, J.D.; Visser, A.E.; Spear, S.K.; Reichert, W.M.; Swatloski, R.P.; Broker, G.A.; Rogers, R.D. Mercury(II) partitioning from aqueous solutions with a new, hydrophobic ethylene-glycol functionalized bis-imidazolium IL. *Green Chem.,* **2003**, *5*, 129-135. [http://dx.doi.org/10.1039/b300971h]

[64] Eber, J.; Wasserscheid, P.; Jess, A. Deep desulfurization of oil refinery streams by extraction with ionic liquids. *Green Chem.,* **2004**, *6*, 316-322. [http://dx.doi.org/10.1039/B407028C]

[65] Zhang, S.; Zhang, Q.; Conrad Zhang, Z. Extractive desulfurization and denitrogenation of fuels using ionic liquids. *Ind. Eng. Chem. Res.,* **2004**, *43*, 614-622.

[http://dx.doi.org/10.1021/ie030561+]

[66] Huang, C.; Chen, B.; Zhang, J.; Liu, Z.; Li, Y. Desulfurization of gasoline by extraction with new ionic liquids. *Energy Fuels,* **2004**, *18*(6), 1862-1864. [http://dx.doi.org/10.1021/ef049879k]

[67] Lo, W-H.; Yang, H-Y.; Wei, G-T. One-pot desulfurization of light oils by chemical oxidation and solvent extraction with room temperature ionic liquids. *Green Chem.,* **2003**, *5*, 639-642. [http://dx.doi.org/10.1039/b305993f]

[68] Berthod, A.; Carda-Broch, S. A new class of solvents for CCC: the room temperature ionic liquids. *J. Liq. Chromatogr. Relat. Technol.,* **2003**, *26*, 1493-1508. [http://dx.doi.org/10.1081/JLC-120021262]

[69] Marszall, M.P.; Kaliszan, R. Application of ionic liquids in liquid chromatography. *Crit. Rev. Anal. Chem.,* **2007**, *37*, 127-140. [http://dx.doi.org/10.1080/10408340601107847]

[70] Shetty, P.H.; Youngberg, P.J.; Kersten, B.R.; Poole, C.F. Solvent properties of liquid organic salts used as mobile phases in microcolumn reversed-phase liquid chromatography. *J. Chromatogr. A,* **1987**, *411*, 61-79. [http://dx.doi.org/10.1016/S0021-9673(00)93960-8]

[71] Waichigo, M.M.; Danielson, N.D. Ethylammonium formate as an organic solvent replacement for ion-pair reversed-phase liquid chromatography. *J. Chromatogr. Sci.,* **2006**, *44*(10), 607-614. [http://dx.doi.org/10.1093/chromsci/44.10.607] [PMID: 17254370]

[72] Zheng, J.; Row, K.H. Effects of ionic liquid on the separation of 2-chlorophenol and 2,4,6-trichlorophenol in RP-HPLC. *J. Chromatogr. Sci.,* **2009**, *47*(5), 392-395. [http://dx.doi.org/10.1093/chromsci/47.5.392] [PMID: 19476708]

[73] Xiao, X.H.; Zhao, L.; Liu, X.; Jiang, S.X. Ionic liquids as additives in high performance liquid chromatography—analysis of amines and the interaction mechanism of ionic liquids. *Anal. Chim. Acta,* **2004**, *519*, 207-211. [http://dx.doi.org/10.1016/j.aca.2004.06.038]

[74] Kaliszan, R.; Marszałł, M.P.; Markuszewski, M.J.; Baczek, T.; Pernak, J. Suppression of deleterious effects of free silanols in liquid chromatography by imidazolium tetrafluoroborate ionic liquids. *J. Chromatogr. A,* **2004**, *1030*(1-2), 263-271. [http://dx.doi.org/10.1016/j.chroma.2003.09.020] [PMID: 15043278]

[75] He, L.; Zhang, W.; Zhao, L.; Liu, X.; Jiang, S. Effect of 1-alkyl-3-methylimidazolium-based ionic liquids as the eluent on the separation of ephedrines by liquid chromatography. *J. Chromatogr. A,* **2003**, *1007*(1-2), 39-45. [http://dx.doi.org/10.1016/S0021-9673(03)00987-7] [PMID: 12924549]

[76] Marszall, M.P.; Baczek, T.; Kaliszan, R. Reduction of silanophilic interactions in liquid chromatography with the use of ionic liquids. *Anal. Chim. Acta,* **2005**, *547*, 172-178. [http://dx.doi.org/10.1016/j.aca.2005.05.045]

[77] Zhang, W.Z.; He, L.J.; Gu, Y.L.; Liu, X.; Jiang, S.X. Effect of ionic liquids as mobile phase additives on retention of catecholamines in reversed-phase highperformance liquid chromatography. *Anal. Lett.,* **2003**, *36*, 827-838. [http://dx.doi.org/10.1081/AL-120018802]

[78] Waichigo, M.M.; Danielson, N.D. Comparison of ethylammonium formate to methanol as a mobile-phase modifier for reversed-phase liquid chromatography. *J. Sep. Sci.,* **2006**, *29*(5), 599-606. [http://dx.doi.org/10.1002/jssc.200500367] [PMID: 16605076]

[79] Martín-Calero, A.; Pino, V.; Ayala, J.H.; González, V.; Afonso, A.M. Ionic liquids as mobile phase additives in high-performance liquid chromatography with electrochemical detection: application to the determination of heterocyclic aromatic amines in meat-based infant foods. *Talanta,* **2009**, *79*(3),

590-597.
[http://dx.doi.org/10.1016/j.talanta.2009.04.032] [PMID: 19576417]

[80] Holbrey, J.D.; Reichert, W.M.; Nieuwenhuyzen, M.; Sheppard, O.; Hardacre, C.; Rogers, R.D. Liquid clathrate formation in ionic liquid-aromatic mixtures. *Chem. Commun. (Camb.),* **2003**, *4*(4), 476-477. [http://dx.doi.org/10.1039/b212726a] [PMID: 12638957]

[81] Zheng, J.; Row, K.H. Separation of guanine and hypoxanthine with some ionic liquids in RP-HPLC. *Am. J. Appl. Sci.,* **2006**, *3*, 2160-2166. [http://dx.doi.org/10.3844/ajassp.2006.2160.2166]

[82] Ruiz-Angel, M.J.; Carda-Broch, S.; Berthod, A. Ionic liquids *versus* triethylamine as mobile phase additives in the analysis of beta-blockers. *J. Chromatogr. A,* **2006**, *1119*(1-2), 202-208. [http://dx.doi.org/10.1016/j.chroma.2005.11.132] [PMID: 16386264]

[83] Pan, L.; LoBrutto, R.; Kazakevich, Y.V.; Thompson, R. Influence of inorganic mobile phase additives on the retention, efficiency and peak symmetry of protonated basic compounds in reversed-phase liquid chromatography. *J. Chromatogr. A,* **2004**, *1049*(1-2), 63-73. [http://dx.doi.org/10.1016/S0021-9673(04)01149-5] [PMID: 15499918]

[84] Berthod, A.; Ruiz-Angel, M.J.; Huguet, S. Nonmolecular solvents in separation methods: dual nature of room temperature ionic liquids. *Anal. Chem.,* **2005**, *77*(13), 4071-4080. [http://dx.doi.org/10.1021/ac050304+] [PMID: 15987112]

[85] Sun, Y.; Cabovska, B.; Evans, C.E.; Ridgway, T.H.; Stalcup, A.M. Retention characteristics of a new butylimidazolium-based stationary phase. *Anal. Bioanal. Chem.,* **2005**, *382*(3), 728-734. [http://dx.doi.org/10.1007/s00216-004-3051-x] [PMID: 15883788]

[86] Van Meter, D.S.; Sun, Y.; Parker, K.M.; Stalcup, A.M. Retention characteristics of a new butylimidazolium-based stationary phase. Part II: anion exchange and partitioning. *Anal. Bioanal. Chem.,* **2008**, *390*(3), 897-905. [http://dx.doi.org/10.1007/s00216-007-1739-4] [PMID: 18040667]

[87] Qiu, H.; Jiang, S.; Liu, X. N-Methylimidazolium anion-exchange stationary phase for high-performance liquid chromatography. *J. Chromatogr. A,* **2006**, *1103*(2), 265-270. [http://dx.doi.org/10.1016/j.chroma.2005.11.035] [PMID: 16325829]

[88] Wang, Q.; Baker, G.A.; Baker, S.N.; Colón, L.A. Surface confined ionic liquid as a stationary phase for HPLC. *Analyst (Lond.),* **2006**, *131*(9), 1000-1005. [http://dx.doi.org/10.1039/B607337A] [PMID: 17047799]

[89] Qiu, H.; Jiang, S.; Liu, X.; Zhao, L. Novel imidazolium stationary phase for high-performance liquid chromatography. *J. Chromatogr. A,* **2006**, *1116*(1-2), 46-50. [http://dx.doi.org/10.1016/j.chroma.2006.03.016] [PMID: 16616175]

[90] Chaves, A.R.; Chiericato Júnior, G.; Queiroz, M.E. Solid-phase microextraction using poly(pyrrole) film and liquid chromatography with UV detection for analysis of antidepressants in plasma samples. *J. Chromatogr. B Analyt. Technol. Biomed. Life Sci.,* **2009**, *877*(7), 587-593. [http://dx.doi.org/10.1016/j.jchromb.2008.12.070] [PMID: 19185550]

[91] Brix, R.; Bahi, N.; Lopez de Alda, M.J.; Farré, M.; Fernandez, J.M.; Barceló, D. Identification of disinfection by-products of selected triazines in drinking water by LC-Q-ToF-MS/MS and evaluation of their toxicity. *J. Mass Spectrom.,* **2009**, *44*(3), 330-337. [http://dx.doi.org/10.1002/jms.1509] [PMID: 19034890]

[92] Endo, Y.; Tagiri-Endo, M.; Seo, H.S.; Fujimoto, K. Identification and quantification of molecular species of diacyl glyceryl ether by reversed-phase high-performance liquid chromatography with refractive index detection and mass spectrometry. *J. Chromatogr. A,* **2001**, *911*(1), 39-45. [http://dx.doi.org/10.1016/S0021-9673(00)01240-1] [PMID: 11269594]

[93] Savaliya, A.A.; Shah, R.P.; Prasad, B.; Singh, S. Screening of Indian aphrodisiac ayurvedic/herbal healthcare products for adulteration with sildenafil, tadalafil and/or vardenafil using LC/PDA and

extracted ion LC-MS/TOF. *J. Pharm. Biomed. Anal.,* **2010**, *52*(3), 406-409. [http://dx.doi.org/10.1016/j.jpba.2009.05.021] [PMID: 19540696]

[94] Silva, M.M.; Bergamasco, J.; Lira, S.P.; Lopes, N.P.; Hajdu, E.; Peixinho, S.; Berlinck, R.G. Dereplication of bromotyrosine-derived metabolites by LCPDA-MS And analysis of the chemical profile of 14 aplysina sponge specimens from the Brazilian coastline. *Aust. J. Chem.,* **2010**, *63*, 886-894. [http://dx.doi.org/10.1071/CH09616]

[95] Varandas, S.; Teixeira, M.J.; Marques, J.C.; Aguiar, A.; Alves, A.; Bastos, M.M. Glucose and fructose levels on grape skin: interference in Lobesia botrana behavior. *Anal. Chim. Acta,* **2004**, *513*, 351-355. [http://dx.doi.org/10.1016/j.aca.2003.11.086]

[96] Qiu, H.; Jiang, S.; Liu, X. N-Methylimidazolium anion-exchange stationary phase for high-performance liquid chromatography. *J. Chromatogr. A,* **2006**, *1103*(2), 265-270. [http://dx.doi.org/10.1016/j.chroma.2005.11.035] [PMID: 16325829]

[97] Qiu, H.; Jiang, S.; Liu, X.; Zhao, L. Novel imidazolium stationary phase for high-performance liquid chromatography. *J. Chromatogr. A,* **2006**, *1116*(1-2), 46-50. [http://dx.doi.org/10.1016/j.chroma.2006.03.016] [PMID: 16616175]

[98] Qiu, H.; Jiang, Q.; Liu, X.; Jiang, S. Comparison of anion-exchange and hydrophobic interactions between two new silica-based long-chain alkylimidazolium stationary phases for LC. *Chromatographia,* **2008**, *68*, 167-171. [http://dx.doi.org/10.1365/s10337-008-0683-6]

[99] Auler, L.M.; Silva, C.R.; Collins, K.E.; Collins, C.H. New stationary phase for anion-exchange chromatography. *J. Chromatogr. A,* **2005**, *1073*(1-2), 147-153. [http://dx.doi.org/10.1016/j.chroma.2004.10.012] [PMID: 15909516]

[100] Peng, J.F.; Liu, J.F.; Hu, X.L.; Jiang, G.B. Direct determination of chlorophenols in environmental water samples by hollow fiber supported ionic liquid membrane extraction coupled with high-performance liquid chromatography. *J. Chromatogr. A,* **2007**, *1139*(2), 165-170. [http://dx.doi.org/10.1016/j.chroma.2006.11.006] [PMID: 17113589]

[101] Liu, S-J.; Zhou, F.; Zhao, L.; Xiao, X.H.; Liu, X.; Jiang, S.X. Immobilized 1,3-Dialkylimidazolium Salts as New Interface in HPLC Separation. *Chem. Lett.,* **2004**, *33*, 496-497. [http://dx.doi.org/10.1246/cl.2004.496]

[102] Stalcup, A.A.; Cabovska, B. Ionic liquids in chromatography and capillary electrophoresis. *J. Liq. Chromatogr. Relat. Technol.,* **2004**, *27*, 1443-1459. [http://dx.doi.org/10.1081/JLC-120030611]

[103] Yanes, E.G.; Gratz, S.R.; Baldwin, M.J.; Robison, S.E.; Stalcup, A.M. Capillary electrophoretic application of 1-alkyl-3-methylimidazolium-based ionic liquids. *Anal. Chem.,* **2001**, *73*(16), 3838-3844. [http://dx.doi.org/10.1021/ac010263r] [PMID: 11534705]

[104] López-Pastor, M.; Simonet, B.M.; Lendl, B.; Valcárcel, M. Ionic liquids and CE combination. *Electrophoresis,* **2008**, *29*(1), 94-107. [http://dx.doi.org/10.1002/elps.200700645] [PMID: 18161692]

[105] Yanes, E.G.; Gratz, S.R.; Stalcup, A.M. Tetraethylammonium tetrafluoroborate: a novel electrolyte with a unique role in the capillary electrophoretic separation of polyphenols found in grape seed extracts. *Analyst (Lond.),* **2000**, *125*(11), 1919-1923. [http://dx.doi.org/10.1039/b004530f] [PMID: 11193077]

[106] Yue, M.E.; Shi, Y.P. Application of 1-alkyl-3-methylimidazolium-based ionic liquids in separation of bioactive flavonoids by capillary zone electrophoresis. *J. Sep. Sci.,* **2006**, *29*(2), 272-276. [http://dx.doi.org/10.1002/jssc.200500292] [PMID: 16524102]

[107] François, Y.; Varenne, A.; Juillerat, E.; Servais, A.C.; Chiap, P.; Gareil, P. Nonaqueous capillary

electrophoretic behavior of 2-aryl propionic acids in the presence of an achiral ionic liquid. A chemometric approach. *J. Chromatogr. A,* **2007**, *1138*(1-2), 268-275. [http://dx.doi.org/10.1016/j.chroma.2006.10.063] [PMID: 17113588]

[108] Tran, C.D.; Mejac, I. Chiral ionic liquids for enantioseparation of pharmaceutical products by capillary electrophoresis. *J. Chromatogr. A,* **2008**, *1204*(2), 204-209. [http://dx.doi.org/10.1016/j.chroma.2008.04.013] [PMID: 18440541]

[109] Tian, Y.; Feng, R.; Liao, L.; Liu, H.; Chen, H.; Zeng, Z. Dynamically coated silica monolith with ionic liquids for capillary electrochromatography. *Electrophoresis,* **2008**, *29*(15), 3153-3159. [http://dx.doi.org/10.1002/elps.200800056] [PMID: 18654980]

[110] Armstrong, D.W.; He, L.; Liu, Y.S. Examination of ionic liquids and their interaction with molecules, when used as stationary phases in gas chromatography. *Anal. Chem.,* **1999**, *71*(17), 3873-3876. [http://dx.doi.org/10.1021/ac990443p] [PMID: 10489532]

[111] Anderson, J.L.; Armstrong, D.W. High-stability ionic liquids. A new class of stationary phases for gas chromatography. *Anal. Chem.,* **2003**, *75*(18), 4851-4858. [http://dx.doi.org/10.1021/ac0345749] [PMID: 14674463]

[112] Letcher, T.M.; Soko, B.; Reddy, P.; Deenadayalu, N. Determinationof activity coefficients at infinite dilution of solutes in the ionic liquid 1-hexyl-3- methylimidazolium tetrafluoroborate using gas-liquid chromatography at the temperatures 298.15 K and 323.15 K. *J. Chem. Eng. Data,* **2003**, *48*, 1587-1590. [http://dx.doi.org/10.1021/je030187k]

[113] Warren, D.; Trevor, L.M.; Deresh, R.; Raal, D.J. Activity coefficients of hydrocarbon solutes at infinite dilution in the IL, 1-methyl-3-octyl-imidazolium chloride from gas-liquid chromatography. *J. Chem. Thermodyn.,* **2003**, *35*, 1335-1341. [http://dx.doi.org/10.1016/S0021-9614(03)00091-0]

[114] Heintz, A.; Kulikov, D.V.; Verevkin, S.P. Thermodynamic properties of mixtures containing ionic liquids: Activity coefficients at infinite dilution of polar solutes in 4-methyl-N-butylpyridinium tetrafluoroborate using gas-liquid chromatography. *J. Chem. Thermodyn.,* **2002**, *34*, 1341-1347. [http://dx.doi.org/10.1006/jcht.2002.0961]

[115] Heintz, A.; Kulikov, D.V.; Verevkin, S.P. Thermodynamic properties of mixtures ccontaining ionic liquids: II. Activity coefficients at infinite dilution of hydrocarbons and polar solutes in 1-methyl-3-ethyl-imidazolium bis(trifluoromethyl-sulfonyl) amide and in 1,2-dimethyl-3-ethyl-imidazolium bis(trifluoromethylsulfonyl) amide using gas-liquid chromatography. *J. Chem. Eng. Data,* **2002**, *47*, 894-899. [http://dx.doi.org/10.1021/je0103115]

[116] Berthod, A.; He, L.; Armstrong, D.W. Ionic liquids as stationary phase solvents for methylated cyclodextrins in gas chromatography. *Chromatographia,* **2001**, *53*(1–2), 63-68. [http://dx.doi.org/10.1007/BF02492429]

[117] Ding, J.; Welton, T.; Armstrong, D.W. Chiral ionic liquids as stationary phases in gas chromatography. *Anal. Chem.,* **2004**, *76*(22), 6819-6822. [http://dx.doi.org/10.1021/ac049144c] [PMID: 15538810]

[118] Planeta, J.; Roth, M. Partition coefficients of low-volatility solutes in the ionic liquid 1-n-buty--3-methylimidazolium hexafluorophosphate-supercritical CO2 system from chromatographic retention measurements. *J. Phys. Chem.,* **2004**, *108*, 11244-11249. [http://dx.doi.org/10.1021/jp049094f]

[119] Berthod, A.; Carda-Broch, S. Use of the ionic liquid 1-butyl-3-methylimidazolium hexafluorophosphate in countercurrent chromatography. *Anal. Bioanal. Chem.,* **2004**, *380*(1), 168-177. [http://dx.doi.org/10.1007/s00216-004-2717-8] [PMID: 15365674]

[120] Endres, F. Ionic liquids: solvents for the electrodeposition of metals and semiconductors. *ChemPhysChem,* **2002**, *3*(2), 145-154.

[http://dx.doi.org/10.1002/1439-7641(20020215)3:2<144::AID-CPHC144>3.0.CO;2-#] [PMID: 12503121]

[121] Sato, T.; Maruo, T.; Marukane, S.; Takagi, K. Ionic liquids containing carbonate solvent as electrolytes for lithium ion cells. *J. Power Sources,* **2004**, *138*(1–2), 253-261. [http://dx.doi.org/10.1016/j.jpowsour.2004.06.027]

[122] Garcia, B.; Lavallee, S.; Perron, G.; Michot, C.; Armand, M. Room temperature molten salts as lithium battery electrolyte. *Electrochim. Acta,* **2004**, *49*, 4583-4588. [http://dx.doi.org/10.1016/j.electacta.2004.04.041]

[123] Li, Z.; Liu, H.; Liu, Y.; He, P.; Li, J. A room-temperature ionic-liquid-templated proton-conducting gelatinous electrolyte. *J. Phys. Chem. B,* **2004**, *108*, 17512-17518. [http://dx.doi.org/10.1021/jp047252x]

[124] Zein El Abedin, S.; Borissenko, N.; Endres, F. Electrodeposition of nanoscale silicon in a room temperature ionic liquid. *Electrochem. Commun.,* **2004**, *6*, 510-514. [http://dx.doi.org/10.1016/j.elecom.2004.03.013]

[125] Wadhawan, J.D.; Schr¨oder, U.; Neudeck, A.; Wilkins, S.J.; Compton, R.G.; Marken, F.; Consorti, C.S.; de Souza, R.F.; Dupont, J. Ionic liquid modified electrodes: Unusual partitioning and diffusion effects of Fe(CN)64-/3- in droplet and thin layer deposits of 1-methyl-3-(2,6-(S)-dimethylocte--2-yl)-imidazolium tetrafluoroborate. *J. Electroanal. Chem.,* **2000**, *493*(1–2), 75-83. [http://dx.doi.org/10.1016/S0022-0728(00)00308-9]

[126] Ding, J.; Zhou, D.; Spinks, G.; Wallace, G.; Forsyth, S.; Forsyth, M.; MacFarlane, D. Use of ionic liquids as electrolytes in electromechanical actuator systems based on inherently conducting polymers. *Chem. Mater.,* **2003**, *15*, 2392-2398. [http://dx.doi.org/10.1021/cm020918k]

[127] Bennett, M.D.; Leo, D.J. Ionic liquids as stable solvents for ionic polymer transducers. *Sens. Actuators A Phys.,* **2004**, *115*, 79-90. [http://dx.doi.org/10.1016/j.sna.2004.03.043]

[128] Wang, R.; Hoyano, S.; Ohsaka, T. O2 gas sensor using supported hydrophobic room-temperature IL membrane-coated electrode. *Chem. Lett.,* **2004**, *33*, 6-7. [http://dx.doi.org/10.1246/cl.2004.6]

[129] Wang, R.; Okajima, T.; Kitamura, F.; Ohsaka, T. Anovel amperometric O2 gas sensor based on supported room-temperature IL porous polyethylene membrane-coated electrodes. *Electroanalysis,* **2004**, *16*, 66-72. [http://dx.doi.org/10.1002/elan.200302919]

[130] Brinz, T.; Simon, U. *Electrochemical gas sensor with ionic liquids as electrolyte.,* **2004**.

[131] Lee, Y.G.; Chou, T.C. Ionic liquid ethanol sensor. *Biosens. Bioelectron.,* **2004**, *20*(1), 33-40. [http://dx.doi.org/10.1016/j.bios.2004.01.026] [PMID: 15142574]

[132] Gaillon, L.; Bedioui, F. Voltammetric analysis of the catalytic reactivity of electrogenerated CoI-salen with organohalogenated derivatives in an ionic liquid at room temperature. *J. Mol. Catal. Chem.,* **2004**, *214*(1), 91-94. [http://dx.doi.org/10.1016/j.molcata.2003.10.062]

[133] Lu, W.; Fadeev, A.G.; Qi, B.; Mattes, B.R. Fabricating conducting polymer electrochromic devices using ionic liquids. *J. Electrochem. Soc.,* **2004**, *151*, H33-H39. [http://dx.doi.org/10.1149/1.1640635]

[134] Appleby, D.; Hussey, C.L.; Seddon, K.R.; Turp, J.E. Room-temperature ionic liquids as solvents for electronic absorption spectroscopy of halide complexes. *Nature,* **1986**, *323*, 614-615. [http://dx.doi.org/10.1038/323614a0]

[135] Seddon, K.R. Electronic Absorption Spectroscopy in Room Temperature Ionic Liquids. In: *Molten Salt Chemistry: NATO ASI Series. Ser. C: Math. Phys. Sci. (ASIC)*; Reidel Publishing Co.: Dordrecht,

1987; 202, pp. 365-381.

[136] Billard, I.; Moutiers, G.; Labet, A.; El Azzi, A.; Gaillard, C.; Mariet, C.; Lützenkirchen, K. Stability of divalent europium in an ionic liquid: spectroscopic investigations in 1-methyl-3-butylimidazolium hexafluorophosphate. *Inorg. Chem.*, **2003**, *42*(5), 1726-1733. [http://dx.doi.org/10.1021/ic0260318] [PMID: 12611545]

[137] Lu, J.; Liotta, C.L.; Eckert, C.A. Spectroscopically probing microscopic solvent properties of room-temperature ionic liquids with the addition of carbon dioxide. *J. Phys. Chem. A,* **2003**, *107*, 3995-4000. [http://dx.doi.org/10.1021/jp0224719]

[138] Alvaro, M.; Ferrer, B.; Garc'ıa, H.; Narayana, M. Screening of an ionic liquid as medium for photochemical reactions. *Chem. Phys. Lett.*, **2002**, *362*, 435-440. [http://dx.doi.org/10.1016/S0009-2614(02)01138-7]

[139] Armstrong, D.W.; Zhang, L-K.; He, L.; Gross, M.L. Ionic liquids as matrixes for matrix-assisted laser desorption/ionization mass spectrometry. *Anal. Chem.*, **2001**, *73*(15), 3679-3686. [http://dx.doi.org/10.1021/ac010259f] [PMID: 11510834]

[140] Zabet-Moghaddam, M.; Heinzle, E.; Tholey, A. Qualitative and quantitative analysis of low molecular weight compounds by ultraviolet matrix-assisted laser desorption/ionization mass spectrometry using ionic liquid matrices. *Rapid Commun. Mass Spectrom.*, **2004**, *18*(2), 141-148. [http://dx.doi.org/10.1002/rcm.1293] [PMID: 14745762]

[141] Carda-Broch, S.; Berthod, A.; Armstrong, D.W. Ionic matrices for matrix-assisted laser desorption/ionization time-of-flight detection of DNA oligomers. *Rapid Commun. Mass Spectrom.*, **2003**, *17*(6), 553-560. [http://dx.doi.org/10.1002/rcm.931] [PMID: 12621617]

[142] Mank, M.; Stahl, B.; Boehm, G. 2,5-Dihydroxybenzoic acid butylamine and other ionic liquid matrixes for enhanced MALDI-MS analysis of biomolecules. *Anal. Chem.*, **2004**, *76*(10), 2938-2950. [http://dx.doi.org/10.1021/ac030354j] [PMID: 15144208]

[143] Romero-Sanz, I.; Bocanegra, R.; Fernandez de la Mora, J.; Gamero-Castano, M. Source of heavy molecular ions based on Taylor cones of ionic liquids operating in the pure ion evaporation regime. *J. Appl. Phys.*, **2003**, *94*, 3599-3605. [http://dx.doi.org/10.1063/1.1598281]

[144] Romero-Sanz, I.; Fernandez de la Mora, J. Energy distribution and spatial structure of electrosprays of ionic liquids in vacuo. *J. Appl. Phys.*, **2004**, *95*, 2123-2129. [http://dx.doi.org/10.1063/1.1639947]

[145] Dyson, P.J.; Khalaila, I.; Luettgen, S.; McIndoe, J.S.; Zhao, D. Direct probe electrospray (and nanospray) ionization mass spectrometry of neat ionic liquids. *Chem. Commun. (Camb.),* **2004**, *19*(19), 2204-2205. [http://dx.doi.org/10.1039/b407217k] [PMID: 15467873]

[146] Jackson, G.P.; Duckworth, D.C. Electrospray mass spectrometry of undiluted ionic liquids. *Chem. Commun. (Camb.),* **2004**, *5*(5), 522-523. [http://dx.doi.org/10.1039/b314754a] [PMID: 14973588]

[147] Karkkainen, J.; Asikkala, J.; Laitinen, R.S.; Lajunen, M. Effect of temperature on the purity of product in the preparation of 1-butyl-3-methylimidazolium-based ionic liquids. *Zeitsch. Naturforsch. B: Chem. Sci.*, **2004**, *59*, 763-770.

[148] Zhao, D. Analysis of ionic liquids and dissolved species by electrospray ionization MS. *Aust. J. Chem.*, **2004**, *57*, 509-510. [http://dx.doi.org/10.1071/CH03327]

[149] Quijano, G.; Couvert, A.; Amrane, A. Ionic liquids: applications and future trends in bioreactor technology. *Bioresour. Technol.*, **2010**, *101*(23), 8923-8930. [http://dx.doi.org/10.1016/j.biortech.2010.06.161] [PMID: 20667722]

[150] Pham, T.P.; Cho, C.W.; Yun, Y.S. Environmental fate and toxicity of ionic liquids: a review. *Water Res.,* **2010**, *44*(2), 352-372.
[http://dx.doi.org/10.1016/j.watres.2009.09.030] [PMID: 19854462]

[151] Azimova, M.A.; Morton, S.A., III; Frymier, P.D. Comparison of three bacterial toxicity assays for imidazolium-derived ionic liquids. *J. Environ. Eng.,* **2009**, *10*, 1388-1392.
[http://dx.doi.org/10.1061/(ASCE)EE.1943-7870.0000092]

[152] Pernak, J.; Sobaszkiewicz, K.; Mirska, I. Anti-microbial activities of ionic liquids. *Green Chem.,* **2003**, *5*, 52-56.
[http://dx.doi.org/10.1039/b207543c]

[153] Romero, A.; Santos, A.; Tojo, J.; Rodríguez, A. Toxicity and biodegradability of imidazolium ionic liquids. *J. Hazard. Mater.,* **2008**, *151*(1), 268-273.
[http://dx.doi.org/10.1016/j.jhazmat.2007.10.079] [PMID: 18063302]

[154] Docherty, K.M.; Kulpa, C.F. Toxicity and antimicrobial activity of imidazolium and pyridinium ionic liquids. *Green Chem.,* **2005**, *7*, 185-189.
[http://dx.doi.org/10.1039/b419172b]

[155] Lee, S.M.; Chang, W.J.; Choi, A.R.; Koo, Y.M. Influence of ionic liquids on the growth of Escherichia coli. *Korean J. Chem. Eng.,* **2005**, *22*, 687-690.
[http://dx.doi.org/10.1007/BF02705783]

[156] Sendovski, M.; Nir, N.; Fishman, A. Bioproduction of 2-phenylethanol in a biphasic ionic liquid aqueous system. *J. Agric. Food Chem.,* **2010**, *58*(4), 2260-2265.
[http://dx.doi.org/10.1021/jf903879x] [PMID: 20095569]

CHAPTER 11

New Trends in Enantioanalysis of Pharmaceutical Compounds using Electrochemical Sensors

Raluca-Ioana Stefan-van Staden*

Laboratory of Electrochemistry and PATLAB Bucharest, National Institute of Research for Electrochemistry and Condensed Matter, 202 Splaiul Independentei 060021, Bucharest, Romania

Abstract: Enantioanalysis became very important in pharmaceutical field due to the different pathways of the enantiomers of the same chiral compound. Therefore, development of reliable methods for enantioanalysis was considered necessary. The utilization of enantioselective electrodes for enantioanalysis simplified enantioanalysis and improved tremendously its reliability. This chapter shows the role and new trends of enantioselective electrodes for pharmaceutical enantioanalysis, the design of these electrodes and their applications in enantioanalysis.

Keywords: Chiral selectors, Enantioselective electrodes, Pharmaceutical enantioanalysis, Stochastic enantioselective sensors.

ENANTIOANALYSIS – A NEED FOR PHARMACEUTICAL INDUSTRY

For many years, all chiral compounds were produced and formulated without care about the ratio between the enantiomers, or type of the enantiomers contained by the final pharmaceutical compound. A simple positive evaluation of the uniformity content test was to give credentials to the market as the resulting product. Because a significant number of the pharmaceutical compounds had a chiral structure, depending on the synthesis conditions, one or more enantiomers were produced in a stochastic manner. Unwanted secondary effects could not have been fully explained until the time when chiral drugs were produced as single enantiomers and the pathway of these in the body was checked. From that point in time, more attention has been paid not only to the purity of the active compounds, but also to the enantiopurity tests, being recognized that not all enantiomers are those for which the drug should be designed, namely in some of the cases, the other enantiomer(s) having a different pathway in body, arriving in unwanted

* **Corresponding author Raluca-Ioana Stefan-van Staden:** Laboratory of Electrochemistry and PATLAB Bucharest, National Institute of Research for Electrochemistry and Condensed Matter, 202 Splaiul Independentei >060021, Bucharest, Romania; Tel: +40751507779; Fax: +4021 3163113; E-mail: ralucavanstaden@gmail.com

Atta-ur-Rahman, Sibel A. Ozkan & Rida Ahmed (Eds.)

organs, and creating critical conditions, or having no effect resulting in no improvement of the state of health of the patient, were poison for the body affecting the general health condition of the patients. Synthesis was also conducted in the direction of the production of the wanted enantiomer to be formulated in the pharmaceutical compound.

Therefore, the enantiomeric drugs can be classified according to these differences in the following classes: (1) only one of the enantiomers presents pharmacological activity the other being considered impurity that may be active with desirable or undesirable activity or inactive (*e.g.*, the enantiomers of antihypertensive active methyldopa), (2) both enantiomers present the same qualitative and quantitative pharmacological activities (*e.g.*, the enantiomers of promethazine), (3) the enantiomers may have activity that is qualitative similar but quantitatively different (*e.g.*, the adrenergic blocking agent propranolol and verapamil used in anginal therapy), (4) both enantiomers have independent therapeutic potencies (*e.g.*, R-propoxyphene is an analgesic agent and S-propoxyphene is an effective antitussive), (5) one of the enantiomers exhibits undesirable side effects (*e.g.*, S(+)-ketamine is an active anesthetic and analgesic, and undesirable side effects like hallucination and agitation is being associated with the R(-)-distomer), (6) some enantiomers exhibit therapeutic advantages (*e.g.*, R(+)-indacrinone is active diuretic agent but possesses the undesirable side effect of uric acid retention where S(-)-indacrinone can antagonize the undesirable side effect of its antipode by acting as uricosuric).

Since one has to produce pharmaceutical compounds based on one enantiomer as active compound, the need for developing methods to perform enantioanalysis is essential. The first methods developed were based on enantioseparation using chromatography, and later the capillary zone electrophoresis. These methods are now standard methods for enantioanalysis and enantiopurity tests performed for each chiral active compound to be formulated. The main disadvantages of these methods remain the laborious sample preparation, the cost per analysis, the time of analysis and also the accuracy of the analysis due to the number of steps used for the sampling step of each analysis. Finding the correct column and maintenance and running costs of the chromatographs are in many cases high, contributing to the increase of the cost per analysis, and also to the increase of the cost of the pharmaceutical compound. Furthermore, in process control, there is an increase of need to develop fast method for *on site* analysis and enantioanalysis in order to be able to get the best results for the active compounds.

Therefore, the development of the enantioselective sensors was a great step in enantioanalysis of chiral drugs. While the chromatographic methods can be mostly used for the qualitative enantioanalysis [1], enantioselective potentiometry and amperometry (enantioselective, potentiometric membrane electrodes and amperometric biosensors) can be used with a high reliability for the quantitative enantioanalysis [2 - 4]. The enantioselective electrodes were also incorporated in the conduits of the flow systems (flow injection analysis and sequential injection analysis) to form a simple low cost enantioanalyzer [5, 6], for the simultaneous enantioanalysis of the S- and R-enantiomer. The advantages of using these electrodes as detectors in such flow systems are the following: no contamination between the samples was recorded; the linear concentration range, the sensitivity, the limit of detection and the response time of the enantioselective electrode are improved in flow systems due to establishment of a small thickness diffusing layer at the electrode surface; the ionic species flowing out of the reference electrode cannot influence the response of the indicator electrode because the reference electrode is usually placed down-stream; the intensity of current or potential is measured in non-equilibrium conditions but always at the same moment after the sample injection.

TYPES OF ENANTIOSELECTIVE, ELECTROCHEMICAL SENSORS USED IN ENANTIOANALYSIS

The classical enantioselective sensors are based on two principles for the enantiomer recognition:

1. Utilization of a molecule with a special architecture that can recognize only one of the enantiomers of the chiral substance; these molecules can be chiral or achiral compounds with special architectures like antibiotics, dextrines, antibodies;
2. Utilization of a catalyst that can favorized only the reaction of one of the enantiomers, *e.g.*, enzymes, or substances immobilized in the membranes used for detection using amperometric techniques.

Accordingly, the classical enantioselective sensors can be divided in the following categories:

- Electrochemical sensors;
- Biosensors;
- Immunosensors.

The electrochemical sensors' class can be divided in (a) Amperometric enantioselective electrodes, and (b) Enantioselective, potentiometric membrane electrodes. A special class of enantioselective stochastic sensors was developed. The principle of molecular recognition was based on the conductivity of the channel situated at the membrane-solution interface.

The type of the electrode selected must also consider the working concentration range and the limit of detection of the electrode which must be in agreement with the concentration of the enantiomer in the sample. It is well-known that the potentiometric electrodes are usually working at concentration between 10^{-2} and 10^{-5} mol/L with limits of detection of 10^{-6} to 10^{-7} mol/L magnitude order. For the enantioselective, potentiometric membrane electrodes the working concentration ranges as well as the limits of detection were brought down to lower concentrations (up to 10^{-8}mol/L) due to selection of the chiral selector as well as of utilization of the KCl solution 0.1Cmol/L as internal solution in the design of the electrode. The amperometric sensors are usually working on concentration ranges of nmol/L to pmol/L and even lower. Therefore, for the pharmaceutical analysis one will prefer to have a potentiometric electrode for the enantioanalysis of the main enantiomer and an amperometric biosensor/immunosensor for the enantioanalysis of its antipode.

The enantioselectivity of these electrodes can be explained using the stability of the complexes formed between the enantiomers and the chiral selector used for the design of the enantioselective, potentiometric membrane electrode: if the stability of the complex formed between the chiral selector and the enantiomer to be determined is higher than the stability of the complex formed between the chiral selector and the antipode of the enantiomer to be determined, than the potentiometric electrode is enantioselective. For the amperometric biosensors, the enantioselectivity is given by the enantioselectivity of the enzyme for the substrate [7]. It is a fact that the advantage of the amperometric biosensors over the enantioselective, potentiometric membrane electrodes is that the amperometric biosensors combines the high enantioselectivity of the enzyme with the high sensitivity of the amperometric transducer. Furthermore, if on is designing amperometric immunosensors for enantioanalysis these can even became enantiospecific due to the high affinity and sensitivity of the immunoreaction as well as to the fact that the antibody will recognize the chiral center.

The lifetime of an electrode is very important for the end-user who will prefer a sensor that can be used for a longer period with the same response characteristics. While the amperometric biosensors must be kept in the fridge at certain temperature and even so they can only be used for maximum three months, most of the enantioselective, potentiometric membrane electrodes can be kept at room temperature and used for more than 6 months or a year. Accordingly, the enantioselective, potentiometric membrane electrodes are selected.

Both types of electrodes, potentiometric and amperometric have got advantages and limitations for enantioanalysis. To select the proper electrode for enantioanalysis there is a need to look for the complexity of the molecule of the enantiomer, matrix from where it must be determined, range of concentration within which it can be found and if necessary requirements like robustness and lifetime must be considered.

DESIGN OF ENANTIOSELECTIVE SENSORS

The reliability of the design of the enantioselective, and potentiometric membrane electrodes is very important for the reliability of the analytical information obtained using these electrodes. The following types of matrices were considered for the design of the sensors: PVC based matrices [8, 9] and carbon based matrices [10 - 43]. A highly reliable design was selected for the imprinted polymers based sensors [44]. The highly reproducible design was the one based on carbon/diamond paste matrix. The non-reproducibility of PVC based matrices is due to non-uniformity and non-reproducibility of the repartition of the electroactive material in the matrix [45].

Chiral Selectors Used in the Design of the Enantioselective, Potentiometric Membrane Electrodes

The selection of the chiral selector is essential for the enantioanalysis using potentiometry, because the potential development is directly correlated with the stability of the complex formed between the chiral selector and enantiomer [46]. Molecular modeling studies were proposed for identification of the best chiral selectors [47]. Unfortunately, the molecular modeling cannot at this stage fully justify the enantioselectivity, but it can show which one of the chiral selectors will form a more stable complex with a given enantiomer.

The most used chiral selectors for the enantioselective, potentiometric membrane electrodes design include crown-ethers [8], cyclodextrins [9 - 23], maltodextrins [24 - 32], antibiotics [33 - 40] and fullerenes [41 - 43].

Design of Carbon Paste Based Enantioselective Electrodes

The carbon paste was prepared by mixing the paraffin oil and graphite powder in a ratio of 1:4 (w/w). The modified paste was obtained by the addition of the solution containing the chiral selector/catalyst (10^{-3} mol/L) to the carbon paste in a ratio 1:1=μL:mg. The unmodified paste was filled into a plastic pipette peak leaving 3 to 4 mm empty in the top to be filled with the modified paste. The diameter of the proposed sensors was 3 mm. Electric contact was obtained using a silver wire inserted into the carbon paste. Before each use, the surface of the electrode was wetted with deionised water and then polished with an alumina paper (polishing strips 30144-001, Orion).

Design of Diamond Paste Based Enantioselective Electrodes

The diamond paste was obtained by mixing the paraffin oil and diamond powder in a ratio of 1:2 (w/w). The chiral selector/catalyst solution (10^{-3} mol/L) was added to the diamond paste in a ratio of 1:2=μL:mg to give the modified diamond paste. The unmodified diamond paste was filled into a plastic pipette peak leaving 3 to 4 mm empty in the top to be filled with the modified diamond paste. The diameter of the sensor was 3 mm. Electric contact was made by inserting a silver wire in the diamond paste. Before each use, the surface of the electrode was wetted with deionized water and then polished with an alumina paper (polishing strips 30144-001, Orion).

Design of PVC Based Enantioselective, Potentiometric Membrane Electrodes

The membrane of the PVC based EPMEs comprise of 1.2% chiral selector, 65.6% plastifiant (orthonitrophenyloctyl ether (oNPOE)/dioctylsebacate (DOS)), 0.4% tetrakis([3,5-bis(trifluoromethyl)phenyl]borate) (TKB) in a tetrahydrofuran solution containing the polyvinyl chloride (PVC) (32.4% from the membrane composition) [8]. A 7 mm diameter circle was cut from the membrane and inserted into an electrode body along with an inner solution containing the enantiomer to be determined in a certain concentration (usually 10^{-3} mol/L).

Design of Molecularly Imprinted Polymers Based Enantioselective Sensors

Of main importance for this class of sensors is the synthesis of polymer [44]. A reliable synthesis will give reliable response characteristics for the sensor as well as reliable analytical information. Polymers can be prepared by either thermal initiation or photolytic initiation with azobisnitriles (AIBN), or herein by Lucirin LR 8893X as photolytic initiator. In a 50 mL tube, 7800 mg, 1,4-butanedienyl diacrylate (BDDA), 3360 mg α-methylacrylic acid (MAA), 165 mg L-phenylalanine and 100 mg photoinitiator, ethyl 2,4,6-trimethylbenzophenyphosphinate (LR 8893 X) in 15 mL methanol were thoroughly mixed. Solubilization was achieved by sonication and the mixture was then cooled on ice. After degassing, the tube was sealed under nitrogen and exposed to a UV-laser beam (366 nm) for 18 h. The polymers obtained were ground in a mortar and wet-sieved in water to the desired size distribution. Particles which passed through a 25 μm sieve constituted a fraction of < 25 μm.

PHARMACEUTICAL ENANTIOANALYSIS USING POTENTIOMETRIC SENSORS

Enantioselective, Potentiometric Membrane Electrodes Based on Polyaniline Films

Yin *et al.* [48] proposed optical active polyaniline films based enantioselective, potentiometric electrodes for the enantioanalysis of chiral amino acids. The slopes were 60mV/decade of concentration. When the sensor was exposed to a mixture of enantiomers, the slope of the response changed along with their proportion. Therefore, this parameter was employed as an index of the enantiomeric purity of the mixture.

Enantioselective, Potentiometric Membrane Electrodes Based on Cyclodextrins

Enantioselective, potentiometric membrane electrodes based on α- and γ-cyclodextrins were proposed for the enantioanalysis of R-baclofen [19]. 59.50 mV/pR-baclofen and 51.00 mV/pR-baclofen were recorded as slopes for the EPME based on α- and γ-cyclodextrin, respectively. The detection limits of the proposed electrodes were: $7x10^{-9}$ mol/L for the EPME based on α-cyclodextrin, and $1.44x10^{-10}$ mol/L for the EPME based on γ-cyclodextrin. Both electrodes were enantioselective. Due to their response characteristics, selectivity and enantioselectivity, the EPMEs based on cyclodextrins can be employed for the assay of R-baclofen raw materials and its pharmaceutical formulation, *e.g.*, Norton-Baclofen® tablets.

S-Captopril was the first angiotensin-converting enzyme inhibitor determined using an enantioselective, potentiometric membrane electrode [11]. The enantioselective, potentiometric membrane electrode based on impregnation of 2-hydroxy-3-trimethylammoniopropyl-β-cyclodextrin (as chloride salt) solution in a carbon paste, was successfully used for S-captopril assay in the linear concentration range 10^{-6} - 10^{-2} mol/L, within a pH range 3.0 - 6.5. The slope recorded was near-Nernstian, of 57.70 mV/decade of concentration, and the limit of detection of $2 x 10^{-7}$ mol/L. S-captopril was recovered with an average recovery of 99.99% (RSD = 0.05%), therefore it was used for the enantioanalysis of S-captopril in pharmaceutical tablet formulations (25 mg S-captopril/tablet), when the average recovery was 99.69% (RSD = 0.39%). The sensor was enantioselective. The response characteristics of this electrode were tested also for R-captopril when a non-Nernstian slope of 27.30 mV/decade of concentration as well as a detection limit of 10^{-6} magnitude order were obtained.

Enantioanalysis of S-Cilazapril [13] was performed using an enantioselective, potentiometric membrane electrode based on modification of carbon paste with 2-hydroxy-3-trimethylammoniopropyl-β-cyclodextrin (as chloride salt) solution in a carbon paste. The slope of the electrode was near-Nernstian (56.85 mV/decade of concentration) on the 10^{-5} - 10^{-2} mol/L concentration range (pH between 3.0 and 5.5), with a limit of detection of 5.0 x 10^{-6} mol/L. The electrode was enantioselective. The average recovery of S-cilazapril was 99.36% with a RSD of 0.27%.

S-Enalapril enantioanalysis was performed using the potentiometric electrode based on impregnation of 2-hydroxy-3-trimethylammoniopropyl-β-cyclodextrin (as chloride salt) solution (10^{-3} mol/L) in a carbon paste, in the 3.6 x 10^{-5} - 6.4 x 10^{-2} mol/L (pH between 3.0 and 6.0) linear concentration range with a detection limit of 1.0 x 10^{-5} mol/L [12]. The slope was near-Nernstian: 55.00 mV/decade of concentration. The average recovery of S-enalapril - raw material was 99.96% (RSD = 0.098%). The potentiometric selectivity coefficient over D-proline (6.5 x 10^{-4}) proved the sensor's enantioselectivity. S-Enalapril was determined from pharmaceutical tablets with an average recovery of 99.59% (RSD = 0.20%).

(-)-Ephedrine is the pharmacologically active enantiomers of ephedrine. The enantiopurity tests for (-)-ephedrine can be performed using a peroctylated α-cyclodextrin based plastic membrane electrode [9]. Bis(1-butylpentyl)adipate (BBPAP) was used as plasticizer and 10^{-3} mol/L NH_4Cl as inner solution. The slope of the electrode was 60mV/decade of concentration, and the potentiometric enantioselectivity coefficient is less than 10^{-4}. The limit of detection is of 10^{-7} mol/L magnitude order.

Four enantioselective, potentiometric membrane electrodes (EPMEs) based on carbon paste impregnated with α-, β-, 2-hydroxyl-3-trimethylammoniopropyl ---(as chloride salt) and γ-cyclodextrins, were proposed as a chiral selectors for assay of S-flurbiprofen raw materials and in its pharmaceutical formulation Froben 100® tablets [21]. The slopes of the electrodes based on α-, β-, 2-hydroxy--3-trimethylammoniopropyl -β-(as chloride salt) and γ-cyclodextrins were 57.90, 59.20, 56.10, and 59.90 mV/decade of concentration within the following concentration ranges: 1.5x10^{-8} – 1.5 x10^{-4}; 1.5x10^{-8} – 1.5 x10^{-4}; 3.0x10^{-7} – 5.5 x10^{-4}; 1.5x10^{-8} – 1.5 x10^{-3} mol/L having the limits of detection of 10^{-9} and 10^{-8} mol/L magnitude order. The electrodes were enantioselective. S-flurbiprofen was assayed in the Froben 100® tablets with an average recovery higher than 99.80%.

The enantioanalysis of L-histidine raw material and in its pharmaceutical formulation was performed using four enantioselective, potentiometric membrane electrodes based on carbon paste impregnated with α-, β-, 2-hydroxyl-

3-trimethylammoniopropyl -β-(as chloride salt) and γ-cyclodextrins [22]. The proposed electrodes showed near-Nernstian response over L-histidine (between 56.36 and 57.76 mV/decade of concentration) but not over D-histidine. The limits of detection were in the 10^{-8} and 10^{-11} mol/L magnitude order with working concentration ranges up to 10^{-7} and 10^{-10} mol/L. The recovery of L-histidine in the presence of D-histidine was higher than 99.10% with RSD lower than 0.1%.

S-pentopril was determined in the presence of R-pentopril using an enantioselective, potentiometric membrane electrode based on impregnation of 2-hydroxy-3-trimethylammoniopropyl-β-cyclodextrin (as chloride salt) solution in a carbon paste. The enantioanalysis can be performed on the 10^{-6} - 10^{-2} mol/L (pH = 3.0 - 5.5) concentration range [13]. The limit of detection was 7.58 x 10^{-5} mol/L. Slope was near-Nernstian: 58.16 mV/decade of concentration. The potentiometric selectivity coefficient over D-proline was 6.9 x 10^{-4}. S-pentopril average recovery is 99.79% (RSD = 0.17%).

The enantioselective, potentiometric membrane electrode based on impregnation of 2-hydroxy-3-trimethylammoniopropyl-β-cyclodextrin (as chloride salt) solution in a carbon paste, can be used for the S-perindopril assay in the 10^{-5} - 10^{-2} mol/L (pH = 2.35 - 6.00) concentration range [15]. The detection limit was 5 x 10^{-6} mol/L, and the slope was near-Nernstian: 54.23 mV/decade of concentration. The selectivity was checked over R-perindopril and D-proline. The 10^{-4} magnitude orders obtained for potentiometric selectivity coefficients proved enantioselectivity of the sensor. For R-perindopril, the potentiometric, enantioselective membrane electrode has the following response characteristics: slope is non-Nernstian (38.00 mV/decade of concentration); linear range between 10^{-4} and 10^{-2} mol/L, and detection limit of 10^{-6} mol/L magnitude order. S-perindopril can be determined with an average recovery of 99.58% (RSD = 0.33%).

Better results for the enantioanalysis of S-perindopril were obtained when the carbon-paste was impregnated with α-, β- and γ-cyclodextrins as chiral selectors [16]. There had been improvements in detection limits (lower with unsubstituted CDs as chiral selectors), enantioselectivity (higher especially when α-CD was used as chiral selector) and the values of the slope of the electrodes (higher especially when α-CD and γ-CD were used as chiral selectors).

An enantioselective, potentiometric membrane electrode based on impregnation of 2-hydroxy-3-trimethylammoniopropyl-β-cyclodextrin (as chloride salt) solution in a carbon paste was proposed for the assay of L-proline [10]. The linear concentration range for the assay of L-proline was between 5.0x 10^{-5} and 1.5 x 10^{-1} mol/L, with a detection limit of 1.0 x 10^{-5} mol/L. The average recovery was

99.90% with a RSD of 0.12% (n=10). The electrode was enantioselective.

Enantioselective, potentiometric membrane electrodes based on carbon paste impregnated with α-, β- and γ-cyclodextrins were proposed for the assay of L-proline [17]. Response characteristics showed that the proposed electrodes could be reliably used for the enantioanalysis of L-proline, with the best enantioselectivity and time-stability exhibited by α-cyclodextrin based enantioselective, potentiometric membrane electrode. The enantioselective, potentiometric membrane electrodes based on the proposed unsubstituted cyclodextrins showed lower detection limits (10^{-10} magnitude order) than the one previously studied, based on β-cyclodextrin derivative [10]. The widest linear concentration range was recorded for the γ-cyclodextrin based EPME (10^{-8} - 10^{-3}mol/L). The recovery tests performed for the assay of L-proline in the presence of D-proline (recoveries higher than 99.65%) proved that the proposed electrodes can be reliable used for enantioselectivity tests of proline raw material.

2-Hydroxy-3-trimethylammoniopropyl-β-cyclodextrin was used as chiral selector for the design of enantioselective, potentiometric membrane electrode based on plastic membrane. The electrodes were used for the determination of S- and R-propranolol [23]. Two plasticizers were used for the design of the electrodes: ortho-nitrophenyl octyl ether (o-NPOE) and bis(t-butyl-phentyl)adipate (BBPA). 10^{-3} mol/L R-propranolol and 10^{-3} mol/L NH_4Cl were used as inner filling solutions for electrodes A (designed for the assay of R-propranolol) and B (designed for the assay of S-propranolol). Electrode A has a slope of 59.1 mV/decade of concentration with a detection limit of 1.20 x 10^{-5} mol/L while electrode B has a slope of 60.0 mV/decade of concentration with a detection limit of 5.24 x 10^{-6} mol/L.

S-Ramipril can be determined as raw material and from its pharmaceutical formulations, using the EPME based on impregnation of 2-hydroxy-3-trimethylammoniopropyl-β-cyclodextrin (as chloride salt) solution in a carbon paste, in the 1.8 x 10^{-5} - 2.3 x 10^{-1} mol/L (pH between 2.5 and 6.0) concentration range with an average recovery of 99.94% (RSD = 0.030%) and 98.98% (RSD = 0.67%), respectively [14]. The detection limit was of 10^{-5} mol/L magnitude order. The slope was near-Nernstian: 52.00 mV/decade of concentration. The electrode was enantioselective.

The enantioselective, potentiometric membrane electrode based on impregnation of 2-hydroxy-3-trimethylammoniopropyl-β-cyclodextrin (as chloride salt) solution in a carbon paste, was reliably used for the enantioanalysis of S-trandolapril with an average recovery of 99.77% (RSD = 0.22%) [13]. The linear concentration range was 10^{-4} - 10^{-2} mol/L on the 2.5 - 5.5 pH range. The detection limit was of

10^{-5} mol/L magnitude order. The slope was near-Nernstian: 52.45 mV/decade. The sensor was enantioselective.

Enantioselective, Potentiometric Membrane Electrodes Based on Maltodextrins

Different maltodextrins with dextrose equivalent (DE) of 4.0 - 7.0, 13.0 - 17.0, and 16.5 - 19.5, were proposed as novel chiral selectors for the construction of enantioselective, potentiometric membrane electrode used for the enantioanalysis of S-captopril [24]. The potentiometric, enantioselective membrane electrodes were used reliably for the assay of S-captopril as raw material and from its pharmaceutical formulation Novocaptopril tablets, using direct potentiometry. The best response was obtained for the EPME based on the maltodextrin with the highest DE. The best enantioselectivity and stability in time was achieved when the EPMS based on the maltodextrin with the lowest value was used. L-proline was found to be the main interferent for the EPMEs proposed for the enantioselectivity of S-captopril. The surface of the electrodes was regenerated by simply polishing with Al paper.

Two enantioselective, potentiometric membrane electrodes based on maltodextrins with different values of dextrose equivalent (DE) (maltodextrin I: DE 4.0-7.0; maltodextrin II: DE 16.5-19.5) were proposed for the enantioanalysis of baclofen enantiomers in baclofen raw materials and from its pharmaceutical formulations (*e.g.*, Norton-Baclofen® tablets) [28]. The slopes of the proposed EPMEs were 55.0 mV/pS-baclofen for maltodextrin I-based EPME and 59.0 mV/pR-baclofen for maltodextrin II-based EPME. The detection limits were 1.34 x 10^{-6} mol/L (S-baclofen) and 2.52 x 10^{-10} mol/L (R-baclofen), respectively.

The enantioselective, potentiometric membrane electrode based on maltodextrin with dextrose equivalence (DE) 16.5-19.5 was used for the enanatioanalysis of R-deprenyl [49]. The linear concentration range for the EPME was 10^{-10}-10^{-3} mol/L. The slope of the electrode was 53.1 mV per decade of concentration. The detection limit was 3.6 x 10^{-11} mol l^{-1}. The proposed electrode could be reliably employed for the assay of R-deprenyl raw material and its pharmaceutical formulation, Lentogesic tablets.

Three enantioselective, potentiometric membrane electrodes based on carbon paste impregnated with different maltodextrins [dextrose equivalent (DE) (DE 4.0 – 7.0 (I), 13 –17 (II), 16.5 -19.5 (III)], were proposed as chiral selectors for the assay of S-Flurbiprofen raw materials and from its pharmaceutical formulation, Froben 100® tablets [32]. The best response and enantioselectivity were obtained when maltodextrin with the lowest DE was used for the electrode design. The three EPMEs showed very low detection limits.

Three enantioselective, potentiometric membrane electrodes based on maltodextrin with different values of dextrose equivalence (DE) (maltodextrin I: DE 4.0-7.0; maltodextrin II: DE 13.0-17.0; maltodextrin III: DE 16.5-19.5) were proposed for the assay of S-ibuprofen [31]. The linear concentration ranges for the proposed EPMEs were 10^{-10}-10^{-3}, 10^{-8}-10^{-3} and 10^{-10}-10^{-3} mol/L for the EPMEs based on maltodextrin I, II and III, respectively. The slopes of the EPMEs were 59.0, 58.4 and 55.6 mV/decade of concentration for maltodextrin I, II and III based EPMEs, respectively, with detection limits of 5.5 x 10^{-11}, 8.0 x 10^{-9} and 4.1 x 10^{-12} mol/L for the electrodes based on maltodextrin I, II and III, respectively. The enantioanalysis of S-ibuprofen raw material and in its pharmaceutical formulations Myprodol capsules and Nurofen tablets was performed with high reliability.

Enantioselective, potentiometric membrane electrodes based on maltodextrins of different dextrose equivalent (DE) (DE: 4.0-7.0 (I), 13.0-17.0 (II), 16.5-19.5 (III)) as chiral selectors were used for the assay of S-perindopril [25]. These electrodes can be successfully used in the assay of S-perindopril in the presence of R-perindopril, D-proline, PVC and inorganic ions such as Na^+, K^+ and Ca^{2+}. The best results were obtained by using maltodextrins I and II in the construction of the EPMEs. Accordingly, these are the electrodes of choice for the enantioselective analysis of S-perindopril. Improvements in detection limits (lower with maltodextrins as chiral selectors), working concentration ranges, slope and enantioselectivity (higher especially with maltodextrin of lower DE as chiral selector) were recorded *vs* the EPMEs based on β-CD for the enantioanalysis of S-perindopril [15].

Enantioselective, potentiometric membrane electrodes based on carbon paste impregnated with the following maltodextrins having dextrose equivalence (DE) 4.0 - 7.0 (I), 13.0 - 17.0 (II) and 16.5 - 19.5 (III) - as chiral selectors were proposed for the enantioanalysis of L-proline [26]. The response characteristics showed that the proposed electrodes could be reliably utilized in the assay of L-proline as raw material, with the best enantioselectivity and time-stability exhibited when the EPME based on maltodextrin I was used. The three EPMEs showed very low detection limits (of 10^{-9} mol/L magnitude order).

Enantioselective, Potentiometric Membrane Electrodes Based on Antibiotics

Two enantioselective, potentiometric membrane electrodes based on macrocyclic glycopeptide antibiotics – vancomycin and teicoplanin - were designed for the assay of acetyl-L-carnitine [37]. The linear concentration ranges for the proposed electrodes were 10^{-5} – 10^{-2} mol/L for the vancomycin-based electrode and 10^{-4} – 10^{-2} mol/L for the teicoplanin-based electrode, with slopes of 58.1 and 55.0

mV/p(acetyl-L-carnitine), respectively. The proposed electrodes were enantioselective.

Two enantioselective, potentiometric membrane electrodes based on macrocyclic glycopeptide antibiotic, teicoplanin as chiral selector were proposed for the assay of R-baclofen [38]. The linear concentration ranges for the proposed enantioselective, membrane electrodes were 10^{-7} – 10^{-4} mol/L for teicoplanin-based electrode and 10^{-6} – 10^{-4} mol/L for the electrode based on teicoplanin modified with acetonitrile. The slopes of the electrodes were 60.0 mV/pR-baclofen and 57.2 mV/pR-baclofen for teicoplanin and teicoplanin modified with acetonitrile - based EPMEs, respectively. The proposed electrodes were enantioselective, and were used reliably for the enantioanalysis of R-baclofen raw material and from its pharmaceutical formulation, *e.g.*, Norton-Baclofen® tablets.

In order to determine the enantiopurity of L-carnitine, three EPMEs were proposed for the assay of L-carnitine [33]. The electrodes were designed using macrocyclic glycopeptide antibiotics – vancomycin and teicoplanin. Acetonitrile was added to the teicoplanine to design a modified teicoplanine based electrode. The linear concentration ranges for the proposed enantioselective membrane electrodes were 10^{-4} – 10^{-2} mol/L for electrodes based on vancomycin and teicoplanin and 10^{-5} – 10^{-2} mol/L for electrode based on teicoplanin modified with acetonitrile. The slopes of the electrodes were 56.5 mV/pL-carnitine; 54.5 mV/pL-carnitine and 58.3 mV/pL-carnitine for vancomycin-, teicoplanin- and teicoplanin modified with acetonitrile – based EPMEs, respectively. The electrodes were enantioselective. They were used for the enantioanalysis of L-carnitine raw material and its pharmaceutical formulation, Carnilean® capsules.

Three enantioselective, potentiometric membrane electrodes based on macrocyclic glycopeptide antibiotics – vancomycin and teicoplanin (modified or not with acetonitrile) – were proposed for the determination of L- and D-enantiomers of methotrexate (Mtx) [35]. The linear concentration ranges for the proposed enantioselective membrane electrodes were between 10^{-6} and 10^{-3} mol/L for L- and D- methotrexate. The slopes of the electrodes were 58.00 mV/pL-Mtx for vancomycin-based electrode; 57.60 mV/pD-Mtx for teicoplanin-based electrode and 55.40 mV/pD-Mtx for teicoplanin modified with acetonitrile - based electrode. The detection limits of the proposed electrodes were of 10^{-8} mol/L magnitude order. All proposed electrodes proved to be successful for the determination of the enantiopurity of Mtx as raw material and of its pharmaceutical formulations (tablets and injections).

An EPME based on carbon paste modified with macrocyclic antibiotic vancomycin as chiral selector was proposed for the enantioanalysis of S-perindopril raw material and from its pharmaceutical formulations, Coversyl® tablets. The limit of detection for the S-perindopril was 15 nmol/L, with a working concentration range of $5.5x10^{-8}$ – $1.0x10^{-3}$ mol/L. A slope of 55.71 mV/decade of concentration was recorded. The proposed electrode was stable and reproducible over two weeks test period. The response of the electrode does not depend on the pH in the range 3.0 – 7.8. The sensors is enantioselective, and selective *vs* PVP, and D-proline but L-proline interfere in the enantioanalysis of S-perindopril.

Enantioselective, Potentiometric Membrane Electrode Based on Crown Ethers

A crown ether (19-[(10-undecen-1-yl)oxy]-4R,14R-(-)4,14-diphenyl-3-6,9,12,5-pentaoxa-21-azabicyclo[15.3.1]heneicosa-1(21),17,19-triene) was used in a plastic membrane as chiral selector for the design of EPMEs [8]. These electrodes can differentiate between S- and R-1-phenylethylammonium ions. The values of the slopes obtained when different plasticizers were used for EPMEs design were between 51.3 and 60.6mV/decade of concentration proving the suitability of these electrodes for potentiometric measurements.

Enantioselective, Potentiometric Membrane Electrode Based on Fullerenes

New enantioselective, potentiometric membrane electrodes based on carbon paste impregnated with C_{60} fullerene (I), (1,2-methanofullerene C_{60})- 61- carboxylic acid (II), diethyl (1,2-methanofullerene C_{60})-61-61-dicarboxylate (III) and tert-butyl (1,2-methanofullerene C_{60})-61-carboxylic acid (IV) were proposed for the enantioanalysis of L-histidine with slopes of 56.0 to 57.7 mV/decade of concentration, within linear concentration ranges up to 10^{-8} to 10^{-10} mol/L and with detection limits up to 10^{-12} mol/L magnitude order [43]. The average recoveries of L-histidine in its pharmaceutical formulations were higher than 96.40%.

[5, 6]-fullerene-C_{70} (I) and diethyl (1,2-methanofullerene C_{70})-71,71-dicarboxylate (II) were proposed as new chiral selectors for the design of EPMEs [41]. L-proline was selected as model analyte for the study of the behavior of these electrodes. Both electrodes exhibited a cation-enantioselective behavior. The best response was obtained when diethyl (1,2-methanofullerene C_{70})-71,71-dicarboxylate was used for the electrode's construction (59.4 mV/decade of concentration). The best enantioselectivity was achieved for the [5, 6]-fullerene-C_{70} based electrode (pK_{pot}=3.85 over D-proline).

PHARMACEUTICAL ENANTIOANALYSIS USING AMPEROMETRIC SENSORS

Amperometric sensors had a very important contribution to the enantioanalysis of chiral drugs, especially for the enantiopurity determination of pharmaceutical compounds with a trace or ultra trace amount of chiral substance [50].

Addition of an enzyme to the design of the amperometric sensors improved the enantioselectivity [51]. *In situ* molecular imprinting based amperometric sensors were proposed for the enantioanalysis [52]: a microchip for enantioanalysis was designed using molecular imprinting technique into a microchannel wall using acrylamide as the functional monomer and ethylene glycol dimethacrylate as the cross-linker.

Imprinted polymer films were designed for amperometric enantioanalysis by co-electropolymerization of phenol with a 3-hydroxyphenyl boronic acid-monosaccharide complex containing D-glucose and D-mannose on an Au support [53].

Screen-printed amperometric sensors for enantioanalysis of D-aminoacids were also designed [54]. D-aminoacid oxidase was immobilized on a graphite working electrode of a screen-printed strip modified with Prussian Blue and Nafion layers.

PHARMACEUTICAL ENANTIOANALYSIS USING STOCHASTIC SENSORS

Stochastic microsensors based on four inulins (IN, IQ, TEX, and HD) immobilized on diamond paste (DP) were proposed for the assay of the enantioanalysis of L-T4 in its pharmaceutical formulations [55]. The best stochastic microsensor was the one based on TEX/DP. For the enantiorecognition of free L-T4 and D-T4 in pharmaceutical samples, while for free L-T3 (400 fmol/L) the best stochastic microsensor was the one based on HD/DP. The best limit of determination for the assay of free L-T4, and D-T4 (1 pmol/L) was obtained using the stochastic microsensor based on TX/DP. For the enantiorecognition of free L-T3 in pharmaceutical samples the best microsensor was the one based on HD/DP (the wider linear concentration range, and the lower limit of quantification – of pmol/L magnitude order). For the enantiorecognition of free L-T3 in pharmaceutical samples the best microsensor was the one based on HD/DP (the wider linear concentration range, and the lower limit of quantification – of pmol/L magnitude order). Free L-T3, L-T4, and D-T4 were enantioanalysed with high reliabilities in pharmaceutical samples (recoveries higher than 95.00% with RSD values lower than 1.00%).

CONCLUSIONS

The pharmaceutical industry and also biomedical analysis needed development of new tools for the enantioanalysis of chiral drugs, due to difference in pathways of these drugs, as well as due to the differences in illness for which the chiral biomarkers are responsible. Enantioselective electrodes are a good alternative to the chromatographic techniques. They are reliable, easy to use, most of the time the samples do not need any pretreatment or this step is reduced to dissolution of the solid sample in water/organic solvent. Furthermore, the method is also fast (response is obtained within 5min) and the cost is less than for the chromatographic methods. They can also be used for multianalyte assay when needed.

CONSENT FOR PUBLICATION

Not applicable.

CONFLICT OF INTEREST

The authors declares no conflict of interest, financial or otherwise.

ACKNOWLEDGEMENTS

Declared none.

REFERENCES

[1] Izake, E.L. Chiral discrimination and enantioselective analysis of drugs: an overview. *J. Pharm. Sci.,* **2007**, *96*(7), 1659-1676.
[http://dx.doi.org/10.1002/jps.20820] [PMID: 17221858]

[2] Wattanakit, C.; Come, Y.B.S.; Lapayre, V.; Bopp, Ph.A.; Heim, M.; Yadnum, S.; Nokbin, S.; Warakulwit, C.; Limtrakul, J.; Kuhn, A. Enantioselective recognition at mesoporous chiral metal surface. *Nature Comm.,* **2014**, *5*, 3325.

[3] Schugerl, K.; Ulber, R.; Scheper, T. Developments of biosensors for enantiomeric analysis. *TrAC,* **1996**, *15*, 56-62.

[4] Trojanowicz, M.; Kaniewska, M. Electrochemical chiral sensors and biosensors. *Electroanalysis,* **2009**, *21*, 229-238.
[http://dx.doi.org/10.1002/elan.200804382]

[5] Stefan, R.I.; van Staden, J.F.; Aboul-Enein, H.Y. An amperometric biosensor/SIA system for the simultaneous determination of S- and R-captopril. *Biosens. Bioelectron.,* **2000**, *15*, 1-5.
[http://dx.doi.org/10.1016/S0956-5663(99)00075-5] [PMID: 10826637]

[6] Stefan, R.I.; van Staden, J.K.; Aboul-Enein, H.Y. Simultaneous detection of S and R captopril using sequential injection analysis. *Talanta,* **2000**, *51*(5), 969-975.
[http://dx.doi.org/10.1016/S0039-9140(00)00282-4] [PMID: 18967928]

[7] Dominiguez, R.; Serra, B.; Reviejo, A.J.; Pingarron, J.M. Electrochemical chiral sensors and biosensors. *Anal. Biochem.,* **2001**, *298*, 275-282.
[PMID: 11700983]

[8] Horvath, V.; Takacs, T.; Horvai, G.; Huszthy, P.; Bradshaw, J.S.; Izatt, R.M. Enantiomer-selectivity of ion-selective electrodes based on a chiral crown ether ionophore. *Anal. Lett.*, **1997**, *30*, 1591-1609. [http://dx.doi.org/10.1080/00032719708001680]

[9] Kataky, R.; Parker, D.; Kelly, P.M. Potentiometric, enantioselective sensors for alkyl and aryl ammonium ions of pharmaceutical significance, based on lipophilic cyclodextrins. *Scand. J. Clin. Lab. Invest.*, **1995**, *55*(5), 409-419. [http://dx.doi.org/10.3109/00365519509104980] [PMID: 8545599]

[10] Stefan, R.I.; van Staden, J.F.; Aboul-Enein, H.Y. A new construction for potentiometric, enantioselective membrane electrode, and use for L-proline assay. *Anal. Lett.*, **1998**, *31*, 1787-1794. [http://dx.doi.org/10.1080/00032719808005262]

[11] Stefan, R.I.; van Staden, J.K.; Aboul-Enein, H.Y. A new construction for a potentiometric, enantioselective membrane electrode--its utilization to the S-captopril assay. *Talanta,* **1999**, *48*(5), 1139-1143. [http://dx.doi.org/10.1016/S0039-9140(98)00335-X] [PMID: 18967558]

[12] Stefan, R.I.; van Staden, J.F.; Aboul-Enein, H.Y. Potentiometric, enantioselective membrane electrode for S-enalapril assay. *Analusis,* **1999**, *27*, 53-56. [http://dx.doi.org/10.1051/analusis:1999105]

[13] Stefan, R.I.; van Staden, J.F.; Aboul-Enein, H.Y. Detection of S-enantiomers of cilazapril, pentopril and trandolapril using a potentiometric, enantioselective membrane electrode. *Electroanalysis,* **1999**, *11*, 192-194. [http://dx.doi.org/10.1002/(SICI)1521-4109(199903)11:3<192::AID-ELAN192>3.0.CO;2-O]

[14] Stefan, R.I.; van Staden, J.F.; Baiulescu, G.E.; Aboul-Enein, H.Y. A potentiometric, enantioselective membrane electrode for S-ramipril assay. *Chem.Anal.(Warsaw),* **1999**, *44*, 417-422.

[15] Stefan, R.I.; Van Staden, J.K.; Aboul-Enein, H.Y. S-perindopril assay using a potentiometric, enantioselective membrane electrode. *Chirality,* **1999**, *11*(8), 631-634. [http://dx.doi.org/10.1002/(SICI)1520-636X(1999)11:8<631::AID-CHIR4>3.0.CO;2-K] [PMID: 10467314]

[16] Ozoemena, K.I.; Stefan, R.I.; van Staden, J.F.; Aboul-Enein, H.Y. Enantioanalysis of S-perindopril using different cyclodextrin-based potentiometric sensors. *Sens. Actuators B Chem.*, **2005**, *105*, 425-429. [http://dx.doi.org/10.1016/j.snb.2004.06.032]

[17] Ozoemena, K.I.; Stefan, R.I. Enantioselective potentiometric membrane electrodes based on α-, β- and γ-cyclodextrins as chiral selectors for the assay of L-proline. *Talanta,* **2005**, *66*(2), 501-504. [http://dx.doi.org/10.1016/j.talanta.2004.11.024] [PMID: 18970013]

[18] Stefan, R.I.; Nejem, R.M. Enantioselective, potentiometric membrane electrodes based on cyclodextrins for the assay of L- and D-2-hydroxyglutaric acid. *Anal. Lett.*, **2005**, *38*, 1847-1855. [http://dx.doi.org/10.1080/00032710500230830]

[19] Stefan van Staden, R.I.; Rat'ko, A.A. Enantioselective, potentiometric membrane electrodes based on cyclodextrins: application for the determination of R-baclofen in its pharmaceutical formulation. *Talanta,* **2006**, *69*(5), 1049-1053. [http://dx.doi.org/10.1016/j.talanta.2005.12.022] [PMID: 18970679]

[20] Stefan, R.I.; Nejem, R.M. Cyclodextrins-based enantioselective, potentiometric membrane electrodes for L-vesamicol assay in serum samples. *Sens. Actuators B Chem.*, **2006**, *117*, 123-127. [http://dx.doi.org/10.1016/j.snb.2005.11.007]

[21] Stefan, R.I.; Bokretsion, R.G.; Ozoemena, K.I.; van Staden, J.F.; Aboul-Enein, H.Y. Enantioselective, potentiometric membrane electrodes based on different cyclodextrin s as chiral selectors for the assay of S-flurbiprofen. *Electroanalysis,* **2006**, *18*, 1718-1721. [http://dx.doi.org/10.1002/elan.200603574]

[22] Stefan, R.I.; Holo, L. Enantioselective, potentiometric membrane electrodes based on cyclodextrins for the determination of L-histidine. *Sens. Actuators B Chem.,* **2007**, *120*, 399-402. [http://dx.doi.org/10.1016/j.snb.2006.02.033]

[23] Sun, C.J.; Sun, X.X.; Aboul-Enein, H.Y. Potentiometric, enantioselective electrode responsive to propanolol enantiomers based on a lipophilic α-cyclodextrin. *Anal. Lett.,* **2004**, *37*, 2259-2266. [http://dx.doi.org/10.1081/AL-200028059]

[24] Stefan, R.I.; van Staden, J.F.; Aboul-Enein, H.Y. Maltodextrins as new chiral selectors in the design of potentiometric, enantioselective membrane electrodes. *Fresenius J. Anal. Chem.,* **2001**, *370*(1), 33-37. [http://dx.doi.org/10.1007/s002160100741] [PMID: 11393233]

[25] Ozoemena, K.I.; Stefan, R.I.; van Staden, J.F.; Aboul-Enein, H.Y. Utilization of maltodextrin based enantioselective, potentiometric membrane electrodes for the enantioselective assay of S-perindopril. *Talanta,* **2004**, *62*(4), 681-685. [http://dx.doi.org/10.1016/j.talanta.2003.08.035] [PMID: 18969348]

[26] Ozoemena, K.I.; Stefan, R.I.; van Staden, J.F.; Aboul-Enein, H.Y. Enantioselective, potentiometric membrane electrodes based on maltodextrins. Their applications for the determination of L-proline. *Sens. Actuators B Chem.,* **2004**, *98*, 97-100. [http://dx.doi.org/10.1016/j.snb.2003.09.029]

[27] Stefan, R.I.; Nejem, R.M.; van Staden, J.F.; Aboul-Enein, H.Y. Enantioselective, potentiometric membrane electrodes for the determination of L-pipecolic acid in serum. *Electroanalysis,* **2004**, *16*, 1730-1733. [http://dx.doi.org/10.1002/elan.200303022]

[28] Rat'ko, A.A.; Stefan-van Staden, R.I. Determination of baclofen enantiomers in pharmaceutical formulations using maltodextrin-based enantioselective, potentiometric membrane electrodes. *Farmaco,* **2004**, *59*(12), 993-997. [http://dx.doi.org/10.1016/j.farmac.2004.08.005] [PMID: 15598435]

[29] Nejem, R.M.; Stefan, R.I.; Staden, J.F.; Aboul-Enein, H.Y. Enantioanalysis of l-2-hydroxyglutaric acid in urine samples using enantioselective, potentiometric membrane electrodes based on maltodextrins. *Talanta,* **2005**, *65*(2), 437-440. [http://dx.doi.org/10.1016/j.talanta.2004.06.040] [PMID: 18969817]

[30] Stefan, R.I.; Nejem, R.M. Enantioanalysis of glyceric acid using enantioselective, potentiometric membrane electrodes: applications for the diagnosis of PH II and D-glyceric academia/acidurias. *Sens. Actuators B Chem.,* **2005**, *106*, 736-740. [http://dx.doi.org/10.1016/j.snb.2004.09.025]

[31] Stefan-van Staden, R.I.; Mashile, T.R. Enantioselective assay of S(+)-ibuprofen using enantioselective, potentiometric membrane electrodes based on maltodextrins. *Sens. Actuators B Chem.,* **2006**, *120*, 295-297. [http://dx.doi.org/10.1016/j.snb.2006.02.044]

[32] Stefan-van Staden, R.I.; Bokretsion, R.G.; Ozoemena, K.I. Utilization of maltodextrin-based enantioselective, potentiometric membrane electrodes for the enantioselective assay of S-flurbiprofen. *Anal. Lett.,* **2006**, *39*, 1065-1073. [http://dx.doi.org/10.1080/00032710600620401]

[33] Rat'ko, A.A.; Stefan, R.I.; van Staden, J.K.; Aboul-Enein, H.Y. Determination of l-carnitine using enantioselective, potentiometric membrane electrodes based on macrocyclic antibiotics. *Talanta,* **2004**, *63*(3), 515-519. [http://dx.doi.org/10.1016/j.talanta.2003.11.036] [PMID: 18969462]

[34] Rat'ko, A.A.; Stefan, R.I.; van Staden, J.F.; Aboul-Enein, H.Y. Enantioselective, potentiometric membrane electrode based on vancomicyn. Its application for the determination of D-pipecolic acid. *Sens. Actuators B Chem.,* **2004**, *99*, 539-543. [http://dx.doi.org/10.1016/j.snb.2004.01.004]

[35] Rat'ko, A.A.; Stefan, R.I.; van Staden, J.K.; Aboul-Enein, H.Y. Macrocyclic antibiotics as chiral selectors in the design of enantioselective, potentiometric membrane electrodes for the determination of l- and d-enantiomers of methotrexate. *Talanta,* **2004**, *64*(1), 145-150. [http://dx.doi.org/10.1016/j.talanta.2004.01.022] [PMID: 18969580]

[36] Ozoemena, K.I.; Stefan, R.I.; van Staden, J.F.; Aboul-Enein, H.Y. Enantioselective, potentiometric membrane electrodes based on vancomycin as chiral selector, for the assay of S-perindopril. *Instrum. Sci. Technol.,* **2004**, *32*, 371-378. [http://dx.doi.org/10.1081/CI-120037669]

[37] Rat'ko, A.A.; Stefan, R.I.; van Staden, J.K.; Aboul-Enein, H.Y. Macrocyclic antibiotics as chiral selectors in the design of enantioselective, potentiometric membrane electrodes for the determination of l- and d-enantiomers of methotrexate. *Talanta,* **2004**, *64*(1), 145-150. [http://dx.doi.org/10.1016/j.talanta.2004.01.022] [PMID: 18969580]

[38] Rat'ko, A.A.; Stefan, R.I. Teicoplanin-based enantioselective, potentiometric membrane electrodes for the determination of R-baclofen in pharmaceutical formulations. *Anal. Lett.,* **2004**, *37*, 3161-3173. [http://dx.doi.org/10.1081/AL-200040315]

[39] Stefan, R.I.; Nejem, R.M.; van Staden, J.F.; Aboul-Enein, H.Y. Determination of D-2-hydroxyglutaric acid in urine samples using enantioselective, potentiometric membrane electrodes based on antibiotics. *Sens. Actuators B Chem.,* **2005**, *106*, 791-795. [http://dx.doi.org/10.1016/j.snb.2004.09.031]

[40] Stefan, R.I.; Nejem, R.M. Determination of L-vesamicol in serum samples using enantioselective, potentiometric membrane electrodes based on antibiotics. *Anal. Lett.,* **2006**, *39*, 675-682. [http://dx.doi.org/10.1080/00032710600609750]

[41] Stefan, R.I. New enantioselective, potentiometric membrane electrodes based on C_{70} fullerenes as chiral selectors. *Sens. Lett.,* **2003**, *1*, 71-74. [http://dx.doi.org/10.1166/sl.2003.014]

[42] Stefan, R.I.; Lal, B. Enantioselective, potentiometric membrane electrodes based on C_{60} derivatives as chiral selectors for the enantioanalysis of S-clenbuterol. *Anal. Lett.,* **2006**, *39*, 1311-1319. [http://dx.doi.org/10.1080/00032710600666396]

[43] Stefan-van Staden, R.I.; Lal, B.; Holo, L. Enantioselective potentiometric membrane electrodes based on C(60) fullerene and its derivatives for the assay of l-Histidine. *Talanta,* **2007**, *71*(3), 1434-1437. [http://dx.doi.org/10.1016/j.talanta.2006.05.088] [PMID: 19071470]

[44] Zhan, S.Z.; Dai, Q.; Yuan, C.W.; Lu, Z.H.; Haeussling, L. Synthesis, recognition and separation of print molecule in molecularly imprinted polymers. *Anal. Lett.,* **1999**, *32*, 677-687. [http://dx.doi.org/10.1080/00032719908542848]

[45] Stefan, R.I.; van Staden, J.F.; Aboul-Enein, H.Y. *Electrochemical Sensors in Bioanalysis*; Marcel Dekker, Inc.: New York, **2001**.

[46] Stefan, R.I.; Aboul-Enein, H.Y. Ion-selective membrane electrodes based on ion-pair complexes: correlation between slopes and stability of ion-pair complexes. *Instrum. Sci. Technol.,* **1999**, *27*, 105-110. [http://dx.doi.org/10.1080/10739149908085838]

[47] Montanari, M.L.; Cass, Q.B.; Andricopulo, A.D.; Leitao, A.; Montanari, C.A. Identification of chiral selectors for improved enantioseparation based on molecularinteraction fields. *Anal. Chim. Acta,* **2005**, *545*, 33-45. [http://dx.doi.org/10.1016/j.aca.2005.04.039]

[48] Yin, X.; Ding, J.; Zhang, S.; Kong, J. Enantioselective sensing of chiral amino acids by potentiometric sensors based on optical active polyaniline films. *Biosens. Bioelectron.,* **2006**, *21*(11), 2184-2187. [http://dx.doi.org/10.1016/j.bios.2005.10.010] [PMID: 16303298]

[49] Stefan-van Staden, R.I.; Mashile, T.R. Determination of R-deprenyl using a maltodextrin based

enantioselective, potentiometric membrane electrode. *Instrum. Sci. Technol.,* **2007**, *35*, 117-123. [http://dx.doi.org/10.1080/10739140601126262]

[50] Budnikov, G.K.; Evtyugin, G.A.; Budnikova, Yu.G.; Al'fonsov, V.A. Chemically modified electrodes with amperometric response in enantioselective analysis. *J. Anal. Chem.,* **2008**, *63*, 2-12. [http://dx.doi.org/10.1134/S1061934808010024]

[51] Fu, Y.; Chen, Q.; Zhou, J.; Han, Q.; Wang, Y. Enantioselective recognition of mandelic acid based on γ-globulin modified glassy carbon electrode. *Anal. Biochem.,* **2012**, *421*(1), 103-107. [http://dx.doi.org/10.1016/j.ab.2011.10.017] [PMID: 22056409]

[52] Qu, P.; Lei, J.; Ouyang, R.; Ju, H. Enantioseparation and amperometric detection of chiral compounds by *in situ* molecular imprinting on the microchannel wall. *Anal. Chem.,* **2009**, *81*(23), 9651-9656. [http://dx.doi.org/10.1021/ac902201a] [PMID: 19883060]

[53] Granot, E.; Tel-Vered, R.; Lioubashevski, O.; Willner, I. Stereoselective and enantioselective electrochemical sensing of monosaccharides using imprinted boronic acid-functionalized polyphenol films. *Adv. Funct. Mater.,* **2008**, *18*, 478-484. [http://dx.doi.org/10.1002/adfm.200700498]

[54] Wcisło, M.; Compagnone, D.; Trojanowicz, M. Enantioselective screen-printed amperometric biosensor for the determination of D-amino acids. *Bioelectrochemistry,* **2007**, *71*(1), 91-98. [http://dx.doi.org/10.1016/j.bioelechem.2006.09.001] [PMID: 17071143]

[55] Mitrofan, G.; Stefan-van Staden, R.I.; Comnea-Stancu, I.R.; van Staden, J.F.; Bazylak, G.; Kapnissi-Christodoulou, C.P.; Aboul-Enein, H.Y. Fast screening of whole blood samples and pharmaceutical compounds for enantiorecognition of free L-T3, L-T4, and D-T4. *Chirality,* **2015**, *27*(12), 973-978. [http://dx.doi.org/10.1002/chir.22537] [PMID: 26447904]

SUBJECT INDEX

A

Atta-ur-Rahman, Sibel A. Ozkan & Rida Ahmed (Eds.)

C

D

G

H

N

O

P

T

U

V

W

Z

www.ingramcontent.com/pod-product-compliance
Lightning Source LLC
LaVergne TN
LVHW070944210926
848482LV00002B/85